SUPERCONDUCTIVITY – THEORY AND APPLICATIONS

Edited by **Adir Moysés Luiz**

Superconductivity – Theory and Applications
Edited by Adir Moysés Luiz

Published by InTech
Janeza Trdine 9, 51000 Rijeka, Croatia

Technical Editor Sonja Mujacic

Cover Designer InTech Design Team

Additional hard copies can be obtained from orders@intechopen.com

Superconductivity – Theory and Applications,
Edited by Adir Moysés Luiz
p. cm.
ISBN 978-953-307-151-0

For EU product safety concerns:
IN TECH d.o.o., Prolaz Marije Krucifikse Kozulić 3, 51000 Rijeka, Croatia,
info@intechopen.com or visit our website at intechopen.com.

Meet the editor

Adir Moysés Luiz is associate Professor at Instituto de Física at Federal University in Rio de Janeiro, (UFRJ) Brazil. His e-mail contact is: adir@if.ufrj.br and alternative e-mail: adir.luiz@oi.com.br He obtained doctor degree in Chemical Engineering (DSc.) at COPPE - "Universidade Federal do Rio de Janeiro", (UFRJ) in Brazil. The year of degree obtaining is 1982. The research areas which were always in the focus of his activities are: thermodynamics, solid state physics, materials science and superconductivity.
His present interests are: superconductivity, superconducting materials, applications of superconductivity, superconducting devices, microscopic theories of superconductivity.
Until now, Dr. Luiz has published 44 articles in scientific journals, 23 books and 62 articles in annals of events. He has also published 22 presentations of events (different congresses, conference or lectures), translated 19 books, and participated in 43 graduate boards. He has already been active as the Editor of 3 published books.

Contents

Preface

Superconductivity was discovered 1911 by Kamerlingh Onnes. During the last century, the history of superconductivity has been full of theoretical challenges and practical developments. In 1986, the discovery of Bednorz and Müller of an oxide superconductor with critical temperature (Tc) approximately equal to 35 K, has given a novel impetus to this fascinating subject. Since this discovery, there are a great number of laboratories all over the world involved in research of superconductors with high Tc values, the so-called "high-Tc superconductors". The discovery of a room temperature superconductor has been a long-standing dream of many scientists. The technological and practical applications of such a discovery should be tremendous. However, the actual use of superconducting devices is limited by the fact that they must be cooled to low temperatures to become superconducting. Until 2011, one hundred years after the first Kamerlingh Onnes' discovery, the highest Tc value is approximately equal to 135 K at 1 atm. The knowledge of the microscopic mechanisms of high-Tc superconductors should be a theoretical guide in the researches of room temperature superconductivity.

This book contains 15 chapters reporting interesting researches about theoretical and experimental aspects of superconductivity. Here you will find a great number of works containing theories and describing properties of high-Tc superconductors (materials with $Tc > 30$ K). In a few chapters there are also discussions about low-Tc superconductors (materials with $Tc < 30$ K).

The plan of this book is:

Chapter 1 contains theoretical discussions about the possibility of room temperature superconductivity.

In the chapters 2, 3, 4 and 5 are discussed interesting theories about physical properties of superconductors.

Chapter 6 presents report about the electronic transport in an NS system with a pure normal channel.

In chapter 7 can be found a theoretical discussion concerning the effects of impurities on a noncentrosymmetric superconductor.

Chapter 8 contains a research about the Meissner effect and the use of superconductors as magnets.

Chapter 9 is a report about the properties of macroscopic quantum effects in superconductors.

Chapter 10 presents theoretical discussions concerning the vortex state of superconductors.

In Chapter 11 the development of Josephson voltage standards is analyzed.

Chapter 12 contains critical state analysis using a SQUID magnetometer.

Chapter 13 is a theoretical discussion about superconducting transistors.

Chapter 14 presents some physical properties of the superconductor photonic crystal.

Finally, in chapter 15 you can find a theoretical discussion about the electrodynamics of high pinning superconductors.

I expect that this book will be useful to encourage further experimental and theoretical research of superconductivity.

Adir Moysés Luiz
Instituto de Física, Universidade Federal do Rio de Janeiro
Brazil

1

Room Temperature Superconductivity

Adir Moysés Luiz
Instituto de Física, Universidade Federal do Rio de Janeiro
Brazil

1. Introduction

Superconductivity was discovered by Kamerlingh Onnes in 1911. For one century superconductivity has been a great challenge to theoretical physics. The first successful set of phenomenological equations for superconducting metals was given by F. London in 1935. Yet, in 1950, almost 40 years after the discovery of this phenomenon, there was not any adequate microscopic theory of superconductivity. However, by 1935, single elements necessary to a successful theory to explain superconductivity was known to theorists. The peculiar condensation of a Bose-Einstein gas was predicted by Einstein in 1925. The idea that pairs of fermions can combine to form bosons has been known since 1931. In 1950 the most relevant ideas of superconductivity has been summarized by F. London in his famous book "Superfluids", volume 1. At last, BCS theory (Bardeen et al., 1957) was the first successful theory to explain the microscopic mechanisms of superconductivity in metals and alloys.

Practical applications of superconductivity are steadily improving every year. However, the actual use of superconducting devices is limited by the fact that they must be cooled to low temperatures to become superconducting. For example, superconducting magnets used in most particle accelerators are cooled with liquid helium, that is, it is necessary to use cryostats that should produce temperatures of the order of 4 K. Helium is a very rare and expensive substance. On the other hand, because helium reserves are not great, the world's supply of helium can be wasted in a near future. Thus, because liquid nitrogen is not expensive and the reserves of nitrogen could not be wasted, it is important to use high-T_c superconductors cooled with liquid nitrogen. Superconductors with critical temperatures greater 77 K may be cooled with liquid nitrogen.

We know that BCS theory (Bardeen et al., 1957) explains the microscopic mechanisms of superconductivity in metals. According to BCS theory, electrons in a metallic superconductor are paired by exchanging phonons. According to many researchers (De Jongh, 1988; Emin, 1991; Hirsch, 1991; Ranninger, 1994), BCS theory is not appropriate to be applied to explain the mechanisms of superconductivity in oxide superconductors. Nevertheless, other models relying on a BCS-like picture replace the phonons by another bosons, such as: plasmons, excitons and magnons, as the mediators causing the attractive interaction between a pair of electrons and many authors claim that superconductivity in the oxide superconductors can be explained by the conventional BCS theory or BCS-like theories (Canright & Vignale, 1989; Tachiki & Takahashi, 1988; Takada, 1993).

Copper oxide superconductors are the most important high-T_c superconductors. The discovery of a room temperature superconductor should trigger a great technological revolution. There are claims of synthesis of a room temperature superconductor (see, for example, www.superconductors.org, 2011). But these claims are not accepted by the scientific community. It is generally accepted in the scientific literature that the highest T_c is approximately equal to 135 K at 1 atm in the Hg-Ba-Ca-Cu-O system (Schilling & Cantoni, 1993). However, T_c in this system can be raised up to 180 K using high external pressures. We believe that the discovery of a room temperature superconductor would be possible only when the microscopic mechanisms of oxide superconductors should be clarified. However, up to the present time, the microscopic mechanisms responsible for high-T_c superconductivity are unclear. In a recent article (Luiz, 2010), we have discussed a simple model to study microscopic mechanisms in high-T_c superconductors. The objective of this chapter is to present new studies in order to give new theoretical support for that simple model. We also discuss the possibility of room temperature superconductivity.

2. Possibility of room temperature superconductivity

It is well known that the superconducting state is characterized by a quantum macroscopic state that arises from a Bose-Einstein condensation (BEC) of paired electrons (Cooper pairs). Initially, it is convenient to clarify some concepts regarding BEC. It is well known that a collection of particles (bosons) that follows the counting rule of Bose-Einstein statistics might at the proper temperature and density suddenly populate the collections ground state in observably large numbers (Silvera, 1997). The average de Broglie wavelength λ_{dB} which is a quantum measurement of delocalization of a particle, must satisfy this condition. We know that $\lambda_{dB} = h/p$, where h is Planck's constant and p is the momentum spread or momentum uncertainty of the wave packet. In the other extreme, for particles in the zero momentum eigenstate, the delocalization is infinite; i.e., the packet is spread over the entire volume V occupied by the system. It is generally accepted that BEC occurs when the interparticle separation is of the order of the delocalization λ_{dB} (Silvera, 1997).
The thermal de Broglie wavelength λ_{dB} is a measure of the thermodynamic uncertainty in the localization of a particle of mass M with the average thermal momentum. Thus, λ_{dB} is given by

$$\lambda_{dB} = h/[3MkT]^{1/2} \tag{1}$$

where k is Boltzmann's constant. Equation (1) shows that at a certain low temperature T or/and for a small mass M, λ_{dB} may be spread over great distances. In order to determine the critical temperature T_c at which the addition of more particles leads to BEC it is sufficient to calculate a certain critical density n = N/V, where N is the number of bosons. This calculation is performed using Bose-Einstein statistics; according to (Silvera, 1997) and considering the mass of the boson (Cooper pair) M = 2m*, where m* is the effective mass of the electron, we obtain

$$T_c = 3.31h^2n^{2/3}/(4\pi^2kM) \tag{2}$$

The first application of BEC theory to explain ^4He superfluidity was realized in 1938 (London, 1938). In an important paper (Blatt, 1962), the BEC approach has been extended to give the same results predicted by BCS theory. Thus, it is reasonable to conclude that the conventional n-type superconductivity in metals (explained by BCS theory) is a special case that can also be considered as a phenomenon of BEC of Cooper pairs.

There are three possibilities of occurrence of BEC: (a) BEC involving just bosons, (b) BEC involving just fermions, and (c) BEC involving bosons and fermions simultaneously. In (a) there is a direct BEC without the need of an interaction to bind the bosons. However, in the cases (b) and (c) BEC is possible only indirectly in two steps: in the first step it occurs the binding between pairs of fermions giving rise to bosons and, in the second step, BEC of these bosons may occur.

Because liquid ^4He is a system of bosons, the condensation of ^4He is a BEC of type (a). Superfluidity of ^3He (Lee, 1997) is an example of BEC of type (b). Because liquid ^3He is a system of fermions, in order to occur BEC, two particles must be binded to form a boson and, in he next step, a BEC of these bosons may occur. Another example of BEC of type (b) is the phenomenon of superconductivity in metals and alloys. In the last case, BCS theory (Bardeen et al., 1957) is a successful theory to explain the microscopic mechanisms of superconductivity in metals, and in this case, Equation (2) is not appropriate to calculate the critical temperature T_c because we cannot predict the density n of the bosons formed exchanging phonons. In BCS theory, the critical temperature T_c is the temperature at which a great number of Cooper pairs are formed by exchanging phonons. When the density n of pairs formed are sufficiently high it is possible to occur a Bose-Einstein condensation. For example, in pure copper the density n of Cooper pairs formed are not sufficiently high, thus pure copper cannot become superconductor even at temperatures in the neighborhood of 0 K.

We study now the possibility of occurrence of a Bose-Einstein condensation in an oxide material. If possible, this phenomenon should be a BEC of type (c) just mentioned, that is, the mechanism should involve bosons and fermions simultaneously. In order to verify if BEC is possible in oxide superconductors, it is sufficient to calculate the order of magnitude of the critical temperature T_c using Equation (2). According to Table 1 in the reference (De Jongh, 1988), in a p-type copper oxide superconductor, a typical order of magnitude of the carrier density is given by $n = 10^{21}/cm^3$. Considering an effective mass $m^* = 12m$, where m is the rest mass of the electron, we obtain by Equation (2) the following approximated value: $T_c = 100$ K. This calculation is very crude because Equation (2) is based on an isotropic model of an ideal Bose gas. However, oxide superconductors are not isotropic; on the other hand, pair of electrons (bipolarons) in oxide materials are not an ideal Bose gas because we must consider Coulomb interactions. But the crude calculation based on Equation (2) is sufficient to show that BEC in oxide superconductors cannot be ruled out. A more appropriate formula to calculate T_c (supposing BEC) has been derived in (Alexandrov & Edwards, 2000).

On the basis of the crude calculation based on Equation (2) we will now discuss the possibility of room temperature superconductivity. Using the same above mentioned values and considering a carrier density greater than $n = 10^{21}/cm^3$ we conclude that the critical temperature T_c could be enhanced. For example, considering $m^* = 12m$ and a carrier density $n = 3 \times 10^{21}/cm^3$, we obtain a critical temperature $T_c = 300$ K. Thus, if we apply Equation (2), it is reasonable to conclude that room temperature superconductivity is possible.

According to the type of charge carriers, superconductors can be classified in two types: n-type superconductors, when the charge carriers are Cooper pairs of electrons and p-type superconductors, when the charge carriers are Cooper pairs of holes.

We claim that only p-type materials should be considered in the researches to synthesize a room temperature superconductor. We claim that n-type materials are not qualified to obtain a room temperature superconductor, because in an n-type material the carriers are electrons and these electrons should be binded in order to form bosons. On the other hand, due to Coulomb interactions, it is very difficult to accept a situation of an ideal Bose gas of electrons pairs in order to apply Equation (2). However, considering p-type materials, there are two types of interactions: Coulomb repulsions between electrons, but Coulomb attractions between electrons and holes. Thus, it is reasonable to accept an approximation considering an ideal Bose gas of Cooper pairs of holes in order to apply Equation (2). On the other hand, Bose-Einstein condensation is not restricted only to an ideal Bose gas of bosons (Blatt, 1962). The phenomenon of Bose-Einstein condensation could also be extended to a real Bose gas. It is worthwhile to develop a complete theory to extend the predictions of BEC to a real Bose gas. A fermion-boson mixture of unpaired electrons coexisting and interacting with Cooper pairs treated as real two-electrons and two-holes has been proposed in a letter (Tolmachev, 2000). Finally, considering the simple calculation based on Equation (2) we conclude that the possibility of room temperature superconductivity cannot be ruled out.

3. Superconductors containing oxygen

The most relevant metallic superconductors are pure metals and alloys. BCS theory is appropriate to explain the microscopic mechanisms of superconductivity in pure metals and alloys. However, a great number of oxide materials may become non-metallic superconductors. It seems that BCS theory is not appropriate to explain the microscopic mechanisms in superconductors containing oxygen. An interesting review about oxide superconductors is found in the references (Cava, 2000). The history of oxide superconductors begins in 1933 with the synthesis of the superconductor NbO; with $T_c = 1.5$ K (Sleight, 1995). In 1975 it was discovered the oxide superconductor $BaPb_{0.7}Bi_{0.3}O_3$ (Sleight et al., 1975) with $T_c = 13$ K.

Superconductor	Year	T_C	Reference
NbO	1933	1.5	Sleight, 1995
K_xWO_3	1967	6.0	Remeika et al., 1967
$LiTi_{2+x}O_4$	1973	1.2	Johnston et al., 1973
$BaPb_{1-x}Bi_xO_3$	1975	13	Sleight et al., 1975
$La_{2-x}Ba_xCuO_4$	1986	30	Bednorz & Müller, 1986
$YBa_2Cu_3O_{7-x}$	1987	90	Wu et al., 1987
$Ba_{1-x}K_xBiO_3$	1988	30	Cava et al.,1988
$BiSrCaCu_2O_{6+x}$	1988	105	Maeda et al., 1988
$Tl_2Ba_2Ca_2Cu_3O_{9+x}$	1988	110	Shimakawa et al., 1988
$HgBa_2Ca_2Cu_3O_{8+x}$	1993	130	Schilling & Cantoni, 1993
$NdFeAsO_{1-x}$	2008	54	Yang et al., 2008

Table 1. List of the most relevant superconductors containing oxygen in chronological order.

In 1986, the oxide superconductor $Ba_{0.15}La_{1.85}CuO_4$ with $T_c = 30$ K has been discovered (Bednorz & Müller, 1986). The expression "high-T_c superconductors" has been generally used in the literature to denote superconductors with critical temperatures higher than 30 K. After this famous discovery many cuprate high-T_c superconductors have been synthesized. The cuprate superconductor $HgBa_2Ca_2Cu_3O_{8+x}$ (Hg-1223) has the highest critical temperature ($T_c = 135$ K) at 1 atm (Schilling & Cantoni, 1993). In 2008, a new type of high-T_c

superconductor containing iron (without copper) has been discovered (Yang et al., 2008). In Table 1, we list in chronological order the most important discoveries of superconductors containing oxygen. In Table 1, T_c is expressed in Kelvin and x is a variable atomic fraction of the doping element.

4. Valence skip and double valence fluctuations

What is the principal feature observed in all superconductors listed in Table 1? It is easy to verify that all metals used in the synthesis of superconductors containing oxygen have mixed oxidation states. For example, we can verify that in the superconductor NbO, Nb may have the oxidation states Nb(+III) and Nb(+V). In the bronze superconductor K_xWO_3, W may have the oxidation states W(+IV) and W(+VI). In the superconductor $LiTi_{2+x}O_4$, Ti may have the oxidation states Ti(+II) and Ti(+IV). In the superconductor $BaPb_{1-x}Bi_xO_3$, Pb may have the oxidation states: Pb(+II) and Pb(+IV) and Bi may have the oxidation states Bi(+III) and Bi(+V). Finally, we verify that in the copper oxide superconductors listed in Table 1, Cu may have the oxidation states Cu(+I) and Cu(+III).

Note also that in the superconductor $NdFeAsO_{1-x}$, an example of the recent discovery of iron-based superconductors (Yang et al., 2008), we can verify that Fe may have the oxidation states Fe(+II) and Fe(+IV) and As may have the oxidation states As(+III) and As(+V).

According to a number of authors the probable existence of double charge fluctuations in oxide superconductors is very likely (Callaway et al., 1987; Foltin, 1988; Ganguly & Hegde, 1988; Varma, 1988). Spectroscopic experiments (Ganguly & Hegde, 1988), indicate that double charge fluctuations is a necessary, but not sufficient, criterion for superconductivity. We argue that these charge fluctuations should involve paired electrons hoping from ions (or atoms) in order to occupy empty levels. That is, our basic phenomenological hypothesis is that the electrons involved in the hopping mechanisms might be paired electrons coming from neighboring ions or neighboring atoms.

The discovery of Fe-based high-T_c superconductors (Yang et al., 2008) has reopened the hypothesis of spin fluctuations for the microscopic mechanisms of high-T_c superconductivity. However, it is interesting to note that Fe may have the oxidation states Fe(+II) and Fe(+IV). Thus, the conjecture of double charge fluctuations cannot be ruled out in the study of the microscopic mechanisms in all Fe-based high-T_c superconductors. It is worthwhile to study the competition between double charge fluctuations and spin fluctuations in order to identify which phenomenon is more important in the microscopic mechanisms responsible for the condensation of the superconducting state of Fe-based materials.

What is valence skip? About fifteen elements in the periodic table skip certain valences in all compounds they form. For example, it is well known that the stable oxidation states of bismuth are Bi(+III) and Bi(+V). The oxidation state Bi(+IV) is not stable. If the state Bi(+IV) is formed, it occurs immediately a disproportionation, according to the reaction: 2Bi(+IV) = Bi(+III) + Bi(+V). In the compound $BaBiO_3$, the formal valence Bi(+IV) is understood as an equilibrium situation involving a mixture of equal amounts of the ions Bi(+III) and Bi(+V). Other important examples of elements with valence skip are As, Pb and Tl. In (Varma, 1988) there is an interesting discussion about the microscopic physics responsible for the phenomenon of valence skip. The electronic states of valence-skipping compounds are described in a conference paper (Hase & Yanagisawa, 2008). Elements with valence skip, like Bi and Pb, are the most appropriate elements to study the hypothesis of double charge fluctuations proposed in this chapter. It has been claimed that all elements with valence skip may be used in the synthesis of superconductors (Varma, 1988).

5. Oxygen doping by diffusion

The most relevant doping procedures used for the synthesis of cuprate superconductors have been described in a review article (Rao et al., 1993). Historically, the first synthesis of a high-T_c oxide superconductor was the copper oxide $Ba_xLa_{2-x}CuO_4$ (Bednorz & Muller, 1986). This superconductor is synthesized by doping the parent material La_2CuO_4 with Ba atoms. Soon after this discovery, it was realized (Schirber et al., 1988) that doping the parent material La_2CuO_4 with oxygen, without the introduction of any Ba atomic fraction x, it is also possible to synthesize the superconductor La_2CuO_{4+x}. Thus, in this case, we conclude that the introduction of oxygen is responsible for the doping mechanism of the parent material La_2CuO_4 (Schirber et al., 1988).

Oxide materials may become superconductors when a parent material is doped by the traditional doping mechanism with cation substitution or by a doping mechanism based on oxygen nonstoichiometry (De Jongh, 1988). If a certain oxide contains a metal with mixed oxidation numbers, by increasing (or decreasing) the oxygen content, the metal may be oxidized (or reduced) in order to maintain charge neutrality. Therefore, the synthesis of p-type superconductors may be obtained by doping the parent materials with an excess of oxygen atoms and the synthesis of n-type superconductors may be obtained by doping the parent materials with a deficiency of oxygen atoms.

The most famous example of oxygen doping is provided by the family of p-type oxide superconductors Y-Ba-Cu-O. It is well known that $YBa_2Cu_3O_{6+x}$, considering compositions x between x = 0.5 and x = 0.9, are superconductors, and with a maximum T_c corresponding to a composition x = 0.9. An important example of n-type superconductor is provided by the recent discovery of the superconductor $GdFeAsO_{1-x}$, a high-T_c superconductor with oxygen-deficiency; it has been shown that oxygen doping is a good and reliable procedure for the synthesis of a new family of iron-based high-T_c superconductors (Yang et al., 2008).

In the present chapter we shall study only oxygen doping of p-type oxide superconductors. It is well known that high-T_c superconductors are generally synthesized when a parent material is doped by the traditional doping mechanism with cation (or anion) substitution. But it seems that doping mechanisms based on oxygen nonstoichiometry are more appropriate than doping mechanisms based on cation (or anion) substitutions. However, doping mechanisms based on oxygen nonstoichiometry are not completely clear. In this chapter we give some ideas to study the microscopic mechanisms associated with oxygen doping of p-type oxide superconductors.

A normal atmosphere contains about 21% of O_2, 78% of N_2, and 1% of other gases. Consider an oxide material. When it is heated in a furnace containing atmospheric air at ambient pressure or containing an oxygen reach atmosphere, due to diffusion, some O_2 molecules may be absorbed in the bulk of the solid material. Thus, oxygen nonstoichiometry is a necessary consequence of heating processes of oxide materials submitted to ambient pressure or submitted to atmospheres containing an oxygen excess. The diffusion of oxygen in binary metal oxides has been studied in a book (Kofstad, 1983).

According to molecular orbital (MO) theory (Petrucci et al., 2000), to obtain the molecular orbital electronic configuration of O_2 molecule it is necessary to combine the atomic orbitals of two O atoms. We obtain the following molecular orbital electronic configuration of O_2 molecules:

[MO configuration of O_2]: [inner electrons]$(\sigma_{2s})^2(\sigma^*_{2s})^2(\pi_{2p})^4(\sigma_{2p})^2(\pi^*_{2p})^2$
Considering the above MO configuration of O_2 we conclude that there are two unpaired electrons in the orbital π^*_{2p}. By diffusion and solid state reaction, O_2 molecules may oxidize metal atoms or ions in the bulk material. Because there are two unpaired electrons, the O_2 molecule may pick two electrons in the neighboring metal ions, becoming a peroxide ion O_2(-II) species. Our hypothesis of formation of peroxide ion species in oxide superconductors is supported by a great number of spectroscopic measurements (Rao et al., 1987; Sarma et al., 1987; Dai et al., 1988; Mehta et al., 1992).
In order to understand the microscopic mechanisms of oxygen doping, we give an example. Consider an oxide material containing bismuth (without copper); for instance, consider $BaBiO_3$. As we have stressed in Section 4, the bismuth stable oxidation states are Bi(+III) and Bi(+V). When an oxygen molecule reacts with a Bi(+III) ion, it is reasonable to suppose that this ion gives two electrons to the O_2 molecule, that is, we may write the following reaction:

$$O_2 + Bi(+III) \rightarrow O_2(-II) + Bi(+V)$$

Considering this oxidation reaction we conclude that the formation of the peroxide species O_2(-II) corresponds to the creation of a double hole. In many oxide compounds the electrons or holes may be considered to be localized at lattice atoms forming lattice defects. In such a case, we may suppose that p-type conduction may involve hopping of electrons from site to site (Kofstad, 1983). Therefore, in the bismuth example just mentioned, we claim that conductivity (and superconductivity) may be explained by hopping of electron pairs that jump from neighboring sites to occupy hole pairs.
It is well known that for p-type superconductors the optimal oxygen doping of high-T_c oxide superconductors corresponds to a certain critical hole content. An under-doped superconductor is synthesized when the hole content is less than this critical value and an over-doped superconductor is synthesized when the hole content is greater than this critical value. The prediction of the optimal doping is an unresolved issue. In the next Section we propose a simple model to estimate the optimal doping of p-type oxide superconductors.

6. Optimal doping of p-type oxide superconductors containing bismuth without copper

Our basic hypothesis is that the existence of double charge fluctuations involving paired electrons may be a key to study the microscopic mechanisms in oxide superconductors. The essential concept in this hypothesis is that the hopping mechanism involves two paired electrons, instead of the hopping of a single electron. Our hypothesis may be easily applied in the oxide superconductors containing Bi (without Cu) because, in this case, it is well known that Bi (+III) and Bi (+V) are the only stable oxidation states for the Bi ions. Thus, double charge fluctuations may occur between the ions Bi (+III) and Bi (+V). In this Section we propose a simple method to calculate the optimal doping of oxide superconductors containing bismuth without copper and in the next Section we consider oxide superconductors containing copper without bismuth.
What should be the ideal chemical doping of oxide p-type superconductors containing bismuth without copper in order to obtain the maximum value of T_c? This optimal doping should be obtained by cation substitution or by increasing the oxygen content in the material.

Let us first suppose equal amounts of the ions Bi (+III) and Bi (+V). In this case the formal oxidation state of bismuth should be Bi (+IV). Thus, charge fluctuations should be balanced and the formation of holes is very difficult.

For example, it is well known that $BaBiO_3$ is an insulator because in this material the formal oxidation state of bismuth is Bi (+IV). However, by doping $BaBiO_3$ by cation substitution or by increasing the oxygen content a superconductor may be obtained. Chemical doping may destroy the balance between the ions Bi (+III) and Bi (+V). If the ions Bi (+V) are increased, it is possible to create holes. It is reasonable to suppose that the maximum concentration for optimal doping should correspond to a ratio [(Bi(+V) ions)/(Bi(+III) ions)] = 2, that is, the optimal concentration of the ions Bi (+V) should be the double of the concentration of the ions Bi (+III). Why we have proposed the ratio [(Bi(+V) ions)/(Bi(+III) ions)] = 2 for optimal doping? It is well known that T_c decreases when the hole concentration is higher than a certain critical concentration (Zhang & Sato, 1993). This important property is the nonmonotonic dependence of T_c on the carrier concentration, a high-T_c characteristic feature of all oxide superconductors. If the concentration of the ions Bi (+V) are further increased (and the concentration of the ions Bi (+III) are further decreased), the material becomes overdoped and T_c decreases. Thus, for optimal doping, the bismuth ion concentrations should be: (2/3)Bi (+V) and (1/3)Bi (+III). In the next section we propose an analogous simple model to estimate the optimal doping of p-type copper oxide superconductors (without bismuth).

Now we apply the above simple model to estimate the optimal doping of p-type oxide superconductors containing bismuth without copper. As an example, we apply our hypothesis to the material $Ba_{0.6}K_{0.4}BiO_x$, a famous oxide superconductor without copper, with T_c approximately equal to 30 K (Cava et al., 1988).

We shall suppose for the Bi ions the proportionality assumed in the model just suggested, that is, for optimal doping, we assume that the relative concentrations should be given by: (2/3)Bi(+V) and (1/3)Bi(+III). Considering the oxidation states Ba(+II) and K(+I) and using the charge neutrality condition, we have:

$$1.2 + 0.4 + (1/3)(3) + (2/3)(5) - 2x = 0 \tag{3}$$

From Equation (3) we obtain the result:

$$x = 2.97 \tag{4}$$

The result (4) is in good agreement with the value (x = 3) reported in the reference (Cava et al., 1988).

In the next section we extend the simple model just proposed to estimate the optimal doping of p-type oxide superconductors containing copper without bismuth.

7. Optimal doping of p-type oxide superconductors containing copper without bismuth

To apply our hypothesis to a copper oxide superconductor (without Bi) it should be necessary to suppose the existence of Cu (+I) because we are assuming double charge fluctuations between the states Cu(+I) and Cu(+III). In p-type Cu oxide superconductors, the existence of the oxidation state Cu(+III) is obvious by the consideration of charge neutrality. Thus, from an experimental point of view, it is very important to verify if the oxidation state Cu(+I) is present in the high-T_c Cu oxide superconductors. The probable existence of the states Cu(+I) and Cu(+III) has been verified in the works (Karppinen et al., 1993; Sarma & Rao, 1988).

It is generally believed that the microscopic mechanisms in a cuprate superconductor depends only on the ions Cu(+II) and Cu(+III), without the presence of the ions Cu(+I). Let us suppose that the hopping mechanism involves just a single electron between Cu(+II) and Cu(+III); if this single charge fluctuation would be responsible for superconductivity, we should conclude that the enhancement of Cu(+III) ions should produce a continuous enhancement of the critical temperature T_c. However, it is well known that T_c decreases when the hole concentration is higher than a certain concentration (Zhang & Sato, 1993). This important property is the nonmonotonic dependence of T_c on the carrier concentration, a high-T_c characteristic feature of all oxide superconductors. Thus, by this reasoning and considering the experimental results (Karppinen et al., 1993; Sarma & Rao, 1988), we can accept the presence of the mixed oxidation states Cu(+I), Cu(+II) and Cu(+III) in the copper oxide superconductors. On the other hand, this conjecture is supported if we consider the copper disproportionation reaction (Raveau et al, 1988): 2Cu(+II) = Cu(+I) + Cu(+III).

What should be the optimal chemical doping of Cu oxide superconductors in order to obtain the maximum value of T_c? Initially, considering the disproportionation reaction (Raveau et al, 1988): 2Cu(+II) = Cu(+I) + Cu(+III), we may suppose an equal probability for the distribution of the copper ions states Cu (+I), Cu (+II) and Cu (+III). Thus, the initial concentrations of these ions should be (1/3)Cu(+I), (1/3)Cu(+II) and (1/3)Cu(+III). However, we may suppose that by oxidation reactions, (1/3)Cu(+II) ions may be completely converted to (1/3)Cu(+III) ions. In this case, the maximum concentration of the Cu(+III) ions should be: (1/3) + (1/3) = (2/3). Thus, the optimal doping should correspond to the following maximum relative concentrations: (1/3)Cu(+I) ions and (2/3)Cu(+III) ions. That is, the optimal doping, should be obtained supposing the following concentration ratio: [(Cu(+III) ions)/(Cu(+I) ions)] = 2.

We apply this hypothesis to estimate the optimal doping of the famous cuprate superconductor $YBa_2Cu_3O_x$, where x is a number to be calculated. Using the relative values: (1/3) for Cu(+I) ions and (2/3) for Cu(+III) ions, we may write the formula unit: $YBa_2Cu_1(+I)Cu_2(+III)O_x$. Considering the oxidation states Y(+III), Ba(+II) and O(-II) and using the charge neutrality condition, we get:

$$3 + (2 \times 2) + (1 \times 1) + (2 \times 3) - 2x = 0 \tag{5}$$

From Equation (5) we obtain:

$$x = 7.0 \tag{6}$$

The result (6) is in good agreement with the result (x = 6.9) reported in (Kokkallaris et al., 1999).

Using the simple model just described, we estimate the necessary oxygen content to obtain the optimal doping of the most relevant p-type cuprate superconductors. It is important to note that the experimental determination of the oxygen content is a very difficult task. In Table 2 we have selected a number of works containing this experimental information.

Superconductor	Predicted	Measured	REFERENCE
$Ba_{0.15}La_{1.85}CuO_x$	x = 4.1	x = 4.0	Bednorz & Müller, 1986
La_2CuO_x	x = 4.2	x = 4.1	Schirber et al., 1988
$YBa_2Cu_3O_x$	x = 7.0	x = 6.9	Kokkallaris et al.,1999
$YBa_2Cu_4O_x$	x = 8.2	x = 8.0	Rao et al., 1993
Sr_2CuO_x	x = 3.17	x = 3.16	Hiroi et al., 1993
$HgBa_2Ca_2Cu_3O_x$	x = 8.5	x = 8.4	Hung et al., 1997

Table 2. Comparison between the values of oxygen content of copper oxide superconductors (not containing Bi) predicted by our simple model and the experimental values reported in the literature.

According to our model, we have used for the copper ions the following relative values: (1/3) for Cu(+I) ions and (2/3) for Cu(+III) ions. For the other elements in Table 2, we have considered the following stable oxidation states: La(+III), Y(+III), Ba(+II), Sr(+II), Ca(+II), Hg(+II) and O(-II). We verify that the results predicted by the simple model proposed here are in good agreement with the experimental results listed in Table 2.

8. Discussion

We believe that the simple model proposed in this paper in the case of p-type oxide superconductors could also be extended to estimate the optimal doping of n-type oxide superconductors. However, in the case of n-type oxide superconductors, the reaction produced by oxygen doping is a reduction reaction instead of an oxidation reaction that occurs in p-type oxide superconductors. Since we have not found in the literature any experimental determination of the oxygen content in the case of n-type oxide superconductors we shall not discuss this issue here. This question will be addressed in a future work.

We have proposed a simple model to estimate the relative concentrations of the ions involved to estimate the oxygen content for optimal doping of p-type oxide superconductors. The predictions based on this model are in good agreement with experimental results reported in the literature (Table 2). However, we emphasize that the model proposed in this chapter is not a complete theoretical model, it is just a simple phenomenological model.

Our conjectures can be used to explain some remarkable properties of high-T_c superconductors: (a) the anisotropy is explained considering that the electrons involved in the hopping mechanisms are 3d-electrons (in the case of copper oxide superconductors); (b) the order of magnitude of the coherence length (the mean distance between two electron pairs) is in accordance with the order of magnitude of the distance between the electron clouds of two neighboring ions; (c) the nonmonotonic dependence of T_c on the carrier concentration is explained by the hypothesis of double charge fluctuations and the optimal doping model proposed in this chapter.

The theory of bipolaronic superconductivity (Alexandrov & Edwards, 2000) is similar to our phenomenological model. In the theory of bipolaronic superconductivity, bipolarons are formed supposing a mechanism to bind two polarons. However, by our hypothesis, it is not necessary to suppose the formation of bipolarons by the binding of two polarons. We have assumed that the preformed pairs are just pairs of electrons existing in the electronic configurations of the ions or atoms involved in double charge fluctuations. These pairs should be, for example, lone pairs in atoms or ions or pairs of electrons in the electronic configurations obtained when Hund's rule is applied.

9. Concluding remarks

In this chapter we have studied the most relevant questions about the microscopic mechanisms of superconductivity in oxide materials. Parts of our arguments may be found in the list of references in the next Section. However, we believe that our ideas have been expressed in a clear form for the questions at hand.

The simple model described here is not a theoretical model and cannot be used to account quantitatively for the microscopic mechanisms responsible for superconductivity in oxide materials. However, we believe that our assumptions are helpful to the investigations of the microscopic mechanisms in oxide superconductors. We expect that this simple model will also be useful to encourage further experimental and theoretical researches in superconducting materials. It is worthwhile to study the details of the role of double charge fluctuations in the microscopic mechanisms responsible for superconductivity in oxide materials.

We have stressed in Section 2 that in p-type materials there are two types of interactions: Coulomb repulsions between electrons, but Coulomb attractions between electrons and holes. Thus, considering the possibility of a Bose-Einstein condensation, we claim that only p-type materials are qualified to be considered in the researches to obtain a room temperature superconductor.

Finally, we suggest some future researches. It is worthwhile to make experiments to verify if this model is correct. Supposing that this simple model works, it would be possible to calculate stoichiometric compositions in order to obtain the optimal doping in the researches to synthesize new oxide superconductors. It is well known that the most important method in semiconductor technology is obtained by ion implantation techniques. Similarly, we believe that ion implantation techniques probably will be important in superconductor technology as well. Thus, we hope that future researches based on ion implantation techniques could open a new route in the synthesis of high-T_c superconductors. These future researches, using ion implantation, should take advantage of the possibility of double charge hoping mechanisms, instead of single charge hoping mechanisms existing in the case of ion implantation in the semiconductor technology.

10. References

Alexandrov, A. S. & Edwards, P. P. (2000). High T_c cuprates: a new electronic state of matter? *Physica C*, 331, pp. 97-112

Bardeen, J.; Cooper, L. N. & Schrieffer, J. R. (1957). Theory of superconductivity. *Phys. Rev.*, 108, 5, pp. 1175-1204

Bednorz, J. G. & Müller, K. A. (1986). Possible high T_c superconductivity in the Ba-La-Cu-O system. *Zeitschrift fur Physik B, Condensed Matter*, 64, 2, pp. 189-193

Blatt, J. M. (1962). Qualitative arguments concerning the Bose-Einstein condensation of fermion pairs. *Progress of Theoretical Physics*, 27, pp. 1137-1142

Callaway, J.; Kanhere, D. G & Misra, P. K. (1987). Polarization-induced pairing in high-temperature superconductivity. *Phys. Rev. B*, 3, 13, pp. 7141-7144

Canright, J. S. & Vignale, G. (1989). Superconductivity and acoustic plasmons in the two-dimensional electron gas. *Phys. Rev. B*, 39, 4, pp. 2740-2743

Cava, R. J.; Batlogg, B.; Krajewski, J. J.; Farrow, R.; Rupp Jr, L. W.; White, A. E.; Peck, W. E. & Kometani, T. (1988). Superconductivity near 30 K without copper: the $Ba_{0.6}K_{0.4}BiO_3$ system. *Nature*, 332, pp. 814-816

Cava, R. J. (2000). Oxide superconductors. *Journal American Ceramic Society*; 83, 1, pp. 5-28

Dai, Y.; Manthiram, A.; Campion, A. & Goodenough, J. B. (1988). X-ray-photoemission-spectroscopy evidence for peroxide in 1:2:3 copper oxides containing disordered or excess oxygen. *Physical Review* B 38, 7, pp. 5091-5094

De Jongh, L. J. (1988). A comparative study of (bi)polaronic (super)conductivity in high- and low-T_c superconducting oxides. *Physica C: Superconductivity*, 152, pp. 171-216

Emin, D. (1991). Large bipolarons and superconductivity. *Physica C: Superconductivity*, 185-189, Part 3, pp. 1593-1594

Foltin, J. (1988). Attractive interaction between electrons: An electron-pairing mechanism for superconductivity. *Phys. Review B*, 38, 15, pp. 10900-10902

Ganguly, P. & Hegde, M. S. (1988). Evidence for double valence fluctuation in metallic oxides of lead. *Phys. Rev. B*; 37, 10, pp. 5107-5111

Hase, I. & Yanagisawa, T. (2008). Electronic states of valence-skipping compounds. *Journal of Physics: Conference Series*, 108, 012011, pp. 1-4

Hiroi, Z.; Takano M.; Azuma, M. & Takeda, Y. (1993). A new family of copper oxide superconductors $Sr_{n+1}Cu_nO_{2n+1+\delta}$ stabilized at high pressure. *Nature*, 364, 6435, pp. 315-317

Hirsch, J. E. (1991). Bose condensation versus pair unbinding in short-coherence-length superconductors. *Physica C: Superconductivity*, 179, pp. 317-332

Hung, K. C.; Lam, C.; Shao, H. M.; Wang, S. D. & Yao, X. X.; (1997). Enhancement in flux pining and irreversibility field by means of a short time annealing technique for $HgBa_2Ca_2Cu_3O_{8.4}$ superconductor. *Superc. Science Technology*, 10, 11, pp. 836-842

Johnston, D. C.; Prakash, H.; Zachariessen, W. H. & Vishvanathan, B. (1973) High temperature superconductivity in the Li-Ti-O ternary system. *Materials Research Bulletin*; 8, 7, pp. 777-784

Karppinen, M.; Fukuoca, A.; Wang, J.; Takano, S.; Wakata, M.; Ikemachi, T. & Yamauchi, H. (1993). Valence studies on various superconducting bismuth and Lead cuprates and related materials. *Physica C: Superconductivity*, 208, pp. 130-136

Kofstad, P. (1983). *Diffusion and Electrical Conductivity in Binary Metal Oxides*. Robert E. Krieger Publ. Co., Malabar, Florida

Kokkallaris, S.; Deligiannis, K.; Oussena, M.; Zhukov, A. A.; Groot, P. A. J.; Gagnon, R. & Taillefer, L. (1999). Effect of oxygen stoichiometry on the out-of-plane anisotropy of $YBa_2Cu_3O_{7-\delta}$ single crystals near optimal doping. *Superconductor Science Technology*, 12, 10, pp. 690-693

Lee, D. M. (1997). The extraordinary phases of liquid ^{3}He. *Reviews of Modern Physics*, 69, pp. 645-666

London, F. (1938). On the Bose-Einstein condensation. *Phys. Rev.*, 54, pp. 947-954

Luiz, A. M. (2010). A model to study microscopic mechanisms in high-Tc superconductors, *Superconductor*, Adir Moysés Luiz (Ed.), ISBN: 978-953-307-107-7, Sciyo, Available from the site: http://www.intechopen.com/articles/show/title/a-model-to-study-microscopic-mechanisms-in-high-tc-superconductors

Maeda, H.; Tanaka, Y.; Fukutomi, M. & Asano, T. (1988). A new high-T_c oxide superconductor without a rare earth element. *Japanese Journal of Applied Physics*; 27, 2, pp. L209-L210

Mehta, A.; DiCarlo, J. & Navrotsky, A. (1992). Nature of hole states in cuprate superconductors. *Journal of Solid State Chemistry.*, 101, pp. 173-185

Petrucci, R. H.; Harwood, W. S. & Herring, F. G. (2000). *General Chemistry - Principles and Modern Applications*, Eighth Edition., Prentice Hall, ISBN 0-13-014329-4, New Jersey

Ranninger, J. (1994). The polaron scenario for high T_c superconductivity. *Physica C: Superconductivity*, 235-240, Part 1, pp. 277-280

Rao , C. N. R.; Ganguly, P.; Hegde, M. S. & Sarma, D. D. (1987). Holes in the oxygen (2p) valence bands and the concomitant formation of peroxide-like species in metal oxides: their role in metallicity and superconductivity. *J. Am. Chem. Soc.*, 109, pp. 6893-6895

Rao, C. N. R.; Nagarajan R. & Vijayaraghavan, R. (1993). Synthesis of cuprate superconductors. *Superconductor Science Technology*, 6, 1, pp. 1-22

Raveau, B.; Michel, C.; Hervieu, M. & Provost, J. (1988). Crystal chemistry of perovskite superconductors. *Physica C: Superconductivity*, 153-155, pp. 3-8

Remeika, J. P.; Geballe, T. H.; Mathias, B. T.; Cooper, A. S.; Hull, G. W. & Kellye, M. (1967). Superconductivity in hexagonal tungsten bronzes. *Physics Letters A*, 24, 11, pp. 565-566

Sarma, D. D. & Rao, C. N. R. (1988). Nature of the copper species in superconducting $YBa_2Cu_3O_{7-\delta}$. *Sol. State Commun.*, 65, pp. 47-49

Sarma, D. D.; Sreedhar, K.; Ganguly, P. & Rao, C. N. R. (1987). Photoemission study of $YBa_2Cu_3O_7$ through the superconducting transition: Evidence for oxygen dimerization. *Physical Review* B 36, 4, pp. 2371-2373

Schilling, A. & Cantoni, M. (1993). Superconductivity above 130 K in the Hg-Ba-Ca-Cu-O system. *Nature*, 363, 6424, pp. 56-58

Schirber, J. E.; Morosin, B.; Merrill, R. M.; Hilava, P. F.; Venturinie L.; Kwak, J. F.; Nigrey, P. J.; Baughman, R. J. & Ginley, D. S. (1988). Stoichiometry of bulk superconducting $La_2CuO_{4+\delta}$: A superconducting superoxide? *Physica c*, 152, 1, 121-123

Shimakawa, Y.; Kubo, Y.; Manako, T.; Nakabayashi, Y. & Igarashi, H. (1988). Ritveld analysis of $Tl_2Ba_2Ca_{n-1}Cu_nO_{4+n}$ (n = 1, 2, 3) by powder X-ray diffraction. *Physica C: Superconductivity*, 156, 1 (1 August 1988), pp. 97-102

Silvera, I. F. (1997). Bose-Einstein condensation. *American Journal of Physics*, 65, 570-574

Sleight, A. W.; Gilson, J. L. & Bierstedt, P. E. (1975). High-temperature superconductivity in the $BaPb_{1-x}Bi_xO_3$ system. *Solid State Commun.*, 17, pp. 27-28

Sleight, A. W. (1995). Room temperature superconductivity. *Acc. Chem. Res.*, 28, pp. 103-108

Tachiki, M. & Takahashi, S. (1988). Pairing interaction mediated by the Cu-O charge-transfer oscillations associated with LO phonons in oxide superconductors and their high-T_c superconductivity. *Phys. Rev. B*, 38, 1, pp. 218-224

Takada, Y. (1993). S- and p-wave pairings in the dilute electron gas: Superconductivity mediated by the Coulomb hole in the vicinity of the Wigner-crystal phase. *Phys. Rev. B*, 47, 9, pp. 5202-5211

Tolmachev, V. V. (2000). Superconducting Bose-Einstein condensation of Cooper pairs interacting with electrons. *Physics Letters* A 266, pp. 400-408

Varma, C. M. (1988). Missing valence states, diamagnetic insulators, and superconductors. *Phys. Rev. Letters*; 61, 23, pp. 2713-2716

Wu, M. K.; Ashburn J. R.; Torng, C. J.; Hor, P.H.; Meng, R. L.; Goa, L.; Huang, Z. J.; Wang, Y. Q. & Chu, C. W. (1987). Superconductivity at 93 K in a new mixed-phase Y-Ba-Cu-O compound system at ambient pressure. *Phys. Rev. Letters*, 58, 9, pp. 908-910

Yang, J.; Li, Z. C.; Lu, W.; Yi, W.; Shen, X. L.; Ren, Z. A.; Che, G. C.; Dong, X. L.; Sun, L. L.; Zhou, F. & Zhao, Z. X. (2008). Superconductivity at 53.5 K in $GdFeAsO_{1-\delta}$. *Superconductor Science Technology*, 21, 082001, pp. 1-3

Zhang, H. & Sato, H. (1993). Universal relationship between T_c and the hole content in p-type cuprate superconductors. *Phys. Rev. Letters*, 70, 11, pp. 1697-1699

2

Unconventional Superconductivity Realized Near Magnetism in Hydrous Compound $Na_x(H_3O)_zCoO_2 \cdot yH_2O$

Yoshihiko Ihara[1] and Kenji Ishida[2]
[1]*Department of Physics, Faculty of Science, Hokkaido university*
[2]*Department of Physics, Graduate School of Science, Kyoto university*
Japan

1. Introduction

Two different ground states, superconductivity and magnetism, were believed to be incompatible, and impossible to coexist in a single compound. The Ce based heavy Fermion superconductor $CeCu_2Si_2$, however, was discovered in the vicinity of magnetic phase Steglich et al. (1979). This new class of superconductors, which are referred to as "unconventional" superconductor, demonstrate various novel properties which are not accounted for in the framework of the BCS theory. Electrons in unconventional superconductors are strongly correlated through the Coulomb interaction, while strong electron-electron correlations are not preferable for the conventional BCS superconductors. Modern theory predicts that the repulsive Coulomb interaction can induce attractive interaction to form superconducting Cooper pairs as the result of many-body effect

Unconventional superconductivity is often observed nearby a quantum critical point (QCP), where magnetic instability is suppressed to $T = 0$ by some physical parameters. It is invoked that the quantum critical fluctuations, which are enhanced around QCP, drive the superconducting pairing interactions, instead of the electron-phonon interaction proposed in the BCS theory. In addition to the novel pairing mechanisms, unconventional superconductivity shows various novel superconducting states, such as Fulde-Ferrell-Larkin-Ovchinnikov state and spin-triplet pairing state. Unveiling novel mechanism and resulting novel properties is the main topic of condensed matter physics.

The cobaltate compound is also classified to an unconventional superconductor when we take the results of nuclear spin-lattice relaxation rate Fujimoto et al. (2004); Ishida et al. (2003), and specific heat Yang et al. (2005) measurements into account. A power-law temperature dependence, observed for both physical quantities in the superconducting state, yields the existence of nodes (zero gap with sign change) on the superconducting gap, and addresses the unconventional pairing mechanism. Besides, a magnetic instability was found in the sufficiently water intercalated cobaltates Ihara, Ishida, Michioka, Kato, Yoshimura, Takada, Sasaki, Sakurai & Takayama-Muromachi (2005). The close proximity of superconductivity to magnetism in cobaltates lead us to consider that the same situation as heavy Fermion superconductors is realized in cobaltates.

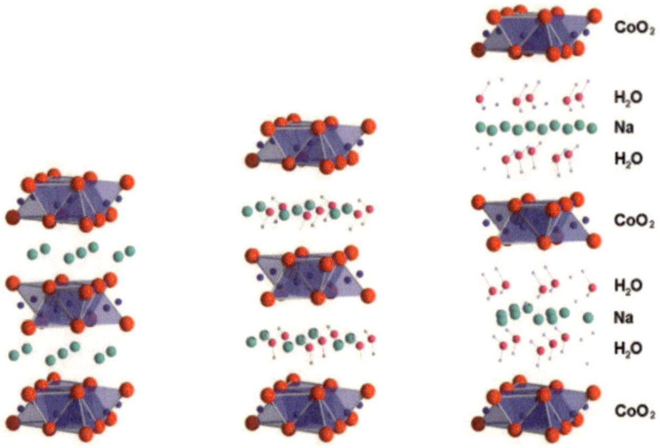

Fig. 1. Crystal structures of non-hydrate, mono-layer hydrate and bilayer hydrate compounds.

In this chapter, the relationship between superconductivity and magnetism will be explored from the nuclear magnetic resonance (NMR) and nuclear quadrupole resonance (NQR) experiments on superconducting and magnetic cobaltate. The principles of experimental technique is briefly reviewed in §3. Then the experimental results are presented in the following sections, §4 and §5. Finally, we will discuss the superconducting paring mechanisms in bilayer-hydrate cobaltate, showing the similarity between cobaltate and heavy Fermion superconductors.

2. Water induced superconductivity in $Na_x(H_3O)_zCoO_2 \cdot yH_2O$

Superconductivity in a cobalt oxide compound was discovered in 2003 Takada et al. (2003). The hydrous cobaltate $Na_x(H_3O)_zCoO_2 \cdot yH_2O$ demonstrates superconductivity when water molecules are sufficiently intercalated into the compound by a soft chemical procedure. In contrast, anhydrous Na_xCoO_2 does not undergo superconducting transition at least above 40 mK Li et al. (2004). A peculiarity of superconductivity in $Na_x(H_3O)_zCoO_2 \cdot yH_2O$ compound is the necessity of sufficient amount of water intercalation between the CoO_2 layers, and depending on the water content, superconducting transition temperatures vary from 2 K to 4.8 K. This compound is the first superconductor which shows superconductivity only in the hydrous phase.

The cobaltate compound has three types of crystal structures with different water concentrations as shown in Fig. 1. The parent compound Na_xCoO_2, which is $y = 0$ and $z = 0$, contains the randomly occupied Na layer between the CoO_2 layers. When Na ions are deintercaleted and water molecules are intercalated between the CoO_2 layers, the crystal structure changes to bilayer hydrate (BLH) structure, in which the Na layer is sandwiched with double water layers to form H_2O-Na-H_2O block layer. Due to the formulation of this thick block layer, the CoO_2 layers are separated by approximately 10 Å, and is considered to have highly two-dimensional nature. Superconductivity is observed in this composition below 5 K. The crystal structure of the superconducting BLH compound changes to mono-layer hydrate (MLH) structure, which forms Na-H_2O mixed layers between the CoO_2 layers containing less water molecules than those of BLH compounds. The water molecules inserted between CoO_2 layers are easily evaporated into the air at an ambient

condition, seriously affecting the physical properties, namely superconductivity, of BLH compounds.
A BLH compound left in the vacuum space for three days becomes a MLH compound, and does not demonstrate superconductivity. Inversely the crystal structure of the MLH compound stored in high-humid atmosphere comes back to the BLH structure, and superconductivity recovers, although the transition temperature of the BLH compound after the dehydration-hydration cycle is lower than that of a fresh BLH compound. Superconducting and normal-state properties in various kinds of samples have been investigated with several experimental methods.
In the normal state, spin susceptibility is almost temperature independent above 100 K, which is a unique behavior irrespective of samples. The sample dependence appears below 100 K, for instance, spin susceptibility increases with decreasing temperature in some samples, but some do not. From the temperature dependence of spin susceptibility below 100 K, temperature independent susceptibility χ_0, effective moment μ_{eff} and Weiss temperature Θ_W were reported to be $\chi_0 = 3.02 \times 10^{-4}$ emu/mol, $\mu_{\text{eff}} \sim 0.3\ \mu_B$ and $\Theta_W = -37$ K by Sakurai et al. (2003). Different values are reported by Chou, Cho & Lee (2004), where the increase of susceptibility toward low temperature is hardly observed, and correspondingly, μ_{eff} is rather small. Although the low temperature behavior of susceptibility is strongly sample dependent, we believe that the slight increase below 100 K is an intrinsic behavior because it is observed in most of the high-quality powder samples and also observed in the Knight shift measured by nuclear magnetic resonance Ihara, Ishida, Yoshimura, Takada, Sasaki, Sakurai & Takayama-Muromachi (2005) and muon spin rotation measurements Higemoto et al. (2004).
In the superconducting state, specific heat is intensively measured by several groups Cao et al. (2003); Chou, Cho, Lee, Abel, Matan & Lee (2004); Jin et al. (2005); Lorenz et al. (2004); Oeschler et al. (2005); Ueland et al. (2004); Yang et al. (2005). The specific heat jump at superconducting transition temperature $\Delta C/\gamma T_c$ is estimated to be approximately 0.7, which is half of the BCS value 1.43. The small jump suggests either the quality of the sample is insufficient, or the superconductivity is an unconventional type with nodes. Below the superconducting transition temperature, C/T does not follow exponential temperature dependence but follows power law behavior, which is universally observed in unconventional superconductor. The power law behavior observed from nuclear-spin-lattice relaxation rate $1/T_1$ measurement also supports unconventional superconductivity Fujimoto et al. (2004); Ishida et al. (2003). The results of $1/T_1$ measurements are presented in § 4. The values of Sommerfeld constant, which are $12 \sim 16$ mJ/molK^2 depending on samples, are comparable to those of anhydrous compound $Na_{0.3}CoO_2$ and less than those of mother compound $Na_{0.7}CoO_2$. It is curious that the BLH compound which has smaller density of state compared to mother compounds demonstrates superconductivity, while a single crystal of $Na_{0.7}CoO_2$ with larger density of state is not a superconductor. The hydrous phases ought to have specific mechanisms to induce superconducting pairs.
The discovery of magnetism in a sufficiently water intercalated BLH compound provides important information to understand the origin of superconductivity. The superconducting BLH is located in the close vicinity of magnetic phase, as in the case for heavy Fermion superconductors. This similarity lets us invoke that the magnetic fluctuations near magnetic criticality can induce superconductivity in BLH system. The magnetic fluctuations are examined in detail with nuclear quadrupole resonance and nuclear magnetic resonance technique in order to unravel the superconducting mechanisms.

3. Nuclear magnetic resonance and nuclear quadrupole resonance

3.1 Nuclear quadrupole resonance measurement

In this section, the fundamental principles of nuclear quadrupole resonance (NQR) are briefly reviewed. Resonance phenomena are observed between split nuclear states and radio frequency fields with energy comparable to the splitting width. To observe the resonance, degenerated nuclear spin states have to be split by a magnetic field and/or an electric-field gradient (EFG). For NQR measurements, a magnetic field is not required because the nuclear levels are split only by the electric-field gradient. Under zero magnetic field, nuclear levels are determined by the electric quadrupole Hamiltonian $\mathcal{H}_Q$, which describes the interaction between the electric quadrupole moment of the nuclei Q and the EFG at the nuclear site. In general, $\mathcal{H}_Q$ is expressed as

$$\mathcal{H}_Q = \frac{\nu_{zz}}{6}\left[(3I_z^2 - \boldsymbol{I}^2) + \frac{1}{2}\eta(I_+^2 + I_-^2)\right], \qquad \left(\nu_{zz} = \frac{6e^2qQ}{4I(2I-1)}\right) \tag{1}$$

where $eq(= V_{zz})$ and $\eta = (V_{yy} - V_{xx})/V_{zz}$ are the EFG along the principal axis (z axis) and the asymmetry parameter, respectively. The resonant frequency is calculated by solving the Hamiltonian. From the measurement of these resonant frequencies, ν_{zz} and η are estimated separately. These two quantities provide information concerning with the Co-3d electronic state and the subtle crystal distortions around the Co site, because the EFG at the Co site is determined by on-site 3d electrons and ionic charges surrounding the Co site. The ionic charge contribution is estimated from a calculation, in which the ions are assumed to be point charges (point-charge calculation). The result of the point-charge calculation indicates that V_{zz} is mainly dependent on the thickness of the CoO_2 layers, because the effect of the neighboring O^{2-} ions is larger than that of oxonium ions and Na^+ ions that are distant from the Co ions. Due to the ionic charge contribution, the resonant frequency was found to increase with the compression of the CoO_2 layers.

When small magnetic fields, which are comparable to EFG, are applied, the perturbation method is no longer valid to estimate the energy level. The resonant frequency should be computed numerically by diagonalizing Hamiltonian, which includes both Zeeman and electric quadrupole interactions. The total Hamiltonian is expressed as

$$\mathcal{H} = \mathcal{H}_Q + \mathcal{H}_Z \tag{2}$$

$$= \frac{\nu_{zz}}{6}\left[(3I_z^2 - \boldsymbol{I}^2) + \frac{1}{2}\eta(I_+^2 + I_-^2)\right] - \gamma\hbar\left(I_zH_z + I_xH_x + I_yH_y\right). \tag{3}$$

The results of a numerical calculation is displayed in Fig. 2, where the parameters ν_{zz}, η are set to be 4.2 MHz and 0.2, respectively. The shift of the resonant frequency depends on the direction of the small magnetic fields. The magnetic fields parallel to the principal axis of EFG (z axis) affect the transition between the largest m states, while spectral shift of this transition is small when the magnetic fields are perpendicular to z axis. In other word, when internal magnetic fields appear in the magnetically ordered state, the direction of the internal fields can be determined from the observation of NQR spectrum with the highest resonant frequency above and below the magnetic transition.

3.2 Nuclear spin-lattice relaxation rate

The nuclear spin system has weak thermal coupling with the electron system. Through the coupling, heat supplied to the nuclear spin system by radio-frequency pulses flows into the

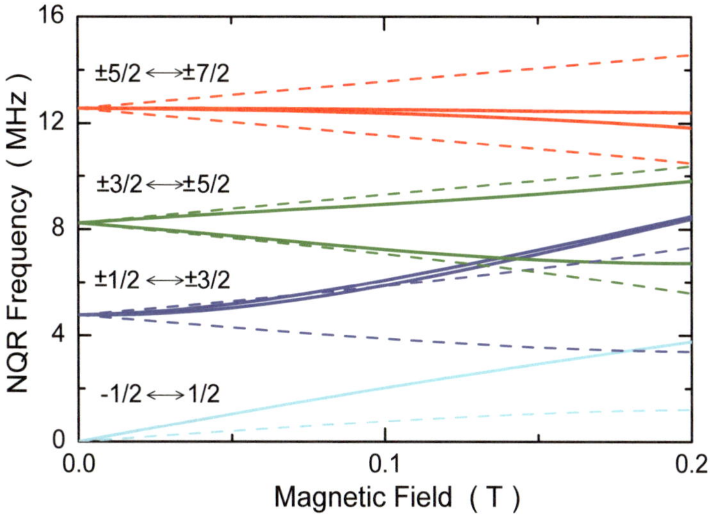

Fig. 2. NQR frequency calculated from equation (3) with small magnetic fields up to 0.2 T. The solid and dashed lines are the resonant frequency in magnetic fields perpendicular and parallel to the principal axis of EFG (*z* axis). The parameters ν_{zz} and η were set to the realistic values of 4.2 MHz and 0.2, respectively

electron system, and the excited nuclear spin system relaxes after a characteristic time scale T_1. The nuclear spin-lattice relaxation rate $1/T_1$ contains important information concerning with the dynamics of the electrons at the Fermi surface.

When the nuclear spin system is relaxed by the magnetic fields induced by electron spins $\delta\boldsymbol{H}$, the transition probability is formulated by using the Fermi's golden rule as

$$W_{m,\nu\to m+1,\nu'} = \frac{2\pi}{\hbar}\left|\langle m,\nu|\gamma_{\rm n}\hbar\boldsymbol{I}\cdot\delta\boldsymbol{H}|m+1,\nu'\rangle\right|^2\delta(E_m+E_\nu-E_{m+1}-E_{\nu'}). \quad (4)$$

With this transition probability, $1/T_1$ is defined as

$$\frac{1}{T_1} = \sum_{\nu,\nu'}\frac{2W_{m,\nu\to m+1,\nu'}}{(I-m)(I+m+1)} \quad (5)$$

$$= \frac{\gamma_{\rm n}^2}{2}\int_{-\infty}^{\infty}dt\cos\omega_0 t\left\langle\frac{\delta H_+(t)\delta H_-(0)+\delta H_-(t)\delta H_+(0)}{2}\right\rangle. \quad (6)$$

Equation (6) is derived by expanding equation (4) with the relation $\delta H_+(T) = \exp(i\mathcal{H}t/\hbar)\delta H_+\exp(-i\mathcal{H}t/\hbar)$. When fluctuation-dissipation theorem is adopted to equation (6), the general representation for $1/T_1$ is obtained.

$$\frac{1}{T_1} = \frac{2\gamma_n^2 k_{\rm B}T}{(\gamma_e\hbar)^2}\sum_{\boldsymbol{q}}A_{\boldsymbol{q}}A_{-\boldsymbol{q}}\frac{\chi''_\perp(\boldsymbol{q},\omega_0)}{\omega_0}. \quad (7)$$

Here, $A_{\boldsymbol{q}}$ and $\chi''_\perp(\boldsymbol{q},\omega)$ are the coupling constant in $\boldsymbol{q}$ space and the imaginary part of the dynamical susceptibility, respectively. ω_0 in the equation is the NMR frequency, which is usually less than a few hundreds MHz. The $\boldsymbol{q}$ dependence of $A_{\boldsymbol{q}}$ is weak, when $1/T_1$ is measured at the site where electronic spins are located.

The temperature dependence of $1/T_1$ has been studied by using the self-consistent renormalization (SCR) theory Moriya (1991). The dynamical susceptibility assumed in this theory is formulated as

$$\chi(\boldsymbol{Q}+\boldsymbol{q},\omega) = \frac{\chi(\boldsymbol{Q})}{1+\boldsymbol{q}^2 A\chi(\boldsymbol{Q}) - i\omega\boldsymbol{q}^{-\theta}C\chi(\boldsymbol{Q})}. \quad (8)$$

The dynamical susceptibility is expanded in $\boldsymbol{q}$ space around $\boldsymbol{Q}$, which represents the ordering wave vector. When the magnetic ordering is ferromagnetic ($\boldsymbol{Q}=0$), $\theta=1$ is used to calculate $\chi(\boldsymbol{q},\omega)$, and when it is anti-ferromagnetic ($\boldsymbol{Q}>0$), θ is zero. The parameters A and C are determined self-consistently to minimize the free energy. The SCR theory is available when the electronic system is close to a magnetic instability. The temperature dependence of $1/T_1$ above T_C and T_N is derived using the renormalized dynamical susceptibility. The results depend on the dimensionality and the θ values, which determine that the magnetic ordering is ferromagnetic (FM) or anti-ferromagnetic (AFM). The temperature dependence anticipated from the SCR theory is listed in Table 1.

	2-D	3-D
FM	$1/T_1 \propto T/(T-T_C)^{3/2}$	$1/T_1 \propto T/(T-T_C)$
AFM	$1/T_1 \propto T/(T-T_N)$	$1/T_1 \propto T/(T-T_N)^{1/2}$

Table 1. Temperature dependence of $1/T_1$ anticipated from the SCR theory.

In the superconducting state, thermally exited quasiparticles can contribute to the Knight shift K and the nuclear spin-lattice relaxation rate $1/T_1$. Since the quasiparticles do not exist within the superconducting gap, the energy spectrum of the quasiparticle density of state is expressed as

$$N(E;\theta,\phi) = \frac{N_0 E}{\sqrt{E^2-\Delta^2(\theta,\phi)}} \qquad \text{for } E>\Delta(\theta,\phi) \quad (9)$$

$$= 0 \qquad \text{for } 0<E<\Delta(\theta,\phi), \quad (10)$$

where E is the quasiparticle energy, which is determined as $E^2 = \varepsilon^2+\Delta^2$, and N_0 is the density of state in the normal state. In order to obtain the total density of state $N(E)$, $N(E;\theta,\phi)$ should be integrated over all the solid angles. $N(E)$ is calculated with considering simple angle dependence for Δ, and the resulting energy spectra are represented in Fig. 3. The angle dependence of the superconducting gap, which we considered for the calculations, are listed below.

$$\Delta(\theta,\phi) = \Delta_0 \qquad \text{for } s\text{-wave}, \quad (11)$$

$$\Delta(\theta,\phi) = \Delta_0 \sin\theta e^{i\phi} \qquad \text{for } p\text{-wave axial state}, \quad (12)$$

$$\Delta(\theta,\phi) = \Delta_0 \cos\theta e^{i\phi} \qquad \text{for } p\text{-wave polar state}, \quad (13)$$

$$\Delta(\theta,\phi) = \Delta_0 \cos 2\phi \qquad \text{for two-dimensional } d\text{-wave}. \quad (14)$$

The temperature dependence of the Knight shift and the nuclear spin-lattice relaxation rate in the superconducting state can be computed from the following equations with using $N(E)$

calculated above. The temperature dependence of the gap maximum Δ_0 was assumed to be that anticipated from the BCS theory for all the gap symmetry considered above.

$$K = \frac{A_{\rm hf}}{N\mu_{\rm B}}\chi_{\rm s} = -\frac{4\mu_{\rm B}A_{\rm hf}}{N}\int N(E)\frac{\partial f(E)}{\partial E}dE, \tag{15}$$

$$\frac{1}{T_1} = \frac{\pi A_{\rm hf}^2}{\hbar N^2}\int\left(1+\frac{\Delta_0^2}{E^2}\right)N(E)^2 f(E)(1-f(E))dE. \tag{16}$$

Here, $f(E)$ is the Fermi distribution function. The term $(1+\Delta_0^2/E^2)$ in equation (16) is referred to as the coherence factor, which is derived from the spin flip process of unpaired electrons through the interactions between the unpaired electrons and the Cooper pairs Hebel & Slichter (1959).

As an initial state $|i\rangle$, it is assumed that one unpaired electron with up spin and one Cooper pair have wave numbers k and k', respectively. After the interaction between these particles, the wave numbers are exchanged, and the electronic spin of the unpaired electron can flip with preserving energy. The final state $|f\rangle$ is chosen to be one electron with down spin and wave number $-k'$, and one Cooper pair with wave number k. This process can contribute to the relaxation rate by exchanging the electronic spins and the nuclear spins. The initial state and the final state are expressed with using the creation and annihilation operators c_k^* and c_k

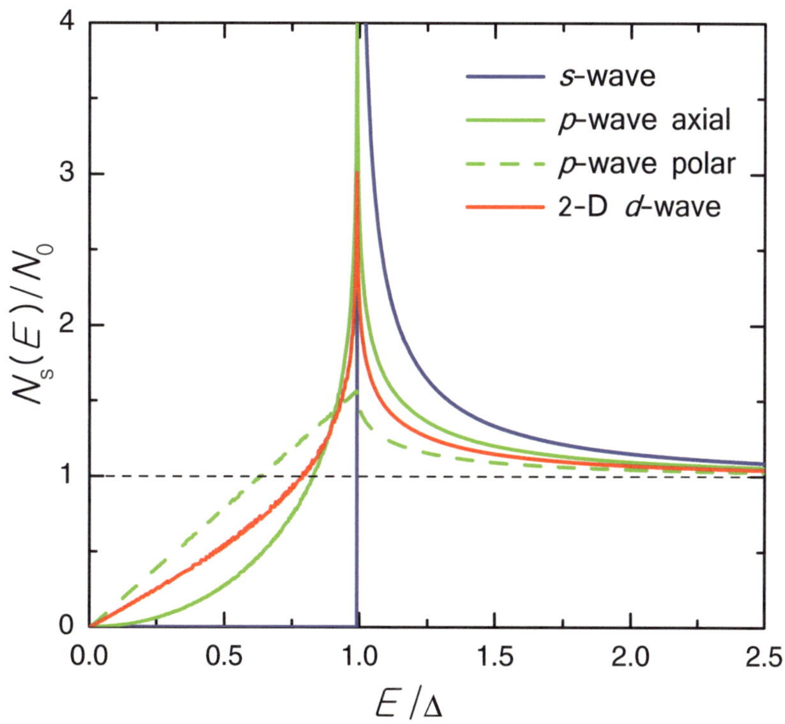

Fig. 3. Quasiparticle density of state in various superconducting states. Finite $N_s(E)$ exists at the energy lower than maximum gap size, when superconducting gap has nodes. The energy dependence of $N_s(E)$ at low energy determines the exponents for the power-law temperature dependence of physical quantities.

as

$$|i\rangle = \left[\sqrt{1-h_{k'}} + \sqrt{h_{k'}} c^*_{k'\uparrow} c^*_{-k'\downarrow}\right] c^*_{k\uparrow} |\psi_0\rangle \tag{17}$$

$$|f\rangle = \left[\sqrt{1-h_k} + \sqrt{h_k} c^*_{k\uparrow} c^*_{-k\downarrow}\right] c^*_{-k'\downarrow} |\psi_0\rangle, \tag{18}$$

where ψ_0 represents the vacuum state, and h_k is determined as $h_k = (1 - \varepsilon_k/E_k)/2$. The perturbation Hamiltonian, which flips one electronic spin, is described as

$$\mathcal{H}_{\rm p} = A \sum_{k,k'} \frac{1}{2} \left(I_+ c^*_{k'\downarrow} c_{k\uparrow} + I_- c^*_{k'\uparrow} c_{k\downarrow} \right). \tag{19}$$

The transition probability from $|i\rangle$ to $|f\rangle$ originating from $\mathcal{H}_{\rm p}$ is derived as

$$\left|\langle f|\mathcal{H}_{\rm p}|i\rangle\right|^2 = \frac{1}{2}\left(\sqrt{1-h_k}\sqrt{1-h_{k'}} + \sqrt{h_k}\sqrt{h_{k'}}\right)^2 \tag{20}$$

$$= \frac{1}{2}\left(1 + \frac{\varepsilon_k \varepsilon_{k'}}{E_k E_{k'}} + \frac{\Delta^2}{E_k E_{k'}}\right). \tag{21}$$

Here, the anticommutation relation of c^*_k and $E_k = \sqrt{\varepsilon_k^2 + \Delta^2}$ are used. The second term of equation (21) is canceled out when this term is integrated over the Fermi surface, because ε_k is the energy from the Fermi energy. The counter process that a down spin flips to an up spin should also be considered. When this process is included, the transition probability becomes twice of $|\langle f|\mathcal{H}_{\rm p}|i\rangle|^2$. The third term $\Delta^2/E_k E_{k'}$ possesses finite value only for an *s*-wave superconductor with an isotropic superconducting gap, such as Al metal Hebel & Slichter (1959). This term vanishes when the superconducting pairing symmetry is *p*-wave and two-dimensional *d*-wave type, which are represented in equations (12), (13) and (14).
In unconventional superconductors with an anisotropic order parameter, the temperature dependence of $1/T_1$ shows power-law behavior, because the quasiparticle density of states exist even below the maximum value of the gap Δ_0. The existence of the density of states in the small energy region originates from the nodes on the superconducting gap. In the *p*-wave axial state, where the superconducting gap possesses point nodes at $\theta = 0, \pi$, $N(E)$ is proportional to E^2 near $E = 0$. As a result, $1/T_1$ is proportional to T^5 far below $T_{\rm c}$. In the *p*-wave polar state and the *d*-wave state, where the superconducting gap possesses line nodes at $\theta = \pi/2$ and $\phi = 0, \pi/2$, respectively, the temperature dependence of $1/T_1$ becomes T^3, which is derived from the linear energy dependence of $N(E)$ in the low energy region. The observation of the power-law behavior in the temperature dependence of $1/T_1$ far below $T_{\rm c}$ is strong evidence for the presence of nodes on the superconducting gap, and therefore, for the unconventional superconductivity.

4. Ground states of $Na_x(H_3O)_zCoO_2 \cdot yH_2O$

4.1 Superconductivity

In order to investigate the symmetry of superconducting order parameter, $1/T_1$ was measured at zero field using NQR signal of the best superconducting sample with $T_c = 4.7$ K. The temperature dependence of $1/T_1$ observed in the superconducting state shows a power-law behavior as shown in Fig. 4 Ishida et al. (2003). This power-law decrease starts just below T_c without any increase due to Hebel-Slichter mechanism, and gradually changes the exponent

from $\sim$ 3 at half of T_c to unity below 1 K. The overall temperature dependence can be sufficiently fitted by the theoretical curve assuming the two-dimensional d-wave pairing state with the gap size $2\Delta/k_BT_c = 3.5$ and residual density of state $N_{res}/N_0 \sim 0.32$. The unconventional superconductivity with nodes on the superconducting gap is concluded for the superconductivity in BLH compounds. The residual density of states are generated by tiny amount of impurities inherent in the powder samples, because superconducting gap diminishes to zero along certain directions in the unconventional superconductors.
Next, the normal-state temperature dependence of $1/T_1T$ for BLH compound is compared with those of non-superconducting MLH and anhydrous compounds in order to identify the origin of superconductivity. Surprisingly, $1/T_1T$ in MLH is nearly identical to that in anhydrous cobaltate, even though the water molecule content is considerably different. In these compositions, gradual decrease in $1/T_1T$ from room temperature terminates around 100 K, below which Fermi-liquid like Korringa behavior is observed. This gradual decrease in $1/T_1T$ at high temperatures is reminiscent of the spin-gap formation in cuprate Alloul et al. (1989); Takigawa et al. (1989). The temperature dependence of the MLH and anhydrous samples are mimicked by

$$\left(\frac{1}{T_1T}\right)_{PG} = 8.75 + 15\exp\left(-\frac{\Delta}{T}\right) \quad (\mathrm{sec}^{-1}\mathrm{K}^{-1}) \tag{22}$$

with $\Delta = 250$ K. The value of Δ is in good agreement with the pseudogap energy $\sim$ 20 meV determined by photoemission spectroscopy Shimojima et al. (2006). The spin-gap behavior observed by relaxation rate measurement indicates the decrease in density of states due to the strong correlations.

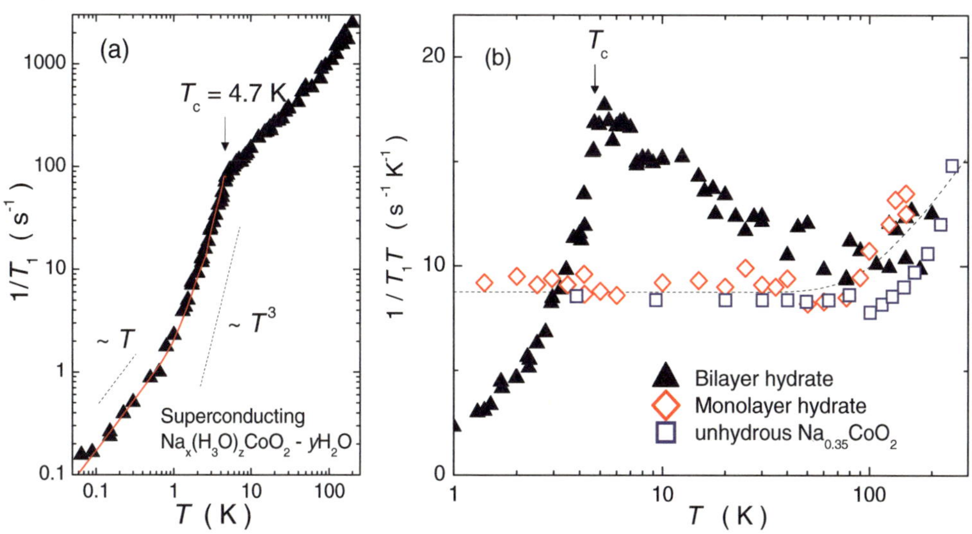

Fig. 4. (a) Temperature dependence of $1/T_1$ measured at zero magnetic field Ishida et al. (2003). Red solid curve is a theoretical fit to the date below $T_c = 4.7$ K, for which line nodes on the superconducting gap are taken into account. (b) Temperature dependence of $1/T_1T$ in superconducting BLH compound, MLH compound, and anhydrous cobaltate $Na_{0.35}CoO_2$. The experimental data of the anhydrous cobaltate was reported by Ning et al. (2004). The dashed line represents the sample-independent pseudogap contribution, which is expressed in equation (22). Ihara et al. (2006)

In contrast to the sample independent high-temperature behavior, the Korringa behavior is not observed in the superconducting BLH samples and, instead, magnetic fluctuations increase approaching to T_c. The sample independent high-temperature behavior and strongly sample dependent low-temperature behavior lead us to conclude that the pseudogap contribution robustly exists in all phases, and the increasing part of $1/T_1T$ detected only in the superconducting BLH sample is responsible for the superconducting pairing interactions. Multi orbital band structure of cobaltate allows to coexist strongly and weakly correlated bands in a uniform system. The sophisticated analyses based on the sample dependent $1/T_1T$ measurements are made in § 5.

4.2 Magnetism

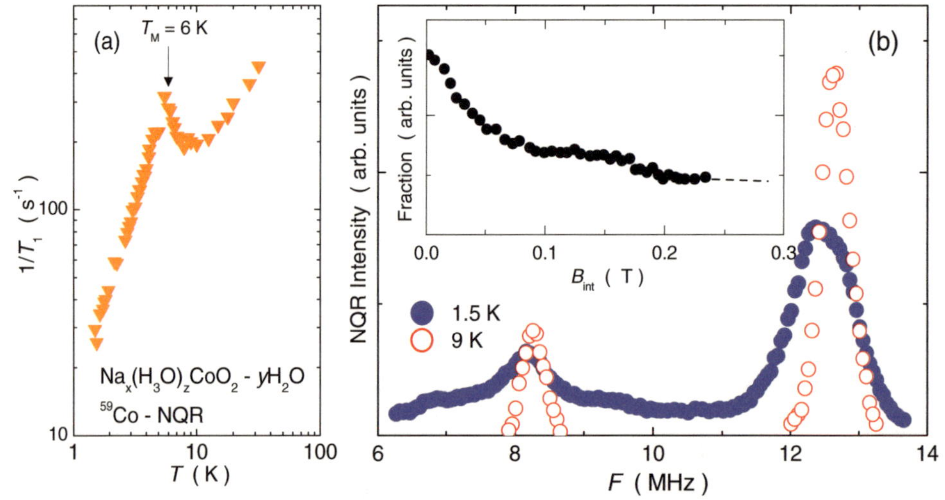

Fig. 5. (a) Temperature dependence of $1/T_1$ for magnetic ordering sample Ihara, Ishida, Michioka, Kato, Yoshimura, Takada, Sasaki, Sakurai & Takayama-Muromachi (2005). Abrupt increase in relaxation rate was observed at the magnetic ordering temperature $T_M = 6$ K. (b) The ^{59}Co NQR spectra above and below T_M. The clear spectral broadening was observed at 1.5 K. The internal field distribution is estimated from the broadened spectra and exhibited in the inset.

Immediately after the water filtration for BLH compounds, some samples do not show superconductivity. Superconductivity appears, even for these samples, after a few days of duration. NQR experiment on freshly hydrated non-superconducting samples has revealed that magnetic ordering sets in below $T_M = 6$ K. Ihara, Ishida, Michioka, Kato, Yoshimura, Takada, Sasaki, Sakurai & Takayama-Muromachi (2005) This magnetism was evidenced from divergence of $1/T_1$ at T_M (Fig. 5(a)) and NQR spectral broadening (Fig. 5(b)).
In a case of conventional Néel state, in which the same size of ordered moments are arranged antiferromagnetically, the internal-field strength could be uniquely determined from the split NQR spectra. The frequency separation between the split spectra is converted to internal field strength through the frequency-field relation derived from the Hamiltonian introduced by equation (3). For the cobaltate, however, the NQR spectrum does not split but just broadens due to the distributing internal fields. We have succeeded in extracting the distribution of the internal fields by taking the process explained below.

First, we determined the direction of the internal field to $H_{int} \perp c$, where the c axis is the principal axis of the electric field gradient, because the resonance peak arising from $m = \pm 5/2 \leftrightarrow \pm 3/2$ transitions become broader than that arising from $m = \pm 7/2 \leftrightarrow \pm 5/2$ transitions. If the internal fields were along the c axis, $\pm 7/2 \leftrightarrow \pm 5/2$ transition line would become the broadest within the three NQR lines. Obviously, this is not a case. The anisotropic broadening of $\pm 7/2 \leftrightarrow \pm 5/2$ transitions is also consistently explained by assuming $H_{int} \perp c$. These two results suggest that the internal fields direct to the ab plane.
Next, the intensity of the internal fields is estimated using frequency-field map derived from equation (3), and shown in Fig. 2. The details of the analyses are described in the separate paper Ihara et al. (2008). For the estimation of the fraction, we used the NQR spectra in the frequency range of 7 $\sim$ 10.5 MHz, because almost linear relationship was observed between the frequency and internal field in this frequency range. The internal field profile is exhibited in the inset of Fig. 5.
It is found that the maximum fraction is at zero field, and that rather large fraction is in the small field region. We also point out a weak hump around 0.15 T, which corresponds to 0.1 μ_B when we adopt 1.47 T/μ_B as the coupling constant Kato et al. (2006). It should be noted that NQR measurements were performed for Co nuclei, where the magnetic moments are located. The distribution of the hyperfine fields at the Co site suggests that the size of the ordered moments has spatial distributions. The magnetic ordering with the modulating ordered moments is categorized to the spin-density-wave type with incommensurate ordering vector. In order to investigate the magnetic structure in detail, neutron diffraction measurements are required.

4.3 Phase diagram

The ground state of $Na_x(H_3O)_zCoO_2 \cdot yH_2O$ strongly depends on the chemical compositions, Na ion (x), oxonium ion H_3O^+ (z) and water molecule (y) contents. The samples evolve drastically after water intercalation, as the water molecules evaporate easily into the air, when the samples were preserved in an ambient condition. This unstable nature causes the sample dependence of various physical quantities. The sample properties of the fragile BLH compounds have to be clarified in detail both from the microscopic and macroscopic measurements before the investigation on ground states.
In order to compare the physical properties of our samples to those of others, the superconducting transition temperatures reported in the literature Badica et al. (2006); Barnes et al. (2005); Cao et al. (2003); Chen et al. (2004); Chou, Cho, Lee, Abel, Matan & Lee (2004); Foo et al. (2005; 2003); Ihara et al. (2006); Jin et al. (2003; 2005); Jorgensen et al. (2003); Lorenz et al. (2004); Lynn et al. (2003); Milne et al. (2004); Ohta et al. (2005); Poltavets et al. (2006); Sakurai et al. (2005); Schaak et al. (2003); Zheng et al. (2006) are plotted against the c-axis length of each sample in Fig. 6(a). A relationship was observed between T_c and the c-axis length. The scattered data points indicate that the c-axis length is not the only parameter that determines the ground state of the BLH compound. The c-axis length, however, behaves as a dominant parameter, and can be a useful macroscopic reference to compare the sample properties of various reports. In the figure, superconductivity seems to be suppressed in some samples with $c \sim 19.75$ Å, probably due to the appearance of magnetism. The magnetism was reported on samples in the red region Higemoto et al. (2006); Ihara, Ishida, Michioka, Kato, Yoshimura, Takada, Sasaki, Sakurai & Takayama-Muromachi (2005); Sakurai et al. (2005). We also found, in some samples, that both the magnetic and superconducting transitions are observed Ihara et al. (2006). It has not been revealed yet how these two transitions coexist in one sample.

The NQR frequency sensitively reflects the crystalline distortions around the Co site, which is induced by the water intercalation. Therefore, in our phase diagram shown in Fig. 6(b), the Co-NQR frequency ν_Q was used as a microscopic reference for the ground states Ihara et al. (2006). Similar phase diagrams, in which NQR frequency is used as the microscopic reference, were reported by subsequent experiments Kobayashi et al. (2007); Michioka et al. (2006), and extended to the higher frequency region. It is empirically shown that the ν_Q detects the sample dependence of parameters which are closely related to the formation of superconductivity. Theoretically, the compression of CoO_2 layer along the c axis has been predicted to have a relation with the formation of superconductivity Mochizuki et al. (2005); Yanase et al. (2005). However, for the full understanding of the relationship between the ground state of the BLH compounds and the NQR frequency, the effect of the hole doping to the CoO_2 layers has to be investigated in detail, because the NQR frequency depends also on the concentration of the on-site $3d$ electrons in addition to the dominant lattice contributions.

5. Magnetic fluctuations near quantum critical point

5.1 Magnetically ordering regime

In this section, we analyze the magnetic fluctuations of BLH phases in detail. First, we analyze the temperature dependence of $1/T_1T$ in the magnetically ordering M1 sample,

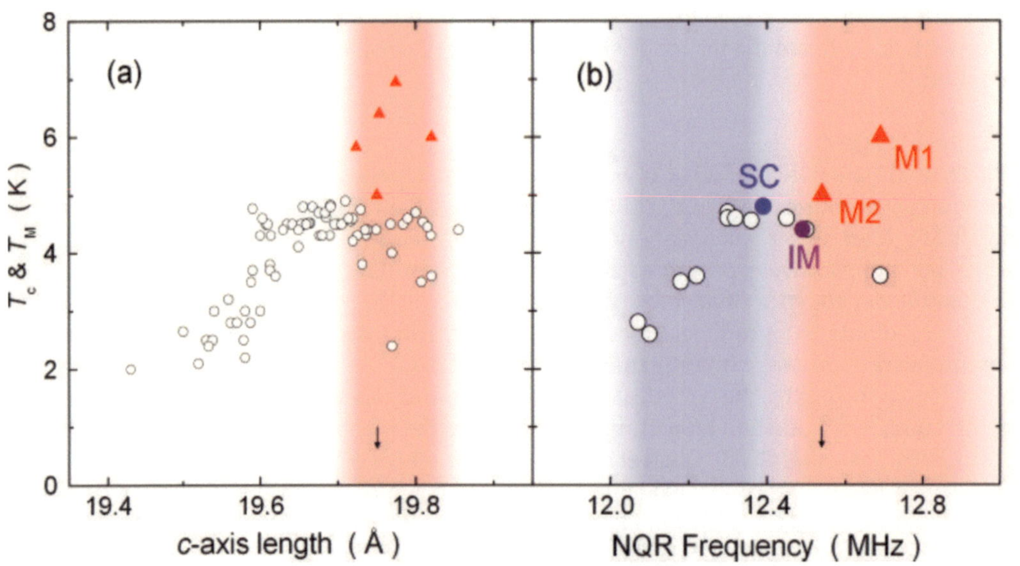

Fig. 6. Superconducting transition temperature T_c and magnetic transition temperature T_M of various samples reported in the literature Badica et al. (2006); Barnes et al. (2005); Cao et al. (2003); Chen et al. (2004); Chou, Cho, Lee, Abel, Matan & Lee (2004); Foo et al. (2005; 2003); Ihara et al. (2006); Jin et al. (2003; 2005); Jorgensen et al. (2003); Lorenz et al. (2004); Lynn et al. (2003); Milne et al. (2004); Ohta et al. (2005); Poltavets et al. (2006); Sakurai et al. (2005); Schaak et al. (2003); Zheng et al. (2006). These values are plotted against (a) the c-axis length and (b) the NQR frequency. The circles and the triangles indicate T_c and T_M, respectively. The down arrow indicates that the superconducting transition was not observed down to 1.5 K. The superconducting transition temperature becomes maximum around $c = 19.69$ Å. The magnetic phase appears in the red region, where the c axis is approximately 19.8 Å.

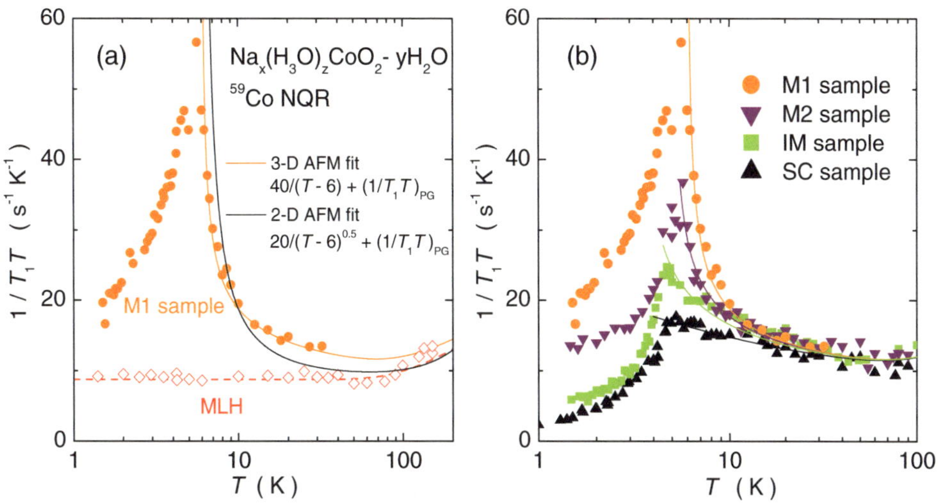

Fig. 7. (a) Temperature dependence of $1/T_1T$ in magnetically ordering sample (M1 in Fig. 6). A prominent peak is observed at $T_M = 6$ K. The full lines are the fitting curves for 2-D antiferromagnetic fluctuations (black) and 3-D antiferromagnetic fluctuations (orange). A good fit of experimental results to the orange line indicates the 3-D nature of magnetic fluctuations in BLH compounds Ihara et al. (2006). (b) Sample dependence of $1/T_1T$ on the samples in the red region in our phase diagram. The full lines are fit to the equation (24). The fitting parameters for each sample are given in Table. 2.

whose position in our phase diagram is located in Fig. 6(b). The strongly enhanced $1/T_1T$ in M1 sample is shown in Fig. 7(a) together with the pseudogap behavior of MLH compound. It is noteworthy that a prominent divergence of $1/T_1T$ at $T_M = 6$ K is observed in M1 sample, which has the highest NQR frequency among thirteen samples examined. We found that the diverging $1/T_1T$ in M1 sample could be fitted to a function consisting of two contributions expressed as

$$\left(\frac{1}{T_1T}\right)_{M1} = \left(\frac{1}{T_1T}\right)_{PG} + \frac{20}{\sqrt{T-6}} \ (\mathrm{sec}^{-1}\mathrm{K}^{-1}). \tag{23}$$

The first term on the right-hand side is the pseudogap contribution expressed in equation (22). The second term represents the magnetic contribution, which gives rise to the magnetic ordering. The functional form of the magnetic-fluctuations contribution, $(T - T_M)^{-1/2}$, is the temperature dependence anticipated for the three-dimensional itinerant antiferromagnet in the framework of the self-consistent renormalization (SCR) theory Moriya (1991), as explained in §3. The functional form for two-dimensional antiferromagnet, $(T - T_M)^{-1}$, would be more appropriate if the two-dimensional crystal structure of the BLH compound were taken into account, but the experimental data cannot be fitted well by the temperature dependence of 2-D antiferromagnet, as shown in Fig. 7(a). The magnetic correlation length is longer than the neighboring Co-Co distance along c axis, which is approximately 10 Å.
We adopted a fitting function with two fitting parameters, a and θ,

$$\frac{1}{T_1T} = \left(\frac{1}{T_1T}\right)_{PG} + \frac{a}{\sqrt{T-\theta}} \tag{24}$$

to inspect the sample-dependent low-temperature magnetic fluctuations. Here, a is a proportionality constant related to the band structure at the Fermi level and to the hyperfine coupling constant. The other parameter θ is the ordering temperature for the magnetically ordering samples and the measure of the closeness to the magnetic instability for the samples without magnetic ordering.

Sample ID	c axis (Å)	ν_Q (MHz)	T_c (K)	T_M (K)	θ (K)	a
SC	19.691	12.39	4.8	–	-1	20
IM	19.739	12.50	4.4	–	4	20
M1	19.820	12.69	3.6	6	6	20
M2	19.751	12.54	–	5	5	20

Table 2. Various parameters for the samples shown in Fig. 6. The θ and a values are determined from the fitting (see text).

Figure 7(b) shows the temperature dependence of $1/T_1T$ in SC, IM, M1, and M2 samples. The microscopic and macroscopic sample properties of these samples are summarized in Table. 2 together with the parameters used for the fitting. As shown in Fig. 7(b), the sample dependent $1/T_1T$ of four samples in red regime in our phase diagram is fitted by tuning only the ordering temperature θ. The sample independent a value indicates that in the red regime, where magnetism appears at low temperatures, the magnetic part of Fermi surface is stably formed. For the samples in superconducting blue region, the a coefficient is also sample dependent, as shown in the next subsection. It should be noted here, that θ for SC sample is -1 K, which indicates that the system is very close to the quantum critical point ($\theta = 0$).

5.2 Superconducting regime

For the superconducting samples, which are located in the blue region in our phase diagram, the magnetic-fluctuations term of equation (24) disappears abruptly so that the a coefficient should also have sample dependence. To better explore the magnetic contribution, the

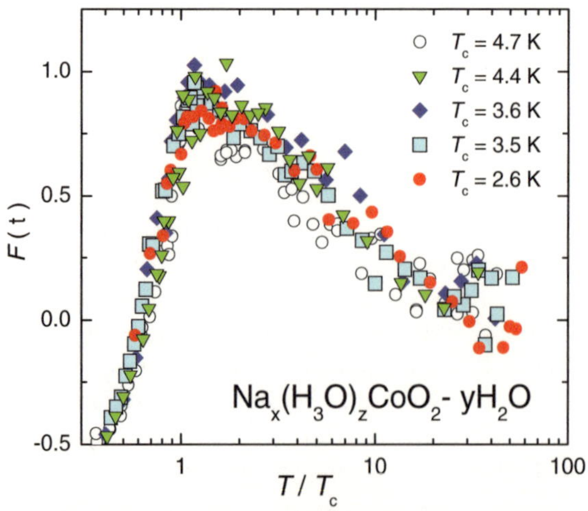

Fig. 8. Scaling plot with respect to T_c in five different samples. The definition of the universal function $F(t)$ and reduced temperature t are found in the text. All the temperature dependence in five samples with various T_c are scaled to single line.

pseudogap contribution is subtracted from the experimental data, and the normalized spin term $F(t)$ is derived from the following formula,

$$F(t) = \frac{1}{T_c}\left[\left(\frac{1}{T_1T}\right)_{\rm BLH} - \left(\frac{1}{T_1T}\right)_{\rm PG}\right]. \tag{25}$$

Here, $t = T/T_c$ is the reduced temperature. A coefficient $1/T_c$ was introduced to normalize the magnetic contribution. The physical interpretation for this coefficient will be given later. As shown in Fig. 8, where $F(t)$ is plotted against the reduced temperature, all the $F(t)$ curves in five different SC samples fall onto a universal line both in superconducting and normal states. A good scaling in superconducting state is not surprising, as the energy scale of superconducting gap determines the character of magnetic excitations. In the normal state, however, a special condition is required to obtain this scaling.

Generally, when the energy and momentum distribution of the spin fluctuations is assumed to be of a Lorentzian form, $1/T_1T$ of a three-dimensional system is proportional to $\chi_Q/\Gamma_Q\xi^3$, where χ_Q, Γ_Q and ξ are the weight of the spin fluctuations at the wave number Q, the characteristic energy of them and the magnetic correlation length, respectively. The temperature dependent $\Gamma(T)$ and $\xi(T)$ give rise to the magnetic contribution of $1/T_1T$. The universal scaling by T_c indicates that superconducting energy scale determines Γ_Q and ξ in the normal state. In addition to Γ_Q and ξ, χ_Q can also be normalized by T_c for the BLH system. This χ_Q normalization was not observed in a similar scaling plot performed for high-T_c cuprate Tokunaga et al. (1997). The weight of spin fluctuations has to be normalized for the multi-band cobaltate superconductor, since magnetic part of the Fermi surface develop only in the superconducting samples, while for the cuprate, a rigid single band is not modified through the carrier doping. The scaling of normal-state magnetic properties by superconducting energy scale clearly evidences that the magnetic fluctuations developed only in the BLH compounds are responsible for the superconducting pairing formation.

5.3 Intermediate regime

In this subsection, let us give an insight into the samples located on the phase boundary, which are labeled as SC and IM in Fig. 6(b). As shown in the Table 2, IM sample with $T_c = 4.4$ K has a positive θ value, indicating a magnetic ground state, while θ for the SC sample with $T_c = 4.8$ K is very close to 0 K. The NQR measurements in zero magnetic field have revealed that the sample inhomogeneity, which is evaluated using the NQR spectral width, is the same for the two samples, and no trace of magnetic anomaly is observed even at the lowest temperature measured. We have carried out NMR measurements, an experiment under magnetic fields, in order to suppress superconducting state and investigate the metallic ground state at low temperatures Ihara et al. (2007; 2009).

The temperature dependence of $1/T_1T$ was measured on SC and IM samples in various fields up to 14 T, and the results for SC sample is displayed in Fig. 9, together with that obtained in zero magnetic field with the NQR measurement. The NQR results are normalized at 10 K in order to eliminate the angle dependence of the coupling constant. For this NMR study, external fields were applied parallel to the EFG principal axis.

In the field-induced normal state, $1/T_1T$ increases at low temperature following the fitting curve defined above 5 K, which is expressed by equation (24) with $a = 30$ and $\theta = -0.7 \pm 0.2$ K. The continuous increase is interrupted only by the onset of superconductivity. The arrows indicate the superconducting transition temperatures in various fields, which were determined as the temperature where $1/T_1T$ starts to deviate from the normal-state temperature dependence. The strongly temperature dependent $1/T_1T$ in the field-induced normal state is a great contrast to the Korringa behavior (T_1T = const.), which would be observed when the electron-electron interactions are weak enough to construct Fermi liquid state. The absence of Korringa behavior down to $T_c(H)$ indicates the existence of the strong electronic correlations. Especially in SC sample, its small θ value indicates that the magnetic fluctuations possess quantum critical nature. As the superconducting transition temperature is the highest in this SC sample, we conclude that the quantum critical fluctuations induce attractive interactions between electrons to cause superconductivity.

The physical properties of the IM sample are quite different from those of the SC sample in magnetic fields, although they are nearly identical in zero field. At the temperatures above 4 K, the relaxation curves were consistently explained by the theoretical function written by

$$\frac{m(\infty) - m(t)}{m(\infty)} = A\left(\frac{3}{14}e^{-\frac{3t}{T_1}} + \frac{50}{77}e^{-\frac{10t}{T_1}} + \frac{3}{22}e^{-\frac{21t}{T_1}}\right), \qquad (26)$$

where t is the time after the saturation pulse Narath (1967). The coefficient A is a fitting parameter, which represents how completely nuclear magnetization is saturated. When IM sample was cooled down below 4 K, the experimental results could not be fitted by the same function. The results of the least square fitting to the whole relaxation curves obtained at 2.5 K and 5.5 K are exhibited in Fig. 10. The insufficient fit at 2.5 K is due to the appearance of short-T_1 components, which originate from the ordered magnetic moments. In Fig. 10, the relaxation curves are plotted against the product of the time and the temperature, in order to represent the existence of the short-T_1 components at low temperatures. If a short-T_1 component contributes to the relaxation curve, the slope would become steep. Obviously, the relaxation curve obtained at 2.5 K indicates the presence of short-T_1 components in the short time regime. In addition, the existence of the long-T_1 components is also observed in the long time regime. The relaxation rate in the magnetically ordered state is distributed in

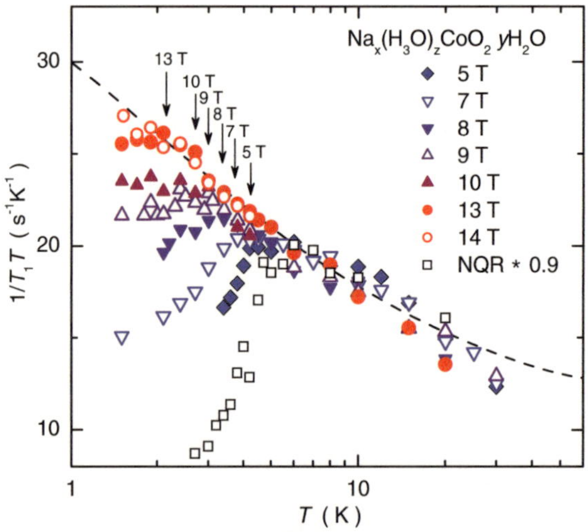

Fig. 9. Temperature dependence of $1/T_1T$ in the SC sample at various magnetic fields Ihara et al. (2007). The arrows indicate the superconducting transition temperatures at each field, which were determined by the deviation of $1/T_1T$ from the normal-state temperature dependence. The dashed line is the fitting function described in the text.

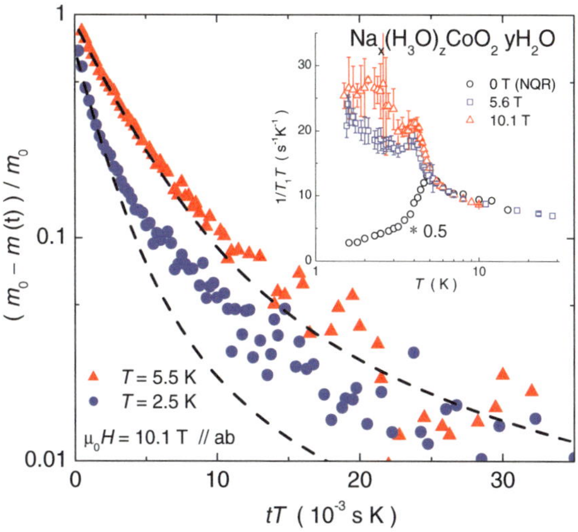

Fig. 10. Relaxation curves of the nuclear magnetization on the IM sample Ihara et al. (2009). The dashed lines represent the theoretical curves for a single T_1. The experimental data cannot be fitted by the single T_1 at low temperatures. Inset shows the temperature dependence of $1/T_1T$ at 5.6 T, 10.1 T, and 0 T(NQR). The absolute value of $1/T_1T$ obtained with NQR measurements were normalized to those with NMR measurement.

space probably due to the distribution of the internal fields at the Co site. More than two distinct components (long and short T_1) are needed for the best fitting, suggesting that the T_1 values are continuously distributed. The spatial distribution could not be resolved by the present NMR experiments on powder samples. We fitted all the relaxation curves to the same theoretical function with the single T_1 component even in the magnetically ordered state to determine the typical T_1 values. The least square fits with using full time range can extract the relaxation rates of the major fraction with large errors below 4 K.
The temperature dependence of $1/T_1T$ is displayed in the inset of Fig. 10, where the normalized NQR results are shown together. Field dependence of $1/T_1T$ is negligible in the normal state, while significant anomaly was observed below 4 K. In the normal state, where the relaxation curves were consistently fitted by the theoretical curve, $1/T_1T$ already has a tendency to diverge toward 4 K, which was foreseen from the positive θ value obtained by the NQR experiment. An anomaly is actually observed at 4 K, when the superconductivity is suppressed by the strong magnetic fields. It should be stressed, here, that the superconductivity of IM sample at zero field is uniform and magnetism is absent at any part of the sample. The magnetic anomaly in IM sample cannot be explained by the sample inhomogeneity, because the anomaly is absent in any fields in SC sample, whose sample inhomogeneity is comparable to that of IM sample.

6. Superconductivity near quantum critical point

Superconductivity in BLH cobaltates appears in the close vicinity of magnetism. We have shown, in the previous sections, that the quantum critical fluctuations bind two electrons to form the Cooper pairs, and induce unconventional superconductivity. The ground state of the samples at the superconductivity-magnetism phase boundary is easily modified by a small perturbation, such as magnetic field. The field-temperature phase diagrams for

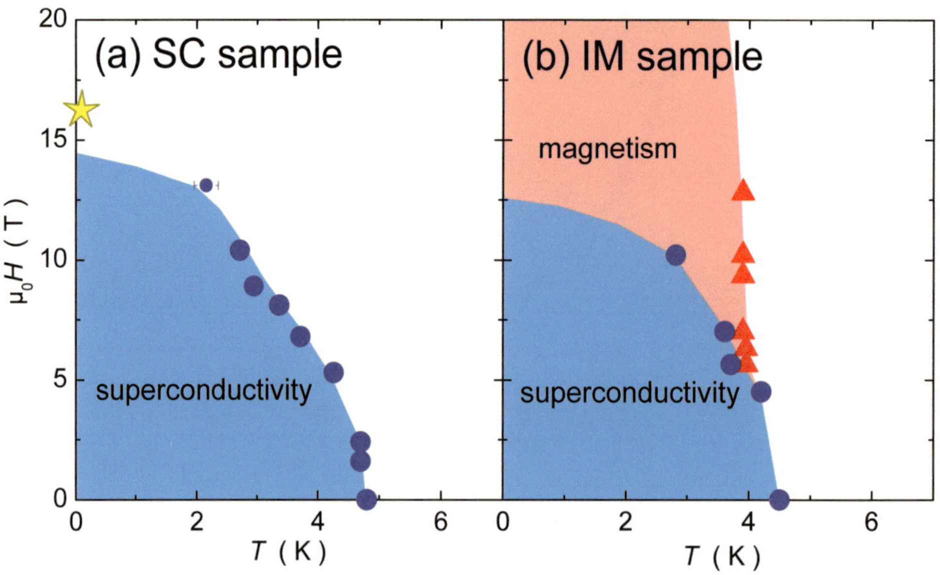

Fig. 11. Field-temperature phase diagram for (a) SC sample, and (b) IM sample. Field induced magnetic anomaly is observed only in the IM sample, which is located just on the superconductivity-magnetism phase boundary.

SC and IM samples, which are located nearby the phase boundary, are shown in Fig. 11. The field-induced normal state for SC sample is paramagnetic with low-energy magnetic fluctuations. The continuous increase in $1/T_1T$, shown in Fig. 9, suggests the existence of quantum critical point at low temperature and high field, where a star is located in Fig. 11(a). In the IM sample, whose chemical composition is slightly different from SC sample, a magnetic phase sets in at high fields, as shown by the red region in Fig. 11(b). The field induced magnetic anomaly suggests that magnetic interactions are strong enough to cause magnetic transition at 4 K, but at zero field in fact, superconductivity disguises magnetism.

A similar phase diagram has been reported for the Ce-based heavy Fermion superconductors, $CeCu_2Si_2$ Steglich et al. (1979), $CeRhIn_5$ Knebel et al. (2006) and $CeCoIn_5$ Young et al. (2007), in which ground states are tuned by pressures of few GPa.

In $CeRhIn_5$, for instance, the optimal pressure p_c for superconductivity is reported to be $\sim$ 2.4 GPa, and a magnetism is induced in pressures lower than p_c. When the pressure is reduced to 2.07 GPa, magnetism is stabilized only in the field-induced normal state with uniform superconductivity in zero field Knebel et al. (2006; 2008). At the optimal pressure, this field-induced magnetism is suppressed to zero K, showing a quantum critical behavior.

In a case of $CeCoIn_5$, the critical point is at an ambient pressure. The continuous increase in the density of state at low temperatures has been reported from the specific heat measurement Ikeda et al. (2001), which is a quantity equivalent to $1/T_1T$. A magnetism was detected by the NMR spectral broadening in strong magnetic fields Young et al. (2007). In this compound, the magnetism can also be induced surrounding a small amount of impurity injected in the sample. Both the field and impurity induced magnetism results from the proximity to quantum critical point.

The interplay between superconductivity and magnetism is universally observed for the superconductivity near quantum critical point. We would refer the interplay as cooperative effects rather than competitive one, as superconductivity can exist only in the vicinity of quantum critical point. The sample quality is a fatal parameter nearby the quantum critical point, where even a small perturbation can drastically modify the ground state. A fine sample control on BLH cobaltates with novel synthetic method, and eventually the single crystal growth will open a path to uncover physics at quantum critical point.

7. Acknowledgement

We would like to appreciate to Hiroya Sakurai for valuable discussion and for his high quality samples. We also acknowledge Kazuyoshi Yoshimura, Chishiro Michioka, Hideo Takeya, and Eiji Takayama-Muromachi for experimental supports and stimulating discussions.
This work was partially supported by the Grants-in-Aid for Scientific Research from the Japan Society for Promotion of Science (JSPS) and from the Ministry of Education, Culture, Sports, Science and Technology (MEXT) of Japan, and by the 21st Century Center-of-Excellence Program "Center for Diversity and Universality in Physics" for MEXT of Japan. Y.I. was financially supported by the JSPS Research Fellowships for Young Scientists.

8. References

Alloul, H., Ohno, T. & Mendels, P. (1989). ^{89}y nmr evidence for a fermi-liauid behavior in $yba_2cu_3o_{6+x}$, *Phys. Rev. Lett.* 63: 1700.

Badica, P., Kondo, T., Togano, K. & Yamada, K. (2006). Anisotropy of $na_{0.35}coo_2 \cdot 1.3h_2o$ superconductor, *J. Phys. Chem. Soli.* 67: 590.

Barnes, P. W., Avdeev, M., Jorgensen, J. D., Hinks, D. G., Claus, H. & Short, S. (2005). Superconductivity and cobalt oxidation state in metastable $na_xcoo_{2-\delta} \cdot yh_2o$ ($x \approx 1/3; y \approx 4x$), *Phys. Rev. B* 72: 134515.

Cao, G., Feng, C., Xu, Y., Lu, W., Shen, J., Fang, M. & Xu, Z. (2003). Superconductivity in a layered cobalt oxyhydrate $na_{0.31}coo_2 \cdot 1.3h_2o$, *J. Phys. Condens. Matter* 15: L519.

Chen, D. P., Chen, H. C., Maljuk, A., Kulakov, A., Zhang, H., Lemmens, P. & Lin, C. T. (2004). Single-crystal growth and investigation of na_xcoo_2 and $na_xcoo_2 \cdot yh_2o$, *Phys. Rev. B* 70: 024506.

Chou, F. C., Cho, J. H., Lee, P. A., Abel, E. T., Matan, K. & Lee, Y. S. (2004). Thermodynamic and transport measurements of superconducting $na_{0.3}coo_2 \cdot 1.3h_2o$ single crystals prepared by electrochemical deintercalation, *Phys. Rev. Lett.* 92: 157004.

Chou, F. C., Cho, J. H. & Lee, Y. S. (2004). Magnetic susceptibility study of hydrated and nonhydrated $na_xcoo_2 \cdot yh_2o$ single crystals, *Phys. Rev. B* 70: 144526.

Foo, M. L., Klimczuk, T. & Cava, R. J. (2005). Hydration phase diagram for sodium cobalt oxide $na_{0.3}coo_2 \cdot yh_2o$, *Mat. Res. Bull.* 40: 665.

Foo, M. L., Schaak, R. E., Miller, V. L., Klimczuk, T., Rogado, N. S., Wang, Y., Lau, G. C., Craley, C., Zandbergen, H. W., Ong, N. P. & Cava, R. J. (2003). Chemical instability of the cobalt oxyhydrate superconductor under ambient conditions, *Soli. Sta. Commun.* 127: 33.

Fujimoto, T., Zheng, G. Q., Kitaoka, Y., Meng, R. L., Cmaidalka, J. & Chu, C. W. (2004). Unconventional superconductivity and electron correlations in the cobalt oxyhydrate $na_{0.35}coo_2 \cdot 1.3h_2o$ from nuclear quadrupole resonance, *Phys. Rev. Lett.* 92: 047004.

Hebel, L. C. & Slichter, C. P. (1959). Nuclear spin relaxation in normal and superconducting aluminum, *Phys. Rev.* 113: 1504.

Higemoto, W., Ohishi, K., Koda, A., Kadono, R., Sakurai, H., Takada, K., Takayama-Muromachi, E. & Sasaki, T. (2006). Possible unconventional superconductivity and weak magnetism in $na_x coo_2 \cdot yh_2o$ probed by μsr, *Physica B* 374-375: 274.

Higemoto, W., Ohishi, K., Koda, A., Ranjan, S., kadono, R., Ishida, K., Takada, K., Sakurai, H., Takayama-Muromachi, E. & Sasaki, T. (2004). Possible unconventional superconductivity in $na_x coo_2 \cdot yh_2o$ porbed by muon spin rotation and relaxation, *Phys. Rev. B* 70: 134508.

Ihara, Y., Ishida, K., Michioka, C., Kato, M., Yoshimura, K., Takada, K., Sasaki, T., Sakurai, H. & Takayama-Muromachi, E. (2005). Weak magnetic order in bilayered-hydrate $na_x coo_2 \cdot yh_2o$ structure probed by co nuclear quadrupole resonance –proposed phase diagram in superconducting $na_x coo_2 \cdot yh_2o$–, *J. Phys. Soc. Jpn.* 74: 867.

Ihara, Y., Ishida, K., Yoshimura, K., Takada, K., Sasaki, T., Sakurai, H. & Takayama-Muromachi, E. (2005). ^{17}o nmr measurements on superconducting $na_{0.35}coo_2 \cdot yh_2o$, *J. Phys. Soc. Jpn.* 74: 2177.

Ihara, Y., Takeya, H., Ishida, K., Ikeda, H., Michioka, C., Yoshimura, K., Takada, K., Sasaki, T., Sakurai, H. & Takayama-Muromachi, E. (2006). Unconventional superconductivity induced by quantum critical fluctuations in hydrate cobalatate $na_x(h_3o)_z coo_2 \cdot yh_2o$ –relationship between magnetic fluctuations and superconductivity revealed by co nuclear quadrupole resonance–, *J. Phys. Soc. Jpn.* 75: 124714.

Ihara, Y., Takeya, H., Ishida, K., Michioka, C., Yoshimura, K., Takada, K., Sasaki, T., Sakurai, H. & Takayama-Muromachi, E. (2007). Quantum critical behavior in superconducting $na_x(h_3o)_z coo_2 \cdot yh_2o$ observed in a high-field co nmr experiment, *Phys. Rev. B* 75: 212506.

Ihara, Y., Takeya, H., Ishida, K., Michioka, C., Yoshimura, Y., Takada, K., Sasaki, T., Sakurai, H. & Takayama-Muromachi, E. (2008). Magnetic ordering in hydrated cobaltate $na_x(h_3o)_z coo_2 \cdot yh_2o$, *Physica B* 403: 1089.

Ihara, Y., Takeya, H., Ishida, K., Yoshimura, K., Takada, K., Sasaki, T., Sakurai, H. & Takayama-Muromachi, E. (2009). Magnetic anomalies of hydrous cobaltate compound $na_x(h_3o)_z coo_2 \cdot yh_2o$ detected by nmr and nqr measurements, *Phys. Rev. B* 79: 024510.

Ikeda, S., Shishido, H., Nakashima, M., Settai, R., Aoki, D., Haga, Y., Harima, H., Aoki, Y., Namiki, T., Sato, H. & Onuki, Y. (2001). Unconventional superconductivity in $cecoin_5$ stuied by the specific heat and magnetization measurements, *J. Phys. Soc. Jpn.* 70: 2248.

Ishida, K., Ihara, Y., Maeno, Y., Michioka, C., Kato, M., Yoshimura, K., Takada, K., Sasaki, T., Sakurai, H. & Takayama-Muromashi, E. (2003). Unconventional superconductivity and nearly ferromagnetic spin fluctuations in $na_x coo_2 \cdot yh_2o$, *J. Phys. Soc. Jpn.* 72: 3041.

Jin, R., Sales, B. C., Khalifah, P. & Mandrus, D. (2003). Observation of bulk superconductivity in $na_x coo_2 \cdot yh_2o$ and $na_x coo_2 \cdot yd_2o$ powder and single crystals, *Phys. Rev. Lett.* 91: 217001.

Jin, R., Sales, B. C., Li, S. & Mandrus, D. (2005). Dependence of the specific heat of $na_x coo_2 \cdot yh_2o/d_2o$ on sodium and water concentrations, *Phys. Rev. B* 72: 060512(R).

Jorgensen, J. D., Avdeev, M., Hinks, D. G., Burley, J. C. & Short, S. (2003). Crystal structure of the sodium cobaltate deuterate superconductor $na_x coo_2 \cdot 4xd_2o$ ($x \approx 1/3$), *Phys. Rev. B* 68: 214517.

Kato, M., Michioka, C., Waki, T., Itoh, Y., Yoshimura, K., Ishida, K., Sakurai, H., Takayama-Muromachi, E., Takada, K. & Sasaki, T. (2006). Possible spin triplet superconductivity in $na_x coo_2 \cdot yh_2o$ – ^{59}co nmr studies, *J. Phys.:Condens. Matter* 18: 669–682.

Knebel, G., Aoki, D., Braithwaite, D., Salce, B. & Flouquet, J. (2006). Coexistence of antiferromagnetismn and superconductivity in $cerhin_5$ under high pressure and magnetic field, *Phys. Rev. B* 74: 020501(R).

Knebel, G., Aoki, D., Brison, J.-P. & Flouquet, J. (2008). The quantum critical point in $cerhin_5$: A resistivity study, *J. Phys. Soc. Jpn.* 77: 114704.

Kobayashi, Y., Yokoi, M., Moyoshi, T. & Sato, M. (2007). $t - 3\nu_Q$ phase diagram of hydrated cobalt oxides $na_x coo_2 \cdot yh_2o$, *J. Phys. Soc. Jpn.* 76: 103705.

Li, S. Y., Taillefer, L., Hawthorn, D. G., Tanatar, M. A., Paglione, J., Sutherland, M., Hill, R. W., Wand, C. H. & Chen, X. H. (2004). Giant electron-electron scattering in the fermi-liquid state of $na_{0.7}coo_2$, *Phy. Rev. Lett.* 93: 056401.

Lorenz, B., Cmaidalka, J., Meng, R. L. & Chu, C. W. (2004). Thermodynamic properties in the normal and superconducting states of $na_x coo_2 \cdot yh_2o$ powder measured by heat capacity experiments, *Physica C* 402: 106.

Lynn, J. W., Huang, Q., Brown, C. M., Miller, V. L., Foo, M. L., Schaak, R. E., Jones, C. Y., Mackey, E. A. & Cava, R. J. (2003). Structure and dynamics of superconducting $na_x coo_2$ hydrate and its unhydrated analog, *Phys. Rev. B* 68: 214516.

Michioka, M., Ohta, H., Itoh, Y. & Yoshimura, K. (2006). ^{59}co nuclear quadrupole resonance studies of superconducting and nonsuperconducting bilayer water intercalated sodium cobalt oxides $na_x coo_2 \cdot yh_2o$, *J. Phys. Soc. Jpn.* 75: 063701.

Milne, C. J., Argyriou, D. N., Chemseddine, A., Aliouane, N., Veira, J., Landsgesell, S. & Alber, D. (2004). Revised superconducting phase diagram of hole-doped $na_x(h_3o)_z coo_2 \cdot yh_2o$, *Phys. Rev. Lett.* 93: 247007.

Mochizuki, M., Yanase, Y. & Ogata, M. (2005). Ferromagnetic and triplet-pairing instabilities controlled by trigonal distortion of coo_6 octahedra in $na_x coo_2 \cdot yh_2o$, *J. Phys. Soc. Jpn.* 74: 1670.

Moriya, T. (1991). Theory of itinerant electron magnetism, *J. Magn. Magn. Mater.* 100: 261.

Narath, A. (1967). Nuclear spin-lattice relaxation in hexagonal transition metals: Titanium, *Phys. Rev.* 162: 320.

Ning, F. L., Imai, T., Statt, B. W. & Chou, F. C. (2004). Spin dynamics in the carrier-doped $s = 1/2$ triangular lattice of $na_x coo_2 \cdot yh_2o$, *Phys. Rev. Lett.* 93: 237201.

Oeschler, N., Fisher, R. A., Phillips, N. E., Gordon, J. E., Foo, M. L. & Cava, R. J. (2005). Heat capacity of $na_{0.3}coo_2 \cdot 1.3h_2o$, a new two-gap superconductor: Comparison with the heat capacity of mgb_2, *Chin. J. Phys.* 43: 574.

Ohta, H., Michioka, C., Itoh, Y. & Yoshimura, K. (2005). Novel phase diagram of superconductor $na_x coo_2 \cdot yh_2o$ in 75 % relative humidity, *J. Phys. Soc. Jpn.* 74: 3147.

Poltavets, V. V., Croft, M. & Greenblatt, M. (2006). Charge transfer, hybridization and local inhomogeneity effects in $na_x coo_2 \cdot yh_2o$: An x-ray absorption spectroscopy study, *Phys. Rev. B* 74: 125103.

Sakurai, H., Takada, K., Sasaki, T. & Takayama-Muromachi, E. (2005). Phase diaram of superconducting $na_x coo_2 \cdot yh_2o$, *J. Phys. Soc. Jpn.* 74: 2909.

Sakurai, H., Takda, K., Yoshii, S., Sasaki, T., Kindo, K. & Takayama-Muromachi, E. (2003). Unconventional upper- and lower-crytical fields and normal-state magnetic susceptibility of the superconducting compound $na_{0.35}coo_2 \cdot 1.3h_2o$, *Phys. Rev. B* 68: 132507.

Schaak, R. E., Klimczuk, T., Foo, M. L. & Cava, R. J. (2003). Superconductivity phase diagram of $na_x coo_2 \cdot 1.3h_2o$, *Nature* 424: 527.

Shimojima, T., Yokoya, T., Kiss, T., Chainani, A., Shin, S., Togano, T., Zhang, C., Chen, C., Watanabe, S., Takada, K., Sasaki, T., Sakurai, H. & Takayama-Muromachi, E. (2006). Laser-excited ultrahigh-resolution photoemission spectroscopy of superconducting $na_{0.35}coo_2 \cdot 1.3h_2o$, *J. Phys. Chem. Soli.* 67: 282.

Steglich, F., Aarts, J., Bredl, C. D., Lieke, W., Mesched, D., Franz, W. & Schäfer, H. (1979). Superconductivity in the presence of strong pauli paramagnetism: $Cecu_2si_2$, *Phys. Rev. Lett.* 43: 1892.

Takada, K., Sakurai, H., Takayama-Muromachi, E., Izumi, F., Dilanian, R. A. & Sasaki, T. (2003). Superconductivity in two-dimensional coo_2 layers, *Nature* 422: 53.

Takigawa, M., Hammel, P. C., Heffner, R. H., Fisk, Z., Ott, K. C. & Thompson, J. D. (1989). ^{17}o nmr study of local spin susceptibility in aligned $yba_2cu_3o_7$ powder, *Phys. Rev. Lett.* 63: 1865.

Tokunaga, Y., Ishida, K., Kitaoka, Y. & Asayama, K. (1997). Novel relation between spin-fluctuations and superconductivity in ni substituted high-t_c cuprate $yba_2cu_3o_7$: Cu nqr study, *Soli. Sta. Commun.* 103: 43.

Ueland, B. G., Schiffer, P., Schaak, R. E., Foo, M. L., Miller, V. L. & Cava, R. J. (2004). Specific heat study of the $na_{0.3}coo_2 \cdot 1.3h_2o$ superconductor: influence of the complex chemistry, *Physica C* 402: 27.

Yanase, Y., Mochizuki, M. & Ogata, M. (2005). Multi-orbital analysis on the superconductivity in $na_x coo_2 \cdot yh_2o$, *J. Phys. Soc. Jpn.* 74: 430.

Yang, H. D., Lin, J.-Y., Sun, C. P., Kang, Y. C., Huang, C. L., Takada, K., Sasaki, T., Sakurai, H. & Takayama-Muromachi, E. (2005). Evidence of nodal superconductivity in $na_{0.35}coo_2 \cdot 1.3h_2o$: A specific-heat study, *Phys. Rev. B* 71: 020504.

Young, B.-L., Urbano, R. R., Curro, N. J., Thompson, J. D., Sarrao, J. L., Vorontsov, A. B. & Graf, M. J. (2007). Microscopic evidence for field-induced magnetism in $cecoin_5$, *Phys. Rev. Lett.* 98: 036402.

Zheng, G. Q., Matano, K., Meng, R. L., Cmaidalka, J. & Chu, C. W. (2006). Na content dependence of superconductivity and the spin correlations in $na_x coo_2 \cdot 1.3h_2o$, *J. Phys.:Condens. Matter* 18: L63–L70.

3

Coherent Current States in Two-Band Superconductors

Alexander Omelyanchouk

B.Verkin Istitute for Low Temperature Physics and Engineering of the National Academy of Science of Ukraine, Kharkov 61103, Ukraine

1. Introduction

To present day overwhelming majority works on theory of superconductivity were devoted to single gap superconductors. More than 50 years ago the possibility of superconductors with two superconducting order parameters were considered by V. Moskalenko

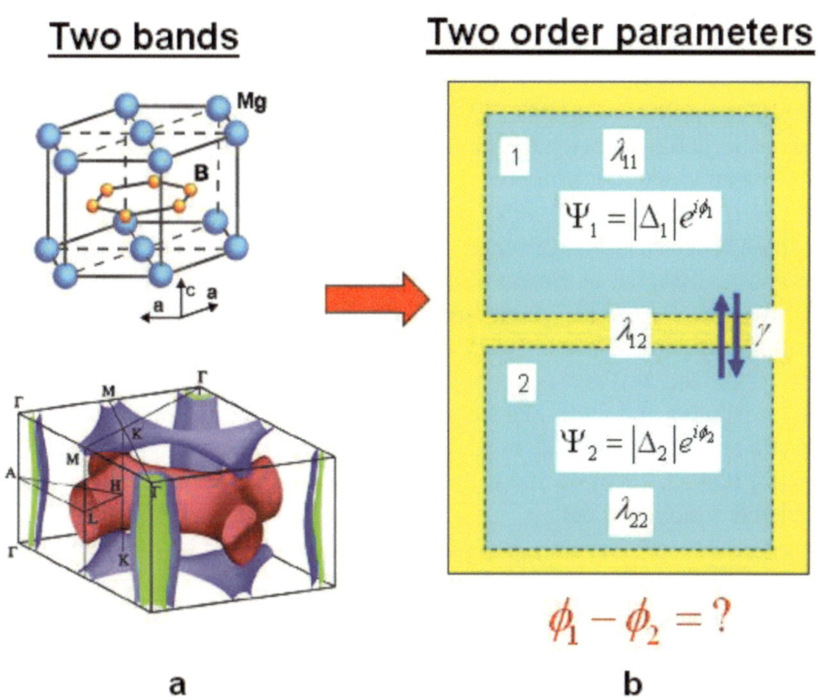

Fig. 1. a. The structure of MgB_2 and the Fermi surface of MgB_2 calculated by Kortus et al. (Kortus et al., 2001).
b. The coexistence of two complex order parameters (in momentum space).

(Moskalenko, 1959) and H. Suhl, B.Matthias and L.Walker (Suhl et al., 1959). In the model of superconductor with the overlapping energy bands on Fermi surface V.Moskalenko has theoretically investigated the thermodynamic and electromagnetic properties of two-band superconductors. The real boom in investigation of multi-gap superconductivity started after the discovery of two gaps in MgB_2 (Nagamatsu et al., 2001) by the scanning tunneling (Giubileo et al., 2001; Iavarone et al., 2002) and point contact spectroscopy (Szabo et al., 2001; Schmidt et al., 2001; Yanson & Naidyuk, 2004). The structure of MgB_2 and the Fermi surface of MgB_2 calculated by Kortus et al. (Kortus et al, 2001) are presented at Fig.1.a. The compound MgB_2 has the highest critical temperature $T_c = 39$ K among superconductors with phonon mechanism of the pairing and two energy gaps $\Delta_1 \approx 7meV$ and $\Delta_2 \approx 2{,}5meV$ at $T = 0$. At this time two-band superconductivity is studied also in another systems, *e.g.* in heavy fermion compounds (Jourdan et al., 2004; Seyfarth et al., 2005), high-T_c cuprates (Kresin & Wolf, 1990), borocarbides (Shulga et al., 1998), liquid metallic hydrogen (Ashcroft, 2000; Babaev, 2002; Babaev et. al, 2004). Recent discovery of high-temperature superconductivity in iron-based compounds (Kamihara et al., 2008) have expanded a range of multiband superconductors. Various thermodynamic and transport properties of MgB_2 and iron-based superconductors were studied in the framework of two-band BCS model (Golubov et al., 2002; Brinkman et al., 2002; Mazin et al., 2002; Nakai et al., 2002; Miranovic et al., 2003; Dahm & Schopohl, 2004; Dahm et al., 2004; Gurevich, 2003; Golubov & Koshelev, 2003). Ginzburg-Landau functional for two-gap superconductors was derived within the weak-coupling BCS theory in dirty (Koshelev & Golubov, 2003) and clean (Zhitomirsky & Dao, 2004) superconductors. Within the Ginzurg-Landau scheme the magnetic properties (Askerzade, 2003a; Askerzade, 2003b; Doh et al., 1999) and peculiar vortices (Mints et al., 2002; Babaev et al., 2002; Gurevich & Vinokur, 2003) were studied.

Two-band superconductivity proposes new interesting physics. The coexistence of two distinctive order parameters $\Psi_1 = |\Psi_1|\exp(i\phi_1)$ and $\Psi_2 = |\Psi_2|\exp(i\phi_2)$ (Fig.1.b.) renewed interest in phase coherent effects in superconductors. In the case of two order parameters we have the additional degree of freedom, and the question arises, what is the phase shift $\delta\phi = \phi_1 - \phi_2$ between Ψ_1 and Ψ_2? How this phase shift manifested in the observable effects? From the minimization of the free energy it follows that in homogeneous equilibrium state this phase shift is fixed at 0 or π, depending on the sign of interband coupling. It does not exclude the possibility of soliton-like states $\delta\phi(x)$ in the ring geometry (Tanaka, 2002). In nonequilibrium state the phases ϕ_1 and ϕ_2 can be decoupled as small plasmon oscillations (Leggett mode) (Legett, 1966) or due to formation of phase slips textures in strong electric field (Gurevich & Vinokur, 2006).

In this chapter we are focusing on the implication of the $\delta\phi$-shift in the coherent superconducting current states in two-band superconductors. We use a simple (and, at the same time, quite general) approach of the Ginsburg–Landau theory, generalized on the case of two superconducting order parameters (Sec.2). In Sec.3 the coherent current states and depairing curves have been studied. It is shown the possibility of phase shift switching in homogeneous current state with increasing of the superfluid velocity v_s. Such switching manifests itself in the dependence $j(v_s)$ and also in the Little-Parks effect (Sec.3). The Josephson effect in superconducting junctions is the probe for research of phase coherent effects. The stationary Josephson effect in tunnel S_1-I-S_2 junctions (I - dielectric) between two- and one- band superconductors have been studied recently in a number of articles (Agterberg et al., 2002; Ota et al., 2009; Ng & Nagaosa, 2009). Another basic type of Josephson junctions are the junctions with direct conductivity, S-C-S contacts (C - constriction). As was shown in (Kulik & Omelyanchouk, 1975; Kulik & Omelyanchouk, 1978; Artemenko et al., 1979) the Josephson behavior of S-C-S structures qualitatively differs from the properties of tunnel junctions. A simple theory (analog of Aslamazov-Larkin theory(Aslamazov & Larkin, 1968)) of stationary Josephson effect in S-C-S point contacts for the case of two-band superconductors is described in Sec.4).

2. Ginzburg-Landau equations for two-band superconductivity.

The phenomenological Ginzburg-Landau (GL) free energy density functional for two coupled superconducting order parameters ψ_1 and ψ_2 can be written as

$$F_{GL} = F_1 + F_2 + F_{12} + \frac{H^2}{8\pi},$$

Where

$$F_1 = \alpha_1 |\psi_1|^2 + \frac{1}{2}\beta_1 |\psi_1|^4 + \frac{1}{2m_1}\left|\left(-i\hbar\nabla - \frac{2e}{c}\vec{A}\right)\psi_1\right|^2 \tag{1}$$

$$F_2 = \alpha_2 |\psi_2|^2 + \frac{1}{2}\beta_2 |\psi_2|^4 + \frac{1}{2m_2}\left|\left(-i\hbar\nabla - \frac{2e}{c}\vec{A}\right)\psi_2\right|^2 \tag{2}$$

and

$$F_{12} = -\gamma\left(\psi_1^*\psi_2 + \psi_1\psi_2^*\right) + \eta\begin{pmatrix}\left(-i\hbar\nabla - \frac{2e}{c}\vec{A}\right)\psi_1\left(i\hbar\nabla - \frac{2e}{c}\vec{A}\right)\psi_2^* + \\ +\left(i\hbar\nabla - \frac{2e}{c}\vec{A}\right)\psi_1^*\left(-i\hbar\nabla - \frac{2e}{c}\vec{A}\right)\psi_2\end{pmatrix} \tag{3}$$

The terms F_1 and F_2 are conventional contributions from ψ_1 and ψ_2, term F_{12} describes without the loss of generality the interband coupling of order parameters. The coefficients γ and η describe the coupling of two order parameters (proximity effect) and their gradients (drag effect) (Askerzade, 2003a; Askerzade, 2003b; Doh et al., 1999), respectively. The microscopic theory for two-band superconductors (Koshelev & Golubov, 2003; Zhitomirsky & Dao, 2004; Gurevich, 2007) relates the phenomenological parameters to microscopic characteristics of superconducting state. Thus, in clean multiband systems the drag coupling term (η) is vanished. Also, on phenomenological level there is an important condition , that absolute minimum of free GL energy exist: $|\eta| < \frac{1}{2\sqrt{m_1 m_2}}$ (see Yerin et al., 2008).

By minimization the free energy F= $\int (F_1 + F_2 + F_{12} + \frac{H^2}{8\pi}) d^3 r$ with respect to ψ_1, ψ_2 and $\vec{A}$ we obtain the differential GL equations for two-band superconductor

$$\begin{cases} \frac{1}{2m_1}\left(-i\hbar\nabla - \frac{2e}{c}\vec{A}\right)^2 \psi_1 + \alpha_1\psi_1 + \beta_1|\psi_1|^2\psi_1 - \gamma\psi_2 + \eta\left(-i\hbar\nabla - \frac{2e}{c}\vec{A}\right)^2 \psi_2 = 0 \\ \frac{1}{2m_2}\left(-i\hbar\nabla - \frac{2e}{c}\vec{A}\right)^2 \psi_2 + \alpha_2\psi_2 + \beta_2|\psi_2|^2\psi_2 - \gamma\psi_1 + \eta\left(-i\hbar\nabla - \frac{2e}{c}\vec{A}\right)^2 \psi_1 = 0 \end{cases} \tag{4}$$

and expression for the supercurrent

$$\begin{aligned} j = &-\frac{ie\hbar}{m_1}\left(\psi_1^*\nabla\psi_1 - \psi_1\nabla\psi_1^*\right) - \frac{ie\hbar}{m_2}\left(\psi_2^*\nabla\psi_2 - \psi_2\nabla\psi_2^*\right) \\ &-2ie\hbar\eta\left(\psi_1^*\nabla\psi_2 - \psi_2\nabla\psi_1^* - \psi_1\nabla\psi_2^* + \psi_2^*\nabla\psi_1\right) - \\ &-\left(\frac{4e^2}{m_1 c}|\psi_1|^2 + \frac{4e^2}{m_2 c}|\psi_2|^2 + \frac{8\eta e^2}{c}\left(\psi_1^*\psi_2 + \psi_2^*\psi_1\right)\right)\vec{A}. \end{aligned} \tag{5}$$

In the absence of currents and gradients the equilibrium values of order parameters $\psi_{1,2} = \psi_{1,2}^{(0)} e^{i\chi_{1,2}}$ are determined by the set of coupled equations

$$\begin{aligned} \alpha_1\psi_1^{(0)} + \beta_1\psi_1^{(0)3} - \gamma\psi_2^{(0)} e^{i(\chi_2-\chi_1)} = 0, \\ \alpha_2\psi_2^{(0)} + \beta_2\psi_2^{(0)3} - \gamma\psi_1^{(0)} e^{i(\chi_1-\chi_2)} = 0. \end{aligned} \tag{6}$$

For the case of two order parameters the question arises about the phase difference $\phi = \chi_1 - \chi_2$ between ψ_1 and ψ_2. In homogeneous zero current state, by analyzing the free energy term F12 (3), one can obtain that for $\gamma > 0$ phase shift $\phi = 0$ and for $\gamma < 0$ $\phi = \pi$. The statement, that ϕ can have only values 0 or π takes place also in a current carrying state, but for coefficient $\eta \neq 0$ the criterion for ϕ equals 0 or π depends now on the value of the current (see below).

If the interband interaction is ignored, the equations (6) are decoupled into two ordinary GL equations with two different critical temperatures T_{c_1} and T_{c_2}. In general, independently of the sign of γ, the superconducting phase transition results at a well-defined temperature exceeding both T_{c_1} and T_{c_2}, which is determined from the equation:

$$\alpha_1(T_c)\alpha_2(T_c) = \gamma^2. \tag{7}$$

Let the first order parameter is stronger then second one, i.e. $T_{c_1} > T_{c_2}$. Following (Zhitomirsky & Dao, 2004) we represent temperature dependent coefficients as

$$\begin{aligned} \alpha_1(T) = -a_1(1 - T/T_{c1}), \\ \alpha_2(T) = a_{20} - a_2(1 - T/T_{c1}). \end{aligned} \tag{8}$$

Phenomenological constants $a_{1,2}, a_{20}$ and $\beta_{1,2}, \gamma$ can be related to microscopic parameters in two-band BCS model. From (7) and (8) we obtain for the critical temperature T_c:

$$T_c = T_{c1}\left(1+\sqrt{\left(\frac{a_{20}}{2a_2}\right)^2+\frac{\gamma^2}{a_1 a_2}}-\frac{a_{20}}{2a_2}\right). \tag{9}$$

For arbitrary value of the interband coupling γ Eq.(6) can be solved numerically. For $\gamma=0$, $T_c = T_{c1}$ and for temperature close to T_c (hence for $T_{c2} < T \le T_c$) equilibrium values of the order parameters are $\psi_2^{(0)}(T)=0$, $\psi_1^{(0)}(T)=\sqrt{a_1(1-T/T_c)/\beta_1}$. Considering in the following weak interband coupling, we have from Eqs. (6-9) corrections $\sim\gamma^2$ to these values:

$$\begin{aligned}
\psi_1^{(0)}(T)^2 &= \frac{a_1}{\beta_1}(1-\frac{T}{T_c})+\frac{\gamma^2}{\beta_1}\left(\frac{1}{a_{20}-a_2(1-\frac{T}{T_c})}-\frac{T}{T_c}\frac{1}{a_{20}}\right),\\
\psi_2^{(0)}(T)^2 &= \frac{a_1}{\beta_1}(1-\frac{T}{T_c})\frac{\gamma^2}{(a_{20}-a_2(1-\frac{T}{T_c}))^2}.
\end{aligned} \tag{10}$$

Expanding expressions (10) over $(1-\frac{T}{T_c})\ll 1$ we have conventional temperature dependence of equilibrium order parameters in weak interband coupling limit

$$\begin{aligned}
\psi_1^{(0)}(T) &\approx \sqrt{\frac{a_1}{\beta_1}}\left(1+\frac{1}{2}\frac{a_{20}+a_2}{a_{20}{}^2 a_1}\gamma^2\right)\sqrt{1-\frac{T}{T_c}},\\
\psi_2^{(0)}(T) &\approx \sqrt{\frac{a_1}{\beta_1}}\frac{\gamma}{a_{20}}\sqrt{1-\frac{T}{T_c}}.
\end{aligned} \tag{11}$$

Considered above case (expressions (9)-(11)) corresponds to different critical temperatures $T_{c_1} > T_{c_2}$ in the absence of interband coupling γ . Order parameter in the second band $\psi_2^{(0)}$ arises from the "proximity effect" of stronger $\psi_1^{(0)}$ and is proportional to the value of γ . Consider now another situation, which we will use in the following as the model case. Suppose for simplicity that two condensates in current zero state are identical. In this case for arbitrary value of γ we have

$$\alpha_1(T)=\alpha_2(T)\equiv\alpha(T)=-a\left(1-\frac{T}{T_c}\right), \beta_1=\beta_2\equiv\beta. \tag{12}$$

$$\psi_1^{(0)}=\psi_2^{(0)}=\sqrt{\frac{|\gamma|-\alpha}{\beta}}. \tag{13}$$

2. Homogeneous current states and GL depairing current

In this section we will consider the homogeneous current states in thin wire or film with transverse dimensions $d \ll \xi_{1,2}(T), \lambda_{1,2}(T)$, where $\xi_{1,2}(T)$ and $\lambda_{1,2}(T)$ are coherence lengths and London penetration depths for each order parameter, respectively. This condition leads to a one-dimensional problem and permits us to neglect the self-magnetic field of the system. (see Fig.2) . In the absence of external magnetic field we use the calibration $\vec{A}=0$.

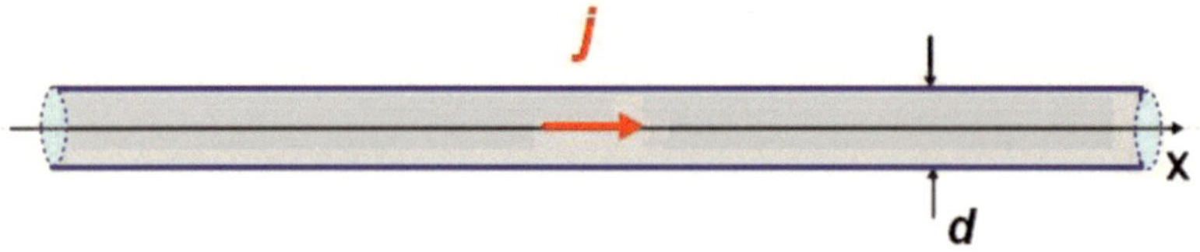

Fig. 2. Geometry of the system.

The current density j and modules of the order parameters do not depend on the longitudinal direction x. Writing $\psi_{1,2}(x)$ as $\psi_{1,2}=|\psi_{1,2}|\exp(i\chi_{1,2}(x))$ and introducing the difference and weighted sum phases:

$$\begin{cases} \phi=\chi_1-\chi_2, \\ \theta=c_1\chi_1+c_2\chi_2, \end{cases} \tag{14}$$

for the free energy density (1)-(3) we obtain

$$\begin{aligned} F &= \alpha_1|\psi_1|^2+\frac{1}{2}\beta_1|\psi_1|^4+\alpha_2|\psi_2|^2+\frac{1}{2}\beta_2|\psi_2|^4+ \\ &+\hbar^2\left(\frac{|\psi_1|^2}{2m_1}+\frac{|\psi_2|^2}{2m_2}+2\eta|\psi_1||\psi_2|\cos\phi\right)\left(\frac{d\theta}{dx}\right)^2+ \\ &+\hbar^2\left(c_2^2\frac{|\psi_1|^2}{2m_1}+c_1^2\frac{|\psi_2|^2}{2m_2}+2c_1c_2\eta|\psi_1||\psi_2|\cos\phi\right)\left(\frac{d\phi}{dx}\right)^2- \\ &-2\gamma|\psi_1||\psi_2|\cos\phi. \end{aligned} \tag{15}$$

Where

$$c_1=\frac{\dfrac{|\psi_1|^2}{m_1}+2\eta|\psi_1||\psi_2|\cos\phi}{\dfrac{|\psi_1|^2}{m_1}+\dfrac{|\psi_2|^2}{m_2}+4\eta|\psi_1||\psi_2|\cos\phi},\quad c_2=\frac{\dfrac{|\psi_2|^2}{m_2}+2\eta|\psi_1||\psi_2|\cos\phi}{\dfrac{|\psi_1|^2}{m_1}+\dfrac{|\psi_2|^2}{m_2}+4\eta|\psi_1||\psi_2|\cos\phi}. \tag{16}$$

The current density j in terms of phases θ and ϕ has the following form

$$j=2e\hbar\left(\frac{|\psi_1|^2}{m_1}+\frac{|\psi_2|^2}{m_2}+4\eta|\psi_1||\psi_2|\cos\phi\right)\frac{d\theta}{dx}. \tag{17}$$

Total current j includes the partial inputs $j_{1,2}$ and proportional to η the drag current j_{12}. In contrast to the case of single order parameter (De Gennes, 1966), the condition $div j=0$ does not fix the constancy of superfluid velocity. The Euler – Lagrange equations for

$\theta(x)$ and $\phi(x)$ are complicated coupled nonlinear equations, which generally permit the soliton like solutions (in the case $\eta = 0$ they were considered in (Tanaka, 2002)). The possibility of states with inhomogeneous phase $\phi(x)$ is needed in separate investigation. Here, we restrict our consideration by the homogeneous phase difference between order parameters $\phi = const$. For $\phi = const$ from equations it follows that $\theta(x) = qx$ (q is total superfluid momentum) and $\cos\phi = 0$, i.e. ϕ equals 0 or π. Minimization of free energy for ϕ gives

$$\cos\phi = sign\left(\gamma - \eta\hbar^2 q^2\right). \tag{18}$$

Note, that now the value of ϕ, in principle, depends on q, thus, on current density j. Finally, the expressions (15), (17) take the form:

$$F = \alpha_1 |\psi_1|^2 + \frac{1}{2}\beta_1 |\psi_1|^4 + \frac{\hbar^2}{2m_1}|\psi_1|^2 q^2 + \alpha_2 |\psi_2|^2 + \frac{1}{2}\beta_2 |\psi_2|^4 + \frac{\hbar^2}{2m_2}|\psi_2|^2 q^2 - \\ -2\left(\gamma - \eta\hbar^2 q^2\right)|\psi_1||\psi_2| sign\left(\gamma - \eta\hbar^2 q^2\right), \tag{19}$$

$$j = 2e\hbar\left(\frac{|\psi_1|^2}{m_1} + \frac{|\psi_2|^2}{m_2} + 4\eta|\psi_1||\psi_2| sign\left(\gamma - \eta\hbar^2 q^2\right)\right)q. \tag{20}$$

We will parameterize the current states by the value of superfluid momentum q, which for given value of j is determined by equation (20). The dependence of the order parameter modules on q determines by GL equations:

$$\alpha_1|\psi_1| + \beta_1|\psi_1|^3 + \frac{\hbar^2}{2m_1}|\psi_1|q^2 - |\psi_2|\left(\gamma - \eta\hbar^2 q^2\right) sign\left(\gamma - \eta\hbar^2 q^2\right) = 0, \tag{21}$$

$$\alpha_2|\psi_2| + \beta_2|\psi_2|^3 + \frac{\hbar^2}{2m_2}|\psi_2|q^2 - |\psi_1|\left(\gamma - \eta\hbar^2 q^2\right) sign\left(\gamma - \eta\hbar^2 q^2\right) = 0. \tag{22}$$

The system of equations (20-22) describes the depairing curve $j(q,T)$ and the dependences $|\psi_1|$ and $|\psi_2|$ on the current j and the temperature T. It can be solved numerically for given superconductor with concrete values of phenomenological parameters.

In order to study the specific effects produced by the interband coupling and dragging consider now the model case when order parameters coincide at $j = 0$ (Eqs. (12), (13)) but gradient terms in equations (4) are different. Eqs. (20)-(22) in this case take the form

$$\begin{gathered} f_1\left(1 - \left(1 + |\tilde{\gamma}|\right) f_1^2\right) - f_1 q^2 + f_2\left(\tilde{\gamma} - \tilde{\eta} q^2\right) sign\left(\tilde{\gamma} - \tilde{\eta} q^2\right) = 0 \\ f_2\left(1 - \left(1 + |\tilde{\gamma}|\right) f_2^2\right) - k f_2 q^2 + f_1\left(\tilde{\gamma} - \tilde{\eta} q^2\right) sign\left(\tilde{\gamma} - \tilde{\eta} q^2\right) = 0 \end{gathered} \tag{23}$$

$$j = f_1^2 q + k f_2^2 q + 2\tilde{\eta} f_1 f_2 q \, sign\left(\tilde{\gamma} - \tilde{\eta} q^2\right) \tag{24}$$

Here we normalize $\psi_{1,2}$ on the value of the order parameters at $j=0$ (13), j is measured in units of $2\sqrt{2}e\frac{|\gamma|+|\alpha|}{\beta}\sqrt{\frac{|\alpha|}{m_1}}$, q is measured in units of $\sqrt{\frac{\hbar^2}{2m_1|\alpha|}}$, $\tilde{\gamma}=\frac{\gamma}{|\alpha|}$, $\tilde{\eta}=2\eta m_1$, $k=\frac{m_1}{m_2}$.
If $k=1$ order parameters coincides also in current-carrying state $f_1=f_2=f$ and from eqs. (23), 24) we have the expressions

$$f^2(q) = \frac{1-q^2+\left|\tilde{\gamma}-\tilde{\eta}q^2\right|}{1+|\tilde{\gamma}|} \tag{25}$$

$$j(q) = 2f^2\left(1+\tilde{\eta}\, sign\left(\tilde{\gamma}-\tilde{\eta}q^2\right)\right)q, \tag{26}$$

which for $\tilde{\gamma}=\tilde{\eta}=0$ are conventional dependences for one-band superconductor (De Gennes, 1966) (see Fig. 3 a,b).

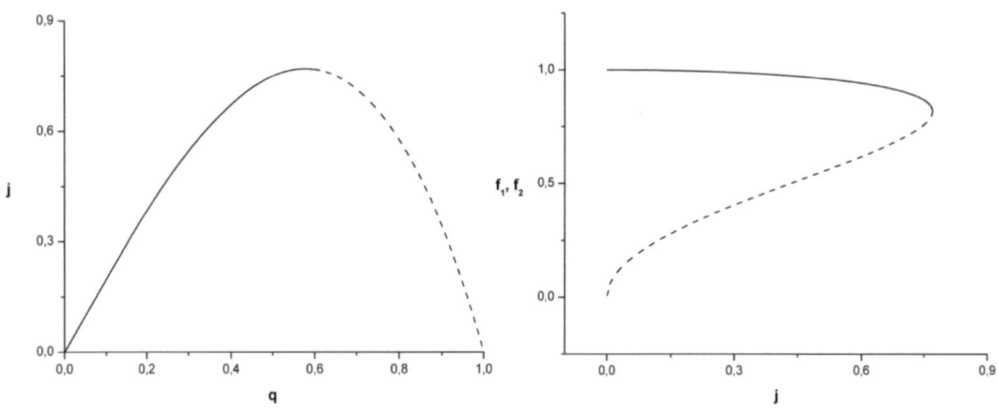

(a) (b)

Fig. 3. Depairing current curve (a) and the graph of the order parameter modules versus current (b) for coincident order parameters. The unstable branches are shown as dashed lines.

For $k \neq 1$ depairing curve $j(q)$ can contain two increasing with q stable branches, which corresponds to possibility of bistable state. In Fig. 4 the numerically calculated from equations (23,24) curve $j(q)$ is shown for $k=5$ and $\tilde{\gamma}=\tilde{\eta}=0$.
The interband scattering ($\tilde{\gamma}\neq 0$) smears the second peak in $j(q)$, see Fig.5.
If dragging effect ($\tilde{\eta}\neq 0$) is taking into account the depairing curve $j(q)$ can contain the jump at definite value of q (for $k=1$ see eq. 34), see Fig.6. This jump corresponds to the switching of relative phase difference from 0 to π.

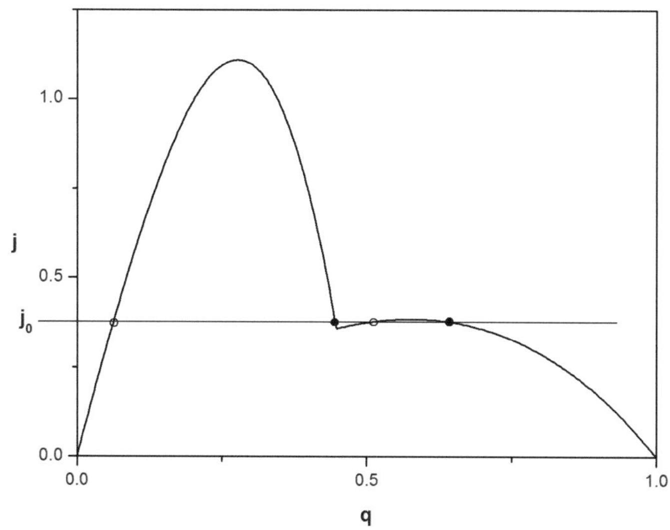

Fig. 4. Dependence of the current j on the superfluid momentum q for two band superconductor. For the value of the current $j = j_0$ the stable ($\circ$) and unstable ($\bullet$) states are depicted. The ratio of effective masses $k = 5$, and $\tilde{\gamma} = \tilde{\eta} = 0$.

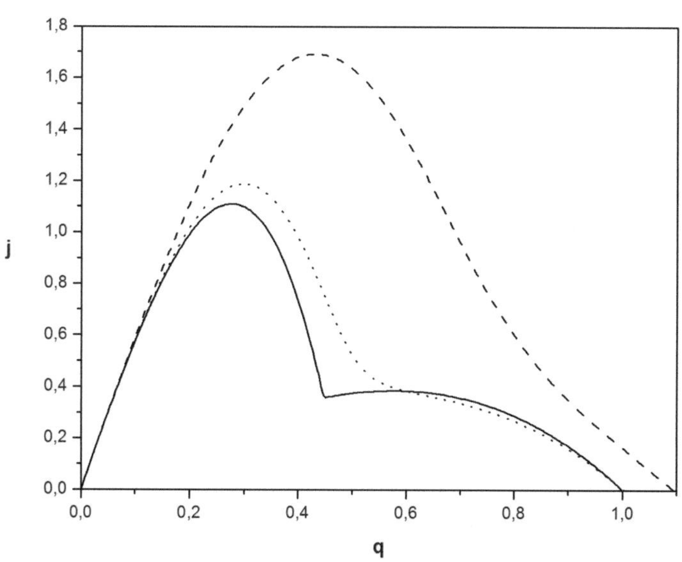

Fig. 5. Depairing current curves for different values of interband interaction: $\tilde{\gamma} = 0$ (solid line), $\tilde{\gamma} = 0.1$ (dotted line) and $\tilde{\gamma} = 1$ (dashed line). The ratio of effective masses $k = 5$, and $\tilde{\eta} = 0$.

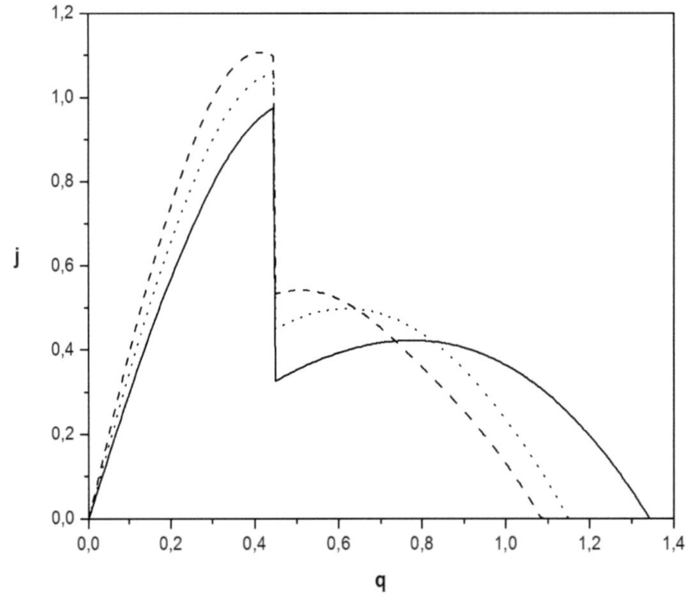

Fig. 6. Depairing current curves for different values of the effective masses ratio $k=1$ (solid line), $k=1.5$ (dotted line) and $k=5$ (dashed line). The interband interaction coefficient $\tilde{\gamma}=0.1$ and drag effect coefficient $\tilde{\eta}=0.5$.

4. Little-Parks effect for two-band superconductors

In the present section we briefly consider the Little–Parks effect for two-band superconductors. The detailed rigorous theory can be found in the article (Yerin et al., 2008). It is pertinent to recall that the classical Little–Parks effect for single-band superconductors is well-known as one of the most striking demonstrations of the macroscopic phase coherence of the superconducting order parameter (De Gennes, 1966; Tinkham, 1996). It is observed in open thin-wall superconducting cylinders in the presence of a constant external magnetic field oriented along the axis of the cylinder. Under conditions where the field is essentially unscreened the superconducting transition temperature $T_{c\Phi}$ (Φ is the magnetic flux through the cylinder) undergoes strictly periodic oscillations (Little–Parks oscillations)

$$\frac{T_c - T_{c\Phi}}{T_c} \propto \min\left(\frac{\Phi}{\Phi_0} - n\right)^2, (n = 0, \pm 1, \pm 2, ...), \tag{27}$$

where $T_{c\Phi} \equiv T_{c\Phi}\big|_{\Phi=0}$ and $\Phi_0 = \pi\hbar c / e$ is the quantum of magnetic flux.

How the Little–Parks oscillations (27) will be modified in the case of two order parameters with taking into account the proximity (γ) and dragging (η) coupling? Let us consider a superconducting film in the form of a hollow thin cylinder in an external magnetic field **H** (see Fig.6).

We proceed with free energy density (19), but now the momentum q is not determined by the fixed current density **j** as in Sec.3. At given magnetic flux $\oint \vec{A} \cdot d\vec{l} = \int \vec{H} \cdot d\vec{\sigma} = \Phi$ the superfluid momentum q is related to the applied magnetic field

$$q = \frac{1}{R}\left(n - \frac{\Phi}{\Phi_0}\right). \tag{28}$$

At fixed flux Φ the value of q take on the infinite discrete set of values for $n = 0, \pm 1, \pm 2, \ldots$. The possible values of n are determined from the minimization of free energy $F[|\psi_1|, |\psi_2|, q]$. As a result the critical temperature of superconducting film depends on the magnetic field. The dependencies of the relative shift of the critical temperature $\Delta t_c \equiv (T_c - T_{c\Phi}) / T_c$ for different values of parameters γ, β, R were calculated in (Yerin et al., 2008). The dependence of $\Delta t_c(\Phi)$ as in the conventional case is strict periodic function of Φ with the period Φ_0 (contrary to the assertions made in Askerzade, 2006). The main qualitative difference from the classical case is the nonparabolic character of the flux dependence $\Delta t_c(\Phi)$ in regions with the fixed optimal value of n. More than that, the term $\left(\gamma - \eta\hbar^2 q^2\right) sign\left(\gamma - \eta\hbar^2 q^2\right)$ in the free energy (19) engenders possibility of observable singularities in the function $\Delta t_c(\Phi)$, which are completely absent in the classical case (see Fig.8.).

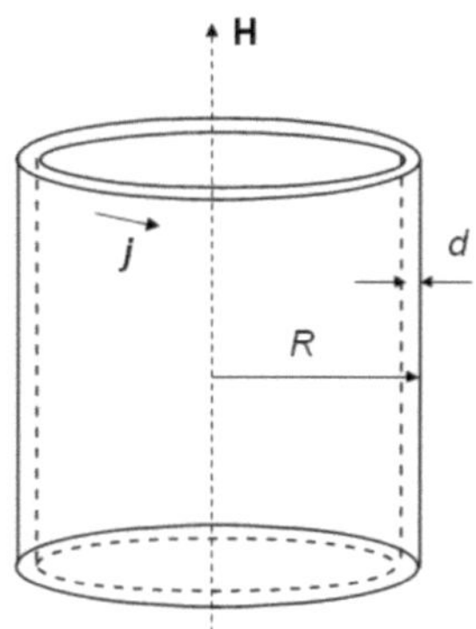

Fig. 7. Geometry of the problem.

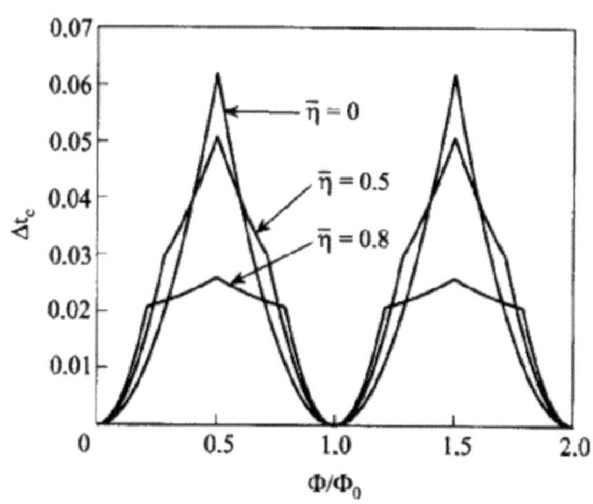

Fig. 8. $\Delta t_c(\Phi)$ for the case where the bands 1 and 2 have identical parameters and values of $\tilde{\eta}$ are indicated.

5. Josephson effect in two-band superconducting microconstriction

In the Sec.3 GL-theory of two-band superconductors was applied for filament's length $L \to \infty$. Opposite case of the strongly inhomogeneous current state is the Josephson microbridge or point contact geometry (Superconductor-Constriction-Superconductor contact), which we model as narrow channel connecting two massive superconductors (banks). The length L and the diameter d of the channel (see Fig. 9) are assumed to be small as compared to the order parameters coherence lengths ξ_1, ξ_2.

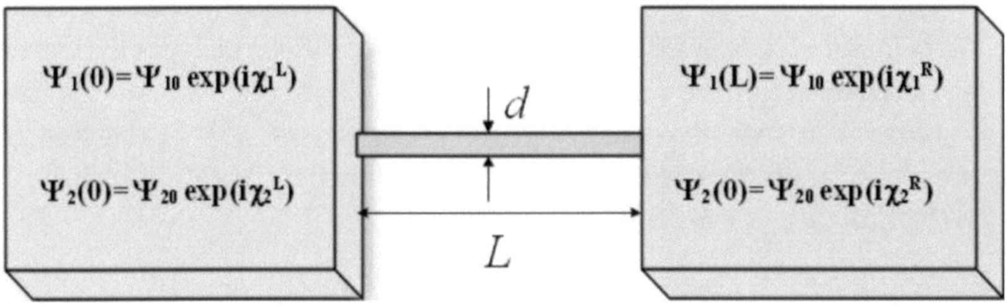

Fig. 9. Geometry of S-C-S contact as narrow superconducting channel in contact with bulk two-band superconductors. The values of the order parameters at the left (L) and right (R) banks are indicated

For $d \ll L$ we can solve one-dimensional GL equations (4) inside the channel with the rigid boundary conditions for order parameters at the ends of the channel.
In the case $L \ll \xi_1, \xi_2$ we can neglect in equations (4) all terms except the gradient ones and solve equations:

$$\begin{cases} \dfrac{d^2\psi_1}{dx^2}=0, \\ \dfrac{d^2\psi_2}{dx^2}=0 \end{cases} \tag{29}$$

with the boundary conditions:

$$\psi_1(0)=\psi_{01}\exp\left(i\chi_1{}^L\right),\ \psi_2(0)=\psi_{02}\exp\left(i\chi_1{}^R\right), \tag{30}$$

$$\psi_1(L)=\psi_{01}\exp\left(i\chi_2{}^L\right),\ \psi_2(L)=\psi_{02}\exp\left(i\chi_1{}^R\right).$$

Calculating the current density j in the channel we obtain:

$$j = j_{11} + j_{22} + j_{12} + j_{21}, \tag{31}$$

$$j_{11} = \frac{2e\hbar}{Lm_1}\psi_{01}^2 \sin\left(\chi_1{}^R - \chi_1{}^L\right),$$

$$j_{22} = \frac{2e\hbar}{Lm_1}\psi_{01}^2 \sin\left(\chi_2{}^R - \chi_2{}^L\right),$$

$$j_{12} = \eta \frac{4e\hbar}{L} \psi_{01} \psi_{02} \sin\left(\chi_1{}^R - \chi_2{}^L\right),$$

$$j_{21} = \eta \frac{4e\hbar}{L} \psi_{02} \psi_{01} \sin\left(\chi_2{}^R - \chi_1{}^L\right).$$

The current density j (31) consists of four partials inputs produced by transitions from left bank to right bank between different bands. The relative directions of components j_{ik} depend on the intrinsic phase shifts in the banks $\delta\phi^L = \chi_1^L - \chi_2^L$ and $\delta\phi^R = \chi_1^R - \chi_2^R$ (Fig.10).

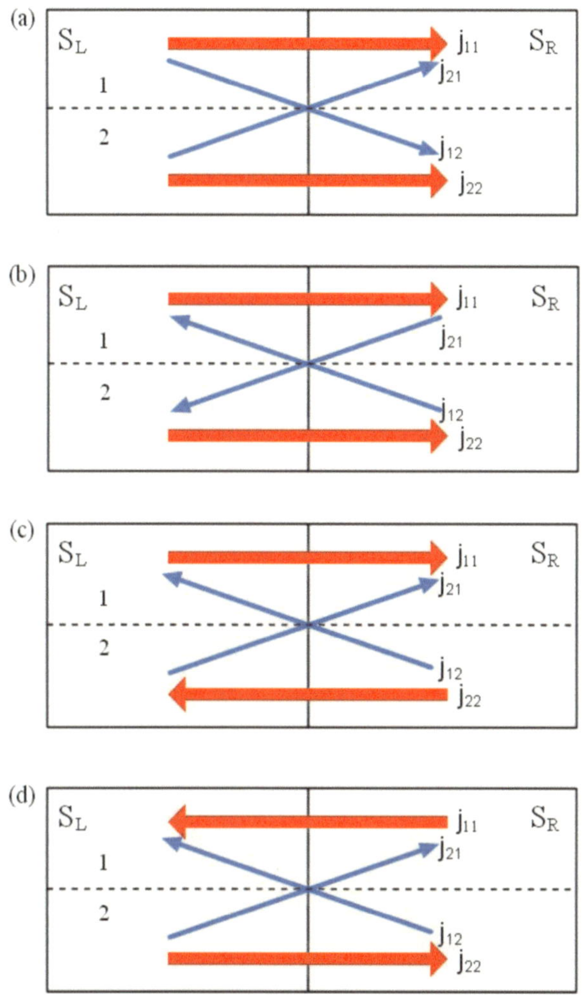

Fig. 10. Current directions in S-C-S contact between two-band superconductors. (a) – there is no shift between phases of order parameters in the left and right superconductors; (b) - there is the π -shift of order parameters phases at the both banks ; (c) – π -shift is present in the right superconductor and is absent in the left superconductor; (d) – π -shift is present in the left superconductor and is absent in the right superconductor .

Introducing the phase difference on the contact $\varphi = \chi_1^R - \chi_1^L$ we have the current-phase relation $j(\phi)$ for different cases of phase shifts $\delta\phi^{R,L}$ in the banks:

a. $\delta\phi^R = \delta\phi^L = 0$

$$j \equiv j_c \sin\phi = \frac{2e\hbar}{L}\left(\frac{\psi_{01}^2}{m_1} + \frac{\psi_{02}^2}{m_2} + 4\eta\psi_{01}\psi_{02}\right)\sin\phi$$

b. $\delta\phi^R = \delta\phi^L = \pi$

$$j \equiv j_c \sin\phi = \frac{2e\hbar}{L}\left(\frac{\psi_{01}^2}{m_1} + \frac{\psi_{02}^2}{m_2} - 4\eta\psi_{01}\psi_{02}\right)\sin\phi$$

c. $\delta\phi^R = \pi, \delta\phi^L = 0$

$$j \equiv j_c \sin\phi = \frac{2e\hbar}{L}\left(\frac{\psi_{01}^2}{m_1} - \frac{\psi_{02}^2}{m_2}\right)\sin(\phi)$$

d. $\delta\phi^R = 0, \delta\phi^L = \pi$

$$j \equiv j_c \sin\phi = \frac{2e\hbar}{L}\left(-\frac{\psi_{01}^2}{m_1} + \frac{\psi_{02}^2}{m_2}\right)\sin(\phi)$$

The critical current j_c in cases a) and b) is positively defined quadratic form of ψ_{01} and ψ_{02} for $|\eta| < \frac{1}{2\sqrt{m_1 m_2}}$. In cases c) and d) the value of j_c can be negative. It corresponds to the so-called π – junction (see *e.g.* (Golubov et. al, 2004)) (see illustration at Fig.11).

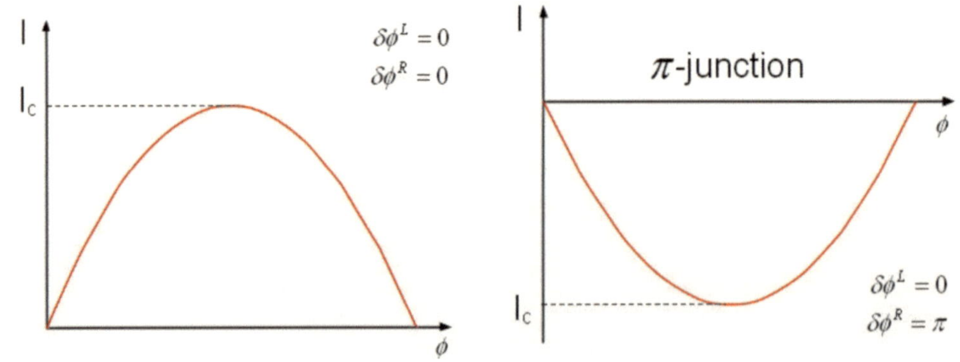

Fig. 11. Current-phase relations for different phase shifts in the banks.

This phenomenological theory, which is valid for temperatures near critical temperature T_c, is the generalization of Aslamazov-Larkin theory (Aslamazov & Larkin, 1968) for the case of two superconducting order parameters. The microscopic theory of Josephson effect in S-C-S junctions (KO theory) was developed in (Kulik & Omelyanchouk, 1975; Kulik &

Omelyanchouk, 1978;) by solving the Usadel and Eilenberger equations (for dirty and clean limits). In papers (Omelyanchouk & Yerin, 2009; Yerin & Omelyanchouk, 2010) we generalized KO theory for the contact of two-band superconductors. Within the microscopic Usadel equations we calculate the Josephson current and study its dependence on the mixing of order parameters due to interband scattering and phase shifts in the contacting two-band superconductors. These results extend the phenomenological theory presented in this Section on the range of all temperatures $0 < T < T_c$. The qualitative feature is the possibility of intermediate between $\sin\phi$ and $-\sin\phi$ behavior $j(\phi)$ at low temperatures (Fig.12).

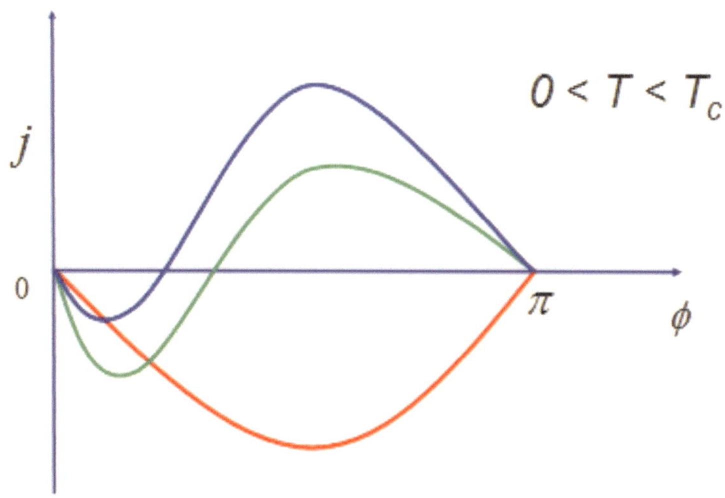

Fig. 12. The possible current-phase relations $j(\phi)$ for hetero-contact with $\delta\phi^R = 0, \delta\phi^L = \pi$.

6. Conclusion

In this chapter the current carrying states in two-band superconductors are described in the frame of phenomenological Ginzburg-Landau theory. The qualitative new feature, as compared with conventional superconductors, consists in coexistence of two distinct complex order parameters Ψ_1 and Ψ_2. It means the appearing of an additional internal degree of freedom, the phase shift between order parameters. We have studied the implications of the $\delta\phi$-shift in homogeneous current state in long films or channels, Little-Parks oscillations in magnetic field, Josephson effect in microconstrictions. The observable effects are predicted. Along with fundamental problems, the application of two band superconductors with internal phase shift in SQUIDS represents significant interest (see review (Brinkman & Rowell, 2007).

7. Acknowledgment

The author highly appreciates S. Kuplevakhskii and Y.Yerin for fruitful collaborations and discussions. The research is partially supported by the Grant 04/10-N of NAS of Ukraine.

8. References

Agterberg D. F., Demler E., & Janko B. (2002). Josephson effects between multigap and single-gap superconductors, *Phys. Rev. B*, V.66, Iss.21, p.214507.

Artemenko S. N., Volkov A. F. & Zaitsev A. V. (1979). Theory of nonstationary Josephson effect in short superconducting junctions, *Zh. Eksp. Teor. Fiz.*, V. 76, No.5, p.1816-1833.

Ashcroft N. W. (2000). The Hydrogen Liquids. *J. Phys.: Condens. Matter*, V.12, No.8A, p. A129-A137.

Askerzade I. N. (2003). Temperature dependence of the London penetration depth of YNi2B2C borocarbids using two-band Ginzburg-Landau theory. *Acta Physica Slovaca*, V.53, No. 4, p.321-327.

Askerzade I. N. (2003). Ginzburg–Landau theory for two-band s-wave superconductors: application to non-magnetic borocarbides LuNi2B2C, YNi2B2C and magnesium diboride MgB2. *Physica C*, V.397, Iss.3-4, p.99-111.

Askerzade I. N. (2006). Ginzburg-Landau theory: the case of two-band superconductors. *Usp. Fiz. Nauk*, V.49, Iss.10, p.1003-1016.

Aslamazov L.G. & Larkin A.I. (1969). The Josephson effect in point superconducting junctions. *Pis'ma Zh. Eksp. Teor. Fiz.*, V.9, No.2, p.150-154.

Babaev E. (2002). Vortices with Fractional Flux in Two-Gap Superconductors and in Extended Faddeev Model. *Phys. Rev. Lett.*, V.89, Iss. 89, p.067001.

Babaev E., Faddeev L. D. & Niemi A. J. (2002). Hidden symmetry and knot solitons in a charged two-condensate Bose system. *Phys. Rev. B*, V.65, Iss.10, p.100512(R).

Babaev E., Sudbo A. & Ashcroft N. W. (2004). A superconductor to superfluid phase transition in liquid metallic hydrogen. *Nature*, V.431, No. 7009, p.666-668.

Brinkman A., Golubov A. A., Rogalla H., Dolgov O. V., Kortus J., Kong Y., Jepsen O. & Andersen O. K. (2002). Multiband model for tunneling in MgB2 junctions. *Phys. Rev.B*, V.65, Iss.18, p.180517(R).

Brinkman A. & Rowell J. (2007). MgB2 tunnel junctions and SQUIDs. *Physica C*, 456, p.188-195.

Dahm T., Graser S. & Schopohl N. (2004). Fermi surface topology and vortex state in MgB2. *Physica C*, V.408-410, p.336-337.

Dahm T. & Schopohl N. (2003). Fermi Surface Topology and the Upper Critical Field in Two-Band Superconductors: Application to MgB2. *Phys. Rev. Lett.*, V.91, Iss. 1, p.017001.

De Gennes P. G. (1966). *Superconductivity of Metals and Alloys*, Benjamin, ISBN 0738201014, New York.

Doh H., Sigrist M., Cho B.K. & Lee S.I. (1999). Phenomenological Theory of Superconductivity and Magnetism in Ho1-xDyxNi2B2C. *Phys. Rev. Lett.*, V.83, Iss.25, p.5350-5353.

Giubileo F., Roditchev D., Sacks W., Lamy R., Thanh D.X., Klein J., Miraglia S., Fruchart D., Marcus J. & Monod P. (2001). Two-Gap State Density in MgB2: A True Bulk Property Or A Proximity Effect? *Phys. Rev. Lett.*, V.87, Iss.17, p.177008.

Golubov A. A., Kortus J., Dolgov O. V., Jepsen O., Kong Y., Andersen O. K., Gibson B. J., Ahn K. & Kremer R. K. (2002). Specific heat of MgB2 in a one- and a two-band model from first-principles calculations. *J. Phys.: Condens. Matter*, V.14, No.6, p.1353-1361.

Golubov A. A. & Koshelev A. E. (2003). Upper critical field in dirty two-band superconductors: Breakdown of the anisotropic Ginzburg-Landau theory. *Phys. Rev. B*, V.68, Iss.10, p.104503.

Golubov A. A., Kupriyanov M. Yu. & Il'ichev E. (2004). The current-phase relation in Josephson junctions. *Rev. Mod. Phys.*, V.76, Iss.2, p. 411–469.

Golubov A. A. & Mazin I. I. (1995). Sign reversal of the order parameter in s wave superconductors. *Physica C*, V.243, Iss.1-2, p.153-159.

Gurevich A. (2003). Enhancement of the upper critical field by nonmagnetic impurities in dirty two-gap superconductors, *Phys. Rev. B*, V.67, Iss.18, p.184515.

Gurevich A. (2007). Limits of the upper critical field in dirty two-gap superconductors. *Physica C*, V. 456, Iss. 1-2, p.160-169.

Gurevich A. & Vinokur V.M. (2003). Interband Phase Modes and Nonequilibrium Soliton Structures in Two-Gap Superconductors. *Phys. Rev. Lett.*, V.90, Iss.4, p.047004.

Gurevich A. & Vinokur V. M. (2006). Phase textures induced by dc-current pair breaking in weakly coupled multilayer structures and two-gap superconductors, *Phys. Rev. Lett.*, V.97, Iss.13, p.137003.

Iavarone M., Karapetrov G., Koshelev A.E., Kwok W.K., Crabtree G.W., Hinks D.G., Kang W.N., Choi E.-M., Kim H.J. & Lee S.I. (2002). Two-Band Superconductivity in MgB2. *Phys. Rev. Lett.*, V.89, Iss.18, p.187002.

Jourdan M., Zakharov A., Foerster M. & Adrian H. (2004). Evidence for Multiband Superconductivity in the Heavy Fermion Compound UNi2Al3. *Phys. Rev. Lett.*, V.93, Iss.9, p.097001.

Kamihara Y., Watanabe T., Hirano M. & Hosono H. (2008). Iron-based layered superconductor La[O(1-x)F(x)]FeAs (x = 0.05-0.12) with T(c) = 26 K. *J. Am. Chem. Soc.*, V.130, Iss.11, p.3296-3297.

Kortus J., Mazin I. I., Belashchenko K. D., Antropov V. P. & L. L. Boyer (2001). Superconductivity of Metallic Boron in MgB2. *Phys. Rev. Lett.*, V.86, Iss.20, p.4656.

Koshelev A. E. & Golubov A. A. (2003). Mixed state of a dirty two-band superconductor: pplication to MgB2, *Phys. Rev. Lett.*, V.90, Iss.17, p.177002.

Kresin V. Z. & Wolf S. A (1990). Multigap structure in cuprates. *Physica C*, V.169, Iss.5-6, p.476–484.

Kulik I. O. & Omelyanchouk A. N. (1975). Microscopic theory of Josephson effect in superconducting microbridges, *Pis'ma Zh. Eksp. Teor. Fiz.*, V.21, No.4, p.216-219.Kulik I. O. & Omelyanchouk A. N. (1978). Josephson effect in superconducting bridges:

microscopic theory, *Fiz. Nizk. Temp.*, V.4, No.3, p.296-311.

Leggett J. (1966). Number-phase fluctuations in two-band superconductors. *Progr. Theor. Phys.*, V.36, No. 5, pp. 901-930.

Mazin I. I., Andersen O. K., Jepsen O., Dolgov O. V., Kortus J., Golubov A. A., Kuz'menko A. B. & van der Marel D. (2002). Superconductivity in MgB2: Clean or Dirty? *Phys. Rev. Lett.*, V.89, Iss.10, p.107002.

Mints R. G., Papiashvili I., Kirtley J. R., Hilgenkamp H., Hammerl G. & Mannhart J. (2002). Observation of Splintered Josephson Vortices at Grain Boundaries in YBa2Cu3O7-δ. *Phys. Rev. Lett.*, V.89, Iss.6, p.067004.

Miranovic P., Machida K. & Kogan V. G. (2003). Anisotropy of the Upper Critical Field in Superconductors with Anisotropic Gaps: Anisotropy Parameters of MgB2. *J. Phys. Soc. Jpn.*, V.72, No.2, p.221-224.

Moskalenko V.A. (1959). Superconductivity of metals within overlapping energy bands. *Fiz. Met. Metallov.*, V.8, p.503-513.

Nagamatsu J., Nakagawa N., Muranaka T., Zenitani Y. & Akimitsu J. (2001), Superconductivity at 39 K in magnesium diboride. *Nature*, V.410, No.6824, p.63-64.

Nakai A., Ichioka M. & Machida K. (2002). Field Dependence of Electronic Specific Heat in Two-Band Superconductors. *J. Phys. Soc. Jpn.*, V.71, No.1, p.23-26.

Ng T. K. & Nagaosa N. (2009). Broken time-reversal symmetry in Josephson junction involving two-band superconductors, *EPL* V.87, No.1, p.17003.

Omelyanchouk A.N. & Yerin Y.S. (2010). Josephson effect in point contacts between two-band superconductors. In: *Physical Properties of Nanosystems*, Bonca J. & Kruchinin S., pp. 111-119, Springer, ISBN: 978-94-007-0043-7, Berlin.

Ota Y., Machida M., Koyama T. & Matsumoto H. (2009). Theory of heterotic superconductor-insulator-superconductor Josephson junctions between single- and multiple-gap superconductors, *Phys. Rev. Lett.*, V.102, Iss.23, p.237003.

Schmidt H., Zasadzinski J.F., Gray K.E. & Hinks D.G. (2001). Evidence for Two-Band Superconductivity from Break-Junction Tunneling on MgB2. *Phys. Rev. Lett.*, V.88, Iss.12, p.127002.

Seyfarth G., Brison J. P., Méasson M.-A., Flouquet J., Izawa K., Matsuda Y., Sugawara H. & Sato H. (2005). Multiband Superconductivity in the Heavy Fermion Compound PrOs4Sb12. *Phys. Rev. Lett.*, V.95, Iss. 10, p.107004.

Shulga S. V., Drechsler S.-L., Fuchs G., Müller K.-H., Winzer K., Heinecke M. & Krug K. (1998). Upper Critical Field Peculiarities of Superconducting YNi2B2C and LuNi2B2C. *Phys. Rev. Lett.*, V.80, Iss.8, p.1730-1733.

Suhl H., Matthias B.T. & Walker L.R. (1959), Bardeen-Cooper-Schrieffer Theory of Superconductivity in the Case of Overlapping Bands. *Phys. Rev. Lett.*, V.3, Iss.12, p. 552-554.

Szabo P., Samuely P., Kacmarcik J., Klein T., Marcus J., Fruchart D., Miraglia S., Mercenat C., & Jansen A.G.M. (2001). Evidence for Two Superconducting Energy Gaps in MgB2 by Point-Contact Spectroscopy. *Phys. Rev. Lett.*, V.87, Iss.13, p.137005.

Tanaka Y. (2002). Soliton in Two-Band Superconductor. *Phys. Rev. Lett.*, V.88, Iss.1, p.017002.Tinkham M. (1996). *Introduction to Superconductivity*, McGraw-Hill, ISBN 0-07-064878-6, New York.

Yanson I. K. & Naidyuk Yu. G. (2004), Advances in point-contact spectroscopy: two-band superconductor MgB2. *Low Temp.Phys.*, V.30, Iss.4, p.261-275.

Yerin Y. S., Kuplevakhskii S. V. & Omelyanchuk A. N. (2008). Little–Parks effect for two-band superconductors. *Low Temp. Phys.*, V.34, Iss.11, p.891-898.

Yerin Y. S. & Omelyanchouk A. N. (2010). Josephson currents in point contacts between dirty two-band superconductors. *Low Temp. Phys.*, V.36, Iss.10, p.969-974.

Zhitomirsky M. E. & Dao V.-H. (2004). Ginzburg-Landau theory of vortices in a multigap superconductor, *Phys. Rev. B*, V.69, Iss.5, p.054508.

4

Nonlinear Response of the Static and Dynamic Phases of the Vortex Matter

S. S. Banerjee[1], Shyam Mohan[1], Jaivardhan Sinha[1],
Yuri Myasoedov[3], S. Ramakrishnan[2] and A. K. Grover[2]
[1]*Department of Physics, Indian Institute of Technology Kanpur, Uttar Pradesh*
[2]*Department of Condensed Matter Physics and Materials Science,*
Tata Institute of Fundamental Research, Mumbai,
[3]*Department of Physics, Weizmann Institute of Science, Rehovot,*
[1,2]*India*
[3]*Israel*

1. Introduction

In the mixed state of type II superconductors, the external magnetic field penetrates the superconducting material in the form of normal cored regions, each carrying a quantum of flux ($\Phi_0 = 2.07\times10^{-7}$ G-cm^2). These normal cores have radii equal to the coherence length (ξ). Surrounding each normal core is a vortex of supercurrent that decays over a characteristic length scale known as the penetration depth (λ). These elastic string-like normal entities (or vortices) mutually repel each other leading to the formation of triangular vortex lattices in ideal superconductors (Blatter et al., 1994; Natterman & Scheidl, 2000). However, real samples always have defects (point defects, dislocations) and inhomogeneities. The superconducting order parameter is preferentially suppressed at these random defect locations, thereby energetically favoring pinning of vortices at these locations. But, pinning also leads to loss of long range order in the vortex lattice. The vortex matter can be considered as a typical prototype for soft materials, where pinning forces and thermal fluctuations are comparable to the elastic energy scale of the vortex lattice. The perennial competition between elastic interactions in the vortex lattice, which establishes order in the vortex state and effects of pinning and thermal fluctuations which try to destabilize the vortex lattice, leads to a variety of pinning regimes, viz., the weak collective pinning regime and the strong pinning regime (Blatter et al., 2004). The competition in different portions of the field-temperature (H,T) phase space leads to the emergence of a variety of vortex phases, like, the Bragg glass, vortex glass, vortex liquid (for review see, Blatter et al., 1994; Natterman & Scheidl, 2000) and transformation amongst them, along with the appearance of significant thermomagnetic history dependent response. The competing effects ever present in the vortex lattice also lead to a quintessential phenomenon called the peak effect (PE), which we shall discuss in the next section.

2. The peak effect phenomenon

Theoretical works in late nineteen eighties and nineties have shown that by taking into account the effects of thermal fluctuations and pinning centers on vortices, the mean field description of a type II superconductor gets substantially modified and new phases and phase boundaries in the vortex matter were predicted. In particular, in a clean pinning free system, it was shown that under the influence of thermal fluctuations, the vortex lattice phase is stable only in the intermediate field range. A new phase was predicted to be present at both very low and at very high fields, viz., the Vortex Liquid State (Nelson, 1988), in which the r.m.s, fluctuation of the vortices about their mean positions become ~10 - 20 % of the intervortex spacing a_0 ($a_0 \propto B^{1/2}$, where B is the field) and the vortex-vortex spatial correlations reduced down length scales of the order ~ a_0. Experimental works on the high temperature superconductors (HTSC) have established the vortex solid to liquid transition at high fields, however, the demonstration of the reentrant behavior of the vortex solid to liquid phase boundary has so far not been vividly elucidated (Blatter et al, 1998; Natterman & Scheidl, 2000). The mean field picture of a perfectly periodic arrangement of vortices in the vortex solid phase is also expected to be modified under the influence of pinning and the vortex solid phase is considered to behave like a vortex glass (Fisher 1989; Fisher, et al. 1989), which is characterized by zero linear resistivity, and could exhibit many metastable states. Further detailed investigations (Giammarchi and P. Le Doussal, 1995), showed, the existence of a novel vortex solid to solid transformation as a function of varying field at a fixed temperature in which a novel Bragg Glass phase (a reasonably well ordered lattice with correlation extending over few hundreds of a_0) at low fields transforms into a Vortex Glass state with spatial correlations surviving over a very short range at high fields. This solid to solid transformation is considered to arise due to a sudden injection / proliferation of dislocations into the Bragg glass phase (for a review see Natterman & Scheidl, 2000).

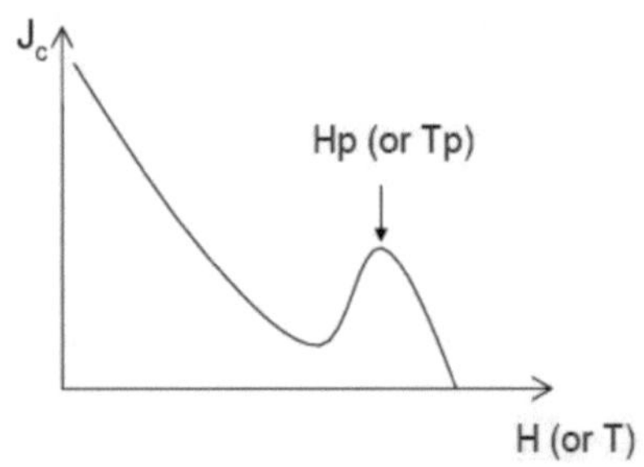

Fig. 1. Schematic representation of the peak effect (PE) in the critical current density, J_c, with applied field (or temperature). The field H_p (or temperature T_p) represents the peak position of the PE.

To experimentally investigate the phases of vortex matter, few popular routes are via ac susceptibility, dc magnetization, transport measurements, all of which provide information on the critical current density (J_c) (the maximum dissipationless current which is carried by a superconductor). Usually a change in the phase of vortex matter is accompanied by a change in the pinning experienced by the vortices. As the J_c is a direct measure of the pinning experienced by a given phase, changes in the behavior of J_c are a good indicator of the transformation/transition in the vortex matter. Usually the J_c of a superconductor is expected to monotonically decrease with increasing values of the temperature or field. However, in a large variety of superconductors it is found that the monotonic decrease in J_c

with increasing field (H) or temperature (T) is interrupted by an anomalous enhancement in J_c just before the superconductor turns normal (Figure.1). This anomaly in the J_c behavior is known as the peak effect (PE) phenomenon and has been observed in many low and high-temperature superconductors (Berlincourt, 1961; Bhattacharya & Higgins, 1993; Higgins and Bhattacharya, 1996; Ling et al., 2001; Ghosh et al., 1996; Banerjee et al., 1998, 1999a, 1999b, 2000a, 2000b, 2001). In electrical transport experiments, from which J_c is deduced, the PE appears as a bump in J_c as in the schematic of Fig.1. Due to the enhancement in pinning, the PE appears as an anomalous increase of the diamagnetic screening or shielding response and a drop in the dissipation response in the ac susceptibility (quadrature signal) measurements before the diamagnetic ac-susceptibility (in -phase signal) crashes to zero at H_{c2} or $T_c(H)$ (Banerjee et al., 1998-2001; Mohan et al., 2007).

Though a complete theoretical description of the PE is lacking, there have been plausible proposals articulating different mechanisms to explain this phenomenon. Pippard (Pippard, 1969) put forth the notion that if the vortex lattice (VL) loses rigidity near H_{c2} at a rate much faster than the pinning force, then the softened vortices would conform more easily to the pinning centers thereby getting strongly pinned, and consequently producing the peak in J_c. The idea acquired a quantitative basis, when a correct statistical summation procedure for the pinning force was proposed by A. I. Larkin and Yu. N. Ovchinnikov (LO) (Larkin, 1970a, 1970b; Larkin and Ovchinnikov, 1979), which took into account the elasticity of the vortex lattice. The basic premise of the LO theory is that the flux lines lower their free energy by passing through the pinning sites, thereby deviating from an ideal periodic arrangement. The deformation of the FLL costs elastic energy despite the lowering in free energy due to the pinning of flux lines. The equilibrium configuration of the flux lines in a deformed state is obtained by minimizing the sum of these two energies. This work of Larkin and Ovchinnikov showed that random distribution of weak pins destroys long range order in the FLL, with short range order being preserved only within a volume bounded by two correlation lengths viz., the radial (R_c, the correlation length across the surface of the sample and perpendicular to the vortex line) and the longitudinal (L_c, the correlation length parallel to the vortex line). These length scales were shown to be related to the elastic modulii of the vortex lattice (Larkin and Ovchinnikov, 1979), and the net pinning force experienced by the VL, viz., $F_p \propto R_c^{-\alpha} L_c^{-\beta}$, where α and β are positive powers. The PE stood explained within the LO theory due to softening of the elastic modulii of the VL, which caused a decrease in R_c and L_c, thereby causing F_p or J_c to anomalously increase at PE. While the LO theory provides an explanation of the PE phenomenon, a quantitative match of the details of the PE with LO theory lacked.

While theoretically some difference persist as regards the origin of the PE phenomenon, the experimental investigations (Banerjee et al., 1998, 1999a, 1999b, 2000a, 2000b, 2001; Bhattacharya & Higgins, 1993; Gammel et al., 1998; Ghosh et al, 1996; Higgins and Bhattacharya, 1996; Marchevsky et al., 2001; Thakur et. al, 2005, 2006; Troyanovski et al., 1999, 2002) are almost concurrent towards in establishing PE as an order to disorder transformation in the vortex lattice. Studies (Banerjee et al, 1998, 1999a, 1999b, 2000a,b, 2001) on different single crystals of 2H-$NbSe_2$, with progressively increasing amounts of the quenched random pinning have revealed that the details of PE phenomenon are significantly affected by level of disorder, amounting to the appearance of significant variation in the metastable response(s) of the vortex lattice. These studies were able to demonstrate the correlation between the thermomagnetic history effects (i.e., difference between the field cooled (FC) and zero - field cooled (ZFC) response exhibited by the FLL in single crystal of a conventional superconductor 2H-$NbSe_2$ and the pinning strength in the samples (Banerjee et al, 1999b). These observations lead to proposals pertaining to the existence of a pinning induced transformation across glassy phases of the vortex matter. In recent times an interesting explanation for PE has been proposed based on a crossover from weak to collective pinning in the vortex matter (Blatter et al. 2004). We shall discuss this work in relation to the experimental findings in section 3.3.

2.1 The effect of disorder on the behavior of critical current (J_c) and the peak effect (PE) phenomenon

2.1.1 Single crystals of different pinning strengths

We are collating here results reported on good quality single crystals of 2H-$NbSe_2$, grown in different laboratories (University of Warwick, UK, NEC research Institute, Princeton, USA and Bell Labs, Murray Hills, USA). On the basis of correlation between pinning strength and the metastability effects in the elastic region of vortex phase diagram, the crystals can be sequentially enumerated in terms of the progressively enhanced pinning. For instance, in 2H-$NbSe_2$ crystals, ranging from nomenclature A to C, the J_c values vary from 10 A/cm^2 to 1000 A/cm^2 (Banerjee et al., 1998, 1999a, 1999b, 2000a, 2000b, 2001; Thakur et al. 2005, 2006).

2.1.2 Identification of different pinning regimes and the behavior of PE as a function of pinning

We extracted J_c(H) (for H//c) in two varieties of single crystals A and B, of 2H-$NbSe_2$, either by directly relating J_c(H) to the widths of the isothermal magnetization hysteresis loops (Bean, 1962, 1964) or by analyzing the in-phase and out-of-phase ac susceptibility data (Bean, 1962, 1964; Angurel et al., 1997). Figure 2 summarizes the J_c vs. H data (H\\c) for the crystals A and B in two sets of log-log plots in the temperature regions close to the respective $T_c(0)$ values (Banerjee, 2000b; Banerjee et al. 2001). The peaks in J_c(H) occur at fields (H_p) less than 1 kOe (see insets in Fig.2(c) and Fig.2(g) for the t_p(H) curves in A and B, viz., locus of the PE in the H - reduced temperature ($t = T/T_c(0)$) space for the two samples, with pinning strength in B > A).

We first focus on the shapes of the J_c(H) curves (cf. Fig.2(a) to 2(d)) in the crystal A. In Fig.2(a), the three regimes (marked I, II and III in the figure) of J_c(H), at a reduced temperature t~0.973, are summarized as follows : (1) At the lowest fields (H $\leq$ 10 Oe), J_c varies weakly with H (region I), as expected in the individual pinning or small bundle pinning regime, noted earlier (Duarte et al. 1996), (2) Above a threshold field value, marked by an arrow, J_c(H) variation (in region II) closely follows the archetypal collective pinning power law (Duarte et al. 1996, Larkin, 1970a, 1970b; Larkin and Ovchinnikov, 1979) dependence (see the linear behaviour in region II of J_c vs. H on log-log scale in Fig.2), (3) This power law regime terminates at the onset (marked by another arrow) position of the PE phenomenon (region III).

On increasing the temperature (see Figs. 2(a) and 2(b) for the data at t=0.973 and 0.994), the following trends are immediately apparent: (1) the peak effect becomes progressively shallower, i.e., the ratio of J_c(H) at the peak position to that at the onset of PE becomes smaller. For instance, the said ratio has a value of about 8 at t=0.973 and it reduces to a value of 3.5 at t=0.994. ; (2) The power law region shrinks; for example, the field interval between the pair of arrows (identifying the power law region) spans from 10 Oe to about 500 Oe at

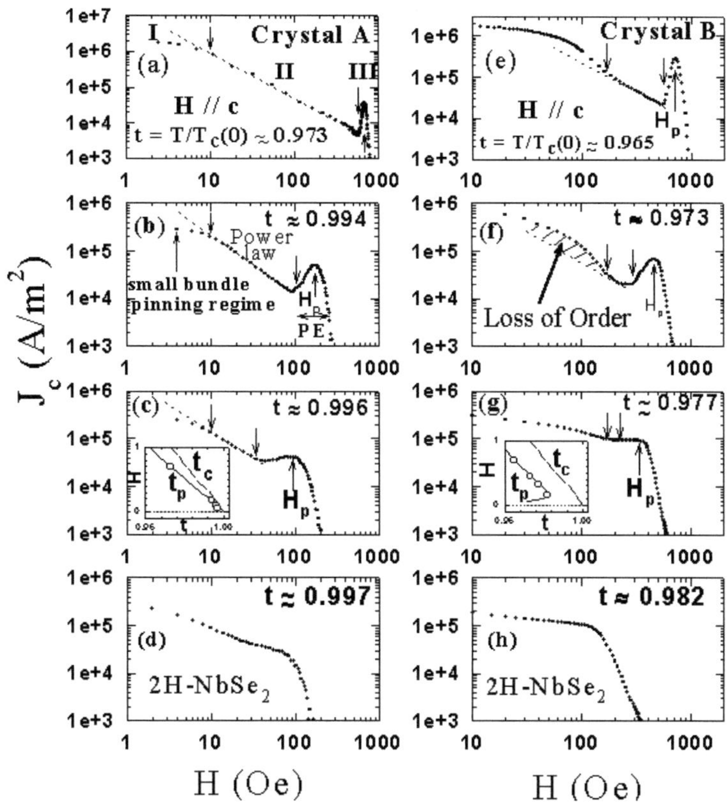

Fig. 2. Log-Log plots of J_c vs. H for H\\c at selected reduced temperatures ($t = T/T_c(0)$) in crystals A and B of 2H- $NbSe_2$. The insets in Fig.2(c) and Fig.2(g) display the locus of PE curve, $t_p(H)(=T_p(H)/T_c(0))$ and the superconductor-normal phase boundary $t_c(H)(=T_c(H)/Tc(0))$ in crystals A and B, respectively. The marked data points on the PE curves in each of these insets identify the reduced temperatures at which $J_c(H)$ data have been displayed in Figs.2(a) to 2(d) and in Figs.2(e) to 2(h). (Ref. Banerjee et al, 2000a)

t=0.973 in Fig.2(a), whereas at t=0.996 in Fig. 2(c), the power law regime terminates near 40 Oe. Also, the slope value of linear variation of log J_c vs. log H in the latter case is somewhat smaller. At still higher temperatures (see, for instance, Fig.2(d) at 0.997), the power law region is nearly invisible and the anomalous PE peak cannot be distinctly identified anymore, as only a residual shoulder survives.

In contrast, the second set of plots (see Figs. 2(e) to 2(h)) in the crystal B shows a different behaviour, although the overall evolution in the shapes of $J_c(H)$ curves is generically the same. In Fig.2(e), at a reduced temperature t~0.965, one can see the same power law regime as in Fig.2(a), but as the extrapolated dotted line shows, $J_c(H)$ departs from the power law behaviour in the low field region (i.e., for H < 200 Oe). As the field decreases below 200 Oe, the current density in crystal B (t=0.965) increases rapidly towards the background saturation limit (i.e., in the single vortex pinning regime). The approach to background saturation limit occurs at much lower field (H < 10 Oe) in crystal A. The smooth crossover to individual or small bundle pinning regime as seen in the crystal A, therefore adds on an additional characteristic in the crystal B. We label the region of rapid rise of $J_c(H)$ at low fields from a power law behaviour in region II into the weakly field dependent $J_c(H)$ behaviour in region I, as the region with "loss of order" (cf. Fig.2(f)). Further, with increasing temperature, the power law regime in the crystal B shrinks faster than that in sample A (cf. Fig.2(e) at t=0.965 and Fig.2(f) at t=0.973), leaving only a rather featureless monotonic $J_c(H)$ behaviour upto the highest fields (cf. Fig.2(g) and Fig.2(h)). Note, also, that the limiting value of the reduced temperature upto which the power law regime along with the PE peak survives in the crystal B is smaller than that in crystal A. In crystal B, the PE peak can be distinctly discerned only upto t=0.977, whereas in crystal A it can be seen even upto t=0.994. Recalling that the crystal B is more strongly pinned than crystal A, the above observation reaffirms the notion that the progressive enhancement in effective pinning (which occurs as we go from sample A to B) shrinks the (H,T) region over which the vortex matter responds like an elastically (ordered) pinned vortex lattice.

Having identified the regime of collective pinning where the vortex matter behaves like an ordered elastic medium and determined its sensitivity to pinning, it is fruitful to explore transformation in the elastic regime for weak collective to strong pinning (Blatter et al., 2004), and investigate if it coincides with the appearance of PE

3. Weak collective pinning, strong pinning and thermal fluctuations dominated regimes for the quasi-static vortex state

3.1.1 AC susceptibility measurements:

It is chosen to focus on A′ type of a crystal of 2H-$NbSe_2$ (cf. section 2.1.1, A′ has pinning inbetween that of samples A and B), has dimensions 1.5 x 1.5 x 0.1 mm^3, $T_c(0)$ ~ 7.2 K and J_c ~ 50 - 100 A/cm^2 (at 4.2 K and 10 kOe). The 2H-$NbSe_2$ system, being a layered material, often has extended defects (dislocations, stacking faults) present along its crystalline *c* axis. If H is applied along the *c* axis, then the vortex lines (also oriented along *c* direction) could be strongly pinned by these extended defects between layers. To reduce the emphasis on the inevitably present strong pinning centers, we have chosen to focus on behaviour obtained for the H ⊥ c orientation (the c-axis of hexagonal crystallographic lattice is aligned along the thickness of the platelet shaped sample) for our measurements. This choice of the field direction also avoids geometric and surface barrier effects, which are known to persist up to the PE in H //c orientation (Zeldov, et al. 1994; Paltiel et al., 1998).

We measured the ac susceptibility response as well as DC magnetization of the vortex state in the weak pinning 2H-$NbSe_2$ sample in the above mentioned orientation. The real (χ') component of the ac susceptibility response (viz., $\chi=\chi'+i\chi''$) is a measure of its diamagnetic shielding response. The maximum value of (normalized) χ'= -1 corresponds to the perfectly shielded, Meissner state of the superconductor. The χ' is related to the shielding currents (= J_c) setup in the sample via (Bean 1962, 1964), $\chi' \propto -J_c/h_{ac}$ for h_{ac} > H*, where h_{ac} is the ac excitation magnetic field used to measure the ac susceptibility response and H* is the penetration field value at which induced screening currents flow through the entire bulk of the sample. (Note, H* $\propto J_c(H,T)$). The quadrature χ'' signal is a measure of energy dissipated by vortices, which maximizes at h_{ac} = H*. If the vortices get strongly pinned then χ''shows a decrease, which is encountered in the PE regime. In the PE region, vortex matter gets better pinned and the χ'' response anomalously decreases.

3.1.2 Typical characteristics of AC susceptibility response

The $\chi'(T)$ behavior in the presence of a dc field (H) of 100 Oe is shown in Fig.3(a). In this figure the various curves correspond to different values of the amplitude of the h_{ac} at a frequency of 211 Hz applied parallel to H ($\perp$ c). Note that at a fixed T, on increasing h_{ac} the χ' (viz., the diamagnetic shielding) response progressively decreases from -1 value (see the dashed arrow marked at 6.8 K in Fig.3(a)). At fixed T, the decrease in χ' is due to h_{ac} approaching close to H* ($\propto J_c$(100 Oe, 6.8 K) and the magnetic flux penetrates the bulk of the sample, leading to a decrease in the screening response. As the h_{ac} penetrates deeper into the superconductor, one begins to clearly observe features associated with the bulk pinning of vortices inside the superconductor, viz., the peak effect (PE) phenomenon. The quintessential PE is easily observed as the anomalous enhancement in χ' between T_{on} (corresponding to the onset of PE at a given H, T) and T_p (the peak of PE at a given H, T) . Notice that due to the enhanced pinning in the PE regime between T_{on} and T_p, the sample attempts to shield its interior better from the penetrating h_{ac} as a consequence the χ' increases. Also notice that as the h_{ac} increases, the PE width between T_{on} and T_p becomes narrower.

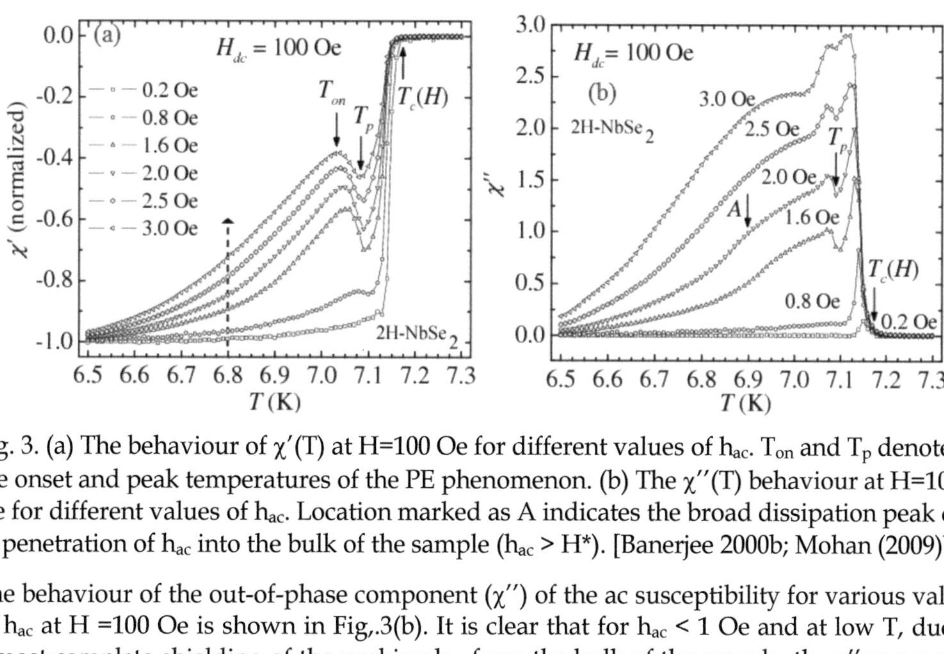

Fig. 3. (a) The behaviour of $\chi'(T)$ at H=100 Oe for different values of h_{ac}. T_{on} and T_p denote the onset and peak temperatures of the PE phenomenon. (b) The $\chi''(T)$ behaviour at H=100 Oe for different values of h_{ac}. Location marked as A indicates the broad dissipation peak due to penetration of h_{ac} into the bulk of the sample (h_{ac} > H*). [Banerjee 2000b; Mohan (2009)b]

The behaviour of the out-of-phase component (χ'') of the ac susceptibility for various values of h_{ac} at H =100 Oe is shown in Fig,.3(b). It is clear that for h_{ac} < 1 Oe and at low T, due to almost complete shielding of the probing h_{ac} from the bulk of the sample, the χ''response is nearly zero. At a fixed T, say T=6.8 K, as h_{ac} increases, χ'' response also increases monotonically. Full penetration of h_{ac} into the bulk of the sample causes a significant rise in dissipation, which in turn leads to a broad maximum in the χ'' response (location marked as A in Fig.3(b) for h_{ac} =2 Oe). On approaching the PE region, due to enhancement in vortex pinning, one observes a drop in χ''response (marked as T_p for h_{ac} = 2 Oe). Beyond T_p, dissipation has a tendency to rise sharply before decreasing close to T_c(H). From Fig.3(b) we note that at H =100 Oe and T=6.8 K, significant flux penetration starts at h_{ac} = 1.6 Oe. Within the Bean's Critical State model (Bean 1962, 1964) the field for flux penetration is given by H*~J_c.d, where d is the relevant dimension in which the critical state is established. Using Fig.3(b), by approximating H*= 1.6 Oe, we estimate the $J_c \sim 130$ A/cm^2 at 6.8 K at 100 Oe (note J_c decreases significantly with increasing H).

3.2 Transformation in the vortex state deep in the elastic regime

From Fig.3 it can be noted that the PE phenomenon is distinctly observed for $h_{ac} \geq 2$ Oe as at these h_{ac}, the ac field fully penetrate the bulk of the superconductor, and one can probe changes in the bulk pinning characteristics of the sample. Choosing h_{ac} = 2 Oe, we measured the $\chi'(T)$ and $\chi''(T)$ for different values of H. Figures 4(a) and 4(c) and Figs. 4(b) and 4(d)

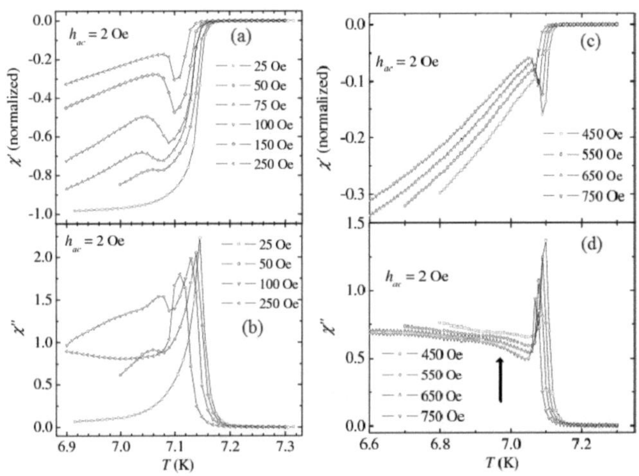

Fig. 4. The real ((a),(c)) and imaginary ((b), (d)) parts of the ac susceptibility as a function of T with h_{ac}=2 Oe and for different H. [Mohan 2009b]

show the $\chi'(T)$ and $\chi''(T)$, respectively. At 7.0 K in Fig.4(a), with increasing H the value of χ'varies from about -1 at 25 Oe to about -0.2 at 250 Oe. This decrease in the diamagnetic shielding response, we believe, arises from the inverse field relation of the critical current density, e.g., $J_c \propto 1/H$ (Kim et al., 1962). In all the curves the location of PE is clearly visible as the anomalous enhancement in χ' due to the anomalous increase in pinning or J_c. However below 100 G the PE is very shallow, and we see an enhancement in χ''which occurs very close to $T_c(H)$. At 100 Oe we see the decrease in χ'' at PE quite clearly, before the χ''increases near $T_c(T)$. At higher fields of 250 Oe (Fig.4(a)) from $\chi'(T)$ we see that the PE gets narrower in temperature width. As one moves to still higher fields (Fig.4(c)), the PE width gets still narrower and sharper. In the $\chi''(T)$ at Fig.4(d), as well as in Fig.4(b) (above 100 Oe) we do not find the drop in χ'' associated with PE as the drop over a narrow temperature window in χ'' due to PE gets merged into the enhancement in χ'' signal one observes in the vicinity of $T_c(H)$. However from Figs.4(c) and 4(d), we see that there is a decrease in χ'' which begins (see an arrow in Fig.4(d)) well before the anomalous enhancement in $\chi'(T)$ sets in at PE.

The fig. 5 provides a glimpse into ac susceptibility data at high fields. Above H = 750 Oe, the signature of PE survives as a subtle change in slope of $\chi'(T)$ at T_P (see locations marked by arrows in Fig.5(a)) just before χ'crashes to zero value at $T_c(H)$. A distinct feature seen at these fields is that the dissipation χ'' behaviour (Fig.5(b)), which is large at lower T, decreases sharply as one approaches T_c. This decrease begins from a region located far below the PE and is similar to the decrease in $\chi''(T)$ found above 450 Oe in fig. 4(d). The sharp increase in the dissipation (on $\chi''(T)$) very close to $T_c(H)$ (as noted in Fig.4), is observed only for 1000 Oe

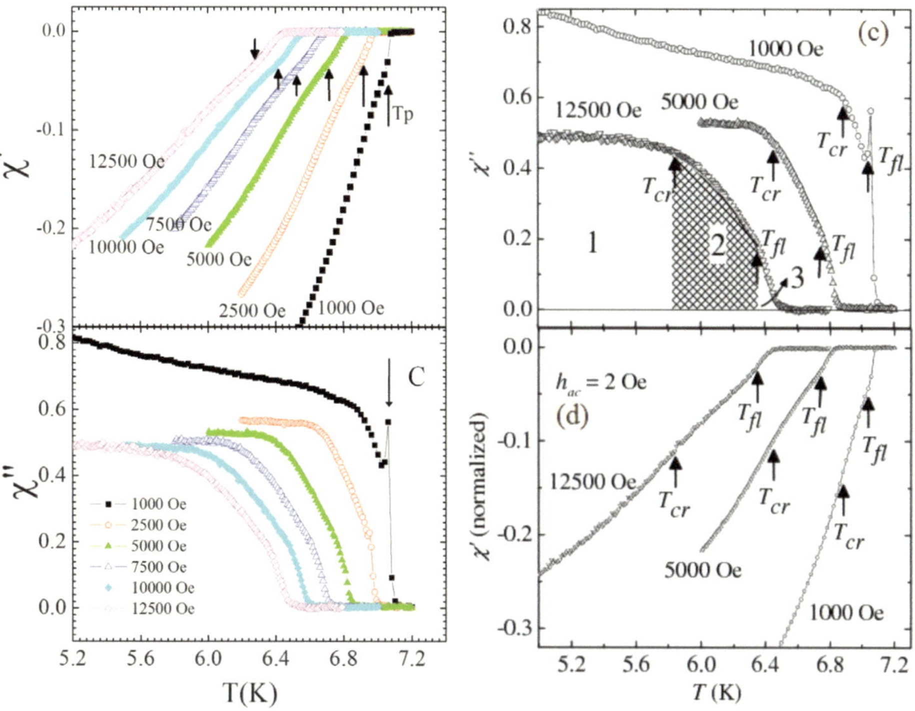

Fig. 5. The real (a) and imaginary (b) parts of the ac susceptibility measured with h_{ac} = 2 Oe and for different dc fields: 1000 Oe ≤ H ≤ 12500 Oe. The arrows in panel (a) mark the peak locations of the PE. (c) The χ'' response for 1000 Oe, 5000 Oe and 12500 Oe. The T_{cr} and T_{fl} locations determine the different regimes of dissipation marked as the regions 1, 2 and 3 (See text for details). (d) The χ' response corresponding to (c). [Mohan et al. 2007; Mohan 2009b].

(position marked C in Fig.5(b)). Above 1000 Oe, instead of a peak in χ''(T), the χ'' response exhibits only a change in slope near T_c(H) before becoming zero on reaching T_c(H). It should be noted that the temperature at which where the χ''response drops sharply from a large value does not correspond to any specific feature in χ'(T) and, also, occurs well before the onset of PE. In Figs.6 (a) to (c) we can identify locations of the drop in dissipation χ'' by detecting the change in slope of through plots dχ''/dT vs T (see Figs.6(f), 6(e) and 6(d)) .
In Figs.6(d)-(f), the onset of the drop in dissipation at lower T is marked with arrows as T_{cr} and the T at which there occurs a change in slope of the dissipation curves close to T_c(H) are marked as T_{fl}. (The nomenclature T_{cr} and T_{fl}, signify the temperature above which, there occur pinning crossover and thermal fluctuation dominated regimes, respectively). The dashed lines are a guide to the eye representing the base line behavior of the dχ''/dT. The onset of deviation in dχ''/dT from the baseline identifies T_{cr} (cf. Figs.6(d) - (f)). In Figs.6 (d)-(f) the base lines for different H have been artificially offset for clarity in the data representation. After the locations of T_{cr} and T_{fl} are identified from dχ''/dT (cf. Figs.6(d) - (f)), their positions are identified and marked on the corresponding χ''(T) curves (Figs.6(a)-(c)). We now consider three representative χ''(T) curves, namely the response for 1000 Oe, 5000 Oe and 12500 Oe in Fig.5(c) to understand the significance of the T_{cr} and T_{fl}.

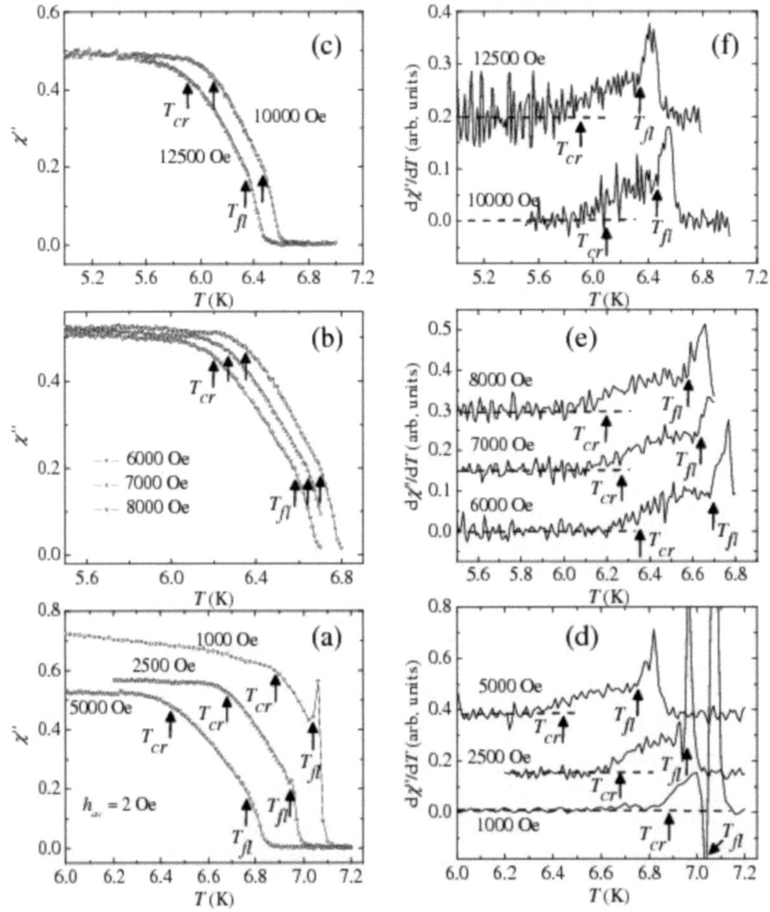

Fig. 6. The panels on the left (a)-(c) show the χ''(T) response for different H. The right hand panels (d)-(f), show the derivative dχ''/dT determined from the corresponding χ''(T) curves on the left panel. (see discussion in the text) [Mohan et al. 2007; Mohan 2009b]

In Fig.5(c), for H= 12500 Oe, three distinct regimes of behaviour in the χ''(T) response have been identified as the regions 1, 2 and 3. Region 1 is characterized by a high dissipation response. As noted earlier, this high dissipation results from full penetration of h_{ac} to the center of the sample, similar to the dissipation peak marked at A in Fig.3(b). As noted earlier in Fig.5(a), at these high fields beyond 1000 G, at T > T_{cr}, χ'(T) response possesses no distinct signature of the PE phenomenon. The absence of any distinct PE feature in χ'(T) should have caused no modulations in the behavior of χ''(T) response, except for a peak in dissipation close to T_c(H). Instead, in the region 2 (cross shaded and located between the T_{cr} and T_{fl} arrows in Fig.5(c)) a new behaviour in the dissipation response is observed, viz., in this region there is a substantial decrease in dissipation.

As seen earlier in the context of PE in Fig.3(b), that any anomalous increase in pinning corresponds to a decrease in the dissipation. The observation of a large drop in dissipation across T_{cr} (Fig.5(c)) indicates there is a transformation from low J_c state to a high J_c state, i.e., a transformation from weak pinning to strong pinning. Subsequent to the drop in χ''(T) in region 2, the dissipation response attempts to show an abrupt increase (see change in slope in $d\chi''/dT$ in Fig.6(d) to (f)) at the onset of region 3 (marked as T_{fl} in Fig.5 and Fig.6). The abrupt increase in dissipation beyond T_{fl} is more pronounced at low H and high T (see behavior in Fig.5(b)). The significance of T_{fl} will be revealed in subsequent sections. In brief, the T_{fl} will be considered to identify the onset of a regime dominated by thermal fluctuations, where pinning effects become negligible and dissipation response goes through a peak. It is interesting to note that the T_{fl} locations are identical to the location of T_p (viz., the peak of PE) in Figs.5(a) and 5(c). For H < 750 Oe, the T_{fl} location can be identified with the appearance of a distinct PE peak at T_p (see Fig.4, where dissipation enhances at $T_p = T_{fl}$). It is important to reiterate that the anomalous drop in dissipation in region 2 near T_{cr} is not associated with the PE phenomenon.

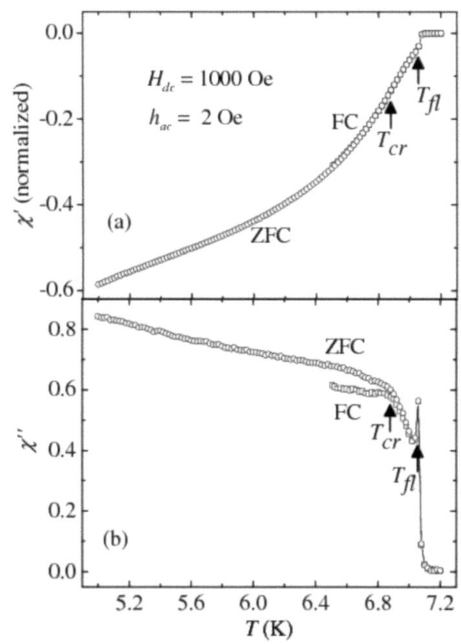

Fig. 7. The real (a) and imaginary (b) parts of the ac susceptibility measured in the ZFC and FC modes, for H = 1000 Oe. Also marked for are the locations of the T_{cr} and T_{fl}. [Mohan et al. 2007; Mohan 2009b]

All the above discussions pertain to susceptibility measurements performed in the zero field cooled (ZFC) mode. Detailed studies of the dependence of the thermomagnetic history dependent magnetization response on the pinning (Banerjee et al. 1999b, Thakur et al., 2006), had shown an enhancement in the history dependent magnetization response and enhanced metastablility developing in the vortex state as the pinning increases across the PE. While the ZFC and field cooling (FC), $\chi'(T)$ response can be identical in samples with weak pinning, the will show that $\chi''(T)$ is a more sensitive measure of small difference in the thermomagnetic history dependent response. Figures 7(a) and 7(b) display $\chi'(T)$ and $\chi''(T)$ measured for a vortex state prepared either in ZFC or FC state in 1000 Oe. Figure 7(a) shows the absence of PE at T_{cr} in the $\chi'(T)$ response at 1000 Oe for vortex state prepared in both FC and ZFC modes. Furthermore, there is no difference between the ZFC and FC $\chi'(T)$

responses (cf.Fig.7(a)). However, the dissipation ($\chi''(T)$) behaviour in the two states (Fig.7(b)) are slightly different. While there are no clear signatures of T_{cr} in the $\chi'(T)$ response, in $\chi''(T)$ response (Fig.7(b)) below T_{cr} one observes that the FC response significantly differs from that of the ZFC state, with the dissipation in the FC state below T_{cr} being lower as compared to that in the ZFC state. The presence of a strong pinning vortex state above T_{cr}, causes the freezing in of a metastable stronger pinned vortex state present above T_{cr}, when the sample is field cooled to $T < T_{cr}$. As the FC state has higher pinning than the ZFC state (which is in a weak pinning state) at the same T below T_{cr}, therefore, the $\chi''(T)$ response is lower for the FC state. Above T_{cr} the behavior of ZFC and FC curves are identical, as both transform into a maximally pinned vortex state above T_{cr}. The behavior of $\chi''(T)$ in the FC state indicates that the pinning enhances across T_{cr}. Beyond T_{cr}, the ZFC and FC curves match and the high pinning regime exists till T_{fl}. This observation holds true for all H_{dc} above 1000 Oe as well.

3.2.1 Transformation in pinning: evidence from DC magnetization measurements

Figure 8 displays measured forward (M_{fwd}) and (M_{rev}) reverse magnetization responses of 2H-$NbSe_2$ at temperatures of 4.4 K, 5.4 K and 6.3 K for $H \perp c$.

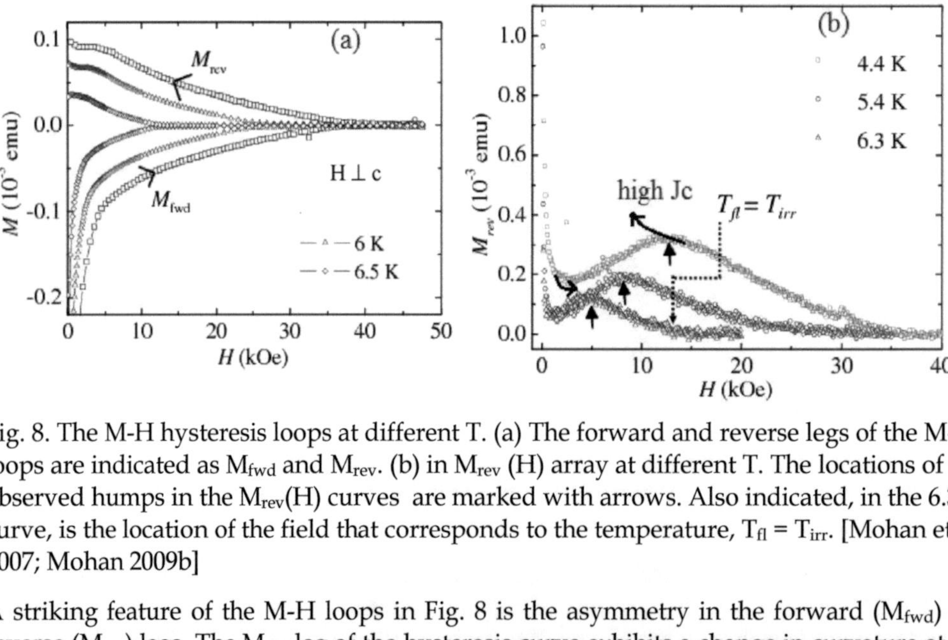

Fig. 8. The M-H hysteresis loops at different T. (a) The forward and reverse legs of the M-H loops are indicated as M_{fwd} and M_{rev}. (b) in M_{rev} (H) array at different T. The locations of the observed humps in the $M_{rev}(H)$ curves are marked with arrows. Also indicated, in the 6.3 K curve, is the location of the field that corresponds to the temperature, $T_{fl} = T_{irr}$. [Mohan et al. 2007; Mohan 2009b]

A striking feature of the M-H loops in Fig. 8 is the asymmetry in the forward (M_{fwd}) and reverse (M_{rev}) legs. The M_{rev} leg of the hysteresis curve exhibits a change in curvature at low fields. In Fig.8(b) we plot only the M_{rev} from the M-H recorded at 4.4 K, 5.4 K and 6.3 K. At low fields, the M_{rev} leg exhibits a hump; the location of the humps are denoted by arrows in Fig.8(b). The characteristic hump-like feature (marked with arrows in Fig.8(b)) can be identified closely with T_{cr} locations identified in Figs.4, 5 and 6. The tendency of the dissipation χ'' to rapidly rise close to $T_{fl}(H)$ (cf. Figs.4, 5 and 6) is a behaviour which is expected across the irreversibility line ($T_{irr}(H)$) in the H-T phase diagram, where the bulk pinning and, hence, the hysteresis in the M(H) loop becomes undetectably small. The decrease in pinning at $T_{irr}(H)$, results in a state with mobile vortices which are free to dissipate. We have confirmed that $T_{fl}(H)$ coincides with $T_{irr}(H)$, by comparing dc magnetization with χ'' response measurements (cf. arrow marked as $T_{fl}= T_{irr}$ in Fig.8 for the 6.3 K curve). Thus $T_{fl}(H)$ coincides with $T_{irr}(H)$, which is also where the peak of the PE occurs, viz., the peak of PE at T_p occurs at the edge of irreversibility (cf. H-T phase diagram in Fig.9).

3.3 The H-T vortex phase diagram and pinning crossover region

Figure 9(a) shows the H - T, vortex matter phase diagram wherein we show the location of the $T_c(H)$ line which is determined by the onset of diamagnetism in $\chi'(T)$, the $T_p(B)$ line which denotes the location of the PE phenomenon, the $T_{cr}(H)$ line across which the $\chi''(T)$ response (shaded region 2 in Fig.5(c)) shows a substantial decrease in the dissipation and the T_{fl} line beyond which dissipation attempts to increase. The PE ceases to be a distinct noticeable feature beyond 750 G and the $T_p(H)$ line (identified with arrows in Fig. 5(a)) continues as the $T_{fl}(H)$ line. Note the $T_{fl}(H)$ line also coincides with $T_{irr}(H)$. For clarity we have indicated only the $T_{fl}(H)$ line in the phase diagram with open triangles in Fig.9(a).

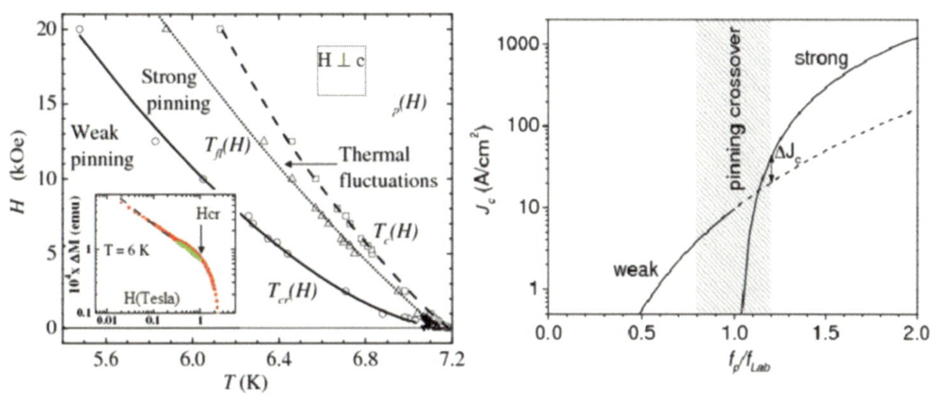

(a) (b)

Fig. 9. (a) The phase diagram showing the different regimes of the vortex matter. The inset is a log-log plot of the width of the hysteresis loop versus field at 6K. (b) An estimate of variation in J_c with f_p/f_{Lab} in different pinning regimes. [Mohan et al. 2007; Mohan 2009b].

We consider the $T_{cr}(H)$ line as a crossover in the pinning strength experienced by vortices, which occurs well prior to the PE. A criterion for weak to strong pinning crossover is when the pinning force far exceeds the change in the elastic energy of the vortex lattice, due to pinning induced distortions of the vortex line. This can be expressed as (Blatter et al, 2004), the pinning force (f_p) ~ Labusch force (f_{Lab}) = $(\varepsilon_0\xi/a_0)$, where $\varepsilon_0 = (\phi_0/4\pi\lambda)^2$ is the energy scale for the vortex line tension, ξ is the coherence length, ϕ_0 flux quantum associated with a vortex, λ is the penetration depth and a_0 is the inter vortex spacing ($a_0 \propto H^{-0.5}$). A softening of the vortex lattice satisfies the criterion for the crossover in pinning. At the crossover in pinning, we have a relationship, $a_0 \approx \varepsilon_0\xi\, f_p^{-1}$. At $H_{cr}(T)$ and far away from T_c, if we use a monotonically decreasing temperature dependent function for $f_p \sim f_{p0}(1-t)^{\beta}$, where $t=T/T_c(0)$ and $\beta > 0$, then we obtain the relation $H_{cr}(T) \propto (1-t)^{2\beta}$. We have used the form derived for $H_{cr}(T)$ to obtain a good fit (solid line through the data Fig.9(a)) for $T_{cr}(B)$ data, giving $2\beta \sim 1.66 \pm 0.03$. Inset of Fig. 9(a) is a log-log plot of the width of the magnetization loop (ΔM)

versus H. The weak collective pinning regime is characterized by the region shown in the inset, where the measured ΔM(H) (red curve) values coincide with the black dashed line, viz., $\Delta M \propto J_c \propto 1/H^p$, with p as a positive integer (discussed earlier). Using expressions for $J_c(f_p/f_{Lab})$ (Blatter, 2004), $a_0 \sim \lambda$ and λ= 2300 A°, ξ= 23 A° for 2H-$NbSe_2$ (Higgins and Bhattacharya, 1996) and parameters like density of pins suitably chosen to reproduce J_c values comparable to those experimentally measured for 2H-$NbSe_2$, Fig.9(b) shows the enhancement in J_c expected at the weak to strong pinning regime, viz., around the shaded region in Fig.9(b) marked ΔJ_c, in the vicinity of $f_p/f_{Lab} \sim 1$. In Fig.9(a), the shaded region in the ΔM(H) ($\propto J_c$(H), Bean, 1962; 1967) plot shows the excess pinning that develops due to the pinning crossover across H_{cr}(T) (≡ T_{cr}(H)). Comparing Figs.9(b) and 9(a) we find $\Delta Jc/J_{c,weak} \sim 1$ compares closely with the (change in ΔM in shaded region ~ 0.6 T in Fig.9(a) inset)/ ΔM (along extrapolated black line ~ 0.6 T) ~ 0.5. In the PE regime, usually $\Delta Jc/J_{c,weak} \geq 10$ (see for example in Fig.2). Note from the above analysis and the distinctness of the T_{cr} and T_p lines in Fig.9(a), shows that the excess pinning associated with the pinning crossover does not occur in the vicinity of the PE, rather it is a line which divides the elastically pinned regime prior to PE. Based on the above discussion we surmise that the T_{cr}(H) line marks the onset of an instability in the static elastic vortex lattice due to which there is a crossover from weak (region 1 in Fig.5(c)) to a strong pinning regime (region 2 in Fig.5(c)). The crossover in pinning produces interesting history dependent response in the superconductor, as seen in the M_{rev} measurements of Fig. 8 and in the χ''(T) response for the ZFC and FC vortex states, in the main panel of Fig.7. In the inset (b) of Fig.8 we have schematically identified the pinning crossover (by the sketched dark curved arrows in Fig.8(b)) by distinguishing two different branches in the M_{rev}(H) curve, which correspond to magnetization response of vortex states with high and low J_c. We reiterate that the onset of instability of the elastic vortex lattice sets in well prior to PE phenomenon without producing the anomalous PE.

As the strong pinning regime commences upon crossing H_{cr}, how then does pinning dramatically enhance across PE? The T_{fl}(H) line in Fig.9 marks the end of the strong pinning regime of the vortex state. Above the T_{fl}(H) line and close to T_c(H), the tendency of the dissipation response to increase rapidly (Figs.1 and 2) especially at low H and high T, implies that thermal fluctuation effects dominate over pinning. We find that our values (H_{fl}, T_{fl}) in Fig.9(a), satisfies the equation governing the melting of the vortex state, viz., $B_{fl} = \beta_m \left(\frac{c_L^4}{G_i}\right) H_{c2}(0) \left(\frac{T_c}{T_{fl}}\right)^2 \left[1 - \frac{T_{fl}}{T_c} - \frac{B_{fl}}{H_{c2}(0)}\right]^2$, where, β_m = 5.6 (Blatter et al, 1994), Lindemann no. $c_L \sim 0.25$ (Troyanovski et al. 1999, 2002), $H_{c2}^{\perp c}(0)$ = 14.5 T, if a parameter, G_i is in the range of 1.5 x 10^{-3} to 10^{-4}. The Ginzburg number, G_i, in the above equation controls the size of the H - T region in which thermal fluctuations dominate. A value of O(10^{-4}) is expected for 2H-$NbSe_2$ (Higgins & Bhattacharya, 1996). The above discussion implies that thermal fluctuations dominate beyond T_{fl}(H). By noting that T_p(H) appears very close to T_{fl}(H), it seems that PE appears on the boundary separating strong pinning and thermal fluctuation dominated regimes.

The above observations (Mohan et al, 2007) imply that instabilities developing within the vortex lattice lead to the crossover in pinning which occurs well before the PE. Infact, PE seems to sit on a boundary which separates a strong pinning dominated regime from a thermal fluctuation dominated regime. These assertions could have significant ramifications pertaining to the origin of PE which was originally attributed to a softening of the elastic modulii of the vortex lattice. Even though thermal fluctuations try to reduce pinning, we believe newer results show that at PE, the pinning and thermal fluctuations effects combine in a non trivial way to dramatically enhance pinning, much beyond what is expected from pinning crossovers. The change in the pinning response deep in the elastic vortex state is expected to lead to nonlinear response under the influence of a drive. It is interesting to ask if these crossovers and transformation in the static vortex state evolve and leave their imprint in the driven vortex state.

4 Nonlinear response of the moving vortex state

4.1 I-V characteristics and the various phases of the driven vortex matter

In the presence of an external transport current (I) the vortex lattice gets set into motion. A Lorentz force, $\mathbf{f_L}=\mathbf{J} \times \phi_0/c$, acting on each vortex due to a net current density J (due to current (I) sent through the superconductor and the currents from neighbouring vortices) sets the vortices in motion. As the Lorentz force exceeds the pinning force, i.e $f_L>f_p$, the vortices begin to move with a force-dependent velocity, v. The motion of the flux lines induces an electric field $\mathbf{E} = \mathbf{B} \times \mathbf{v}$, in the direction of the applied current causing the appearance of a longitudinal voltage (V) across the voltage contacts (Blatter et al, 1994). Hence, the measured voltage, V in a transport experiment can be related to the velocity (v) of the moving vortices via V=Bvd, where d is the distance between the voltage contacts. Measurements of the V (equivalent to vortex velocity v) as a function of I, H, T or time (t) are expected to reveal various phases and their associated characteristics an nonlinear behavior of the driven vortex state.

When vortices are driven over random pinning centers, broadly, four different flow regimes have been established theoretically and through significantly large number of experiments (Shi & Berlinski 1991; Giammarchi & Le Doussal, 1996; Le Doussal & Giammarchi, 1998; Giammarchi & Bhattacharya, 2002). These are: (a) depinning, (b) elastic flow, (c) plastic flow, and (e) the free-flow regime. At low drives, the depinning regime is first encountered, when the driving force just exceeds the pinning force and the vortices begin moving. As the vortex state is set in motion near the depinning regime, the moving vortex state is proliferated with topological defects, like, dislocations (Falesky et al, 1996). As the drive is increased by increasing the current through the sample, the dislocations are found to heal out from the moving system and the moving vortex state enters an ordered flow regime (Giammarchi & Le Doussal, 1996; Yaron et al., 1994; Duarte, 1996). The depinning regime is thus followed by an elastically flowing phase at moderately higher drives, when all the vortices are moving almost uniformly and maintain their spatial correlations. The nature and characteristics of this phase was theoretically described as the moving Bragg glass phase (Giammarchi & Le Doussal, 1996; Le Doussal & Giammarchi, 1998). In the PE regime of the H- T phase diagram, it is found that as the vortices are driven, the moving vortex state is proliferated with topological defects and dislocations, thereby leading to loss of correlation amongst the moving vortices (Falesky et al, 1996; Giammarchi & Le Doussal, 1996; Le Doussal & Giammarchi, 1998; Giammarchi & Bhattacharya 2002). This is the regime of plastic flow. In the plastic flow regime, chunks of vortices remain pinned forming islands of localized vortices, while there are channels of moving vortices flowing around these pinned islands, viz., different parts of the system flow with different velocities

(Bhattacharya & Higgins, 1993, Higgins & Bhattacharya 1996; Nori, 1996; Tryoanovski et al, 1999). The effect of the pins on the moving vortex phase driven over random pinning centers is considered to be equivalent to the effect of an effective temperature acting on the driven vortex state. This effective temperature has been theoretically considered to lead to a driven vortex liquid regime at large drives (Koshelev & Vinokur, 1994). At larger drives, the vortex matter is driven into a freely flowing regime. Thus, with increasing drive, interplay between interaction and disordering effects, causes the flowing vortex matter to evolve between the various regimes.

The plastic flow regime has been an area of intense study. The current (I) - voltage (V) characteristics in the plastic flow regime across the PE regime are highly nonlinear (Higgins & Bhattacharya, 1996), where a small change in I is found to produce large changes in V. Investigations into the power spectrum of V fluctuations revealed significant increase in the noise power on entering the plastic flow regime (Marley, 1995; Paltiel et al., 2000, 2002). The peak in the noise power spectrum in the plastic flow regime was reported to be of few Hertz (Paltiel et al., 2002). The glassy dynamics of the vortex state in the plastic flow regime is characterized by metastability and memory effects (Li et al, 2005, 2006; Xiao et al, 1999). An edge contamination model pertaining to injection of defects from the nonuniform sample edges into the moving vortex state can rationalise variety of observations associated with the plastic flow regime (Paltiel et al., 2000; 2002). In recent times experiments (Li et al, 2006) have established a connection between the time required for a static vortex state to reach steady state flow with the amount of topological disorder present in the static vortex state. By choosing the H-T regime carefully, one finds that while the discussed times scales are relatively short for a well ordered static vortex state, the times scales become significantly large for a disordered vortex state set into flow, especially in the PE regime. The discovery of pinning transformations deep in the elastic vortex state (Mohan et al, 2007), motivated a search for nonlinear response deep in the elastic regime as well as to investigate the time series response in the different regimes of vortex flow (Mohan et al, 2009).

4.2 Identification of driven states of vortex matter in transport measurements

The single crystal of 2H-$NbSe_2$ used in our transport measurements (Mohan et al, 2009) had pinning strength in between samples of A and B variety (see section 2.1.1). The dc magnetic field (H) applied parallel to the c-axis of the single crystal and the dc current (I_{dc}) applied along its 'ab' plane (Mohan et al, 2009). The voltage contacts had spacing of d ~ 1 mm apart. Figure 10(a) shows the plots of resistance ($R=V/I_{dc}$) versus H at 2.5 K, 4 K, 4.5 K, 5 K, 5.8 K and 6 K measured with I_{dc}=30 mA. With increasing H, all the R-H curves exhibit common features viz., nearly zero R values at lowest H, increasing R after depinning at larger H, an anomalous drop in R associated with onset of plastic flow regime and finally, a transition to the normal state at high values of H. To illustrate in detail these main features, and to identify different regime of driven vortex state, we draw attention only to the 5 K data in Figure 10(b).

At 5 K, for H < 1.2 kOe, R < 0.1 mΩ, which implies an immobile, pinned vortex state. Beyond 1.2 kOe (position marked as H_{dp} in Fig.10(b)), the FLL gets depinned and R increases to mΩ range. From this we estimate the critical current I_c to be 30 mA (at 5 K, 1.2 kOe). The enhanced pinning associated with the anomalous PE phenomenon leads to a drop in R starting at around 6 kOe (onset location marked as H_{pl}) and continuing up to around 8 kOe (location marked as H_p). The PE (≡ plastic flow) region is shaded in Fig.10(b). As

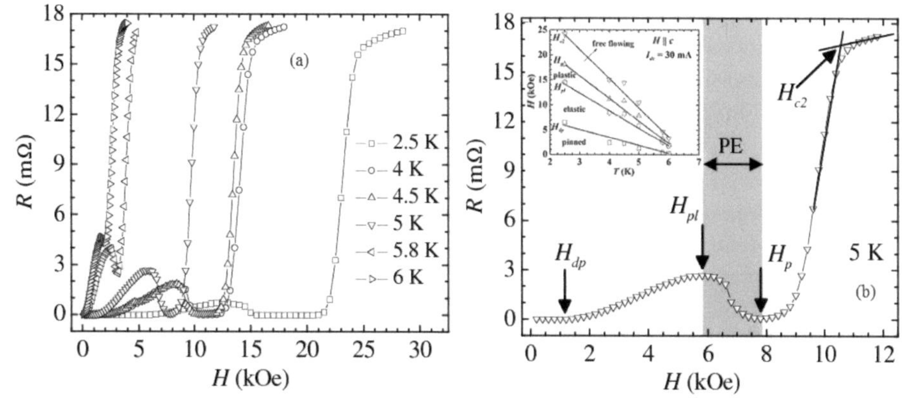

Fig. 10. (a) R versus H (H \\ c) of the vortex state, measured at different T with I_{dc}=30 mA. (b) R-H at 5 K only, with the different driven vortex state regimes marked with arrows. The arrows marks the locations of, depinning (H_{dp}), onset of plastic deformations (H_p), peak location of PE (H_p) and upper critical field (H_{c2}) at 5 K, respectively. The inset location of above fields (Fig.10(b)) on the H-T diagram. [Mohan et al. 2009a; Mohan 2009b].

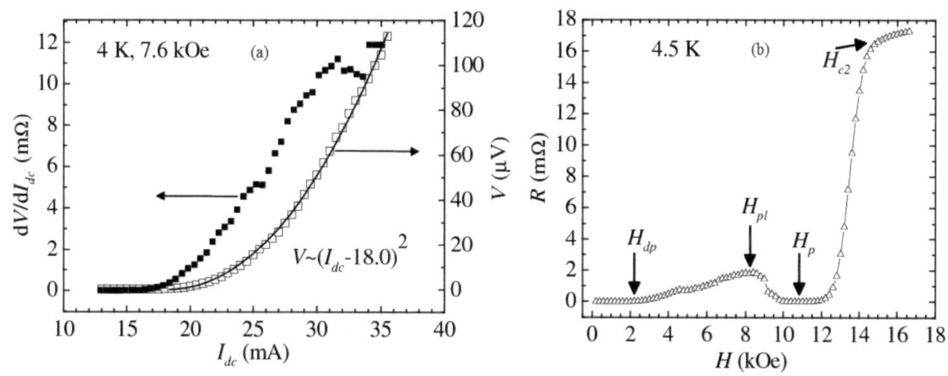

Fig. 11. (a) The V-I_{dc} characteristics and dV/dI_{dc} vs I_{dc} in the elastic phase at 4 K and 7.6 kOe. The solid line is a fit to the V-I_{dc} data, (cf. text for details). (b) R-H curve at 4.5 K and I_{dc}= 30 mA. [Mohan et al. 2009a; Mohan 2009b]

discussed earlier (Fig.9), beyond H_p, thermal fluctuations dominate causing large increase in R associated with pinning free mobile vortices until the upper critical field H_{c2} is reached. We determine $H_{c2}(T)$ as the intersection point of the extrapolated behaviour of the R-H curve in the normal and superconducting states, as shown in Fig.10(b). By identifying these features from the other R-H curves (Fig.10(a)), an inset in Fig.10(b) shows the H-T vortex phase diagram for the vortex matter driven with I_{dc} = 30 mA.

Figure 11 shows the V-I_{dc} characteristics at 4 K and 7.6 kOe; this field value lies between $H_{dp}(T)$ and $H_{pl}(T)$ (see inset, Fig.10(b)), i.e. in the elastic flow regime. It is seen that the data fits (see solid line in Fig.11(a)) to $V\sim(I_{dc} - I_c)^{\beta}$, where $\beta \sim 2$ and I_c = 18 mA ($I = I_c$, when $V \geq 5$ μV, as V develops only after the vortex state is depinned), which inturn indicates the onset of an elastically flow. Experiments indicate the concave curvature in I-V coincides with ordered elastic vortex flow (Duarte et al, 1996; Yaron et al.,1994; Higgins and Bhattacharya 1996). Unlike the elastic flow regime, the plastic flow regime is characterized by a convex curvature in the V-I_{dc} curve alongwith a conspicuous peak in the differential resistance (Higgins and Bhattacharya, 1996), which is absent in Fig.11 (see dV/dI_{dc} vs I_{dc} in Fig.11(a)). All the above indicate ordered elastic vortex flow regime at 4 K, 7.6 kOe and I = 30 mA. The dV/dI_{dc} curve also indicates a nonlinear V-I_{dc} response deep in the elastic flow regime.

4.3 Time series measurements of voltage fluctuations and its evolution across different driven phases of the vortex matter

Figure 11(b), shows the R-H curve for 4.5 K. Like Fig.10(b), in Fig. 11 (b), the H_{dp}, H_{pl}, H_p and H_{c2} locations are identified by arrows, which also identify the field values, at which time series measurements were performed. The protocol for the time series measurements was as follows: At a fixed T, H and I_{dc}, the dc voltage V_0 across the electrical contacts of the sample was measured by averaging over a large number of measurements ~ 100. The V_0 measurement prior to every time series measurement run, ensures that we are in the desired location on R-H curve, viz., the V_0/I value measured before each time series run should be almost identical to the value on the R(H) curve at the given H,T, like the one shown in Figs.10(b) or 11(b). After ensuring the vortex state has acquired a steady flowing state, viz., by ensuring the mean V,i.e., <V> ~ V_0, the time series of the voltage response (V(t)) is measured in bins of 35 ms for a net time period of a minute, at different H, T.

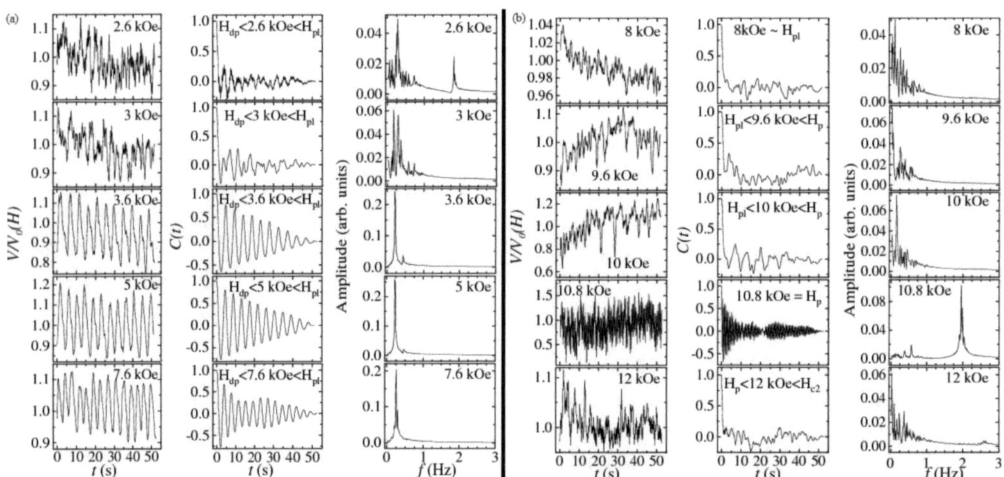

Fig. 12. (a) The left most vertical column of panels represent the fluctuations in voltage V(t)/V_0 measured at different fields at 4.5 K with I_{dc} of 30 mA. Note: V_0(2.6 kOe) = 1.4 μV, V_0(3 kOe) = 3.7 μV, V_0(3.6 kOe) = 9.5 μV, V_0(5 kOe) = 21.1 μV, V_0(7.6 kOe) = 50.7 μV. The middle set of panels are the C(t) calculated from the corresponding V(t)/V_0 panels on the left. The right hand set of panels show the amplitude of the FFT spectrum calculated from the corresponding C(t) panels. In Fig.12 (b), the organization of panels is identical to that in Fig.12 (a) with, V_0(8 kOe) = 54.5 μV, V_0(9.6 kOe) = 9.8 μV, V_0(10 kOe) = 1.0 μV, V_0(10.8 kOe) = 0.2 μV, V_0(12 kOe) = 3.2 μV. [Mohan et al. 2009a; Mohan 2009b]

The time series V(t) measurements at T=4.5 K are summarised in Figs.12 (a), Fig.12 (b), Fig. 13 (a) and Fig.13 (b). The stack of left hand panels in Figs. 12(a), 12(b), 13(a) and 13(b) show the normalized V(t)/V_0 versus time (t) for different driven regimes, viz., the just depinned state (H ~ H_{dp}), the freely flowing elastic regime (H_{dp} <H < H_{pl}), above the onset of the plastic regime ($H > H_{pl}$), deep inside the plastic regime ($H \sim H_p$) and above PE regime ($H > H_p$) (cf. Fig.11(b)). A striking feature in these panels is the amplitude of fluctuations in V(t) about the V_0 value are significantly large, varying between 10-50% of V_0, depending on the vortex flow regime. As one approaches very near to the normal regime, the fluctuations in V(t) are about 1% of V_0 (see bottom most plot at 16 kOe the left stack of panels in Fig.13(a)) and is about 0.02% deep inside the normal state (see Fig. 13(b), left panel). Near H_{dp} (2.6 kOe and 3 kOe, Fig.12(a)) the fluctuations are not smooth, but on entering the elastic flow regime, one can observe spectacular nearly-periodic oscillations of V(t) (see at 3.6 kOe, 5 kOe and 7.6 kOe in panels of Fig.12(a)). Such conspicuously large amplitude, slow time period fluctuations of the voltage V(t), which are sustained within the elastically driven state of the vortex matter (up to 7.6 kOe), begin to degrade on entering the plastic regime (above 8 kOe, see Fig.12(b)).

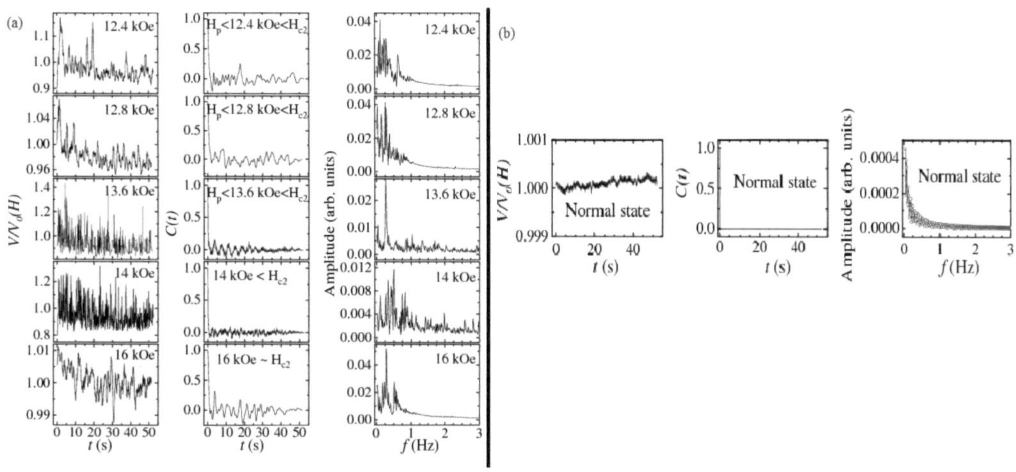

Fig. 13. (a) consists of three columns representing $V(t)/V_0$, C(t) and power spectrum of fluctuations (see text for details) measured with I_{dc} of 30 mA. Note: V_0(12.4 kOe) = 13.6 μV, V_0(12.8 kOe) = 49.6 μV, V_0(13.6 kOe) = 284.9 μV, V_0(14 kOe) = 404.5 μV, V_0(16 kOe) = 513.7 μV. (b)Panels show similar set of panels as (a) in the normal state at T = 10 K and H = 10 kOe with I_{dc} of 30 mA (V_0 = 539. 6 μV). [Mohan et al. 2009a; Mohan 2009b]

Considering that the voltage (V) developed between the contacts on the sample is proportional to the velocity (v) of the vortices (see section 4.1, V=Bvd), therefore to investigate the velocity - velocity correlations in the moving vortex state, the voltage-voltage (≡ velocity - velocity) correlation function: $C(t) = \frac{1}{V_0^2}\langle V(t'+t)V(t)\rangle$, was determined from the $V(t)/V_0$ signals (see the middle sets of panels in Figs.12 (a) and 12 (b) and Fig. 13 for the C(t) plots). In the steady flowing state, if all the vortices were to be moving uniformly, then the velocity - velocity correlation (C(t)) will be featureless and flat. While if the vortex motion was uncorrelated then they would lose velocity correlation within a short interval of time after onset of motion, then the C(t) would be found to quickly decay. Note an interesting evolution in C(t) with the underlying different phases of the vortex matter. While there are almost periodic fluctuations in C(t) at 3.6 kOe, 5 kOe and 7.6 kOe (at $H < H_{pl}$) sustained over long time intervals, there are also intermittent quasi-periodic fluctuations sustained for a relatively short intervals even at H > H_{pl}, viz., at 10.8 kOe and 13.6 kOe (see Fig.12 and Fig.13). The periodic nature of C(t) indicates that in certain regimes of vortex flow, viz., even deep in the driven elastic regime (viz., 3.6 kOe, 5 kOe and 7.6 kOe in Fig.12(a) panels) the moving steady state of the vortex flow, the vortices are not always perfectly correlated. Instead their velocity appears to get periodically correlated and then again drops out of correlation.

Once can deduce the power spectrum of the fluctuations by numerically determining the fast Fourier transform (FFT) of C(t). The FFT results are presented in the right hand set of panels in Figures 12(a), 12(b), 13(a) and 13(b). A summary of the essential features of the power spectrum are as follows. At 2.6 kOe where the vortex array is just above the depinning limit for I_{dc} = 30 mA, one finds two peak-like features in the power spectrum centered around 0.25 Hz and 2 Hz (Fig.12(a)). With increasing field, the peak feature at 2 Hz vanishes, and with the onset of freely flowing elastic regime (>3 kOe), a distinct sharp peak located close to 0.25 Hz survives. This low-frequency peak, which exists up to H = 7.6 kOe, has an amplitude nearly five times that at 0.25 Hz for 2.6 kOe. In the plastic flow regime, viz., H > H_{pl} ~ 8 kOe, the amplitude of the 0.25 Hz frequency starts diminishing (Figs.12(b), the right most panel). At the peak location of the PE (H_p=10.8 kOe), the 0.25 Hz frequency is absent but there is now a well defined peak in the power spectrum close to 2 Hz (see Fig.12(b)). Close to the vortex state depinning out of the plastic regime (i.e., close to the termination of PE (e.g., at 12.4 kOe and beyond, in Fig.13(b)), the 2 Hz peak dissappears and a broad noisy feature, which seems to be peaked, close to mean value ~ 0.25 Hz makes a reappearance (cf. right hand panels set in Fig.13(a)).

Close to 13.6 kOe and 14 kOe, one finds that the fluctuations begin to appear at multiple frequencies, indicating a regime of almost random and chaotic regime of response. Features related to a chaotic regime of fluctuations are being described later in section 4.6. As one begins to approach close to H_{c2}, i.e., at 16 kOe, one observes a broad spread out spectrum with weak amplitude. For the sake of comparison, in the panels in Fig.13(b), the measured and analyzed V(t)/V_0, C(t) and the power spectrum of voltage fluctuations in the normal state of the superconductor at 10 K and 10 kOe stand depicted. Note that the V(t) is just abut 0.02% of V_0, which is far lower than that present in the superconducting state. The C(t) is featureless and the power spectrum of the fluctuations in the normal state also does not show any characteristics peak in the vicinity of 0.25 Hz or 2 Hz.

The evolution in the fluctuations described above at T=4.5 K is also found at other temperatures. Similar to 4.5 K measurements of the voltage - time series were done at 2.5 K, 5 K, 5.8 K, 6 K (Mohan, 2009b). Figure 14 shows the power spectrum of the fluctuations in V recorded at 2.5 K in different field regimes (Mohan, 2009a). Panel (a) of Fig.14 shows the R-H behavior plot for T=2.5 K, where the field locations of H_{dp}, H_{pl}, H_p and H_{c2} have been marked with arrows. By comparing the power spectrum of fluctuations at 2.5 K (Figs.14 (a) and 14(b)) with those at 4.5 K (the left most set of panels in Figs.12(a), 12(b) and 13(a)), one can find similarity in overall features, along with some variations as well. For example, note that like at 4.5 K, in 2.5 K also, just after depinning, the vortex state viz., at 6. 5 kOe at 2.5 K (Fig.14) and 2.6 kOe at 4.5 K (Fig.12(a)), one can observe the presence of two discernable features in the power spectrum located in the vicinity of the 0.2 Hz and 2.25 Hz. However, unlike at 4.5 K where the peak at 2 Hz quickly disappeared by 3 kOe (Fig.12(a)) at 2.5 K on moving to fields away from the H_{dp}, the two peak structure (one close to 0.2 Hz and another close to 2.25 Hz) in the power spectrum persists upto field of 12. 5 kOe (see Fig.14(b)). At 2.5 K the peak located near 2.25 Hz in the power spectrum progressively decreases with increasing H until it dissapears at 13.5 kOe and only a broad feature with peaks in the sub- Hertz regime remains (see, 13.5 kOe and 14.5 kOe data in the panels of Fig.14(c)). Unlike at 4.5 K, where the periodic nature of the fluctuations in the ordered elastic flow regime was clearly discernable, at 2.5 K the fluctuations in V(t) are not as periodic (perhaps due to the admixture of the two characteristic frequencies). Here one can argue that both drive and thermal fluctuation effects play a significant role in generating the characteristic fluctuations. At 2.5 K, on entering the PE regime, similar to 4.5 K data, one finds only find a lone peak surviving near 2 Hz in the power spectrum of fluctuations (compare 18 kOe data at 2.5 K in Fig.14(c) panel with the 10.8 kOe data in Fig.12(b)). Beyond the PE regime at 22 kOe at 2.5 K only the broad feature in the sub-Hertz regime survives. At other higher T (> 4.5 K and close to $T_c(H)$) the features in the power spectrum are almost identical to those seen for 4.5 K with the difference being that features in the sub-Hertz regime become dominant compared to the Hertz regime (Mohan, 2009).

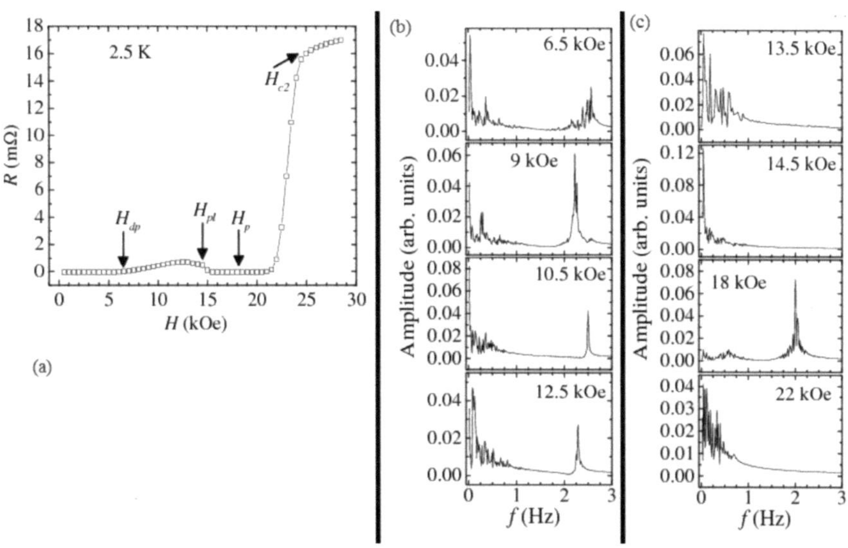

Fig. 14. (a) R–H behavior at 2.5 kOe measured with I_{dc} = 30 mA. Panels (b) and (c) represent the power spectrum of fluctuations at 2.5 K at different H. [Mohan 2009b]

4.4 Excitation of resonant like modes of fluctuations in the driven vortex phase

The above measurements have revealed that a dc drive (with I_{dc}) excites large fluctuations in voltage (equivalent to velocity) in the range of 10 – 40% of the mean voltage level (V_0) at characteristic frequencies (f_0 and f'_0) located in the range of 0.2 Hz and 2 Hz, respectively. The observation that low-frequency modes can get excited in the driven (by I_{dc}) vortex lattice had led Mohan et al, (2009) to explore the effect of a small ac current (I_{ac}) superimposed on I_{dc}, where the external periodic drive with frequencies (f) close to f_0 and f'_0 may result in a resonant like response of the driven vortex medium. The vortex lattice was driven with a current, $I = I_{dc} + I_{ac}$, where $I_{ac} = I_0 \mathrm{Cos}(2\pi ft)$ is the superposed ac current on I_{dc}. At 4 K at different H, the vortex state is driven with I(f), and the dc voltage V of the sample was measured while varying the f of $I_{ac}(f)$. Figure 15 shows the measured V against f at different values of H, where I_{dc} = 22 mA and I_0 = 2.5 mA ($I_{ac} = I_0 \mathrm{Cos}(2\pi ft)$), where the I_0 is chosen to ensure that $I_{dc}+I_0$ gives the same V as with only I_{dc} = 30 mA, at the given H,T.

In the elastic regime (7.6 kOe, cf. Fig. 15(a)) one observes spectacular oscillations in V(f). Significantly large oscillations are observed in V at low f , viz., f < 3 Hz, where the oscillations can exceed (by nearly 100%) of the mean V level determined by the I_{dc}. Shown in Fig.15(b) is an enlarged view of the low-f region of the V(f) data at 7.6 kOe presented in Fig.15(a). An important feature to note in Fig. 15(b) is the enhanced regimes of fluctuations in V(f) occurring at the harmonics of 0.25 Hz (see arrows in bold in Fig.15(b)).

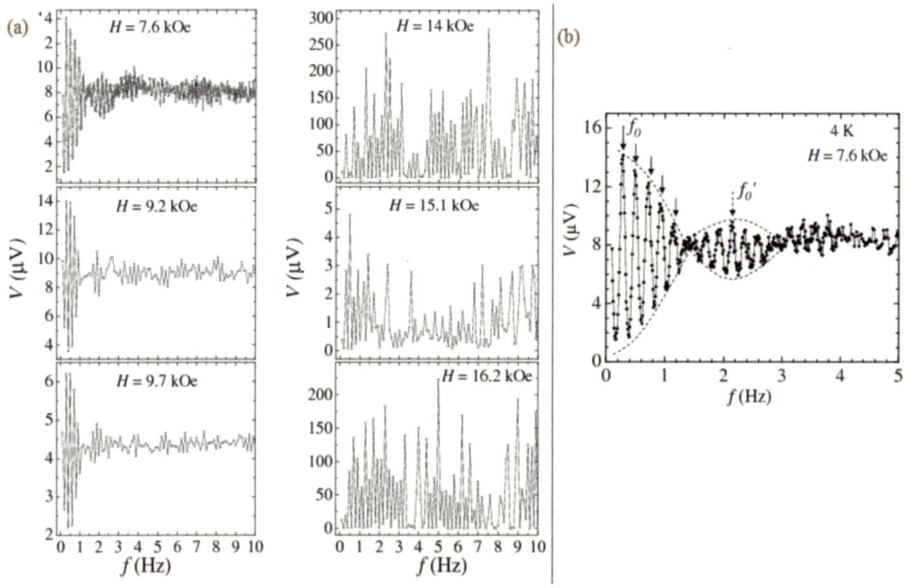

Fig. 15. (a) The measured dc voltage V against frequency f of I_{ac} at different values of H at 4 K and with a current $I = I_{dc} + I_{ac}$, where I_{dc} = 22 mA and I_0 = 2.5 mA. (b) An enlarged view of V(f) at 4 K and 7.6 kOe (panel (a)). The arrows in 'bold' mark the location of the resonant peaks in V(f). [Mohan et al. 2009a; Mohan 2009b]

Note that the peak of the fluctuations in V(f) at the harmonics of 0.25 Hz appears to follow an envelope curve, which has a frequency of 2 Hz (see envelope curve in Fig.15(b)), though the envelope of fluctuation at $f_0' \sim 2$ Hz damps out faster than that at $f_0 \sim 0.25$ Hz. However, one can see that f of I_{ac} matches with the characteristic frequencies f_0 and f'_0 (cf Figs. 12 and 13), which are excited with I_{dc}, viz., ~ 0.25 Hz and ~ 2 Hz, where one observes resonant oscillations in the V. Note that by increasing H as one enters the plastic regime, for example at 9.2 kOe (Fig.15(a)), the enhanced resonant like fluctuations in V(f) at the harmonics of 0.25 Hz seem to rapidly diminish. At 7.6 kOe, while one observes resonant like fluctuations in V(f) upto $6f_0$, f_0 = 0.25 Hz, at 9.2 kOe, one observes the same till only about $4f_0$. Notice that above the peak of the PE, viz., at 14 kOe and beyond, one observes no resonant like behavior in V(f), instead the system seems to exited at all frequencies, which is indicative of a chaotic regime of fluctuations. It is interesting to note similar behavior was also observed in the power spectrum of fluctuations in the vortex velocity excited at 14 kOe in Fig.13(a). Thus, the observation of large (~100%) excursions in the measured V_{dc} signal at harmonics of 0.25 Hz indicates a significantly large *nonlinear* response in the traditionally assumed *linear*, weakly disordered - driven vortex solid prior to the PE. The above chaotic behavior continues well above the onset of the PE regime. Though from the earlier discussion of Figs.

12 and 13, it may have appeared that f_0 ~ 0.2 - 0.25 Hz makes a comeback above the PE, leading one to propose a similarity of driven phases before and above the PE, yet the present measurements indicate that above PE, the f_0 does not excite the resonant like feature which are characteristic of f_0 deep in the elastic regime (viz., see Fig.15).

It has been proposed (Mohan et al, 2009a) that the resistance of the sample varies as, $R = R_0 \pm \sum_{n=1}^{m}[R_n Cos(n2\pi f_0 t) + R_n^{/} Sin(n2\pi f_0 t)]$,under the influence of current, $I = I_{dc} + I_{ac.}$ Here, R_0 is the resistance of the sample in response to the I_{dc} alone, R_n and $R_n/$ are the f dependent coefficients of the in-phase and out-of-phase responses, and f_0 is the characteristic frequency of fluctuations. The f_0 (= 0.25 *Hz*) corresponds to the peak value in the power spectrum for H = 7.6 kOe and T = 4.5 K in Fig.15(b). Taking the time average of the expression, $V = IR$, yields, $V_{dc} = I_{dc}R_0 \pm (I_0R_n)/2$, at $f = nf_0$. From the very large fluctuations (~100%) seen in Fig.15, it is clear that $(I_0R_1)/2 \approx I_{dc}R_0$ or R_1 ~ 20 R_0, is a substantially large component excited at f = f_0. Similarly, at f = 2 f_0, R_2 ~ 15 R_0. Notice from Fig.15, that the nonlinear response can be easily seen upto f = 5 to $6f_0$ (see the positions of solid arrows in Fig.15(b)). The envelope of the amplitude of fluctuations in Fig.15(b) appears to decrease upto 5 f_0; thereafter, the envelope regenerates itself into second and third cycles of oscillations, but, with progressively, reduced intensities. Thus, a small perturbation with I_{ac} ~ 0.1 I_{dc} triggers large fluctuations along with a higher-harmonic generation indicating a highly nonlinear nature of the dynamics. It is noteworthy that the envelope of the resonant oscillations at nf_0 seen at 7.6 kOe with a frequency of 2 Hz (= f'_0) is damped out in the plastic regime. Thus, the peak in the vicinity of f_0'=2 Hz as seen in Figs.12, 13 and 14, have properties different from f_0 ~ 0.2 - 0.25 Hz. Unlike f_0, the $I_{ac}(f_0')$ does not excite resonant like modes of fluctuations especially in the plastic regime, and even in the elastic regime as noted earlier the envelope (dotted curve in Fig.15(b)) with frequency f_0' = 2 Hz damps out very quickly. Thus f_0 and f_0' are associated with distinct behavior of different states of the driven vortex matter.

4.5 Evolution in the characteristic frequencies observed in the power spectrum with vortex velocity

It is known that the periodically spaced vortices when driven over pins, lead to a specific variety of vortex-velocity fluctuations, called the washboard frequency (Fiory 1971; Felming & Grimes 1979; Harris et al., 1995; Kokubo et al, 2005), which are in the range of 0.1-1 MHz. The wash board frequency is far larger than the frequencies, elucidated above. It has also been reported that the nonlinear *I-V* characteristics in the PE regime is accompanied with low frequency noise (<< washboard frequency) in the range of few Hz (Higgins and Bhattacharya 1996; Paltiel et al, 2000; 2002; Gordeev et al 1997; Marley et al 1995; Merithew et al. 1996). The peak in the noise power density in the vicinty of 3 Hz in the PE regime in 2H-$NbSe_2$ was rationalized within the edge contamination framework (Paltiel et al., 2000;2002). Qualitatively, as per the edge contamination model (Paltiel et al, 2000; 2002), the disordered vortices injected from irregularities on the sample boundaries lead to a slow down of the ordered vortices driven inside the sample. This causes a reduction in the injection rate of the disordered vortices. As the fraction of the injected disordered vortices decreases, the velocity of the driven state inturn increases and the entire process repeats. This is the source of velocity fluctuations via the edge contamination picture. It has been argued that edge contamination should result in velocity fluctuations, which are proportional to the rate of injection of vortices which typically are in the range few Hz. In our case, vortices need about 0.1 s to traverse the typical width of our sample of ~ 0.1 cm, with a vortex velocity , v = <V(t)>/(d.B) ~ 10^{-2} m/s(= 1 cm/s), where V ~ 10 μV observed at 30 mA, B =μ_0H = 1 Tesla, and d is the distance between the electrical contacts = 10^{-3} m. Therefore, the injection rate of disordered vortices into the moving vortex medium from irregularities at the sample edges is at the rate of ~ 10 Hz. The observation of a peak in the velocity fluctuation spectrum centered around 2 Hz (cf. Figs. 12, 13 and 14) in the PE region could be termed as consistent with earlier reported observations of peak in noise power in similar frequency range in the PE regime of $NbSe_2$ (Paltiel et al., 2002; Merithew et al., 1996) and $YBa_2CuO_{7-\delta}$ (Gordeev et al., 1997). However, in the ordered elastic driven vortex state prior to PE, one also notes a much lower frequency of 0.25 Hz (cf. Figs.12, 13 and 14), which as per the edge contamination model would imply an effective sample width of 4 cm (with u = 1 cm/s), which would be >> actual sample width (~ 0.1 cm). This implies a deviation from the edge contamination picture.

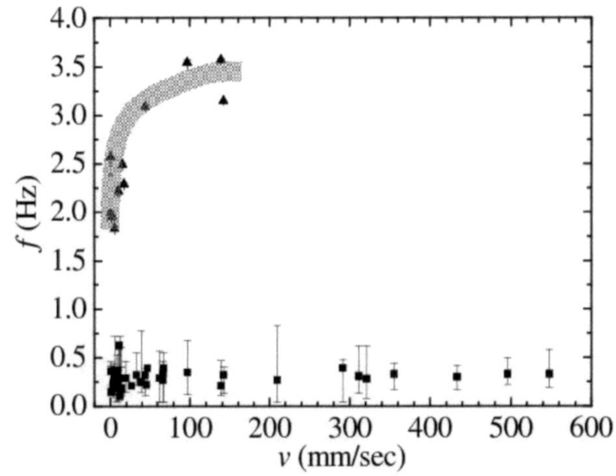

Fig. 16. The evolution of the characteristic frequencies associated with fluctuations in vortex motion as a function of velocity of vortices. The shaded band represents the behaviour of the higher characteristic frequency. [Mohan et al. 2009a; Mohan 2009b]

Figure16 shows an evolution in f_0 (~ 0.05 H_z, solid squares) and f'_0 (~ 2 Hz, solid triangles) with velocity (v) of the vortices (Mohan et al, 2009). This compilation is based on measurements at different H, T, and I_{dc}. One can see that the higher characteristic frequency f_0' increases with v, varying from around 1.75 Hz to 3.5 Hz, while the lower f_0 is v independent. This is consistent with the impression from the I_{dc}+I_{ac} experiments that f_0 and f_0' have distinct behavior and do not correspond to part of the same behavior repeating at different frequencies. From the conventional noise mechanism, based on edge contamination model, one would expect the frequency of v fluctuations (equivalent to the disorder injection rate) to increase with v without showing any tendency to saturate with v. However, this is not the case as seen in Fig.16. While the higher frequency f_0'does seem to increase with v at lower values (see shaded region in Fig.16), it shows a much more slower change with v at higher values, with a tendency to saturate. The lower frequency appears to be nominally v independent, which is unexpected within edge contamination model. One may clarify that in certain v regimes only one of the two frequencies survives. It can be stated that the detailed richness of

the fluctuations in the descriptions presented here do not find a rationalization within the present models relating to noise in the driven vortex state.
The current understanding of the nature of the flowing vortex state and transitions within it are inadequate. This is best illustrated by the nonlinear nature of the response within the steady state of driven elastic vortex medium (cf. discussion pertaining to Figs.12 - 16 above), which is far from the conventional notion that the elastic medium is almost a benign medium, which responds almost linearly to drive. Infact though a lot has been understood regarding the plastic flow regime (see discussion relating to the plastic flow regime in section 4.1), newer works (Olive & Soret, 2006; 2008) have indicated that the vortices in this regime exhibit chaotic regimes of flow, where the velocity fluctuations of the vortices may show intermittent velocity bursts which can be a route for the emergence of chaos in the vortex state.

4.6 Intermittent voltage bursts in driven vortex state

The nature of voltage fluctuations and the associated power spectrum of fluctuations at 4.5 K (cf. Fig.13(a), 14 kOe data) reveal that in the regime just after PE the vortices driven by a dc drive (I_{dc}) begin to exhibit v fluctuations at all possible frequencies. This behavior is further corroborated by the V(f) data in Fig.15, which shows that the vortex state at 14 kOe (just above the PE regime) when driven with I_{dc} and perturbed with $I_{ac.}$ The driven vortex state at 14 kOe begins to show large nonlinear excursion in v (equivalent to V) at all f in the range over which f is varied. Such a behavior, where the nonlinear fluctuations in v exists uniformly over a large frequency interval is indicative of the onset of a chaotic regime of flow in the vortex state.

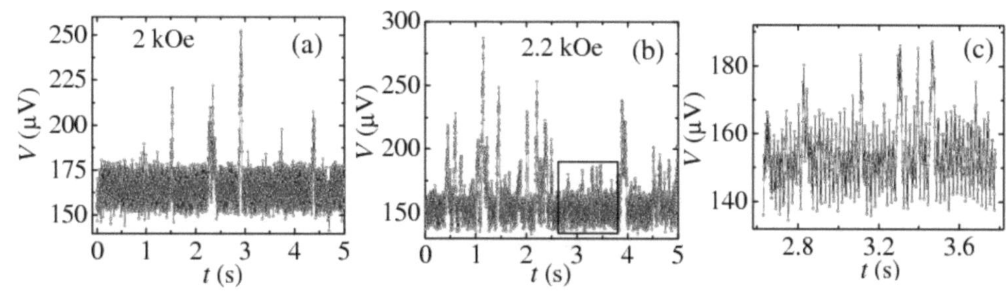

Fig. 17. Panels (a) and (b) show the measured temporal response of the dc voltage (V(t)) at 6 K in the plastic flux-flow regime. Panel (c) is a blow-up of the rectangular region marked in (b). [Mohan et al. 2009a; Mohan 2009b]

One can capture the time resolved voltages (V(t)) in smaller time intervals of 1.25 ms (as against the 35 ms interval in the earlier data) using the data storage buffer of the ADC in a lock-in amplifier. At a higher T and deep in the plastic phase, one observe, the development of an interesting fluctuation behaviour in the time domain, viz., that of intermittency (Mohan et al, 2009a). The panels (a) and (b) in Fig.17 show the measured V(T) data at 6 K in the plastic regime with H=2 kOe and H=2.2 kOe (see phase diagram in the inset of Fig.10). At 2 kOe, one observes nearly-periodic fluctuations about a mean level 160 μV. But, these V fluctuations are interrupted by large, sudden voltage bursts. On entering deeper into the plastic regime, i.e. at 2.2 kOe, these chaotic voltage bursts become much more prominent (see Fig.17(b)). The intermittent large V (equivalent to v) bursts are almost twice as large as the mean V level. In terms of the vortex velocity (v=V/Bd), the mean velocity level at 6 K and 30 mA, is 750 mm/sec whereas during the intermittent bursts the voltage shoots up to a maximum v ~ 1500 mm/sec. Such bursts are followed by time intervals, when the fluctuations are nearly periodic, as can be clearly seen in panel (c) of Fig.17. Here, it is useful to mention that from simulations studies Olive and Soret (2006, 2008) have proposed that in the plastic regime of flow the vortex motion within the channels periodically synchronizes with the fluctuating vortices trapped in the pinned islands leading to periodic fluctuations. This periodic regime can become unstable and give way to a chaotic burst, with large velocity fluctuations. The intermittent velocity bursts indicate the onset of disordered trajectories of the moving vortices, which is symptomatic of the onset of chaotic motion of vortices. Apart from observing intermittency features in the plastic flow regime (Fig.17) at 6 K there are indirect evidences at 14 kOe at 4.5 K close to $T_c(H)$ (see Fig.13(a)) and Fig.15(a), which indicate the onset of chaotic behavior at these T, H. Perhaps onset of such intermittant velocity bursts appear closer to a regime where thermal fluctuations also begin to play a significant role in the behavior of the vortices in the driven state especially after the onset of plastic flow.

5. Epilogue and future directions

The nonlinear response deep within the driven elastic medium is presumably related to a possible transformation into a heterogenous vortex configuration observed deep within the elastic phase (Mohan et al, 2007). Complex nonlinear systems under certain conditions can produce slow spontaneous organization in its dynamics. Under the influence of a sufficient driving force, the system can exhibit coherent dynamics, with well-defined one or more frequencies (Ganapati & Sood 2006; Ganapati et al., 2008). The evolution of fluctuations, such as those illustrated in Figures 12, 13 and 14, can be viewed as the complex behavior of a nonlinear driven vortex state with multiple attractors (stable cycles). The appearance of stable cycles are characteristic of a particular phase of the driven vortex state. Underlying phase transformations in the driven vortex state induce the system to fluctuate between different stable cycles, leading to a typical spectrum of fluctuations discussed in Figs. 12, 13 and 14. The above nature may lead to extreme sensitivity of the driven vortex system to the low amplitude perturbations, as is shown in Fig.15. We believe that the fluctuations with characteristic frequencies with the nonlinear response discussed above are indicative of phase transformations in the driven vortex state. Figures 15 and 16 have shown that the behavior the characteristic low frequencies f_0 and f'_0 are distinct and cannot be completely attributed to irregular edge related effects of the superconductor. Infact f_0 can be attributed to the due to the elastic fraction of the vortices, where its response is found to be maximum in Figs.12-16, while the 2 Hz represents to disordered fraction in the driven vortex state.
To summarise, we have dwelled the nature of transformations deep in the quasi static elastic vortex state. As the vortex state is driven in the steady state, exploration of vortex-velocity fluctuations in the time domain have uncovered signatures of complex nonlinear dynamics even deep in the elastic driven vortex state prior to the onset of plastic flow. These pertain to new regimes of coherent driven dynamics in the elastic phase with distinct frequencies of fluctuations. These regimes are a precursor to chaotic fluctuations, which can germinate deep in the plastic regime. In ongoing experiments pertaining to more detailed time series measurements on systems other than $NbSe_2$, novel interesting signatures of critical behaviour at dynamical phase transition in driven mode of plastically deformed vortex matter have recently been identified (Banerjee et al, 2011, unpublished).

6. Acknowledgements

We first acknowledge Prof. A.K. Sood of I.I.Sc., Bangalore for sharing his insights on non-linear response in soft condensed matter and motivating our recent investigations in vortex state studies. The author acknowledge Prof. Shobo Bhattacharya and Prof. Eli Zeldov for collaborative works in the past. A.K. Grover thanks C.V. Tomy, Geetha Balakrishnan, M.J. Higgins and P.L. Gammel for the crystals of 2H-$NbSe_2$ for vortex state studies at TIFR. We thank Ulhas Vaidya for his help during experiments at TIFR. Satyajit S. Banerjee (S. S. Banerjee) acknowledges funding from DST, CSIR, DST Indo-Spain S &T forum, IIT Kanpur.

7. References

Angurel, L. A., Amin, F., Polichetti, M., Aarts, J. & Kes, P. H. (1997). Dimensionality of Collective Pinning in 2*H*-$NbSe_2$ Single Crystals. *Physical Review B* Vol. 56, No. 6, pp. 3425-3432

Banerjee, S. S., Patil, N. G., Saha, S., Ramakrishnan, S., Grover, A. K., Bhattacharya, S., Ravikumar, G., Mishra, P. K., Chandrasekhar T. V. Rao, Sahni, V. C., Higgins, M. J., Yamamoto, E., Haga, Y., Hedo, M., Inada, Y. & Onuki, Y. (1998). Anomalous Peak Effect in $CeRu_2$ and 2*H*-$NbSe_2$: Fracturing of a Flux Line Lattice. *Physical Review B* Vol. 58, No. 2, pp. 995-999

Banerjee, S. S., Patil, N. G., Ramakrishnan, S., Grover, A. K., Bhattacharya, S., Ravikumar, G., Mishra, P. K., Rao, T. V. C., Sahni, V. C. & Higgins, M. J. (1999a). Metastability and Switching in the Vortex State of 2*H*-$NbSe_2$. *Applied Physics Letters* Vol. 74, No. 1, pp. 126-128

Banerjee, S. S., Patil, N. G., Ramakrishnan, S., Grover, A. K., Bhattacharya, S., Ravikumar, G., Mishra, P. K., Rao, T. V. C., Sahni, V. C., Higgins, M. J., Tomy, C. V., Balakrishnan, G. & Mck Paul, D. (1999b). Disorder, Metastability, and History Dependence in Transformations of a Vortex Lattice. *Physical Review B* Vol. 59, No. 9, pp. 6043-6046

Banerjee, S. S., Ramakrishnan, S., Grover, A. K., Ravikumar, G., Mishra, P. K., Sahni, V. C., Tomy, C. V., Balakrishnan, G., Paul, D. Mck., Gammel, P. L., Bishop, D. J., Bucher, E., Higgins, M. J. & Bhattacharya, S. (2000a). Peak Effect, Plateau Effect, and Fishtail Anomaly: The Reentrant Amorphization of Vortex Matter in 2*H*-$NbSe_2$. *Physical Review B* Vol. 62, No. 17, pp. 11838-11845

Banerjee, S. S., (2000b). In: Vortex State Studies In Superconductors. *Thesis. Tata Institute of Fundamental Research, Mumbai – 400005. University of Mumbai. India*

Banerjee, S. S., Grover, A. K., Higgins, M. J., Menon, Gautam I., Mishra, P. K. , Pal, D., Ramakrishnan, S., Chandrasekhar Rao, T.V., Ravikumar, G., Sahni, V. C., Sarkar S. and Tomy C.V. (2001) Disordered type-II superconductors: a universal phase diagram for low-Tc systems, Physica C Vol 355, pp. 39 - 50.

Bean, C. P. (1962). Magnetization of Hard Superconductors. *Physical Review Letters* Vol. 8, No. 6, pp. 250-253

Bean, C. P. (1964). Magnetization of High-Field Superconductors. *Reviews of Modern Physics* Vol. 36, No. 1, pp. 31-39

Berlincourt, T. G., Hake, R. R. & Leslie, D. H. (1961). Superconductivity at High Magnetic Fields and Current Densities in Some Nb-Zr Alloys. *Physical Review Letters* Vol. 6, No. 12, pp. 671 -674

Bhattacharya, S. & Higgins, M. J. (1993). Dynamics of a Disordered Flux Line Lattice. *Physical Review Letters* Vol. 70, No. 17, pp. 2617-2620

Blatter G., Feigel'man, M. V., Geshkenbein, V. B., Larkin, A. I. & Vinokur, V. M. (1994). Vortices in High-Temperature Superconductors. *Reviews of Modern Physics* Vol. 66, No. 4, pp. 1125-1388

Blatter, G., Geshkenbein, V. B. & Koopmann, J. A. G. (2004). Weak to Strong Pinning Crossover. *Physical Review Letters* Vol. 92, No. 6, pp. 067009(1)-067009(4)

Le Doussal, P. & Giamarchi, T. (1998). Moving Glass Theory of Driven Lattices with Disorder. *Physical Review B* Vol. 57, No. 18, pp. 11356-11403

Duarte, A., Righi, E. F., Bolle, C. A., Cruz, F. de la, Gammel, P. L., Oglesby, C. S., Bucher, E., Batlogg, B. & Bishop, D. J. (1996). Dynamically Induced Disorder in the Vortex Lattice of 2*H*-$NbSe_2$. *Physical Review B* Vol. 53, No. 17, pp. 11336-11339

Faleski, M. C., Marchetti, M.C. & Middleton, A. A. (1996). Vortex Dynamics and Defects in Simulated Flux Flow. *Physical Review B* Vol. 54, No. 17, pp. 12427-12436

Fiory, A. T. (1971). Quantum Interference Effects of a Moving Vortex Lattice in Al Films. *Physical Review Letters* Vol. 27, No. 8, pp. 501-503

Fisher, M. P. A. (1989). Vortex Glass Superconductivity: A Possible New Phase in Bulk High-T_c Oxides. *Physical Review Letters* Vol. 62, No. 12, pp. 1415-1418

Fisher, D. S., Fisher, M. P. A. & Huse, D. A. (1990). Thermal Fluctuations, Quenched Disorder, Phase Transitions, and Transport in Type-II Superconductors. *Physical Review B* Vol. 43, No. 1, pp. 130-159

Fleming, R. M. & Grimes, C. C. (1979). Sliding-Mode Conductivity in $NbSe_3$: Observation of a Threshold Electric Field and Conduction Noise. *Physical Review Letters* Vol. 42, No. 21, pp. 1423-1426

Gammel, P. L., Yaron, U., Ramirez, A. P., Bishop, D. J., Chang, A. M., Ruel, R., Pfeiffer, L. N., Bucher, E., D'Anna, G., Huse, D. A., Mortensen, K., Eskildsen, M. R. & Kes, P. H. (1998). Structure and Correlations of the Flux Line Lattice in Crystalline Nb Through the Peak Effect. *Physical Review Letters* Vol. 80, No. 4, pp. 833-836

Ganapathy, R. & Sood, A. K. (2006). Intermittent Route to Rheochaos in Wormlike Micelles with Flow-Concentration Coupling. *Physical Review Letters* Vol. 96, No. 10, pp. 108301(1)-108301(4)

Ganapathy, R., Mazumdar, S. & Sood, A. K. (2008). Spatiotemporal Nematodynamics in Wormlike Micelles Enroute to Rheochaos. *Physical Review E* Vol. 78, No. 2, pp. 021504(1)- 021504(6)

Ghosh, K., Ramakrishnan, S., Grover, A. K., Menon, G. I., Chandra, G., Rao, T. V. C., Ravikumar, G., Mishra, P. K., Sahni, V. C., Tomy, C. V., Balakrishnan, G., Mck Paul, D. & Bhattacharya, S. (1996). Reentrant Peak Effect and Melting of a Flux Line Lattice in 2H-$NbSe_2$. *Physical Review Letters* Vol. 76, No. 24, pp. 4600-4603

Giamarchi T. & Le Doussal, P. (1995). Elastic Theory of Flux Lattices in the Presence of Weak Disorder. *Physical Review B* Vol. 52, No. 2, pp. 1242-1270

Giamarchi, T. & Le Doussal, P. (1996). Moving Glass Phase of Driven Lattices. *Physical Review Letters* Vol. 76, No. 18, pp. 3408-3411

Giamarchi, T. & Bhattacharya, S. (2002). Vortex Phases, In: *High Magnetic Fields: Applications in Condensed Matter Physics and Spectroscopy,* C. Berthier, L. P. Levy and G. Martinez (Eds.), *Springer*, 314-360, ISBN: 978-3-540-43979-0

Ghosh, K., Ramakrishnan, S., Grover, A. K., Menon, G. I., Chandra, Girish, Chandrasekhar Rao, T. V., Ravikumar, G., Mishra, P. K., Sahni, V. C., Tomy, C. V., Balakrishnan, G.,

Mck Paul, D., Bhattacharya, S. (1996) Reentrant peak effect and melting of flux line lattice in 2H-$NbSe_2$. *Physical Review Letters* Vol. 76, pp. 4600 - 4603.

Gordeev, S. N., de Groot, P. A. J., Ousenna, M., Volkozub, A. V., Pinfold, S., Langan, R., Gagnon, R. & Taillefer, L. (1997). Current-Induced Organization of Vortex Matter in Type-II Superconductors. *Nature* Vol. 385, pp. 324-326

Harris, J. M., Ong, N. P., Gagnon, R. & Taillefer, L.(1995). Washboard Frequency of the Moving Vortex Lattice in $YBa_2Cu_3O_{6.93}$ Detected by ac-dc Interference. *Physical Review Letters* Vol. 74, No. 18, pp. 3684-3687

Higgins, M. J. & Bhattacharya, S. (1996). Varieties of Dynamics in a Disordered Flux-Line Lattice. *Physica C* Vol. 257, pp. 232-254

Kim, Y. B., Hempstead, C. F. & Strnad, A. R. (1962). Critical Persistent Currents in Hard Superconductors. *Physical Review Letters* Vol. 9, No. 6, pp 306-309

Kokubo, N. , Kadowaki, K. and Takita, K. (2005). Peak Effect and Dynamic Melting of Vortetex matter in $NbSe_2$ Crystals. *Physical Review Letters* Vol. 95, No. 17, pp 177005(1)-177005(4)

Larkin, A. I. & Ovchinnikov, Yu. N. (1979). Pinning in Type II Superconductors. *Journal of Low Temperature Physics*, Vol. 34, No. 3/4, pp 409-428

Larkin, A.I. (1970a). Vliyanie neodnorodnostei na strukturu smeshannogo sostoyaniya sverkhprovodnikov. *Zh. Eksp. Teor.* Fiz, Vol. 58, No. 4, pp. 1466-1470

Larkin, A.I. (1970b). Effect of inhomogeneities on the structure of the mixed state of superconductors, *Sov. Phys. JETP* Vol. 31, No. 4, pp 784

Li, G., Andrei, E. Y. , Xiao, Z. L., Shuk, P. & Greenblatt. M. (2005). Glassy Dynamics in a Moving Vortex Lattice. *J. de Physique IV,* Vol. 131, pp 101-106 (and references therein to their earlier work)

Li, G., Andrei, E. Y., Xiao, Z. L., Shuk, P. & Greenblatt, M. (2006). Onset of Motion and Dynamic Reordering of a Vortex Lattice. *Physical Review Letters* Vol. 96, No. 1, pp 017009(1)-017009(4)

Ling X. S., Park, S. R., McClain, B. A., Choi, S. M., Dender D. C. & Lynn J. W. (2001). Superheating and Supercooling of Vortex Matter in a Nb Single Crystal: Direct Evidence of a Phase Transition at the Peak Effect from Neutron Diffraction. *Physical Review Letters* Vol. 86, No. 4, pp 712-715

Marchevsky, M., Higgins, M. J. & Bhattacharya, S. (2001). Two Coexisting Vortex Phases in the Peak Effect Regime in a Superconductor. *Nature,* Vol. 409, pp 591-594

Marley, A. C., Higgins, M. J. and Bhattacharya, S. (1995). Flux Flow Noise and Dynamical Transitions in a Flux Line Lattice. *Physical Review Letters,* Vol. 74, No. 15, pp 3029-3032

Merithew, R. D., Rabin, M. W., Weissman, M. B., Higgins, M. J. and Bhattacharya, S. (1996). *Physical Review Letters* Vol. 77, No. 15, pp 3197-3199

Mohan, S., Sinha, J., Banerjee, S. S. & Myasoedov, Y., (2007). Instabilities in the vortex matter and the peak effect phenomenon. *Physical Review Letters* Vol. 98, No. 2, pp 027003 (1)-027003(4)

Mohan, S., Sinha, J., Banerjee, S. S., Sood, A. K., Ramakrishnan, S.& Grover, A. K. (2009a). Large Low-Frequency Fluctuations in the Velocity of a Driven Vortex Lattice in a Single Crystal of 2H-$NbSe_2$ Superconductor. *Physical Review Letters* Vol. 103, No. 16, pp 167001(1)-167001(4)

Mohan, S., (2009b). In: Instabilities in the vortex state of type II superconductors. *Thesis. Department of Physics. Indian Intitute of Technology – Kanpur, India*

Natterman, T and Scheidl, S (2000), Vortex - Glass phases in type II superconductors, *Advances in Physics* 1460-6967, 49, 607-705.

Nelson, D. R. (1988). Vortex Entanglement in High *Tc* Superconductors. *Physical Review Letters* Vol. 60, No. 19, pp 1973-1976

Nori, F. (1996). Intermittently Flowing River of Magnetic Flux. *Science* Vol. 271, pp 1373-1374

Olive, E. & Soret, J. C. (2006) Chaotic Dynamics of Superconductor Vortices in the Plastic Phase. *Physical Review Letters* Vol. 96, No. 2, pp 027002(1)-027002(4)

Olive, E. & Soret, J. C. (2008). Chaos and Plasticity in Superconductor Vortices: Low-Dimensional Dynamics. *Physical Review B* Vol. 77, No. 14, pp 144514(1)-144514(8)

Pippard A. B, (1969). A Possible Mechanism for the Peak Effect in Type II Superconductors. *Philosophical Magazine* Vol. 19, No. 158, pp 217-220

Paltiel, Y., Fuchs, D. T., Zeldov, E., Myasoedov, Y., Shtrikman, H., Rappaport, M. L. & Andrei, E. Y. (1998). Surface Barrier Dominated Transport in $NbSe_2$, *Physical Review B* Vol. 58, No. 22, R14763-R14766

Paltiel, Y., Zeldov, E., Myasoedov, Y. N., Shtrikman, H., Bhattacharya, S., Higgins, M. J., Xiao, Z. L., Andrei, E. Y., Gammel, P. L. & Bishop, D. J. (2000). Dynamic Instabilities and Memory Effects in Vortex Matter. *Nature*, Vol. 403, pp 398-401

Paltiel, Y., Jung, G., Myasoedov, Y., Rappaport, M. L., Zeldov, E., Higgins, M. J. & Bhattacharya, S. (2002). Dynamic Creation and Annihilation of Metastable Vortex Phase as a Source of Excess Noise. *Europhysics Letters* Vol. 58, No. 1, pp 112-118

Shi, A. C. & Berlinsky, A. J. (1991). Pinning and *I-V* characteristics of a two-dimensional defective flux-line lattice. *Physical Review Letters* Vol. 67, No. 14, pp 1926-1929

Thakur, A. D., Banerjee, S. S., Higgins, M. J., Ramakrishnan, S. and Grover, A. K. (2005) Exploring metastability via third harmonic measurements in single crystals of 2H-$NbSe_2$ showing an anomalous peak effect. *Physical Review B* Vol.72, pp 134524-134529.

Thakur, A. D., Banerjee, S. S., Higgins, M. J., Ramakrishnan, S. and Grover, A. K. (2006) Effect of pinning and driving force on the metastability effects in weakly pinned superconductors and the determination of spinodal line pertaining to order-disorder transition. *Pramana Journal of Physics* Vol. 66, pp. 159 – 177.

Troyanovski, A. M., Aarts, J. & Kes, P.H. (1999). Collective and plastic vortex motion in superconductors at high flux densities, *Nature* Vol. 399, pp 665-668

Troyanovski, A. M., van Hecke, M., Saha, N., Aarts, J. & Kes, P. H. (2002). STM imaging of flux line arrangements in the peak-effect regime. *Physical Review Letters* Vol. 89, No. 14, pp 147006(1)-147006(4)

Xiao, Z. L., Andrei, E. Y. & Higgins, M. J. (1999). Flow Induced Organization and Memory of a Vortex Lattice. *Physical Review Letters* Vol.83, No. 8, pp 1664-1667.

Yaron, U., Gammel, P. L., Huse, D. A., Kleiman, R. N., Oglesby, C. S., Bucher, E., Batlogg, B., Bishop, D. J., Mortensen, K., Clausen, K., Bolle, C. A. & de la Cruz, F. (1994). Neutron Diffraction Studies of Flowing and Pinned Magnetic Flux Line Lattices in $2HNbSe_2$. *Physical Review Letters* Vol. 73, No. 20, pp 2748-2751 (1994)

Zeldov, E., Larkin, A. I., Geshkenbein, V. B., Konczykowski, M., Majer, D., Khaykovich, B., Vinokur, V. M. & Shtrikman, H. (1994). Geometrical Barriers in High-Temperature Superconductors, *Physical Review Letters* Vol. 73, No. 10, pp 1428-1431.

5

Energy Dissipation Minimization in Superconducting Circuits

Supradeep Narayana[1] and Vasili K. Semenov[2]
[1]*Rowland Institute at Harvard, Harvard University, Cambridge*
[2]*Department of Physics and Astronomy, Stony Brook University, Stony Brook USA*

1. Introduction

Low energy dissipation and ability to operate at low temperatures provide for Josephson junction circuits a niche as a support for low temperature devices. With high speed operation (Chen W. et al., 1999) capability the Josephson junction circuits make a prime candidate for applications which are difficult to engineer with existing CMOS technology. The development of Josephson junction technology took a major turn for the better with the invention of the Rapid-Single-Flux-Quantum (RSFQ) devices (Likharev K.K. et al., 1991), an improvement over voltage biased Josephson Junctions logic which were plagued with the junction switching and reset problems. The modern applications of SFQ circuits extend to a larger range of temperature operation and the applications vary from low temperature magnetic sensor, to high speed mixed signal circuits, voltage and current standards (Turner C.W. et al., 1998), and auxiliary components for quantum computing circuits. Most of the SFQ circuits are fabricated with Niobium, but Aluminum based circuits are being used for quantum gates (Nielsen M.A. et al., 2000) and qubit operations. SFQ circuits based on Magnesium di-Boride junctions are being developed for higher temperature operations (Tahara S. et al., 2004). Predominantly most of the Josephson junction circuits today are operated at around 4K. All the circuits are optimized usually for liquid helium temperatures, so the circuits operated in helium bath Dewars or cryostat's do not experience any temperature gradients or drift effects which can affect the operating margins.

With the improvement in the fabrication technology and soft-wares for SFQ circuit technology, the designing complex circuits have become easier. Complex Circuits with over 20K junctions such digital synthesizer and digital RF Trans-receiver have already been demonstrated (Oleg M.A. et al., 2011). Development of circuits over 100K junctions are actively under progress. However, with enormously large circuits, power requirements also increase.

Looking at the range of applications and complexity of the problems of energy minimization, we try to look at the problem in two approaches. One for large complex circuits, we try to reduce the power bias itself, or the overall load of current that is supplied to the chip. And secondly, we try to improve the operation of the circuit blocks by designing components that can be operated in power independent mode.

The proposals made here should be applicable to all operations to make the maximum benefit of the advantage of the design. To operate at lower temperatures such as in milli-Kelvin ranges, required for quantum computing, the junctions and circuit components have to be scaled. The cells, modules or blocks used in design of building larger parts of circuits, are modified in a way such that the cells are capable of maintaining the state of the logic even when the power bias is switched off.

The second and larger energy dissipation source, which can be directly, reduced by lowering the bias current supply. One of the simplest methods of reducing the DC bias current is recycling the bias from one part of the circuit to bias the other parts. This technique called current recycling is a method for serially biasing the circuits. Small scale demonstrations of the technique have been demonstrated a few years ago (Johnson M.W. et al., 2003). We present here some of the results for techniques for over 1k junctions in a single chip and also discuss some of the limitations of these techniques.

2. Background and related work

The problem of power dissipation has been attempted by several groups over the last two decades and the problem has gained more attention based on the new developments of applications into quantum computing technology and wireless technology applications (Tahara S. et al., 2004, Narayana S. 2011). If Josephson SFQ technology has to be extended to quantum computers, which require far fewer junctions but must be operated at much lower temperatures to maintain longer quantum coherence, the issue of power dissipation comes to the forefront.

Despite the numerous advantages, over its semiconductor counterparts, the power dissipation is high in the conventional digital Josephson technology. If the application revolves around quantum computation, the size of circuit is small but power dissipation could seriously disrupt the quantum operations. On the other hand, if the circuits being designed are large power dissipation in the bias lines could be larger by several orders of magnitude compared to the power dissipation in single block or cell.

Early efforts of reducing power dissipation were using large inductances connected to the bias resistors. This method was demonstrated for moderate size circuits in (Yoshikawa N. et al., 2001), but operating margins were reported to be reduced at higher frequencies due limitation of L/R time constant compared to the switching frequency. But reducing R also reduces the maximum clock frequency, which limits the high circuit design.

Static power dissipation, largest source of power dissipation, was eliminated by eliminating resistive biasing elements in circuit design (Polonsky S. 1999). An effort to mimic CMOS logic, also to eliminate static power dissipation was presented by (Silver A.H. et al., 2001), but was harder to integrate into SFQ circuits. But the method successfully was designed for high speed circuits. A new RQL logic has been presented which involves multi phase AC bias, which has been known to cause AC crosstalk (Silver A.H. et al., 2006). Another method for static power dissipation was presented in (Kirichenko D.E. et al., 2011), where the JTL is used to a digital controller to supply bias current to the circuits under operation.

3. Power dissipation in RSFQ circuits

Before we go into methods and experiments results, we can go to present a simplified model as which are well studied in Detail (Rylyakov A. 1997). We will just recap some of the main purpose with a numerical example so as to provide a continuation and feel for the value of method presented. Let us begin with a simple model as to get an idea of the power estimated without going into detailed mathematical models. The most power is dissipated in the bias resistors and second source of power dissipation is the shunt resistors in the junctions when junctions are in the resistive state.

If the clock operation has a frequency f , the power dissipation due to the switching of the Josephson junction is

$$P = FE = f\Phi_0 Ic \tag{1}$$

Where, E is the total energy dissipated, I_C is the critical current of the junction and Φ_0 is the quantum flux constant. Now for a critical current of 100μA, and the clock frequency of 50GHz the power dissipated for a single junction by switching is 10 nW.

Now let us look at the second source of power dissipation, in figure 1, is a Josephson junction network, the junctions are usually biased to a lower value than I_C, about 0.7 I_C, the junction can switch when a correct SFQ pulse arrives.

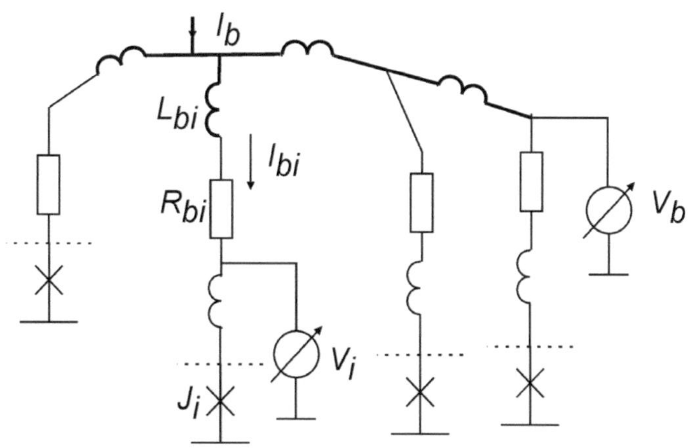

Fig. 1. Josephson junction network

The inductances can L_{bi} and L_i, ratios influences the order of switching events and the effects have been studied well in (Chen W. et al., 1999 and references therein) power dissipation. So for a typical power dissipation based on V = 2.6mv, which is the sub band gap of the niobium superconductors, and critical current 100uA. The power dissipation is P = VI, so the power dissipation is P = 260nW, which is nearly 25 times higher in the bias resistors compared to the junctions. In broader context, one can say that, the power dissipation is at least one order of magnitude higher in bias resistors.

3.1 Temperature scaling

With the growing interest in quantum computing and the favorable choice of single flux quantum circuits for its application, we do have to modify a few parameters for better design. Since most of the quantum computing circuits are operate in milli- Kelvin range, we present a simple scaling method to avoid errors in design of SFQ circuits. The resistance of the shunt resistance (R_{Sh}) and the sub-band gap resistance of the junction is $R_{m,}$ and both the resistances can be seen in parallel and can be calculated as in equation 2.

$$R = \frac{Rm.Rsh}{Rm + Rsh} \tag{2}$$

The principle governing factors for the Josephson junction with the combined resistance R is such that disparity must be avoided so the scaling of resistance should avoid errors due to quantum fluctuations and these quantum fluctuations must be smaller than the thermal fluctuations. So, the ratio of resistances must be smaller to the ratio of thermal noise and critical current contributions and resistance scaling must be smaller. So for scaling conditions to be satisfied we must have,

$$\frac{I_T}{I_C} > \frac{R}{R_Q} \tag{3}$$

Where $R_Q = \frac{\pi T}{e^2}$, I_C is the critical current, I_T is the thermal noise. Bias voltage cannot be scaled similarly as resistance. However, for all conventional reasons the bias voltage is fairly independent of temperature. But in reduction to the crosstalk reductions and circuit designs specifics, the bias voltage can be reduced by a factor of 2 to 5 (Narayana S. 2011, Salvin A. et al., 2006).

4. Power independent RSFQ logic

Superconducting structures have been known to keep circulating currents for unlimited time. If this current or magnetic field caused by this persistent current then one can use this phenomenon to perform useful functions without any energy dissipation. Unfortunately, the list of such functions is quite small (Tahara S. et al., 2004). This are because most of functions or blocks using the persistent currents such as RSFQ cells/latches lose their state when the power is turned off. However, below we would like to show that RSFQ cells could be modified for Power Independent (PI) operation. Let us remind that power independence means an ability of circuits to be un-powered without any loss of stored information. As a result, power independent circuits should be powered only when logic operations should be performed.

The simplest power independent circuit with memory is a well-known single junction SQUID as shown in figure 2a. The single junction SQUID is a superconducting loop with sufficiently large loop inductance L interrupted by a single Josephson junction. The dynamics of single Josephson junction SQUID has been well known for many years now and will not be discussed in detail here. But, it may be sufficient to recap the flux modulation as a function of the bias current to the SQUID as shown in figure 2b. From the figure 2b we can see that, we can write "1" or "0" by applying large enough positive ($I_b > I_{th1}$) or negative ($I_b > I_{th0}$) bias current I_b.

The device continues to remember any of these states if bias current is switched off ($I_b = 0$) as there is no dissipation in the superconducting loop.

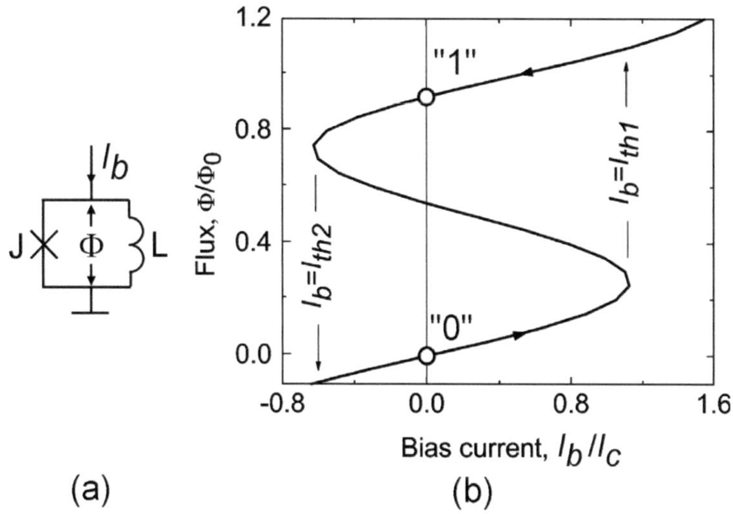

Fig. 2. (a) Single Junction SQUID ;(b) Current vs Flux characteristics. A single junction SQUID as the simplest power independent cell.

The introduced memory cell of single junction SQUID could be incorporated into RSFQ flip-flops and logic gates. The construction of power independent RS flip flop is presented in the next section.

4.1 Power Independent RSFQ RS flip-flop

The transformation of the RSFQ cell into Power Independent RSFQ cell is explained here. In the case of regular RSFQ RS Flip Flop (Turner C.W. et al., 1998), as shown in figure 3a, the Josephson junctions J3, J4 and loop inductance L, form a two junction interferometer with $I_C L = 1.25\Phi_0$, so that a flux quantum can be stored in it. The current in the loop can be expressed as the sum of the bias current equally divided between the two junctions and circulating current $I_P = \pm\Phi / 2L$. Initially, the circulating current is counterclockwise, representing a stored "0". The currents when the bias is applied are $I_{J3} = (I_b / 2) + I_P$ and $I_{J4} = (I_b / 2) - I_P$.

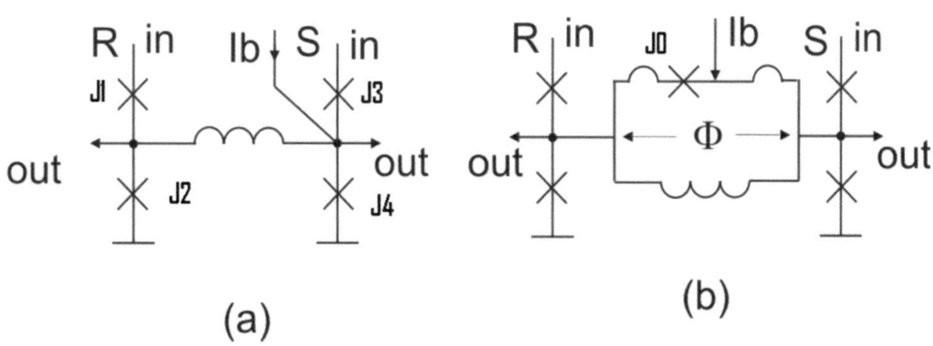

Fig. 3. (a) RSFQ RS- Flip Flop (b) RSFQ Power Independent Flip Flop. Transformation of a conventional RSFQ RS Flip flop into power independent RSFQ RS Flipflop.

When input pluses are applied to the input (set) and clock (reset) terminals, this causes circulating current to reverse polarity. When pulse arrives on the input, its current passes through the J2 (nearly biased at $\Phi = 0$) and causes J3 to switch and the circulating current is transferred to J4. The clockwise circulating current is representative of a stored "1". Then, when a clock pulse (reset) is applied, it passes through L1 and J1 and into J4, thus causing it to switch. The voltage pulse developed during the switching reverses the circulating current, so again a "0" is stored in the loop; it simultaneously applies this SFQ voltage pulse to the output inductor L3.

The junctions J1 and J2 have lower critical currents than J3 and J4 and to protect the inputs from back reaction of the interferometer if pulses come under the wrong circumstances.

In the RSFQ RS Flip Flop cell as shown in figure 3a, the magnetic bias is created by asymmetrically applying bias current I_b . This magnetic bias disappears if the bias current is switched off. As a result the circuit keeps its internal state only as long as the bias current remains applied.

In figure 3b the transformation for the RS flip-flop into Power Independent cell is shown. The operation of the Power independent RS flip flop operates in the similar manner as the conventional RSFQ RS flip flop, the junction and inductance parameters have to be adjusted accordingly. In contrast PI cell (Figure 3b) holds its magnetic bias inside its SQUID, instead of a single quantizing inductance L. To activate the SQUID and the circuit one should apply large enough bias current I_b. (Note that this "activating" current is slightly greater than its nominal value for regular logic operation.) Being activated the circuit remains magnetically biased (presumably by flux about $\Phi_0/2$) even if bias current is off. Note that even power independent circuits should be powered to perform logic operations, in order to provide the needed additional magnetic flux bias.

4.2 Investigations of power independent circuits

In order to investigate the power independent RSFQ flip flop we simulated the RSFQ flip flop. The clocked RS flip flop, where the reset terminal used as the clock terminal can be used as the RSFQ D flip-flop. The circuits were designed based on the simulations to be fabricated for Hypres 1KA/cm^2 Nb trilayer technology.

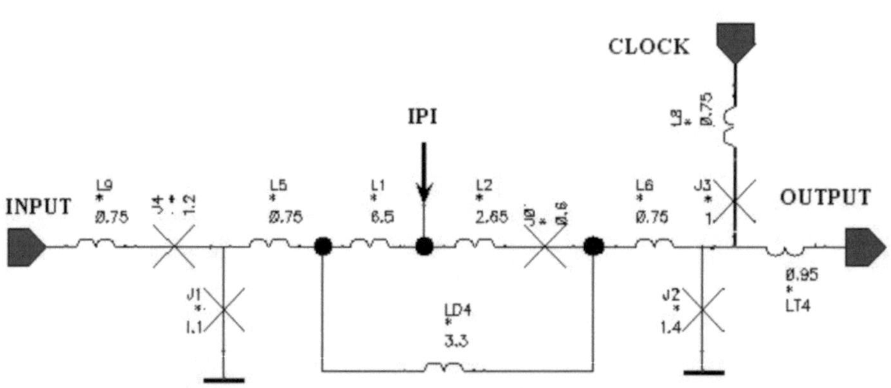

Fig. 4. Power independent D flip-flop(Clocked RS Flip-flop). Bias current IPI does 2 things: it "activates" the SQUID with junction J0 and the power the cell during normal operation. Values of paramerters are shown in dimensionless 'PSCAN' units (Polonsky S. et al., 1991).

4.3 Design of 6-bit shift register with PI cells

Figure 4 shows schematics of a D flip-flop (Clocked RS Flip flop), re-optimized for operation in power independent mode at 4 K. The power independent D flip-flop has the single junction interferometer that can be identified by schematic components L1, L2, LD4 and J0. The interferometer is biased by current IPI. The components in the figure 4 are represented by dimensionless PSCAN units, which are easier for computation.

Figure 5 illustrate current and input data patterns used for a numerical circuit optimization with PSCAN software package. The new feature of the simulation is a more complex shape of applied bias current IPI. During the simulation it was required that junction J0 is switched only one time and when IPI current it applied for the first time. No other junctions switched when bias current goes down.

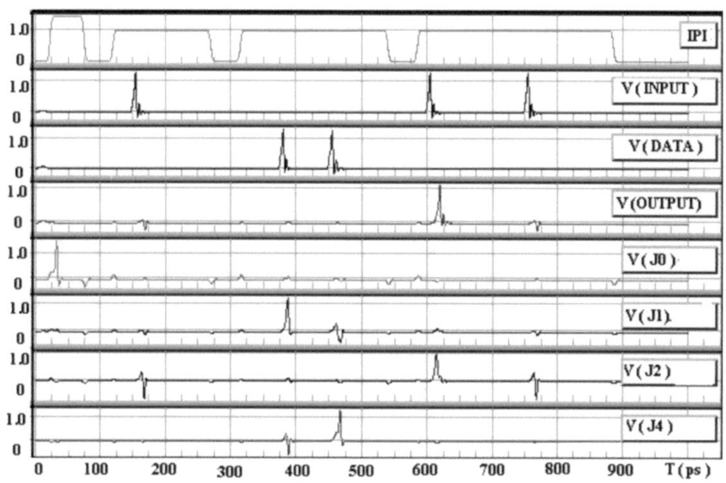

Fig. 5. Current (upper trace) and voltage waveforms illustrating the power independent operation of D cell. Note that the initial "activation" procedure could require larger current IPI than those during the regular circuit operation

A 6-bit shift register with PI D cells has been designed and laid out for HYPRES fabrication technology. The shift register has been incorporated into a benchmark test chip developed for a comparative study of flux trapping sensitivities (Polyakov Y.A. et al., 2007) of different D cells. (The earlier revisions of the test circuits are documented in Narayana S. 2011.)

Figure 6 shows a microphotograph of a fully operational circuit (as shown in Figure 4) fabricated at HYPRES (1 KA/cm2 technology). Bias current margins (±16%) for the only measured chip are about 2 times below our numerical estimations (35%). The figure 7 the shift register was tested with Octupux (Zinoviev D. et al.,) setup where the low speed testing was use to confirm the correct operation of the shift registers. Since the shift register is a counter shift register, the clock and the data pulses travel in the opposite direction and this can be confirmed by the traces in figure 5. The chip was not tested for high speed operation

These measurements show a complete operation of the circuit but with about ±16% bias current margins that are more than 2 times below of our numerical estimations (35%). We believe that the discrepancy is mostly because of the large number (over a dozen) of corners in the SQUID loop. This is because we believe that the inductances for the corners in the loop have been overestimated (Narayana S 2011).

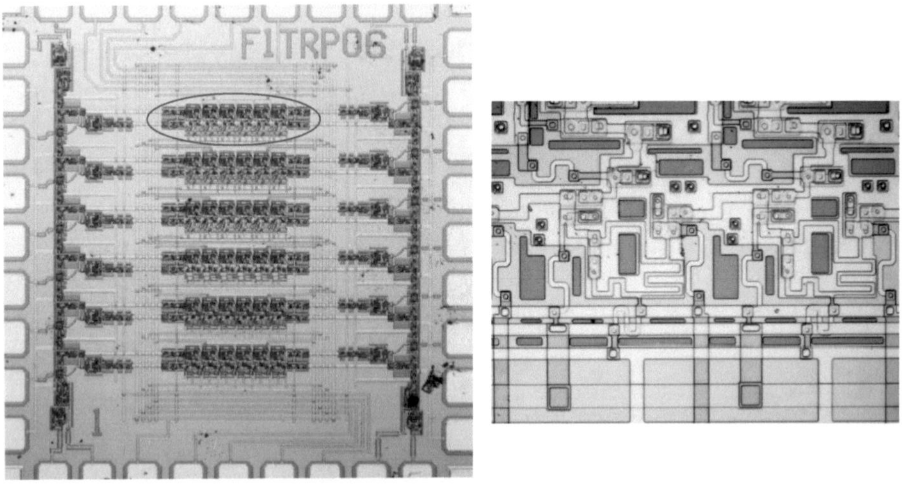

Fig. 6. The microphotographs of the chip and 2 cell fragment of the shift register (encircled on the left) consisting of power independent RSFQ D cells. Horizontal pitch of the 2 cells is 270μm.

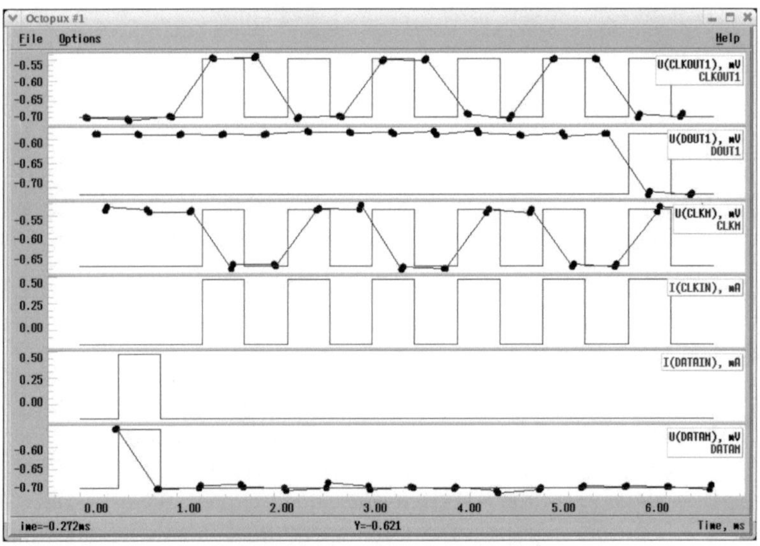

Fig. 7. The low frequency operation of counter flow shift register. The traces 3(clock), 5(data) are the inputs and 1, 2 are the clock and data output traces respectively.

5. Current recycling

One of the main advantages of RSFQ circuit is that only dc bias is needed. It eliminates the cross-talk problems caused by ac biasing and makes designing larger circuits easier. However, in larger circuits the total dc bias current could add up to a few amperes and such large bias currents cause large heat dissipation, which is not preferred (for larger modular designs the bias currents could add up several amperes).One of the techniques that has been proposed (Kang J.H. 2003) is biasing the circuits serially otherwise commonly known as 'current recycling'. Biasing large circuit blocks in series (referred to as current recycling or current re-use) will essentially reduce the total current supply for the superconducting IC to a manageable value. Both capacitive (Teh C.K., et al., 2004) and inductive methods (Kang J.H., et al., 2003) of coupling for current recycling have been demonstrated at a small scale. It is difficult to estimate the impact of the technique based on single gate operation as in (Johnson MW et al, 2003). Current recycling becomes easier at higher current density of the superconducting IC (Narayana S 2011). Current recycling however has its limitations; it does not reduce the on-chip static power dissipation by the circuit blocks and also due to additional structures, the area occupied by the circuit increases.

To demonstrate the method of current recycling, we have designed a Josephson junction transmission line (JTL) as shown in figure 8, which represents one module. The module consists of three parts, the driver, receiver and the payload. The payload is usually the circuit block that is used for operation, in this case to keep matters simple a JTL has been used, as its operating margins are very high. The payload can otherwise be replaced by flip-flops, filters, or logic gates.

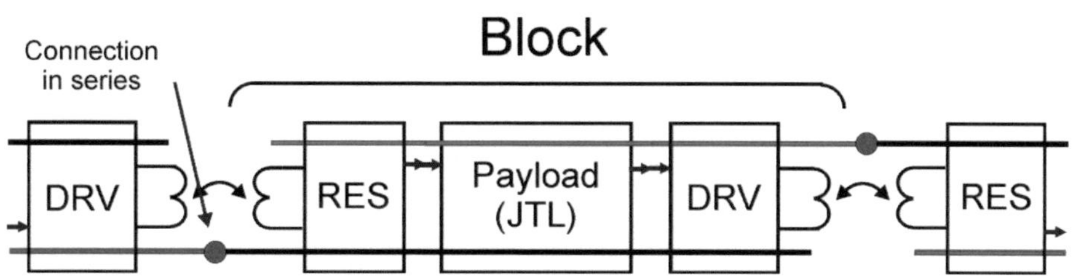

Fig. 8. Block diagram of current recycling digital transmission line

Before we explain the operation of the driver-receiver circuit, it is important to remember a few thumb rules for the current recycling design. For current recycling, the ground planes under adjacent circuit blocks must be separated and subsequent blocks biased in series. It will also be necessary to isolate SFQ transients between adjacent blocks. This may be achieved by low pass filters, but will need to avoid power dissipation in the filters. Series inductance could provide high frequency isolation; the inductors could be damped by shunting with suitable resistance, such that there is no large DC power dissipation. Capacitive coupling between adjacent blocks can be used for current recycling however they are not discussed here and also capacitors (Teh C.K., et al, 2004) used for this method also occupy larger space compared to the inductive filtering method.

5.1 Current recycling basics

The fundamental requirement for serial biasing of circuits is that current drawn from (supplied to) each circuit must be equal and the input/output must not add current to the serially biased circuits. The inputs and outputs are connected via galvanic connection to satisfy the above requirements.

In figure 9, the complete schematic of the driver- receiver is shown. The driver and receiver circuits are completely different electrical grounds. An inductor connecting two Josephson junctions momentarily stores a single flux quantum while an SFQ pulse propagates from one junction to another. Typically, this duration time is about 5picoseconds, depending on the circuit parameters. Between the time when J13 and J14 generate a voltage pulse, the magnetic flux stored in the inductor that connects J13 and J14 induces a current in the inductor, L1u, connecting J1 and J2. With proper circuit parameters, this induced current causes a voltage pulse to be created on J1 and this pulse then propagates through J2 to be further processed. In this way, an SFQ signal pulse is transferred from one ground to another plane.

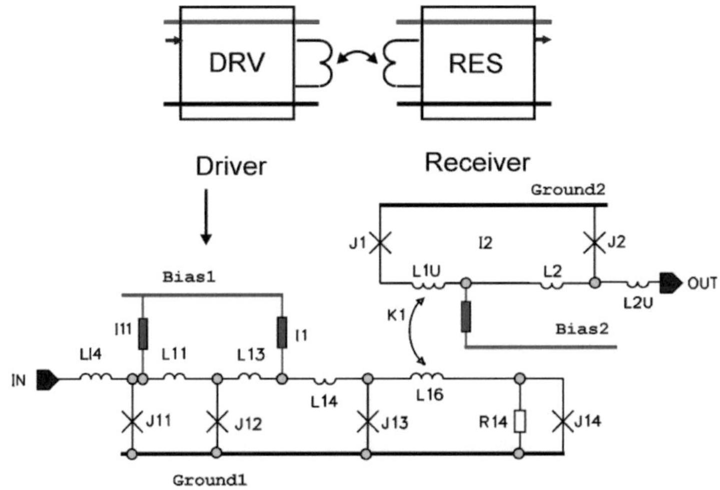

Fig. 9. Circuit schematic for magnetic coupled SFQ pulse transfer between driver and receiver

5.2 Current recycling experimental demonstration

The complete block diagram for the circuit is shown in figure 11a along with the connection scheme for the 80 blocks to be biased serially. Figure 11b shows the microphotograph of the chip which was fabricated for the circuit schematics discussed in of figure 8 and 9. The bias Current for the junction on the input side is passed to one ground plane while the ground for the junctions on the output side is isolated from the other ground by a ground plane moat. The Josephson junction J13 and J14 are damped more heavily than other junctions to guarantee that minimum reflections take place at the end of the input JTL. Tight magnetic coupling is required between the pulse transmitting the JTL and the pulse receiving JTL to obtain a robust circuit with excellent operational margins. To ensure higher coupling holes were opened in both the upper and lower ground planes as shown in figure 10.

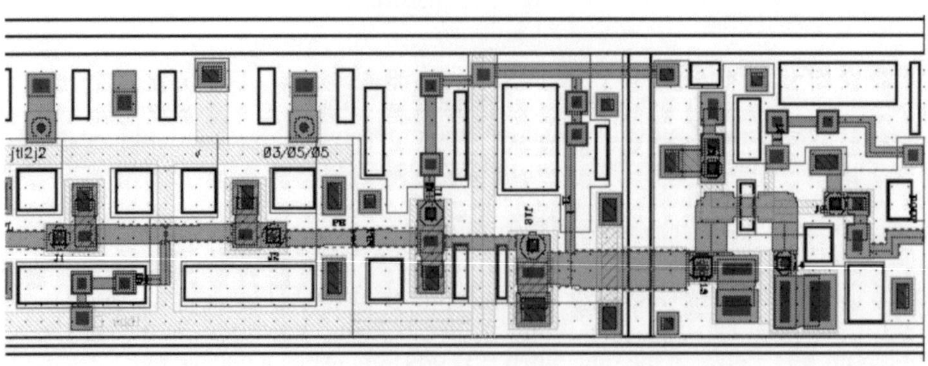

Fig. 10. The layout showing the JTL-driver-receiver connections is shown.

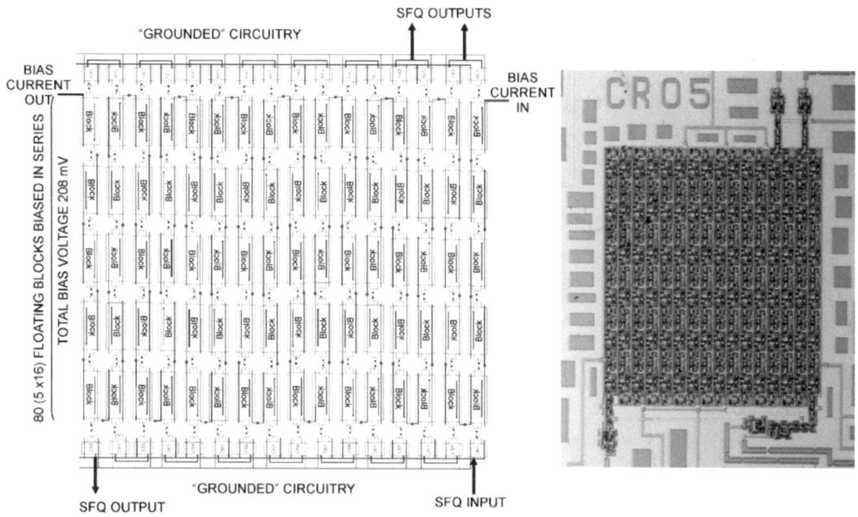

(a) (b)

Fig. 11. (a) The block diagram for serial biasing.(b)The microphotograph of the chip for demonstrating current recycling.

The digital traces for the correct operation of the circuit are shown in figure 12. The measurements were carried out at low frequency using Octupux setup (Zinoviev 1997). The circuit tested used the standard I/O blocks of SFQ/DC converters to measure the operating margins of the circuits. The circuits were fabricated for both 1kA and 4.5KA/cm^2 Hypres tri-layer Niobium technology. The circuit has margins of ±15%. The bias current to obtain correct operation was reduced to 1.7mA by current recycling method; otherwise the operation of 80 blocks with parallel biasing would require nearly 200mA.

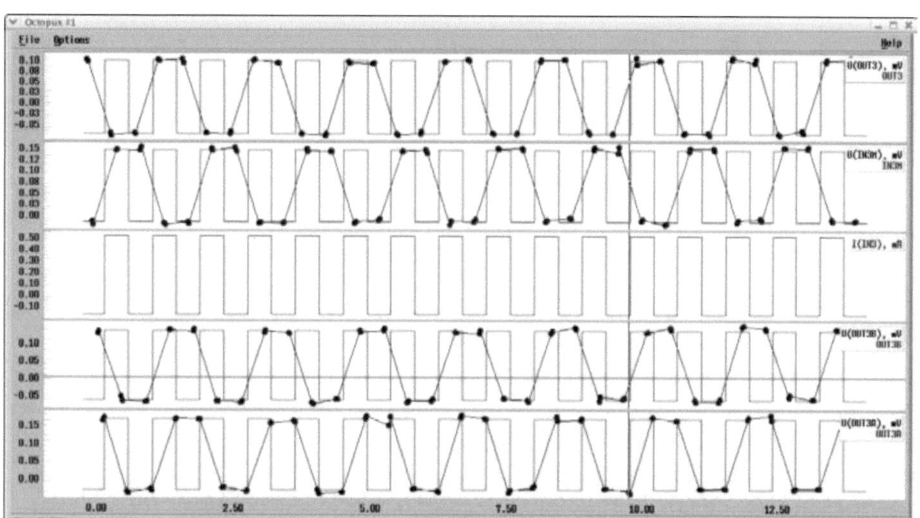

Fig. 12. The digital waveform of the input and three outputs as shown in figure 5.4. The traces 1,3,5 are the outputs and 2 is the input trace

6. Discussions and summary

In sections 4 and 5, we have demonstrated the operation of power independent RSFQ cell and current recycling technique for over 1k junctions in a single chip.. Now let us consider the latter case. If N blocks were parallel biased (that is individually biased), the power dissipated would be

$$Pp = \sum_{i=1}^{N} Ib_i^{\ 2}.R_b \tag{4}$$

For the serially biased case

$$Ps = Ib_i^2.R.N \tag{5}$$

Comparing the two cases for N uniform cells, the ratio of Pp/Ps is N. So essentially, one can reduce the bias current by maximum of N times by serial biasing scheme. However, one should note that, this scheme cannot reduce the on chip power dissipation but only reduce the total bias current load, which could prove very significant in designing large circuits.

In the power independent mode the cells can be turned on only when the cells have to be operated and can be turned off, rest of the time. Also they retain the logic state of the circuit, when they are switched off so one can eliminate static power dissipation by this method. In both the schemes discussed, we note that there is a significant increase in area overhead (about 30%).

In this chapter, we have presented solutions for energy minimization in single flux quantum circuits. We have also presented a method for scaling resistances to modify existing SFQ based circuits to fit designs for quantum computation. We have also presented some of the short comings of the proposed methods, giving us an avenue for further research in the areas to make the proposed methods more widely acceptable for application in quantum computing and high performance mixed signal circuits.

7. Acknowledgements

The authors would like to thank Yuri. A. Polyakov for his assistance with the experiments. The authors would like to thank HYPRES, Inc for fabrication of the chips and we would also like to thank Sergey Tolpygo, Oleg Mukahnov and Alex Kirichenko, all of Hypres Inc, for useful comments related with the fabrication and low power designs. We would also like to thank J. Pekkola for fruitful discussions related to heat dissipation at low temperatures.

8. References

Chen W., Rylyakov A.V., Patel V., Lukens J.E. and Likharev K.K. (1999).Rapid single flux quantum T-flipflop operating up to 770 GHz, *IEEE Trans. Appl. Supercond.,* Vol. 19, No. 1,(Jun 1999),pp. 3212-3215 , ISSN: 1051-8223.

Johnson M.W., Herr Q.P., Durand D.J. and Abelson L.A. (2003). Differential SFQ transmission using Either inductive or capacitive coupling, *IEEE Trans. Appl. Supercond.,*Vol. 13,No. 2,(Jun 2003),pp. 507-510 , ISSN: 1051-8223.

Kang J.H. and Kaplan S.B. (2003). Current recycling and SFQ signal transfer in Large scale RSFQ circuits, *IEEE Trans. Appl. Supercond.*,Vol 13, No. 2,Jun 2003,pp. 547-551 , ISSN: 1051-8223

Kirichenko D.E., Sarawana S. and Kirichenko A.E. (2011). No static power dissipation biasing of RSFQ circuits, *IEEE Trans. Appl. Supercond.(in press)*, ISSN: 1051-8223.

Likharev K.K. and Semenov V.K. (1991).RSFQ Logic/Memory Family: A new Josephson Junction Technology for sub-Terahertz-clock-frequency digital systems, *IEEE Trans. Appl. Supercond.*, Vol 1, No. 1, (Mar 1991),pp. 3-28, ISSN: 1051-8223.

Narayana S., Polyakov Y.A. and Semenov V.K. (2009). Evaluation of Flux trapping in superconducting circuits, *IEEE Trans. Appl. Supercond.*, Vol. 6, No. 3,(Jun 2009),pp. 640-643 , ISSN: 1051-8223.

Narayana S. (2011). Large scale integration issues in superconducting circuits, Proquest dissertations and theses, ISBN 9781124042114.

Nielsen M.A. and Chuang I.L. (2000).*Quantum Computation and Quantum Information*, Cambridge University Press, Cambridge, ISBN 0521635039, USA.

Oleg A.M., Kirichenko D, Vernik I.G, Filippov T. Kirichenko A, Webber R. Dotsenko V., Talalaevskii A., Tang J.C., Sahu. A, Shevchenko P., Miller R., Kaplan S., and Gupta D (2008)., Superconductor Digital RF Receiver Systems., IEICE Trans. Electron., Vol. E91-C, No. 3, (Mar 2008), pp 306-317, ISSN 0916-8516

Polonsky S., Semenov V.K. and Shevchenko P. (1991). PSCAN: Personal superconductor circuit analyzer, *Supercond. Sci. Technol.*, Vol.4, No. 4, pp 667-670, ISSN 0953-2048

Polonsky S. (1999).Delay insensitive RSFQ circuits with zero static power dissipation, *IEEE Trans. Appl. Supercond.*,Vol. 19, No. 2, (Jun 1999), pp. 3535-3539 , ISSN: 1051-8223

Polyakov Y.A., Narayana S. and Semenov V.K. (2007). Flux trapping in superconducting circuits. *IEEE Trans. Appl. Supercond,*, Vol. 6, No. 3,(Jun 2007),pp. 520-525 , ISSN: 1051-8223

Ren J., Semenov V. K., Polyakov Y.A. , Averin D.V. and Tsai J.S. (2009). Progress Towards Reversible Computing With nSQUID Arrays. *IEEE Trans. Appl. Supercond.*,Vol. 6, No. 3,(Jun 2009),pp. 961-967 , ISSN: 1051-8223

Rylyakov R.V. (1997). Ultra-low-power RSFQ Devices and Digital Autocorrelation of Broadband Signals. PhD dissertation State University at New York at Stony Brook.

Savin, A.M. Pekola, J.P. Holmqvist, T. Hassel, J. Gronber L., Helisto P. and Kidiyarova-Shevchenko A (2006). High-resolution superconducting single-flux quantum comparator for sub-kelvin temperatures, *Appl. Phys. Lett.*, Vol 89, pp. 133505, ISSN 0003-6951.

Silver A.H. and Herr Q.P. (2001).A New concept for ultra-low power and ultra-high clock rate circuits, *IEEE Trans. Appl. Supercond.*, Vol 11, No. 1,(Mar 2001),pp. 333-337 , ISSN: 1051-8223

Silver A.H., Bunyk P., Kleinsasser A. and Spargo J. (2006).Vision for single flux quantum very large scale integrated technology, *Supercond. Sci. Technol.*, Vol.12, No. 15, pp 307-311,ISSN 0953-2048

Tahara S., Yorozu S., Kameda Y., Hashimoto Y., Numata H., Satoh T., Hattori W., and Hidaka M. (2004).Superconducting digital electronics," *Proceeds. IEEE* , Vol. 92, No 10,(Oct 2004) ,pp 1549-58, ISSN: 1051-8223.

Teh C.K., Enomoto Y. and Okabe Y. (2003). Current recycling technique using capacitive grounding, *Supercond. Sci. Technol.*, Vol.16, No. 12, pp 1513-1517,ISSN 0953-2048

Turner C. W. and Van Duzer T. (1998). *Princples of Superconducting devices and circuits* (2nd Edition), Prentice Hall, ISBN-10: 0132627426, USA

Yoshikawa N and kato Y (1999). Reduction of power consumption of RSFQ circuits by inductance-load biasing, *Supercond. Sci. Technol.*, vol.12., No. 15,pp 918-920,ISSN 0953-2048

Zinoviev DY and Polyakov YA (1997). Octopux: An advanced automated setup for testing superconductor circuits, *IEEE Trans. Appl. Supercond.*,Vol 7, No. 2, (Jun 1997),pp. 3240-3243, ISSN: 1051-8223.

6

Electronic Transport in an NS System With a Pure Normal Channel. Coherent and Spin-Dependent Effects

Yu. N. Chiang (Tszyan)
B.Verkin Institute for Low Temperature Physics and Engineering, National Academy of Sciences of Ukraine, Kharkov 61103
Ukraine

1. Introduction

The proved possibility really to observe quantum-interference phenomena in metals of various purity, in conditions when the scattering occurs mainly at static defects and the electron mean free path, l_{el}, is much less than the size L of the investigated sample, convinces that the phase φ of electron wave functions does not break down at elastic (without changing electron energy) scattering. Even in very pure metals with $l_{el} \approx 0.1$ mm, at $T \leq 4$ K, collisions of electrons with phonons occur much less frequently than those with static defects occur, so that the role of the former in electron kinetics becomes minor. In other words, the inequality $L_\varphi > l_{el}$ is usually satisfied in a metal at sufficiently low temperatures, where L_φ is the phase-breaking length. This condition, however, is not sufficient to observe coherent phenomena. For example, the phenomena of interference nature such as the oscillations in conductance in a magnetic flux ϕ in normal - metal systems (Sharvin & Sharvin, 1981) or coherent effects in hybrid systems "normal metal/superconductor" (NS) (Lambert & Raimondi, 1998), can become apparent given certain additional relations are fulfilled between the parameters, which define the level of the effects: $l_{el} \leq L \leq \xi_T \leq L_\varphi$. Here, ξ_T is the thermal coherence length. Otherwise, at reverse inequalities, a fraction of the coherent phenomena in total current is expected to be exponentially small. It follows from the general expression for the phase-sensitive current: $J(\phi) \sim F(\phi)(l_{el}/L)\exp(-L/\xi_T)\exp(-L_\varphi/\xi_T)$ ($F(\phi)$ is a periodic function).
The majority of coherent effects, as is known, is realized in experiments on mesoscopic systems with typical parameters $L \sim 1$ μm $\gg l_{el} \sim 0.01$ μm, i. e., under conditions where the level of the effects is exponentially small, but still supposed to be detected (Lambert & Raimondi, 1998; Washburn & Webb, 1986). Really, the maximum possible (in the absence of scattering) spatial coherence length, ξ_0, according to the indeterminacy principle, is of the order of $\xi_0 = (\hbar v_F/k_B T) \sim 1$ μm in value, for a pair of single-particle excitations in a normal metal (for example, for $e-h$ hybrids resulting from the Andreev reflection (Andreev, 1964)). That value is the same as a typical size of mesoscopic systems. As a consequence, from the ratio $\xi_T \sim \sqrt{(1/3)\xi_0 l_{el}}$ it follows that in these systems, coherent effects are always realized under conditions $L \gg \xi_T \gg l_{el}$ and are exponentially small on the L scale.

From the above typical relations between the basic microscopic parameters in mesoscopic systems one can see that such systems cannot give an idea about a true scale of the most important parameter - the phase-breaking length L_φ for electron wave functions. It is only possible to say that its value is greater than $L \sim 1$ μm. In mesoscopic systems, a value of the thermal coherence length ξ_T remains not quite clear as well, since, under strong inequality $\xi_T \gg l_{el}$, the scale of this parameter should be additionally restricted: At high frequency of elastic scattering processes on impurity centers (at short l_{el}), the portion of inelastic scattering on the same impurities, which breaks down a phase of wave functions, should be also significant.

It is clear that due to spatial restrictions, mesoscopic systems are also of little use for experimental investigation of non-local coherent effects; a keen interest in those effects has recently increased in connection with the revival of general interest in non-local quantum phenomena (Hofstetter et al., 2009).

Our approach to the investigation of phase-breaking and coherence lengths in metals is based on an alternative, macroscopical, statement of the experiment. We presume that in order to assess the real spatial scale of the parameters L_φ, ξ_T, and ξ_0 in metals, the preference should be given to studying coherent effects, first, in pure systems, where the contribution from inelastic scattering processes is minimized due to the lowered concentration of impurities, and, second, at such sizes of systems, which would certainly surpass physically reasonable limiting scales of the specified spatial parameters. The listed requirements mean holding the following chain of inequalities: $L > L_\varphi > l_{el} \gg \xi_0$ $(l_{el}/\xi_0 \geq 10)$. They can be satisfied by increasing the electron mean free path l_{el} and the system size L by several orders of magnitude in comparison with the same quantities for mesoscopic systems.

At first glance, such changing in the above parameters should be accompanied by the same, by several orders, reduction in the value of registered effects. Fortunately, this concerns only normal-metal systems, where coherent effects have a weak-localization origin (Altshuler et al., 1981). The remarkable circumstance is that the value of coherent effects in normal (N) and NS systems can differ by many orders of magnitude in favor of the latter. Thus, oscillation amplitude of the conductance in a magnetic field in a normal-metal ring (the Aharonov - Bohm effect in a weak - localization approach (Altshuler et al., 1981; Washburn & Webb, 1986)) can be m times less than that in a ring of similar geometry with a superconducting segment (NS - ring) due to possible resonant degeneration of transverse modes in the Andreev spectrum arising in the SNS system (Kadigrobov et al., 1995). For example, for mesoscopic rings $m \sim 10^4 \div 10^5$. As it will be shown below, coherent effects in macroscopical formulation of experiments remain, nevertheless, rather small in value, and for their observing, the resolution of a voltage level down to 10^{-11} V is required. Unlike mesoscopic statement of the experiment, it makes special non-standard requirements for measuring technique of such low signals. To satisfy the requirements, we have developed the special superconducting commutator with picovolt sensitivity (Chiang, 1985).

Here, we describe the results of our research of quantum coherent phenomena in NS systems consisting of normal metals with macroscopical electron mean free path and having macroscopical sizes, which fact allows us to regard our statement of the experiment as macroscopical. The phenomena are considered, observed in such systems of different connectivity: for both simply connected and doubly connected geometry.

In Section 2, phase-sensitive quantum effects in the "Andreev conductance" of open and closed macroscopic SNS systems are briefly considered. The open SNS systems contain segments, up to 350 μm in length, made of high-pure ($l_{el} \sim 100$ μm) single-crystal normal metals Cu and Al which are in contact with In, Sn, and Pb in the superconducting or intermediate state. The phase-sensitive magnetoresistive oscillations are described, with a period equal to the flux quantum $hc/2e$, which were found in hybrid quasi-ballistic doubly connected SNS structures with single-crystal normal segments of macroscopic sizes ($L = 100 \div 500$ μm) and elastic electron mean free path on the same scale. The description of resistive oscillations of a resonance shape for the structure In(S) - Al(N) - In(S) of the similar size is provided. The oscillations undergo a phase inversion (a shift by π) with respect to the phase of the nonresonance oscillations.

In Section 3, the results of studying coherent and spin-dependent effects in the conductance of macroscopic heterosystems "magnetic (Fe, Ni) - superconductor (In)" are presented. The first proof of the possibility of observing, with adequate resolution, the characteristic coherent effect in the conductivity of sufficiently pure ferromagnets was given on the example of nickel. The effect consists in an interference decrease in the conductivity on the scale of the very short coherence length of Andreev $e - h$ hybrids. It was shown that this length did not exceed the coherence length estimated using the semiclassical theory for ferromagnetic metals with high exchange energy. Additional proof was obtained for spin accumulation on F/S interfaces. This accumulation comes from the special features of the Andreev reflection under the conditions of spin polarization of the current in a ferromagnet.

The first observation of the $hc/2e$-oscillations (solid-state analogue of the Aharonov-Bohm effect (Aharonov & Bohm, 1959)) is described in the conductance of a ferromagnet, Ni, as a part of the macroscopical S(In) - F(Ni) - S(In) interferometer ($L^{\mathrm{Ni}} \sim 500$ μm). A physical explanation is offered for the parameters of the oscillations observed. We have found that the oscillation amplitude corresponds to the value of the positive resistive contribution to the resistance from a ferromagnetic layer, several nanometers thick, adjacent to the F/S interface. We have demonstrated that the scale of the proximity effect cannot exceed that thickness. The oscillations observed in a disordered conductor of an SFS system, about 1 mm in length, indicate that the diffusion phase-breaking length is macroscopical in sufficiently pure metals, including ferromagnetic ones, even at not too low helium temperatures. The analysis of the non-local nature of the effect is offered.

Section 4 is the Conclusion.

2. Macroscopical NS systems with a non-magnetic normal-metal segment

While trying to detect possible manifestations of quantum coherent phenomena in conductivity of normal metals, the main results were obtained in experiments on the samples of mesoscopic size under diffusion transport conditions, $L \geq \xi_T \gg l_{el}$. In such case, the contribution from the coherent electrons is exponentially small relative to the averaged contribution from all electrons in all distributions. The portion of coherent electrons can be increased due to weak-localization effect. For example, in a doubly connected sample, so-called self-intersecting coherent trajectory of interfering electrons is artificially organized (Washburn & Webb, 1986). If the length of the loop, L, covering the cavity of the doubly connected system does not exceed the phase-breaking length, L_φ, then introducing a magnetic field into the cavity may lead to a synchronous shift of the phase of wave functions of all electrons. As a result, the conductance of the system, determined by a superposition of these functions, will oscillate periodically in the magnetic field. The amplitude of the oscillations will be defined by the weak-localization contribution from interfering reversible self-intersecting transport trajectories (Aharonov & Bohm, 1959; Sharvin & Sharvin, 1981), and the period will be twice as small as that for the conventional Aharonov-Bohm effect in a

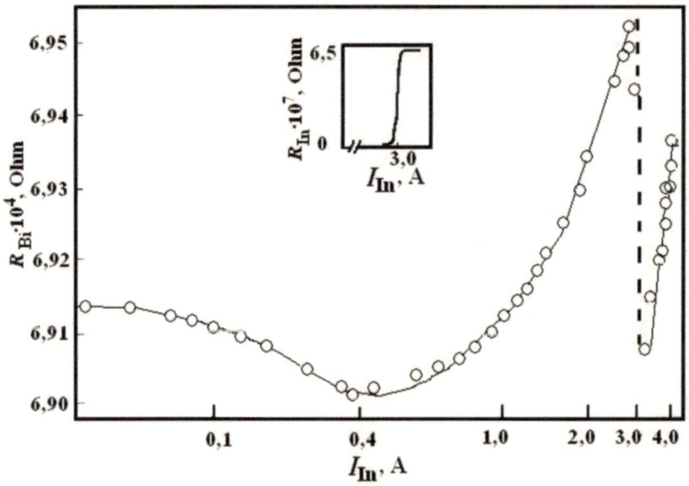

Fig. 1. Positive jump of resistance of bimetallic system Bi/In at converting indium into the superconducting state. Inset: Superconducting transition of In.

normal-metal ring, where the reversibility of the trajectories in the splitted electron beam is not provided.

2.1 Singly connected *NS* systems

2.1.1 Artificial *NS* boundary

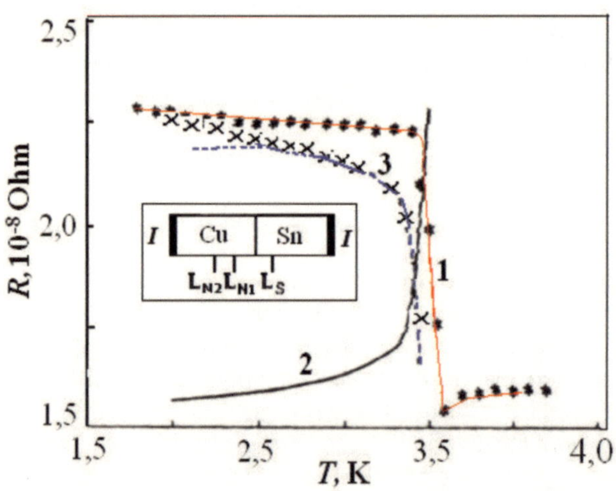

Fig. 2. Positive jump of resistance of bimetallic system Cu/Sn measured on the probes L_{N1}; L_{N2} at converting tin into the superconducting state (curve *1*); calculated portion of the boundary resistance (curve 2); crosses are experimental points excluding curve 2; dotted curve *3* is calculated in accordance with Eq. (4). Inset: Schematic view of the sample.

As it has been noted in Introduction, in comparison with weak-localization situation, the role of coherent interference repeatedly increases in the normal metal, which is affected by Andreev reflection - the mechanism naturally generating coherent quasiparticles. In a hybrid "normal metal/type I superconductor" system (*NS* system), the contribution from the coherent excitations into normal conductivity dominates over the distance scale of the order of a ballistic path from the *NS* boundary ($\sim 1\mu$m), irrespective of the system size, its connectivity, and, generally, the electron mean free path; due to Andreev reflection, the spectrum of coherent excitations is always resolved on this scale. From the discussion in Introduction it clearly follows that in macroscopical statement of experiments, maximum level of the coherent effects should be reached in conditions, where the electron mean free path, l_{el}, the greatest possible coherence length, ξ_0, and the sample length, L (the separation between potential probes), are of the same order of magnitude. In these conditions, when studying even singly connected *NS* systems in 1988, we first revealed an unusual behavior of the normal conductivity of the heterosystem Bi(N)/In(S) (Chiang & Shevchenko, 1988): The resistance of the area containing the boundary between the two metals unexpectedly decreased rather than increased at the transition of one of the metals (In) from the superconducting into normal state (Fig. 1).

Further theoretical (Herath & Rainer, 1989; Kadigrobov, 1993; Kadigrobov et al., 1995) and our experimental research have shown that the effect is not casual but fundamental. It accompanies diffusive transport of electrons through non-ballistic *NS* contacts. Figure 2 presents some experimental data revealing the specific features of coherent excitation scattering in the vicinity of the *NS* boundaries (see more data in (Chiang & Shevchenko, 1998)) for Cu(N)/Sn(S) system (schematic view of the sample is shown in the Inset). The basis of the bimetallic *NS* system under investigation was a copper single crystal with a "macroscopically" large elastic mean free path $l_{el}^{N} \simeq 10 - 20$ μm. The single crystal was in contact with a type I superconductor (tin) ($l_{el}^{S} \simeq 100$ μm). The transverse size of contact areas under probes was 20-30 μm so that tunnel properties were not manifested in view of the large area of the junction. Separation of the N-probes from the boundary L_{N1}, L_{N2}, and L_S were 13, 45, and 31 μm, respectively. The curve *1* shows a general regularity in the behavior of the resistance of

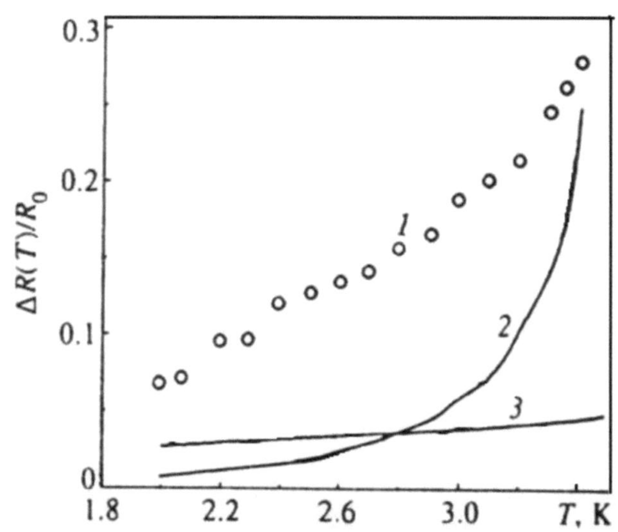

Fig. 3. Resistance of the region of the Cu/In system incorporating the *NS* boundary, below T_c of the superconductor (In): Experimental points (curve *1*) and calculated contributions of the boundary resistance ($z \neq 0$, curve 2) and of the proximity effect (curve *3*).

normal regions adjoining the common boundary of two contacting metals - occurrence of the positive contribution to the resistance closely related to the temperature dependence of the superconducting gap at transition of one of the metals into the superconducting state. This effect is most pronounced just in the macroscopical statement of experiment in the formulated above optimum conditions $l_{el}^N \sim \xi_{0(N)} \sim L_N$.
The considered effect was predicted in (Herath & Rainer, 1989) and (Kadigrobov, 1993). It has been shown that there exists a correction to the normal resistance (hereafter, $\delta R_N^{\mathrm{Andr}}$) leading to an increase in the metal resistance within ballistic distances from the NS boundary upon cooling. The correction may occur due to increased cross section of electron scattering by impurities during multiple interaction of phase-coherent electron and Andreev excitations with impurities and with the NS boundary. According to (Kadigrobov et al., 1995), the relative

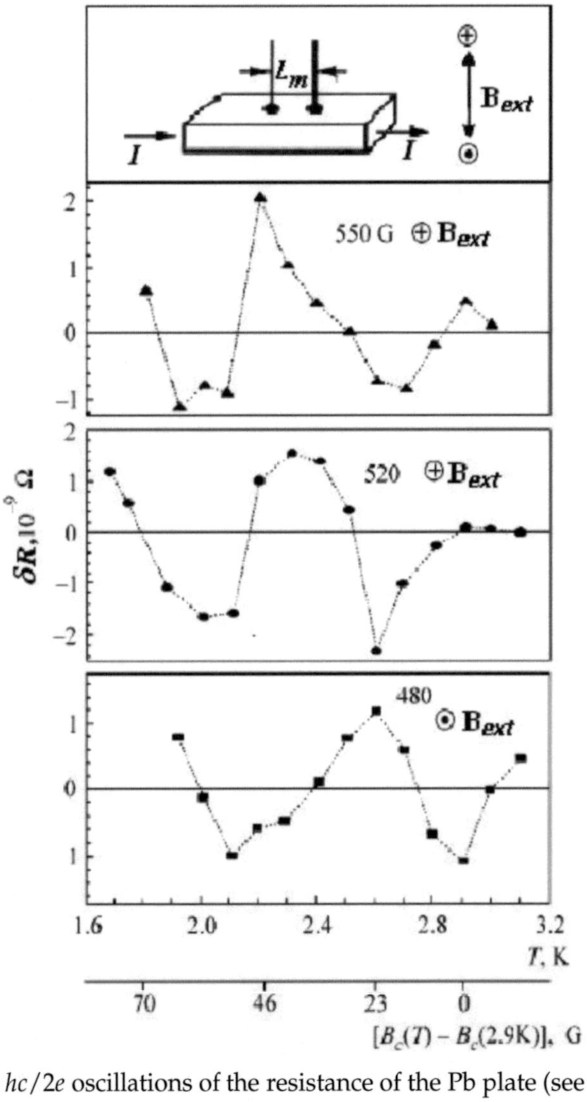

Fig. 4. Temperature $hc/2e$ oscillations of the resistance of the Pb plate (see the outline above) in the intermediate state.

increase in the resistance of a layer, L_N in thickness, measured from the NS boundary and having a resistance R_N prior to the formation of this boundary, should be equal to

$$\frac{\delta R_N^{\text{Andr}}}{R_N} = (l_{el}^N/L_N)\{T_p\}, \tag{1}$$

where $\{T_p\}$ is the effective probability of electron scattering by a layer of thickness of the order of "coherence length" ξ_T taking into account Andreev reflection and the conditions $l_{el}^N \sim \xi_N^* \sim L_N$. The quantity $\{T_p\}$ can be obtained by integrating $T_p = \hbar v_{\text{F}}/\varepsilon l_{el}^N$, viz., the probability that the particle is scattered by an impurity and reflects as an Andreev particle with energy ε (measured from the Fermi level), thus contributing to the resistance over the length l_{el}^N; the integration is over the entire energy range between the minimum energy $\varepsilon_{min} = \hbar v_{\text{F}}/l_{el}^N$ and the maximum energy of the order of the gap energy $\Delta(T)$:

$$\{T_p\} = \int_{\varepsilon_{min}}^{\Delta(T)} (-\frac{\partial f_0}{\partial \varepsilon})T_p d\varepsilon. \tag{2}$$

Integration to a second approximation gives the following analytical result for the correction to the resistance of the layer L_N under investigation as a function of temperature:

$$\frac{\delta R_N^{\text{Andr}}}{R_N} = \frac{\xi_T}{L_N}F(T), \tag{3}$$

where $F(T)$ is of order of unity with $\xi_T \sim l_{el}^N$. For a pair of probes $(L_{N1}; L_{N2})$ (see Fig. 2) with $L_{N1,2} > l_{el}^N$ it provides

$$\frac{\delta R_{L_{N1};L_{N2}}^{\text{Andr}}}{R_{L_{N1};L_{N2}}} = \frac{\xi_T}{L_{N1} - L_{N2}} \ln(L_{N1}/L_{N2})F(T). \tag{4}$$

Using Eq. (4) we have estimated the data received for different samples, including those presented in Fig. 2. The analysis reveals not only qualitative but also quantitative agreement between the experiment and the concept of increasing the dissipative scattering contribution due to Andreev reflection (dotted curve *3* in Fig. 2). It is thus important to emphasize once again that optimum conditions for observing this effect are realized by setting measuring probes at a distance of several ballistic coherence lengths ξ_0 from the NS boundary, i. e., in the macroscopical statement of experiment.

Curve 2 in Fig. 2 gives an idea of the portion of the boundary resistance which arises due to dissipative quasiparticle current flowing in the areas close to NS borders, where the order parameter $\Delta = \Delta(x)$ is less than $\Delta(\infty) = 1$. Since the condition $eV \ll k_{\text{B}}T$ was satisfied in our experiment we calculated this curve using the CESST-HC theory (Clarke et al., 1979; Hsiang & Clarke, 1980). In accordance with the theory, the boundary resistance, R_b^{NS}, caused by the potential V_b^{NS} extending into the near-boundary region of a superconductor where $\Delta(x) < 1$, should be of the order of

$$R_b^{\text{NS}} = V_b^{\text{NS}}/I = Y(z,T)R^{\text{Cu}}; \qquad R^{\text{Cu}} = \lambda_Q \cdot \rho^{\text{Cu}}/A, \tag{5}$$

where λ_Q is the distance from the boundary on the side of the superconductor, over which the potential decays that arises from the imbalance between the charges of the pair current and

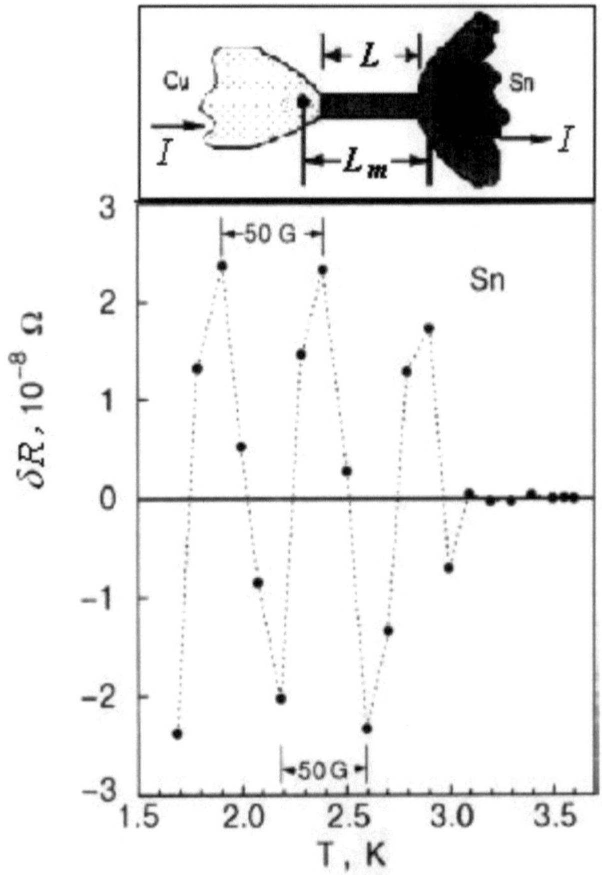

Fig. 5. Temperature $hc/2e$ oscillations of the resistance of the intermediate-state Sn constriction (see the outline above) in a self-current magnetic field of the measuring current $I = 1$ A as the critical magnetic field (50G - intervals are specified).

that of independent quasiparticles; $\rho^{\mathrm{Cu}} = 3/2e^2 N(0) l_{el}^{\mathrm{Cu}} v_{\mathrm{F}}$; $N(0)$ is the density of states per spin at the Fermi level, e is the electron charge, v_{F} is the Fermi velocity, and

$$Y(z,T) = (1+z^2)\frac{k_{\mathrm{B}}T}{\Delta}\sqrt{\frac{2\pi\Delta}{k_{\mathrm{B}}T}}\exp(-\Delta/k_{\mathrm{B}}T). \tag{6}$$

Measurements across the probes $L_{N1}; L_S$, with the NS boundary between them and $L_{N1} > l_{el}^{\mathrm{Cu}}$, and estimation of the boundary resistance of Eqs. (5; 6) indicate a manifestation of such coherent transport mechanism, which leads to an increase in conductivity in these non-ballistic conditions (Fig. 3).

According to the Landauer concept (Landauer, 1970), complete thermalization of an electron is the result not of momentum relaxation but of relaxation of the wave function phase due to inelastic scattering in regions with an equilibrium distribution, called "reservoirs" (regions that are sinks and sources of charges). The simulation of a continuous random walk of an elastically scattered particle in a three-dimensional normal layer of the metal showed the need

to consider the trajectories with multiple Andreev reflections. As shown in (Van Wees et al., 1992), the mean diffusional path length $\langle L \rangle$ depends linearly on the width d of the normal layer, where d is the distance between the boundary and the region of equilibrium distribution (reservoir), which is of the order of magnitude of the inelastic mean free path l_{inel}, i. e., $d \sim l_{inel}$. Since the probability τ_r for excitations to pass from the boundary to the reservoir is inversely proportional to the layer width, $\tau_r \sim (l_{el}/d)$, it follows from the linear relation between $\langle L \rangle$ and d that $\langle L \rangle \sim \tau_r^{-1} \propto l_{inel}$.
It is also necessary to take into account the probability of realizing a diffusional trajectory by equating its length to the elastic mean free path, or, equivalently, equating the length $\langle L \rangle \sim \sqrt{\hbar D/k_B T}$ (D is the diffusion coefficient) to the real length L of the trajectory. On the other hand, $\langle L \rangle = \sqrt{Dt}$ with $t = L/v_F$. Therefore, $\langle L \rangle / L = l_{el}/\langle L \rangle$. In addition, we assume that the only temperature-dependent cause of inelastic scattering is inelastic electron-phonon collisions, with a corresponding mean free path $l_{inel} \sim l_{inel}^{e-ph} \sim T^3$.
As a result, the effective probability for coherent excitations to pass through the phase coherence region in elastic scattering can be written in the form

$$\tau_r = \frac{l_{el}}{l_{inel}} \cdot \frac{l_{el}}{\langle L \rangle} = \beta T^{3.5}, \qquad \beta = l_{el}^{3/2} (\hbar v_F)^{-1/2} (l_{inel}^* T^{*3})^{-1}. \tag{7}$$

In accordance with the Landauer concept, we find the relative contribution to the conductance G in the phase coherence region by calculating the proportion $F(m)$ of coherent trajectories (those that return to the reservoir after m reflections from the boundary, starting with the trajectory with $m = 1$) and their contribution to the current and summing over all trajectories:

$$\frac{\delta G}{G_0} = \sum_{m=1}^{\infty} F(m) I(m), \tag{8}$$

where $\delta G = G - G_0$, $G_0 \equiv G_{T=0}$; $F(m) = \tau_r^2 (1 - \tau_r)^{m-1}$ $(m \neq 0)$.
The probabilistic contribution to the current from a charge on trajectory with reflections is (Blonder et al., 1982; Van Wees et al., 1992)

$$I(m) = 1 + |r_{eh}(m)|^2 |r_{ee}(m)|^2, \qquad |r_{eh}(m)|^2 + |r_{ee}(m)|^2 = 1,$$

where $|r_{ee}(m)|^2$ and $|r_{eh}(m)|^2$ are the probabilities for an electron incident on the *NS* boundary, to leave the boundary after m reflections in the form of an electron wave or a hole (Andreev) wave, respectively. The expression for $I(m)$ shows that for a large enough number of reflections, which increases the probability of Andreev reflection to such a degree that $|r_{eh}(m)|^2 \to 1$, the contribution of the corresponding trajectory to the current increases by a factor of 2. If all of those trajectories reached the reservoir, the dissipation would be increased by the same factor. Formally this is a consequence of the same fundamental conclusion of the theory which was mentioned above: in coherent Andreev reflection the efficiency of the elastic scattering of the electron momentum increases as a result of the interference of the e and h excitations. Actually, the fraction of the coherent trajectories that returns to the reservoir decreases rapidly with increasing distance to the reservoir from the boundary and with increasing number of reflections, which determines the length of the trajectory; thus we have the directly opposite result. In fact, assuming that for low electron energies $(eV/(\hbar v_F/l_{el}) \ll 1)$ and a large contact area the main contribution to the change in conductivity is from coherent trajectories with large numbers of reflections, so that $I(m) \approx 2$, and converting the sum in Eq. (8) to an integral, we find to a second approximation:

$$\frac{\delta G}{G_0} \approx 2 \int_1^{m^*} F(m) dm \approx -\tau_r^2 (m^{*2} \tau_r - 2m^*). \tag{9}$$

The upper limit of integration m^* is the number of reflections corresponding to a certain critical length for a coherent trajectory L that reaches the reservoir. This limit can be introduced as $m^* = \gamma \tau_r^{-1}$ with a certain coefficient γ that is to be determined experimentally. Substituting m^* into Eq. (9), we finally obtain

$$\frac{\delta G}{G_0} \approx -(\gamma^2 - 2\gamma)\tau_r = -AT^{3.5}, \quad A = \beta(\gamma^2 - 2\gamma). \tag{10}$$

The nature of the effect consists in the fact that the number of trajectories leaving from the number of attainable reservoirs increases in the long-range phase coherence region, i.e., an ever greater number of trajectories appear on which the phase of the coherent wave functions does not relax; this decreases the dissipation. Thus, in accordance with Eq. (9), one expects that the temperature dependence of the relative effective resistance measured at the probes located within the phase coherence region will be in the form of a function that decreases with decreasing temperature below T_c as curve *1* in Fig. 3:

$$R/R_N = (R_0/R_N)(1 + AT^{3.5}). \tag{11}$$

2.1.2 Natural *NS* boundary

The discovery of an unusual increase in the resistance of normal conductors upon the appearance of an *NS* boundary (Figs. 1, 2) pointed to the need to deeper understand the properties of the systems with such boundaries. Since then study of the unconventional behavior of the electron transport in such systems has been taken on a broader scope.

As it was noted in Introduction, the early experiments detecting the phase-coherent contribution of quasiparticles to the kinetic properties of normal metals were carried out on samples that did not contain *NS* boundaries. In such a case this contribution, due solely to the mechanism of weak localization of electrons, appears as a small quantum interference correction to the diffusional contribution. Nevertheless, the existence of coherent transport under those conditions was proved experimentally. Study of the *NS* structures containing singly connected type I superconductors in the intermediate state, with large electron elastic mean free path, l_{el}, revealed resistance quantum oscillations of a type similar to the Aharonov-Bohm effect (Tsyan, 2000). The temperature - dependent resistances of Pb and In plates and Sn constriction were studied. The intermediate state was maintained by applying a weak external transverse magnetic field $\mathbf{B}_{ext}$ to the plates and by a self-current field $\mathbf{B}_I$ in the constriction. Thickness of Pb plates was 20 μm, with a separation $L_m \approx 250$ μm between the measuring probes in the middle part of the samples. A rolled In slab with dimensions $L \times W \times t = 1.5$ mm $\times$ 0.5 mm $\times$ 50 μm was soldered at its ends to one of the faces of a copper single crystal and was separated from this face by an insulating spacer. Measurements of the system with In were carried out at a direct current 0.7 A, which self-current magnetic field at a surface of the slab with the specified sizes made 5 G. The tin constriction was $t \approx 20$ μm in diameter and $L \approx 50$ μm in length, with $L_m \approx 100$ μm. At the constriction surface, B_I amounted to ≈ 100 G at $I = 1$ A. The bulk elastic mean free path in the

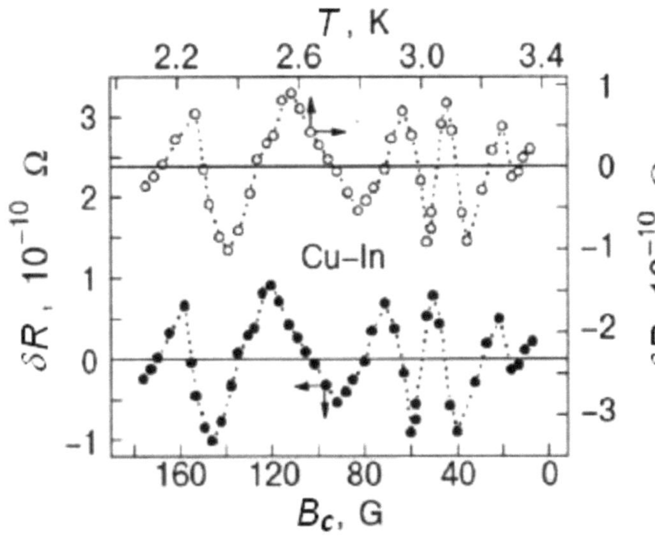

Fig. 6. R - oscillations for the doubly connected Cu-In system as a function of temperature (upper curve) and of the critical magnetic field (lower curve) at temperatures below $T_c^{\rm In}$ in the self-magnetic field ($\sim$ 5 G) of the measuring direct current.

workpieces from which the samples were fabricated was $l_{el} \sim 100$ μm. Such macroscopical value of elastic mean free path in macroscopical statement of experiment makes it necessary to measure samples of $(0.1 \div 1)$ l_{el}^3 in volume, which resistance may make down to $10^{-8} \div 10^{-9}$ Ohm.

Figures 4, 5, and 6 show the oscillatory parts of the current-normalized potential difference δR (hereinafter referred to as R - oscillations), obtained by subtracting the corresponding mean monotonic part for each of the samples. It follows from these graphs that the resistances of the samples oscillate in temperature in the fields maintaining the intermediate state. As is seen, the oscillation amplitude $(\delta R)_{max}$ weakly depends on the temperature and the external magnetic field (although the monotonic resistance components vary over no less than two orders of magnitude). The character of oscillations in the Pb plate at various B_{ext} values (Fig. 4) indicates that the oscillation phase ϕ depends on the strength and sign of the external magnetic field: ϕ(480 G) is shifted from ϕ(550 G) by approximately π, while ϕ(520 G) and ϕ(550 G) coincide.

Constructing the critical-field scale for the oscillation region according to the equation $B_c(T) \simeq B_c(0)[1-(T/T_c(0))^2]$ (T_c is the superconducting transition temperature), one finds that the oscillation period Δ_B in a magnetic field is constant for any pair of points one period apart and is equal to the difference in the absolute values of the critical field (see Figs. 4 and 5) for each of the samples. Here, we used $B_c^{\rm Pb}(0) = 803$ G and $B_c^{\rm Sn}(0) = 305$ G ((Handbook, 1974-1975)). This suggests that the $\Delta_B(B_c)$ period is a function of the direct rather than inverse field. The temperature T^* corresponding to the onset of R-oscillations in the Sn constriction is equal to the temperature for which $B_c^{\rm Sn}(T^*) = B_I(\approx 100$ G), viz., the temperature of the appearance of the intermediate state. The conditions for the confident resolution of the oscillations were fully satisfied for this sample up to 3.5 K. With the values of B_{ext} used for the Pb plate, T^* should lie outside the range of helium temperatures.

It is known that in the intermediate state of a type-I superconductor in a magnetic field, a laminar domain structure arises, with alternating normal and superconducting regions. The observed dependence of the magnitude of the effect on the critical field in the intermediate state, first, provides direct evidence for the presence of a laminar domain NS structure and, second, indicates that the mechanism responsible for the R-oscillations occurs in the normal areas of domains, where, as is known, the magnetic field is equal to the superconductor critical field $B_c(T)$ (De Gennes, 1966). The use of the phenomenological theory of superconductivity (De Gennes, 1966; Lifshitz & Sharvin, 1951) for estimating the number of domains between the measuring probes brought about the values of approximately 12 at 3 K and 16 at 1.5 K for the Pb plate, 1 or 2 for the Sn constriction, and the value of 15-22 mm for the distance d_n between the NS boundaries in the oscillation region of interest. These data suggest the lack of any correlation between the indicated numbers and the number of observed oscillation periods.

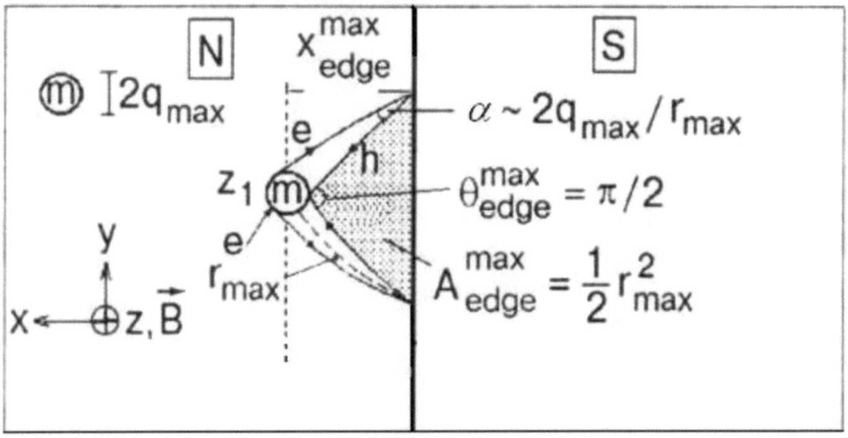

Fig. 7. The criterion of coherent interaction of an electron e and an Andreev hole h with the same elastic scattering center (see Eq. (12) in the text) establishes a distribution of areas A of quantization of the flux of the magnetic vector potential. The maximum admissible area A^{max}_{edge}, bounded by the ballistic trajectories passing through the impurities m of maximum cross section $\sim q^2_{max}$ at the positions $[x^{max}_{edge}, y, z]$ for $\theta^{max}_{edge} = \pi/2$ is separated.

As is known, the direct dependence of the oscillation phase on the field strength arises when the quantization is associated with a real-space "geometric" factor, i. e., with the interference of coherent excitations on the geometrically specified closed dissipative trajectories in a magnetic vector-potential field (Aharonov & Bohm, 1959; Altshuler et al., 1981). At distances of the order of the thermal length $\xi_T \sim \xi_0 \approx \hbar v_F / k_B T$ from the NS boundary, where $l_{el} \gg \xi_0$, the main type of dissipative trajectories are those coherent trajectories on which the elastic-scattering center (impurity) interacts simultaneously with the coherent e (usual) and h (Andreev) excitations (Herath & Rainer, 1989; Kadigrobov, 1993). It was demonstrated in (Herath & Rainer, 1989) that, owing to the doubled probability for the h excitations to be scattered by the impurity, the interference on these trajectories generated the R-oscillations. In the presence of an electric field alone, neither the impurity nor the relevant coherent - trajectory size are set off, so that the oscillations do not arise (Kadigrobov et al., 1995; Van Wees et al., 1992).

Since the e and h trajectories spatially diverge in a magnetic field, the distance r from the impurity to the outermost boundary point, from which the particle can return to the same impurity after being Andreev-reflected, is bounded, according to the simple classical geometric considerations, by the value

$$r = \sqrt{2qR_{\mathrm{L}}[B_c(T)]}. \tag{12}$$

In Eq. (12), R_{L} is the Larmor radius and q is the parameter (of the order of a screening radius) characterizing the impurity size. For instance, in fields of several hundred Gauss, $R_{\mathrm{L}} \approx 1.5 \cdot 10^{-2}$ cm and r does not exceed (1-2) μm at $q \approx (2-5) \cdot 10^{-8}$ cm; i. e., $\xi_0 \leq r \leq \xi_T \sim 10^{-2} l_{el}$ ($l_{el} \gg d_n$, ξ_T; $\xi_T \approx 3$ μm). Therefore, for every impurity with coordinate z, the magnetic field separates in the z = const plane a finite region of possible coherent trajectories passing through the impurity and closing two arbitrary reflection points on the *NS* boundary between the two most distant points which positions are determined by Eq. (12) (see Fig. 7). After averaging over all impurities, only a single trajectory (or a group of identical trajectories) specified by the edge of integration over the quantization area A makes an uncompensated contribution to the wave-function phase. The integration edge $A_{\mathrm{edge}} = (1/2)r^2_{\max}$ corresponds to the area bounded by the trajectory passing through the most efficient (with $\sim q^2_{\max}$) impurity situated at a maximum distance from the boundary, as allowed by criterion (12). One can easily verify that in our samples with $l_{el} \leq 0.1$ mm, every layer of impurity-size thickness parallel to the *NS* boundary comprises no less than 10^3 impurities; i. e., the coherent trajectories corresponding to the integration edge continuously resume upon shifting or the formation of new *NS* boundaries, so that A_{edge} is a continuously defined constant accurate to $\sim q_{\max}/r_{q_{\max}} \sim 10^{-4}$. According to (Aronov & Sharvin, 1987; Chiang & Shevchenko, 1999), the wave-function phase of the excitations with energy $E = eU$ in the field B should change along a coherent trajectory of length Λ as follows

$$\phi = \phi_e + \phi_h = 2\pi[(1/\pi)(E/\hbar v_{\mathrm{F}})\Lambda + BA/(\Phi_0/2)], \tag{13}$$

where $\Phi_0 = hc/e = 4.14 \cdot 10^{-7}$ G·cm^2. The first term in Eq. (13) can be ignored because, in our samples, it does not exceed 10^{-5} at $U \leq 10^{-8}$ V. One can thus expect that the interference contribution coming from the elastic-scattering centers to the conductivity oscillates as $\delta R \propto \delta R_{\max} \cos\phi$ (Chiang & Shevchenko, 2001), where $\delta R_{\max}$ is the amplitude depending on the concentration of the most efficient scattering centers and, hence, proportional to the total concentration c.

The maximum number of oscillation periods $\Delta B_{ext;I}$ that can be observed in a magnetic field upon changing the temperature clearly depends on the B_c variation scale. It varies from the value $B_c(T_O) = B_{ext;I}$ at the temperature T_O at which the *SNS* structure with the intermediate state arises, to the value $B_c(T)$ at a given temperature. Therefore, the phase of the oscillations at a given temperature should depend on the values of $B_{ext;I}$ in a following way:

$$\phi = 2\pi \frac{[B_c(T) - B_{ext;I}]A_{\max}}{\Phi_0/2}. \tag{14}$$

To estimate the interval of B_c values within which the change in A_{edge} with varying B_c may be neglected, we used Eq. (14) and the differential of the parameter r from Eq. (12). This yields $\Delta B_c \approx 3\Delta B_{ext;I}$.

From the condition $\Delta B \cdot A_{\mathrm{edge}} = \Phi_0/2$ and $\Delta B \approx (45;50)$ G, we obtain $r \approx 1$ μm, in accordance with the above-mentioned independent estimation. The ratio of the oscillation amplitudes also conforms to its expected value: $[(\delta R_{osc})^{\mathrm{Sn}}/(\delta R_{osc})^{\mathrm{Pb}}] \sim (c^{\mathrm{Sn}}/c^{\mathrm{Pb}}) \sim (l^{\mathrm{Pb}}_{el}/l^{\mathrm{Sn}}_{el}) \sim 10$. One can expect that a change in the number of domains in the plate from 12 to 16 alters the oscillation amplitude by no more than 40%; i. e., it only modifies the

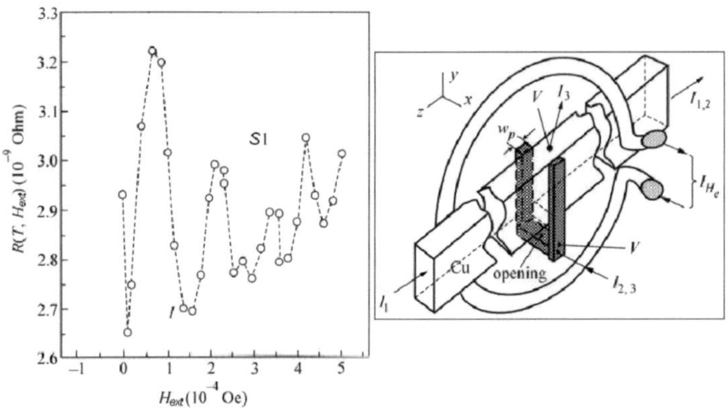

Fig. 8. Oscillations of the generalized resistance of the interferometer $s1$ (In-Cu-Sn) vs magnetic field at $T = 3.25$ K. The oscillation period is $\Delta H = (hc/2e)/A_{\max}$, where$A_{\max}$ is the area of the *abcd* contour in the (xy) cross section of the interferometer (see right panel).

oscillations but does not disturb the overall periodicity pattern (Fig. 4). It also follows from Eq. (14) that the number of periods between the temperature of oscillation onset, T_O, and an arbitrary temperature depends on the value of $B_c(T_O) = B_{ext}$. This makes understandable the relation between the phases of oscillations observed for the Pb plate in different fields: $(\phi_{550G} - \phi_{480G}) \approx 3\pi$ and $(\phi_{520G} - \phi_{480G} + \pi) \approx 3\pi$ (it is taken into account that **B**[520 G] = -**B**[480 G]). Such a relation is a result of the different number of periods, ΔB, measured from $B_{ext;I}$.

The significant distortions of the shape of the oscillation curves in the Pb sample is most likely due to variations in the value of $q_{\max}$ when the number of domains varies in the investigated temperature interval, thereby changing the position of the *NS* boundaries.

In the sample containing In, the values of T_O and $T_c(B = 0)$ are extremely close to each other because of the small $B_I \approx 5$ G. As a result, in the same temperature interval as for Sn and Pb one can observe more than three oscillation periods (compare Figs. 5 and 6). In such case, the change in $A_{\max}$ and, hence, in the oscillation period is hardly noticeable (see the estimation above).

It is appropriate here to compare (although qualitatively) the order of magnitude of the interference contributions to the conductance in the absence of an *NS* boundary, in the approximation of a weak-localization mechanism, and in the presence of an NS boundary. According to the theory of weak localization (Altshuler et al., 1980), the probability of an occurrence of self-intersecting trajectories is of the order of $(\lambda_B/l_{el})^2 \sim 10^{-7}$ ($\lambda_B(\sim q)$ is the de Broglie wavelength; $l_{el} \approx 100$ μm), while as the probability that coherent trajectories will arise in the case of an *NS* boundary in a layer with a characteristic size of the order of the mean free path is larger by a factor of $(r/\lambda_B)^2 \sim 10^8$ than the probability of formation of self-crossing trajectories. The existence of coherent trajectories in the *NS* system is determined by the area of the base of the cone formed by accessible coherent trajectories arising as a result of Andreev reflection, the base of the cone resting on the superconductor and the vertex at an impurity (see Fig. 7). Hence, the expected relative interference contribution to the resistance of an *NS* system is as follows

$$(\delta R/R) \sim (r/\lambda_B)^2 (\lambda_B/l_{el})^2 \gg 1, \quad (15)$$

and agrees completely with the amplitude of the oscillations we observed in a Pb slab.

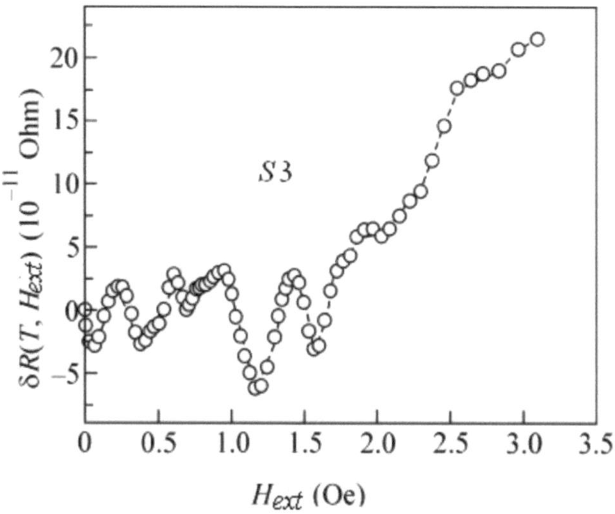

Fig. 9. Oscillations of "Andreev resistance" $\delta R = R(H_{ext}) - R(H_{ext} = 0)$ in interferometer $s3$ (Pb/Cu/Pb) as a function of magnetic field at $T = 4.125$ K. The oscillation period is $\Delta H = (hc/2e)/A_{min}$, where$A_{min}$ is the area of xy projection of the stretched opening of the interferometer onto the direction of magnetic field upon the deviation from the z axis; $R(H_{ext} = 0) = 2.6245 \times 10^{-8}$ Ohm.

2.2 Doubly connected *SNS* structures

Below we present the results from the study of the conductance of doubly connected *NS* systems. Similarly to singly connected ones considered above, they meet the same "macroscopic" conditions, namely, L, $l_{el} \gg \xi_T = \xi_0$ $(L/\xi_T \approx 100)$. This means that a spatial scale of the possible proximity effect is much less than the size of a normal segment of the system, this effect can be therefore neglected completely when considering phenomena of the interferential nature in such "macroscopical" systems.

Macroscopical hybrid samples $s1$ [In(S)/Cu(N)/Sn(S)], $s2$ [Sn(S)/Cu(N)/Sn(S)], $s3$ [Pb(S)/Cu(N)/Pb(S)], and [In(S)/Al(N)/In(S)] were prepared using a geometry of a doubly connected *SNS* Andreev interferometer with a calibrated opening. Figures 8 (right panel) and 10 (upper panel) show schematically (not to scale) the typical construction of the samples, together with a wire turn as a source of an external magnetic field $H_{\rm ext}$ for controlling the macroscopic phase difference in the interferometer formed by a part of a normal single crystal (Cu, Al) and a superconductor connected to it. Interferometers varied in size, type of a superconductor, and area of the *NS* interfaces. The field $H_{\rm ext}$ was varied within a few Oersteds in increments of 10^{-5} Oe. An error of field measurement amounted to no more than 10 %. To compensate external fields, including the Earth field, the container with the sample and the turn was placed into a closed superconducting shield.

Conductance of all the systems studied oscillated while changing the external magnetic field which was inclined to the plane of the opening. It has thus appeared that the areas of extreme projections, S_{extr}, onto the plane normal to the vector of the external magnetic field are related to the periods of observed oscillations, ΔH, by the expression $S_{extr}\Delta H = \Phi_0$, where Φ_0 is the magnetic flux quantum $hc/2e$. The values of $S_{\rm min}$ and $S_{\rm max}$ differed from each other by more than an order of magnitude for each of the interferometers, allowing the corresponding oscillation periods to be resolved (see Figs. 8 - 10).

The sensitivity of dissipative conduction to the macroscopic phase difference in a closed SNS contour is a direct evidence for the realization of coherent transport in the system and the role played by both NS interfaces in it. In turn, at $L \gg \xi_T$, the coherent transport can be caused by only those normal-metal excitations which energies, $\varepsilon \ll T < \Delta$, fill the Andreev spectrum that arises due to the restrictions on the quasiparticle motion because of the Andreev reflections (Zhou et al., 1995). It follows from the quasiclassical dimensional quantization (Andreev, 1964; Kulik, 1969) that the spacing between the levels of the Andreev spectrum should be $\varepsilon_{\rm A} \approx \hbar v_{\rm F}/L_x \approx 20$ mK for the distance between NS interfaces $L_x \simeq$ 0.5 mm. It corresponds to the upper limit for energies of the $e-h$ excitations on the dissipative (passing through the elastic scattering centers) coherent trajectories in the normal region. To zero order in the parameter $\lambda_{\rm B}/l$, only these trajectories can make a nonaveraged phase-interference contribution to conductance, often called the "Andreev" conductance $G_{\rm A}$ (Lambert & Raimondi, 1998). Accordingly, it was supposed that the modulation depth for the normal conductance G_N (or resistance R_N) in our interferometers in the temperature range measured would take the form

$$1 - \frac{G_{\rm A}}{G_N} \equiv \frac{\delta R_{\rm A}}{R_N} \approx \frac{\varepsilon_{\rm A}}{T} \simeq 10^{-2}. \tag{16}$$

In the approximation of noninteracting trajectories, the macroscopic phase, ϕ_i, which coherent excitations with phases ϕ_{ei} and ϕ_{hi} is gaining while moving along an i-th trajectory closed by a superconductor, depends in an external vector-potential field $\mathbf{A}$ on the magnetic flux as follows

$$\phi_i = \phi_{ei} + \phi_{hi} = \phi_{0i} + 2\pi\frac{\Phi_i}{\Phi_0}, \tag{17}$$

where ϕ_{0i} is the microscopic phase related to the length of a trajectory between the interfaces by the Andreev-reflection phase shifts; $\Phi_i = \mathbf{H}_{ext} \cdot \mathbf{S}_i$ is the magnetic flux through the projection S_i onto the plane perpendicular to $\mathbf{H}_{ext}$; $\mathbf{H}_{ext} = \nabla \times \mathbf{A}$ is the magnetic field vector; $\mathbf{S}_i = \mathbf{n}_{S_i} \cdot S_i$; $\mathbf{n}_{S_i}$ is the unit normal vector; S_i is the area under the trajectory; and Φ_0 is the flux quantum $hc/2e$.

The evaluation of the overall interference correction, $2{\rm Re}(f_e f_h^*)$, in the expression for the total transmission probability $|f_e + f_h|^2$ ($f_{e,h}$ are the scattering amplitudes) along all coherent trajectories can be reduced to the evaluation of the Fresnel-type integral over the parameter $\mathbf{S}_i$ (Tsyan, 2000). This results in the separation of the $\mathbf{S}$-nonaveraged phase contributions at the integration limits. As a result, the oscillating portion of the interference addition to the total resistance of the normal region in the SNS interferometer, in particular, for $\mathbf{H}_{ext}||z$, takes the form

$$\frac{\delta R_{\rm A}}{R_N} \sim \frac{\varepsilon_{\rm A}}{T}\sin[2\pi(\phi_0 + \frac{H_{ext}S_{extr}}{\Phi_0})], \tag{18}$$

where S_{extr} is the minimal or maximal area of the projection of doubly connected SNS contours of the system onto the plane perpendicular to $\mathbf{H}$, and $\phi_0 \sim (1/\pi)(L/l_{el}) \sim 1$ (Van Wees et al., 1992). Our experimental data are in good agreement with this phase dependence of the generalized interferometer resistance and the magnitude of the effect. Since all doubly connected SNS contours include $e-h$ coherent trajectories in the normal region with a length of no less than $\sim L \approx 10^2\xi_T$, one can assert that the observed oscillations are due to the long-range quantum coherence of quasiparticle excitations under conditions of suppressed proximity effect for the major portion of electrons.

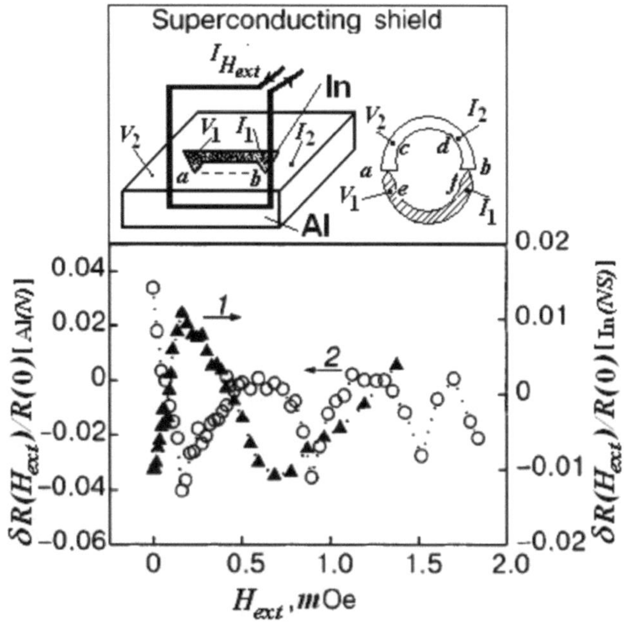

Fig. 10. Non-resonance oscillations of the phase-sensitive dissipative component of the resistance of the indium narrowing (curve *1*) at $T = 3.2$ K and the resonance oscillations of this component in the aluminum part (curve *2*) at $T = 2$ K for the interferometer with $R_a \ll R_b$, as functions of the external magnetic field.

3. Macroscopical NS systems with a magnetic N - segment

The peculiarities of electron transport arising due to the influence of a superconductor contacted to a normal metal and, particularly, to a ferromagnet (F) have been never deprived of attention. Recently, a special interest in the effects of that kind has been shown, in connection with the revived interest to the problem of nonlocal coherence (Hofstetter et al., 2009). Below we demonstrate that studying the coherent phenomena associated with the Andreev reflection, in the macroscopical statement of experiments, may be directly related to this problem. As is known, even in mesoscopic *NS* systems, the coherent effects has been noted in a normal-metal (magnetic) segment at a distance of $x \gg \xi_{\text{exch}}$ from a superconductor (ξ_{exch} is the coherence length in the exchange field of a magnetic) (Giroud et al., 2003; Gueron et al., 1996; Petrashov et al., 1999). That fact gave rise to the intriguing suggestion that magnetics could exhibit a long-range proximity effect, which presumed the existence of a nonzero order parameter $\Delta(x)$ at the specified distance. Such a suggestion, however, contradicts the theory of FS junctions, since $\xi_{\text{exch}} \ll \xi_T \sim v_{\text{F}}/T$, and v_{F}/T is the ordinary scale of the proximity effect in the semiclassical theory of superconductivity (De Gennes, 1966). This assumption, apparently, is beneath criticism, because of the specific geometry of the contacts in mesoscopic samples. As a rule, these contacts are made by a deposition technology. Consequently, they are planar and have the resistance comparable in value with the resistance of a metal located under the interface. A shunting effect arises, and the estimation of the value and even sign of the investigated transport effects becomes ambiguous (Belzig et al., 2000; Jin & Ketterson, 1989; De Jong & Beenakker, 1995).

Influence of the shunting effect is well illustrated by our previous results (Chiang & Shevchenko, 1999); one of them is shown in Fig. 11. The conductance measured outside the *NS* interface (see curve *1* and Inset 1) behaves in accordance with the fundamental ideas of the semiclassical theory (see Sec. 2. 1): Because of "retroscattering", the cross section for elastic scattering by impurities in a metal increases at the coherence length of $e-h$ hybrids formed in the process of Andreev reflection, i. e., the conductivity of the metal decreases rather than increases. Additional scattering of Andreev hole on the impurity is completely ignored in case of a point-like ballistic junction (Blonder et al., 1982). At the same time, the behavior of the resistance of the circuit which includes a planar interface (see Inset 2) may not even reflect that of the metal itself (curve *2*; see also (Petrashov et al., 1999)), but it is precisely this type of behavior that can be taken as a manifestation of the long-range proximity effect.

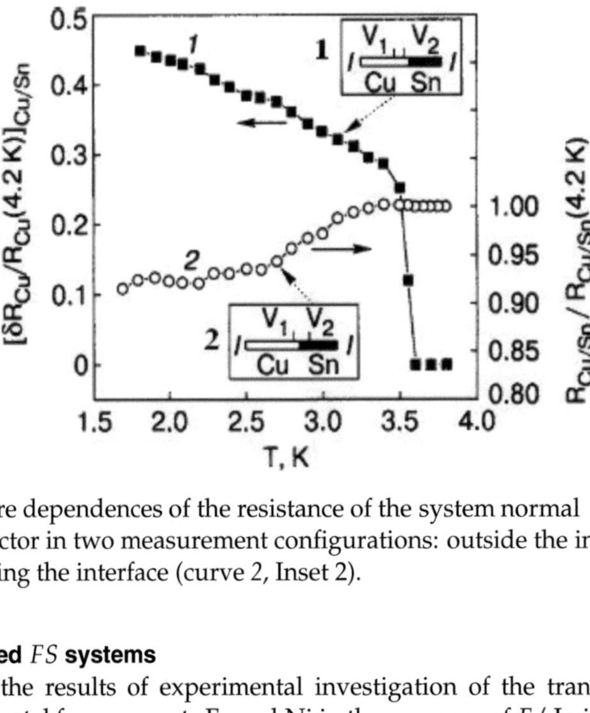

Fig. 11. Temperature dependences of the resistance of the system normal metal/superconductor in two measurement configurations: outside the interface (curve *1*, Inset 1) and including the interface (curve 2, Inset 2).

3.1 Singly connected *FS* systems

Here, we present the results of experimental investigation of the transport properties of non-film single - crystal ferromagnets Fe and Ni in the presence of *F*/ In interfaces of various sizes (Chiang et al., 2007). We selected the metals with comparable densities of states in the spin subbands; conducting and geometric parameters of the interfaces, as well as the thickness of a metal under the interface were chosen to be large in comparison with the thickness of the layer of a superconductor. In making such a choice, we intended to minimize the effects of increasing the conductivity of the system that could be misinterpreted as a manifestation of the proximity effect.

The geometry of the samples is shown (not to scale) in Fig. 12. The test region of the samples with *F*/*S* interfaces *a* and *b* is marked by a dashed line. After setting the indium jumper, the region *abdc* acquired the geometry of a closed "Andreev interferometer", which made it possible to study simultaneously the phase-sensitive effects. Both point (p) and wide (w) interfaces were investigated. We classify the interface as "point" or "wide" depending on the ratio of its characteristic area to the width of the adjacent conductor (of the order of 0.1 or 1, respectively).

3.1.1 Doubling the cross section of scattering by impurities

Figure 13 shows in relative units $\delta R/R = [R(T) - R(T = T_c^{\rm In})]/R(T = T_c^{\rm In})$ the resistance of the ferromagnetic segments with point (Fe, curve *1* and Ni, curve *2*) and wide (Ni, curve *3*) F/S interfaces measured with current flow *parallel* to the interfaces [for geometry, see Insets (a) and (b)]. In this configuration, with indium in the superconducting state, the interfaces, as parts of the potential probes, play a passive role of "superconducting mirrors". It can be seen that for $T \leq T_c^{\rm In}$ (after Andreev reflection is actuated), the resistance of Ni increases abruptly by 0.04% ($\delta R_p \approx 1 \times 10^{-8}$ Ohm) in the case of two point interfaces and by 3% ($\delta R_w \approx 7 \times 10^{-7}$ Ohm) in the case of two wide ones. In Fe with point interfaces, a negligible effect of opposite sign is observed, its magnitude being comparable to that in Ni, $\delta R_p^{\rm Ni}$.
Just as in the case of a nonmagnetic metal (Fig. 11), the observed decrease in the conductivity of nickel when the potential probes pass into the "superconducting mirrors" state, corresponds to an increase in the efficiency of the elastic scattering by impurities in the metal adjoining the superconductor when Andreev reflection appears. (We recall that the shunting effect is small). In accordance with Eq. (3), the interference contribution from the scattering of a singlet pair of $e - h$ excitations by impurities in the layer, of the order of the coherence length ξ in thickness, if measured at a distance L from the N/S interface, is proportional to ξ/L. From this expression one can conclude that the ratio of the magnitude of the effect, δR, to the resistance measured at an arbitrary distance from the boundary is simply the ratio of the corresponding spatial scales. It is thereby assumed that the conductivity σ is a common parameter for the entire length, L, of the conductor, including the scale ξ. Actually, we find from Eq. (3) that the magnitude of the positive change in the resistance, δR, of the layer ξ in whole is

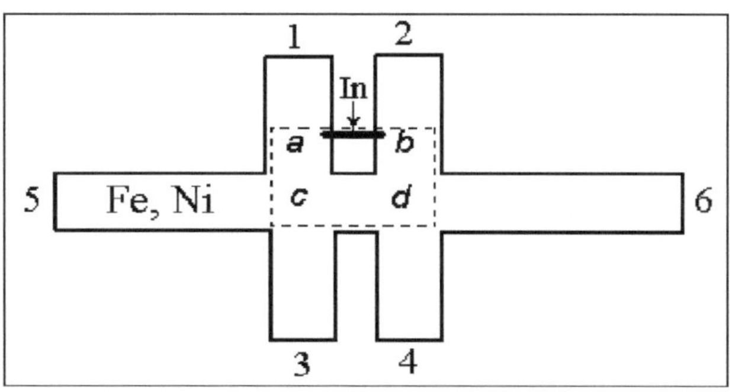

Fig. 12. Schematic view of the F/S samples. The dashed line encloses the workspace. $F/$In interfaces are located at the positions a and b. The regimes of current flow, parallel or perpendicular to the interfaces, were realized by passing the feed current through the branches 1 and 2 with disconnected indium jumper $a - b$ or through 5 and 6 when the jumper was closed (shown in the figure).

$$\delta R^{\,\xi} = (\xi/\sigma_\xi A_{\rm if})\bar{r} \equiv \sum_{i=1}^{N_{\rm imp}} \delta R_i^{\,\xi}. \tag{19}$$

Here, σ_ξ is the conductivity in the layer ξ; $A_{\rm if}$ is the area of the interface; $N_{\rm imp}$ is the number of Andreev channels (impurities) participating in the scattering; δR_i^{ξ} is the resistance resulting from the $e - h$ scattering by a single impurity, and $\bar{r}$ is the effective probability for elastic scattering of excitations with the Andreev component in the layer ξ as a whole. Control

measurements of the voltages in the configurations included and not included interfaces showed that in our systems, the voltages themselves across the interfaces were negligibly small, so that we can assume $\bar{r} \approx 1$. It is evident that the Eq. (19) describes the resistance of the ξ-part of the conductor provided that $\sigma_\xi = \sigma_L$ i. e., for $\xi > l_{el}$. For ferromagnets, $\xi \ll l_{el}$ and $l_{el}^L \neq l_{el}$. In this case, to compare the values of δR measured on the length L with the theory, one should renormalize the value of R_N from the Eq. (3).
In the semiclassical representation, the coherence of an Andreev pair of excitations in a metal is destroyed when the displacement of their trajectories relative to each other reaches a value of the order of the trajectory thickness, i. e., the de Broglie wavelength λ_B. The maximum possible distance ξ_m (collisionless coherence length) at which this could occur in a ferromagnet with nearly rectilinear e and h trajectories (Fig. 14a) is

$$\xi_m \sim \frac{\lambda_B}{\varepsilon_{exch}/\varepsilon_F} = \frac{\pi\hbar v_F}{\varepsilon_{exch}}; \qquad \varepsilon_{exch} = \mu_B H_{exch} \sim T_{exch} \tag{20}$$

(μ_B is the Bohr magneton, H_{exch} is the exchange field, and T_{exch} is the Curie temperature). However, taking into account the Larmor curvature of the e and h trajectories in the field H_{exch}, together with the requirement that both types of excitations interact with the same impurity (see Fig. 14b), we find that the coherence length decreases to the value (De Gennes, 1966) $\xi^* = \sqrt{2qr} = \sqrt{2q\xi_m}$ (compare with Eq. (12)). Here, r is the Larmor radius in the field H_{exch} and q is the screening radius of the impurity $\sim \lambda_B$. Figure 14 gives a qualitative idea of the scales on which the dissipative contribution of Andreev hybrids can appear, as a result of scattering by impurities ($N_{imp} \gg 1$), with the characteristic dimensions of the interfaces $y, z \gg l_{el}$.

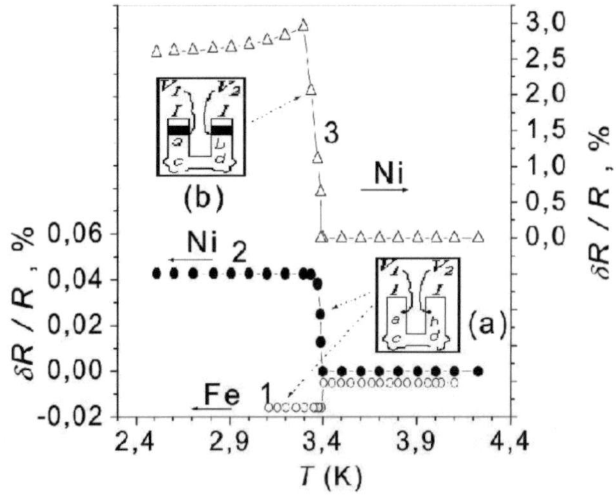

Fig. 13. Temperature dependences of the resistance of Fe and Ni samples in the presence of F/In interfaces acting as "superconducting mirrors" at $T < T_c^{\text{In}}$. Curves *1* and 2: Fe and Ni with point interfaces, respectively; curve *3*: Ni with wide interfaces. Insets: geometry of point (a) and wide (b) interfaces.

For Fe with $T_{exch} \approx 10^3$ K and Ni with $T_{exch} \approx 600$ K, we have $\xi^* \approx 0.001$ μm. It follows that in our experiment with $l_{el} \approx 0.01$ μm (Fe) and $l_{el} \approx 1$ μm (Ni), the limiting case $l_{el} \gg \xi^*$ and $l_{el}^L \neq l_{el}^\xi$ is realized. From Fig. 14b it can be seen that for $y, z \gg l_{el} \gg \xi^*$ in the normal state of the interface, the length l_{el}^{ξ} within the layer ξ^* corresponds to the shortest distance between the impurity and the interface, i. e., $l_{el}^{\xi} \equiv \xi^*$ $(\sigma_L \neq \sigma_{\xi^*})$. Note that for an equally probable distribution of the impurities, the probability of finding an impurity at any distance from the interface in a finite volume, with at least one dimension greater than l_{el}, is equal to unity. Renormalizing Eq. (3), with ξ_T replaced by ξ^*, we obtain the expression for estimating the coherence correction to the resistance measured on the length L in the ferromagnets:

$$\frac{\delta R^{\xi^*}}{R_L} = \frac{\xi^*}{L} \cdot \frac{l_{el}}{l_{el}^{\xi^*}} \bar{r} \approx \frac{l_{el}}{L} \bar{r}; \; \delta R^{\xi^*} = \frac{\xi^*}{\sigma_{\xi^*} A_{if}} \bar{r} \equiv \sum_{i=1}^{N_{imp}} \delta R_i^{\xi^*}. \tag{21}$$

Here, σ_{ξ^*} is the conductivity in the layer ξ^*; $\delta R_i^{\xi^*}$ is the result of $e - h$ scattering by a single impurity. Equation (21) can serve as an observability criterion for the coherence effect in ferromagnets of different purity. It explains why no positive jump of the resistance is seen on curve *1*, Fig. 13, in case of a point Fe/In interface: with $l_{el}^{\mathrm{Fe}} \approx 0.01$ μm, the interference increase in the resistance of the Fe segment with the length studied should be $\approx 10^{-9}$ Ohm and could not be observed at the current $I_{acdb} \leq 0.1$ A, at which the measurement was performed, against the background due to the shunting effect.
Comparing the effects in Ni for the interfaces of different areas also shows that the observed jumps pertain precisely to the coherent effect of the type studied. Since the number of Andreev channels is proportional to the area of an N/S interface, the following relation should be met between the values of resistance measured for the samples that differ only in the area of the interface: $\delta R_w^{\xi^*}/\delta R_p^{\xi^*} = N_{\mathrm{imp}}^w/N_{\mathrm{imp}}^p \sim A_w/A_p$ (the indices p and w refer to point and wide interfaces, respectively). Comparing the jumps on the curves 2 and 3 in Fig. 13 we obtain: $\delta R_w/\delta R_p = 70$, which corresponds reasonably well to the estimated ratio $A_w/A_p = 25 - 100$.
In summary, the magnitude and special features of the effects observed in the resistance of magnetics Fe and Ni are undoubtedly directly related with the above-discussed coherent effect, thereby proving that, in principle, it can manifest itself in ferromagnets and be observed provided *an appropriate instrumental resolution*. Although this effect for magnetics is somewhat surprising, it remains, as proved above, within the bounds of our ideas about the scale of the coherence length of Andreev excitations in metals, which determines the dissipation; therefore, this effect cannot be regarded as a manifestation of the proximity effect in ferromagnets.

3.1.2 Spin accumulation effect

The macroscopic thickness of ferromagnets under F/S interfaces made it possible to investigate the resistive contribution from the interfaces, R_{if}, in the conditions of current flowing perpendicular to them, through an indium jumper with current fed through the contacts 5 and 6 (see Fig. 12 and Inset in Fig. 15).
Figure 15 presents in relative units the temperature behavior of R_{if}^p for point Fe/In interfaces (curve *1*) and R_{if}^w for wide Ni/In interfaces (curve 2) as $\delta R_{if}/R_{if} = [R_{if}(T) - R_{if}(T_c^{In})]/R_{if}(T_c^{In})$. The shape of the curves shows that with the transition of the interfaces from the F/N state to the F/S state the resistance of the interfaces abruptly *increases* but compared with the increase due to the previously examined coherent effect it increases by an incomparably larger amount. It is also evident that irrespective of the interfacial geometry the behavior of the function $R_{if}(T)$ is qualitatively similar in both systems. The value of $R_{if}(T_c^{In})$ is the lowest resistance of the interface that is attained when the current is displaced

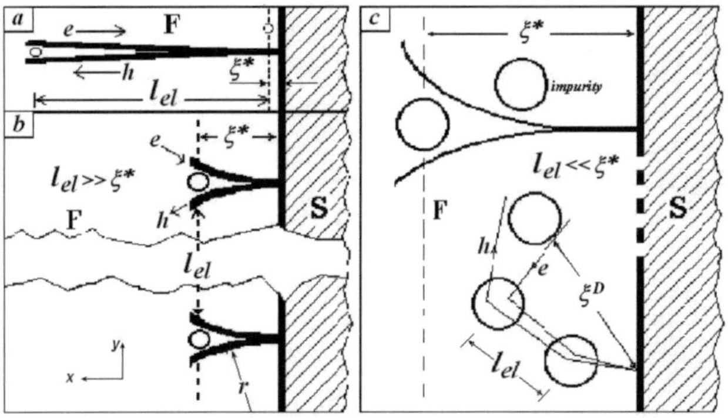

Fig. 14. Scattering of Andreev $e-h$ hybrids and their coherence length ξ^* in a normal ferromagnetic metal with characteristic F/S interfacial dimensions greater than l_{el}. Panels $a, b:\ \xi^* \ll l_{el}$; panel $c:\ \xi^* \gg l_{el}$; $\xi^D \sim \sqrt{l_{el}\xi^*}$.

to the edge of the interface due to the Meissner effect. The magnitudes of the positive jumps with respect to this resistance, $\delta R_{if}/R_{if}(T_c^{In}) \equiv \delta R_{F/S}/R_{F/N}$, are about 20% for Fe (curve *1*) and about 40% for Ni (curve *2*). The values obtained are more than an order of

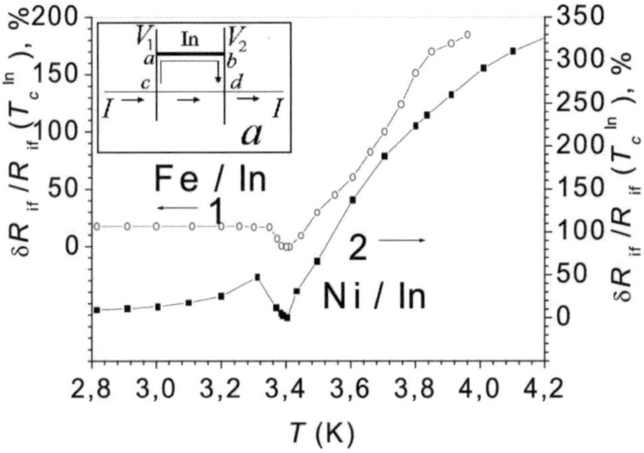

Fig. 15. Spin accumulation effect. Relative temperature dependences of the resistive contribution of spin-polarized regions of Fe and Ni near the interfaces with small (Fe/In) and large (Ni/In) area.

magnitude greater than the contribution to the increase in the resistance of ferromagnets which is related with the coherent interaction of the Andreev excitations with impurities (as is shown below, because of the incomparableness of the spatial scales on which they are manifested). This makes it possible to consider the indicated results as being a direct manifestation of the mismatch of the spin states in the ferromagnet and superconductor, resulting in the accumulation of spin on the F/S interfaces, which decreases the conductivity of the system as a whole. We suppose that such a decrease is equivalent to a decrease in the conductivity of a certain region of the ferromagnet under the interface, if the exchange spin splitting in the ferromagnetic sample extends over a scale not too small compared to the size of this region. In other words, the manifestation of the effect in itself already indicates that the dimensions of the region of the ferromagnet which make the effect observable are comparable to the spin relaxation length. Therefore, the effect which we observed should reflect a resistive contribution from the regions of ferromagnets on precisely the same scale. The presence of such nonequilibrium regions and the possibility of observing their resistive contributions using a four-contact measurement scheme are due to the "non-point-like nature" of the potential probes (finiteness of their transverse dimensions). In addition, the data show that the dimensions of such regions near Fe/S and Ni/S interfaces are comparable in our experiments. Indeed, the value of $\delta R_{Ni/S}/R_{Ni/N}$ corresponding according to the configuration to the contribution from only the nonequilibrium regions and the value of $\delta R_{Fe/S}/R_{Fe/N}$ obtained from a configuration which includes a ferromagnetic conductor of length obviously greater than the spin-relaxation length, are actually of the same order of magnitude. In addition, according to the spin-accumulation theory (Hofstetter et al., 2009; Lifshitz & Sharvin, 1951; Van Wees et al., 1992), the expected magnitude of the change in the resistance of the F/S interface in this case is of the order of

$$\delta R_{F/S} = \frac{\lambda_s}{\sigma A} \cdot \frac{P^2}{1-P^2}; \qquad P = (\sigma_\uparrow - \sigma_\downarrow)/\sigma; \qquad \sigma = \sigma_\uparrow + \sigma_\downarrow. \tag{22}$$

Here, λ_s is the spin relaxation length; P is the coefficient of spin polarization of the conductivity; σ, $\sigma_\uparrow$, $\sigma_\downarrow$, and A are the total and spin-dependent conductivities and the cross section of the ferromagnetic conductor, respectively. Using this expression, substituting the data for the geometric parameters of the samples, and assuming $P^{\mathrm{Fe}} \sim P^{\mathrm{Ni}}$, we obtain $\lambda_s(\mathrm{Fe}/S)/\lambda_s^*(\mathrm{Ni}/S) \approx 2$. This is an additional confirmation of the comparability of the scales of the spin-flip lengths λ_s for Fe/S and λ_s^* for Ni/S, indicating that the size of the nonequilibrium region determining the magnitude of the observed effects for those interfaces is no greater than (and in Fe equal to) the spin relaxation length in each metal. In this case, according to Eq. (22), the length of the conductors, with normal resistance of which the values of $\delta R_{F/S}$ must be compared, should be set equal to precisely the value of λ_s for Fe/S and λ_s^* for Ni/S. This implies the following estimate of the coefficients of spin polarization of the conductivity for each metal:

$$P = \sqrt{(\delta R_{F/S}/R_{F/N})/[1 + (\delta R_{F/S}/R_{F/N})]}. \tag{23}$$

Using our data we obtain $P^{\mathrm{Fe}} \approx 45\%$ for Fe and $P^{\mathrm{Ni}} \approx 50\%$ for Ni, which is essentially the same as the values obtained from other sources (Soulen et al., 1998). If in Eq. (22) we assume that the area of the conductor, A, is of the order of the area of the current entrance into the jumper (which is, in turn, the product of the length of the contour of the interface by the width of the Meissner layer), then a rough estimate of the spin relaxation lengths in the metals investigated, in accordance with the assumption of single-domain magnetization of the samples, will give the values $\lambda_s^{\mathrm{Fe}} \sim 90$ nm and $\lambda_s^{\mathrm{Ni}} > 50$ nm. Comparing these values with the value of coherence length in ferromagnets $\xi^* \approx 1$ nm we see that although the coherent effect leads to an almost 100% increase in the resistance, this effect is localized within a layer which thickness is two orders of magnitude less than that of the layer responsible for the appearance of the spin accumulation effect, therefore it does not mask the latter.

3.2 Doubly connected *SFS* systems

The observation of the coherent effect in the singly connected *FS* systems raised the following question: Can effects sensitive to the phase of the order parameter in a superconductor be manifested in the conductance of ferromagnetic conductors of macroscopic size? To answer

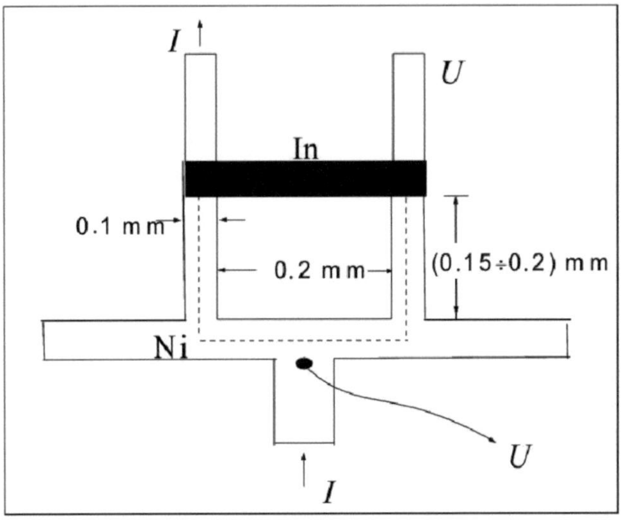

Fig. 16. Schematic diagram of the F/S system in the geometry of a doubly connected "Andreev interferometer". The ends of the single-crystal ferromagnetic (Ni) segment (dashed line) are closed by a superconducting In bridge.

this question we carried out direct measurements of the conductance of Ni conductors in a doubly connected SFS configuration (in the Andreev interferometer (AI) geometry shown in Fig. 16).

Figures 17 and 18 show the magnetic-field oscillations of the resistance of two samples in a doubly connected $S/\mathrm{Ni}/S$ configuration with different aperture areas, measured for the arrangement of the current and potential leads illustrated in Fig. 16. The oscillations in Fig. 17 are presented on both an absolute scale ($\delta R_{\mathrm{osc}} = R_H - R_0$, left axis) and a relative scale ($\delta R_{\mathrm{osc}}/R_0$, right axis). R_0 is the value of the resistance in zero field of the ferromagnetic segment connecting the interfaces in the area of a dashed line in Fig. 16. Such oscillations in SFS systems in which the total length of the ferromagnetic segment reaches the values of the order of 1 mm (along the dashed line in Fig. 16), were observed for the first time. Figures 17 and 18 were taken from two samples during two independent measurements, for opposite directions of the field, with different steps in H and are typical of several measurements, which fact confirms the reproducibility of the oscillation period and its dependence on the aperture area of the interferometer.

The period of the resistive oscillations shown in Fig. 17 is $\Delta B \approx (5-7) \times 10^{-4}$ G and is observed in the sample with the geometrical parameters shown in Fig. 16. It follows from this figure that the interferometer aperture area, enclosed by the midline of the segments and the bridge, amounts to $A \approx 3 \times 10^{-4}\mathrm{cm}^2$. In the sample with twice the length of the sides of the interferometer and, hence, approximately twice the aperture area, the period of the oscillations turned out to be approximately half as large (solid line in Fig. 18). From the values of the periods of the observed oscillations it follows that, to an accuracy of 20%, the periods are proportional to a quantum of magnetic flux $\Phi_0 = hc/2e$ passing through the corresponding area $A: \Delta B \approx \Phi_0/A$.

Obviously, the oscillatory behavior of the conductance is possible if the phases of the electron wave functions are sensitive to the phase difference of the order parameter in the

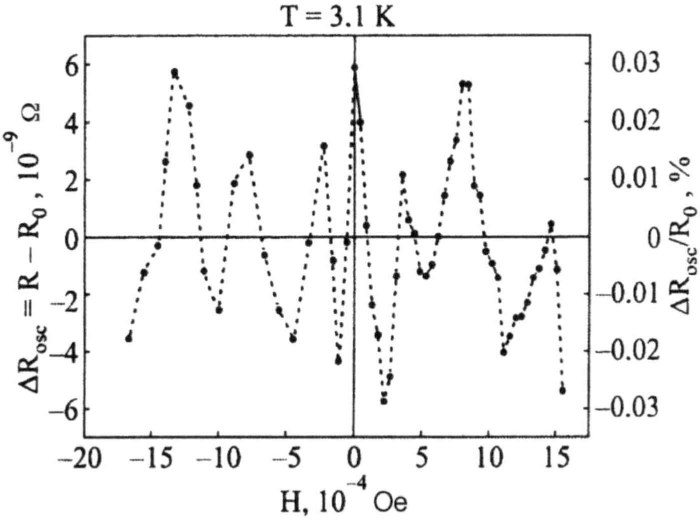

Fig. 17. The $hc/2e$ magnetic-field oscillations of the resistance of a ferromagnetic (Ni) conductor in an AI system with the dimensions given in Fig. 16, in absolute (left-hand scale) and relative (right-hand scale) units. $R_0 = 4.12938 \times 10^{-5}$ Ohm. $T = 3.1$ K.

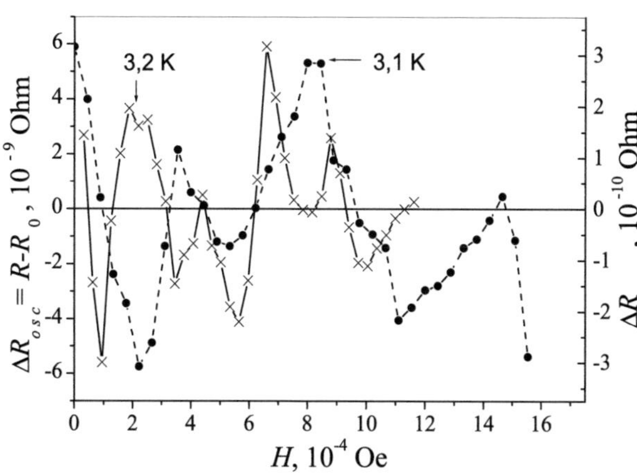

Fig. 18. The $hc/2e$ magnetic-field oscillations of the resistance of a ferromagnetic (Ni) conductor in an AI system with an aperture area twice that of the system illustrated in Fig. 16 (solid curve, right-hand scale). $R_0 = 3.09986 \times 10^{-6}$ Ohm. $T = 3.2$ K. The dashed curve shows the oscillations presented in Fig. 17.

superconductor at the interfaces. Consequently, this parameter should be related to the diffusion trajectories of the electrons on which the "phase memory" is preserved within the whole length L of the ferromagnetic segment. This means that the oscillations are observed in the regimes $L \leq L_\varphi = \sqrt{D\tau_\varphi} \gg \xi_T$ (D is the diffusion coefficient, ξ_T is the coherence length of the metal, over which the proximity effect vanishes, and τ_φ is the dephasing time). It is well known that the possibility for the Aharonov-Bohm effect to be manifested under these conditions was proved by Spivak and Khmelnitskii (Spivak & Khmelnitskii, 1982), although the large value of L_φ coming out of our experiments is somewhat unexpected.

3.2.1 The entanglement of Andreev hybrids

The estimated value of L_φ raises a legitimate question of the nature of the observed effect and the origin of the dephasing length scale evaluated. Since, as discussed in the Introduction, L_φ is determined by the scale of the inelastic mean free path, the main candidates for the mechanism of inelastic scattering of electrons in terms of their elastic scattering on impurities remain electron-electron $(e-e)$ and electron-phonon $(e-ph)$ interactions.

Direct measurement of the temperature-dependent resistance of the ferromagnetic (Ni) segment in the region below T_c^{In} found that $(\delta R_{e-ph}/R_{el}) \sim (l_{el}/l_{e-ph}) \approx 10^{-3} - 10^{-4}$. It follows that for our Ni segment with $l_{el} > 10^{-3}$ cm and $D \sim 10^5$ cm^2/s, the electron-phonon relaxation time should be $\tau_{e-ph} \sim (10^{-7} - 10^{-8})$ s, which value coincides, incidentally, with the semiclassical estimate $\tau_{e-ph} \sim (\hbar/T)(T_{\mathrm{D}}/T)^4$ (T_{D} is the Debye temperature). On the other part, $\tau_{e-e} \sim \hbar\mu_e/T^2$ (μ_e is the chemical potential) at 3 K has the same order of magnitude. Thereby, the dephasing length in the studied systems can have a macroscopical scale of the order of $L_\varphi = \sqrt{D\tau_\varphi} \sim 1$ mm, which corresponds to the length of F segments of our interferometers.

Under these conditions the nature of the observed oscillations can be assumed as follows. According to the arguments offered by Spivak and Khmelnitskii (Spivak & Khmelnitskii, 1982), in a metal, regardless of the sample geometry (the parameters $L_{x,y,z}$), there always exists a finite probability for the existence of constructively interfering transport trajectories, the oscillatory contribution of which does not average out. Such trajectories coexist with destructively interfering ones, the contributions from which average to zero. An example would be the Sharvin's experiment (Sharvin & Sharvin, 1981). In the doubly connected geometry, the probability for the appearance of trajectories capable of interfering constructively increases.

Consider the model shown in Fig. 19. Cooper pairs injected into the magnetic segment are split due to the magnetization and lose their spatial coherence over a distance $\xi^* = \sqrt{2\lambda_{\mathrm{B}} r_{\mathrm{exch}}}$ from the interface (see Sec. 3. 1. 1). r_{exch} is the Larmor radius in the exchange field $H_{\mathrm{exch}} \approx k_{\mathrm{B}}T_{\mathrm{C}}$; $r_{\mathrm{exch}} \sim 1\ \mu$m. (Recall that ξ^* is the distance at which simultaneous interaction of e and h quasiparticles with the same impurity is still admissible.)

The phase shifts acquired by (for example) an electron 3 and hole 2 on the trajectories connecting the interfaces are equal, respectively, to

$$\begin{aligned}\phi_e &= (k_{\mathrm{F}} + \varepsilon_T/\hbar v_{\mathrm{F}})L_e + 2\pi\Phi/\Phi_0 = \phi_{0e} + 2\pi\Phi/\Phi_0,\\ \phi_h &= -(k_{\mathrm{F}} - \varepsilon_T/\hbar v_{\mathrm{F}})L_h + 2\pi\Phi/\Phi_0 = \phi_{0h} + 2\pi\Phi/\Phi_0.\end{aligned} \tag{24}$$

Here ε_T and k_{F} are the energy, measured from the Fermi level and the modulus of the Fermi wave vector, respectively. Since the trajectories of an $e-h$ pair are spatially incoherent, their oscillatory contributions, proportional to the squares of the probability amplitudes, should combine additively:

$$|f_{h(2)}|^2 + |f_{e(3)}|^2 \sim \cos\phi_h + \cos\phi_e \sim \cos(\phi_0 + 2\pi\Phi/\Phi_0), \tag{25}$$

where ϕ_0 is the relative phase shift of the independent oscillations, equal to

$$\phi_0 = (1/2)(\phi_{0e} + \phi_{0h}) \approx (\varepsilon_T/\varepsilon_L)(L_e + L_h)/2L, \tag{26}$$

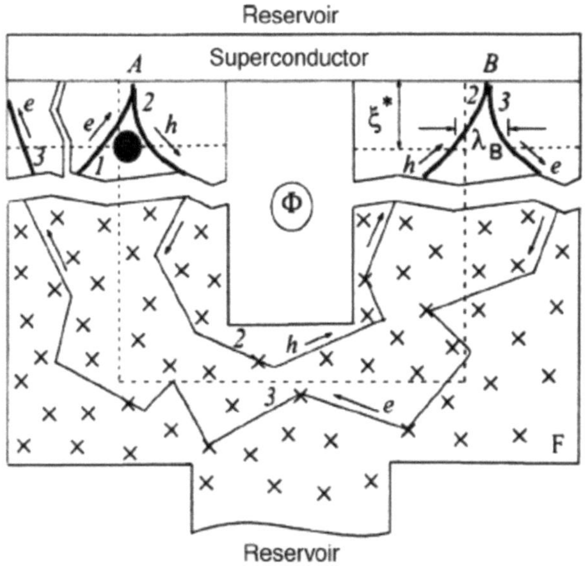

Fig. 19. Geometry of the model.

where $\varepsilon_L = \hbar v_F/L$; $\varepsilon_T = k_B T = \hbar D/\xi_T^2$. Hence it follows that any spatially separated e and h diffusion trajectories with $\phi_0 = 2\pi N$, where N is an integer, can be phase coherent. Clearly this requirement can be satisfied only by those trajectories whose midlines along the length coincide with the shortest distance L connecting the interfaces. In this case, $(L_e + L_h)/2L$ is an integer, since $L_{i(e,h)}, L \propto l_{el}$ and $(L_{i(e,h)}/L) = m(1+\alpha)$, where $\alpha \ll 1$. Furthermore, $(\varepsilon_T/\varepsilon_L)/2\pi$ is also an integer n to an accuracy of $n(1+\gamma)$, where $\gamma \approx (d/L) \ll 1$ (d is the transverse size of the interface). In sum, considering all the foregoing we obtain

$$\cos(\phi_0 + 2\pi\Phi/\Phi_0) \sim \cos(2\pi\Phi/\Phi_0). \tag{27}$$

This means that the contributions oscillatory in magnetic field from all the trajectories should have the same period. Taking into consideration the quasiclassical thickness of a trajectory, we find that the number of constructively interfering trajectories with different projections on the quantization area, those that must be taken into account, is of the order of (l_{el}/λ_B). However, over the greater part of their length, except for the region ξ^*, all (l_{el}/λ_B) trajectories are spatially incoherent. They lie with equal probability along the perimeter of the cross section of a tube of radius l_{el} and axis L, and therefore outside the region ξ^* they average out. Constructive interference of particles on these trajectories can be manifested only over the thickness of the segment ξ^*, reckoned from the interface, where the particles of the $e-h$ pairs are both phase- and spatially coherent. In this region the interaction of pairs with an impurity, as mentioned in the Introduction, leads to a resistive contribution. When the total length of the trajectories is taken into account, the value of this contribution for one pair should be of the order of ξ^*/L. Accordingly, one can expect that the amplitude of the constructive oscillations will have a relative value of the order of

$$\delta R^{\xi^*}/R_L \approx (\xi^*/L)(l_{el}/l_{el}^{\xi^*}) \sim l_{el}^L/L, \tag{28}$$

($l_{el}^{\xi^*} \sim \lambda_B$, see sec. 2.1.1), i. e., the same as the value of the effect measured with the superconducting bridge open. Our experiment confirms this completely: For the samples with the oscillations shown in Figs. 17 and 18, $\delta R^{\xi^*}/R_L \approx 0.03\%$ and 0.01%, respectively. This is much larger than the total contribution from the destructive trajectories, which in the weak-localization approximation is of the order of $(\lambda_B/l_{el})^2$ and which can lead to an increase in the conductance (Altshuler et al., 1981). One should also note that the property of the oscillations under discussion described by Eq. (27) presupposes that the resistance for $H = 0$ will decrease as the field is first introduced, and this, as can be seen in Figs. 17 and 18, agrees with the experiment.

4. Conclusion

Here, we presented the results of the study of Andreev reflection in a macroscopic formulation of experiments, consisting in increasing simultaneously the diffusion coefficient in normal segments of NS hybrid systems and the size of these segments by a factor of $10^3 - 10^4$ as compared with those characteristics of mesoscopic systems. Our data prove that at temperatures below 4 K, the relaxation of the electron momentum, at least at sufficiently rare collisions of electrons with static defects, are not accompanied by a break of the phase of electron wave functions. Hence, the electron trajectories in the classical approximation may be reversible on a macroscopic length scale of the order of several millimeters, both in a nonmagnetic and in a sufficiently pure ferromagnetic metal. In this situation, there appears a possibility to observe conductance oscillations in doubly connected NS systems in Andreev-reflection regime, with a period $hc/2e$ in a magnetic field, which indicates that the interference occurs between singlet bound quasiparticles rather than between triplet bound electrons, as in the Aharonov-Bohm ring. With the current flowing perpendicular to the $N(F)S$ interfaces in singly connected samples, a nonequilibrium resistive contribution of the interfaces was found. We associate this with the spin polarization of a certain region of a ferromagnet under the interface. The observed increase in the resistance corresponds to the theoretically predicted magnitude of the change occurring in the resistance of a single-domain region with spin-polarized electrons as a result of spin accumulation at the F/S interface under the conditions of limiting Andreev reflections.

5. References

Aharonov, Y. & Bohm, D. (1959). Significance of electromagnetic potentials in the quantum theory. *Phys. Rev.*, Vol. 115, No. 3, (1959) 485-491.

Altshuler, B. L., Aronov, A. G. , & Spivak, B. Z. (1981). The Aharonov-Bohm effect in disordered conductors. *JETP Lett.*, Vol. 33, No. 2, (1981) 94-97.

Altshuler, B. L., Khmelnitsky, D. E., Larkin, A. I., et al. (1980). Magnetoresistance and Hall effect in a disordered two-dimensional electron gas. *Phys. Rev. B*, Vol. 22, No. 11, (1980) 5142-5153.

Andreev, A. F. (1964). Thermal conductivity of the intermediate state of superconductors. *Sov. Phys. JETP*, Vol. 19, No. 6, (1964)1228-1234.

Aronov, A. G. & Sharvin, Yu. V. (1987). Magnetic flux effects in disordered conductors. *Rev. Mod. Phys.*, Vol. 59, No. 3, (1987) 755-779.

Belzig, W., Brataas, A., Nazarov, Yu. V., et al. (2000). Spin accumulation and Andreev reflection in a mesoscopic ferromagnetic wire. *Phys. Rev. B*, Vol. 62, No. 14, (2000) 9726-9739.

Blonder, G. E., Tinkham, M., & Klapwijk, T. M. (1982). Transition from metallic to tunneling regimes in superconducting microconstrictions: Excess current, charge imbalance, and supercurrent conversion. *Phys. Rev. B*, Vol. 25, No. 7, (1982) 4515-4532.

Chiang, Yu. N. (1985). Superconducting modulator of the measuring circuit conductance over a wide range of helium temperatures. *Prib. Tekhn. Eksp.*, No. 1, (1985) 202-204.

Chiang, Yu. N. & Shevchenko, O. G. (1988). Measurement of resistance of the normal metal-superconductor (N-S) boundary. *Sov. J. Low Temp. Phys.*, Vol. 14, No. 5, (1988) 299-301.

Chiang, Yu. N. & Shevchenko, O. G. (1998). Contribution of Andreev reflection to the increase in the resistance of the normal metal in a bimetallic N-S structure. *JETP*, Vol. 86, No. 3, (1998) 582-585.

Chiang, Yu. N. & Shevchenko, O. G. (1999). Conductivity of normal metal with phase-coherent excitations in the presence of NS boundary. *Low Temp. Phys.*, Vol. 25, No. 5, (1999) 314-326.

Chiang, Yu. N. & Shevchenko, O. G. (2001). Mesoscopic quantum scillations of the resistance in the intermediate state of type-I superconductors. *Low Temp. Phys.*, Vol. 27, No. 12, (2001) 1000-1009.

Chiang, Yu. N., Shevchenko, O. G., & Kolenov, R. N. (2007). Manifestation of coherent and spin-dependent effects in the conductance of ferromagnets adjoining a superconductor. *Low Temp. Phys.*, Vol. 33, No. 4, (2007) 314-320.

Clarke, J., Eckern, U. , Schmid, A., et al. (1979). Branch-imbalance relaxation times in superconductors. *Phys. Rev. B*, Vol. 20, No. 9, (1979) 3933-3937.

Falko, V. I., Volkov, A. F., & Lambert, C. J. (1999). Andreev reflections and magnetoresistance in ferromagnet-superconductor mesoscopic structures. *JETP Lett.*, Vol. 69, No. 7, (1999) 532-537.

De Gennes, P. G. (1966). *Superconductivity of Metals and Alloys* , W.A. Benjamin, New York.

Giroud, M., Hasselbach, K., Courtois, H., et al. (2003). Electron transport in a mesoscopic superconducting/ferromagnetic hybrid conductor. *Eur. Phys. J. B*, Vol. 31, No. 1, (2003) 103-109.

Gueron, S., Pothier, H., Birge, N. O. et al. (1996). Superconducting proximity effect probed on a mesoscopic length scale. *Phys. Rev. Lett.*, Vol. 77, No. 14, (1996) 3025-3028.

Handbook of Chemistry and Physics, Chem. Pub., Cleveland.

Herath, J. & Rainer, D. (1989). Effects of impurities on Andreev reflection in an NS bilayer. *Physica C*, Vol. 161, No. 2, (1989) 209-218.

Hofstetter, L., Csonka, S., Nygard, J., et al. (2009). Cooper pair splitter realized in a two-quantum-dot Y-junction. *Nature*, Vol. 461, No. 10, (2009) 960-963.

Hsiang, T. Y. & Clarke, J. (1980). Boundary resistance of the superconducting - normal interface. *Phys. Rev. B*, Vol. 21, No. 3, (1980) 945-955.

Jedema, F. J., van Wees, B. J., Hoving, B. H., et al. (1999). Spin-accumulation-induced resistance in mesoscopic ferromagnet/superconductor junctions. *Phys. Rev. B*, Vol. 60, No. 24, (1999) 16549-16552.

Jin, B. Y. & Ketterson, J. B. (1989). Artificial metallic superlattices. *Adv. Phys.*, Vol. 38, No. 3, (1989) 189-366.

De Jong, M. J. M. & Beenakker, C. W. J. (1995). Andreev reflection in ferromagnet-superconductor junctions. *Phys. Rev. Lett.*, Vol. 74, No. 9, (1995) 1657-1660.

Kadigrobov, A. M. (1993). Multiple electron scattering by an impurity and at the NS boundary. Negative differential conductivity of the NSN system. *Low Temp. Phys.*, Vol. 19, No. 8, (1993) 671-672.

Kadigrobov, A., Zagoskin, A., Shekhter, R. I., et al. (1995). Giant conductance oscillations controlled by supercurrent flow through a ballistic mesoscopic conductor. *Phys. Rev. B*, Vol. 52, No. 12, (1995) R8662-R8665.

Kulik, I. O. (1969). Macroscopic quantization and proximity effect in S-N-S junctions. *Sov. Phys. JETP*, Vol. 30, No. 5, (1969) 944-958.

Lambert, C. J. & Raimondi, R. (1998). Phase-coherent transport in hybrid superconducting nanostructures. *J. Phys.: Condens. Matter*, Vol. 10, (1998) 901-941.

Landauer, R. (1970). Electrical resistance of disordered one-dimensional lattices. *Phil. Mag.*, Vol. 21, No. 172, (1970) 863-867.

Lifshitz, E. M. & Sharvin, Yu. V. (1951). On the intermediate state of superconductors. *Dokl. Akad. Nauk Ukr. SSR*, Vol. 79, No. 5, (1951) 783-786.

Petrashov, V. T., Sosnin, I. A., Cox, I., et al. (1999). Giant mutual proximity effects in ferromagnetic/superconducting nanostructures. *Phys. Rev. Lett.*, Vol. 83, No. 16, (1999) 3281-3284.

Sharvin, D. Yu. & Sharvin, Yu. V. (1981). Magnetic flux quantization in a cylindrical film of a normal metal. *JETP Lett.*, Vol. 34, No. 5, (1981) 272-275.

Soulen, R. J., Jr., Byers, J. M., Osofsky, M. S., et al. (1998). Measuring the spin polarization of a metal with a superconducting point contact. *Science*, Vol. 282, No. 5386, (1998) 85-88.

Spivak, B. Z. & Khmelnitskii, D. E. (1982). Influence of localization effects in a normal metal on the properties of SNS junction. *JETP Lett.*, Vol. 35, No. 8, (1982) 412-416.

Tsyan (Chiang), Yu. N. (2000). Resistance quantum oscillations in the intermediate state of singly connected Pb and Sn samples. *JETP Lett.*, Vol. 71, No. 8, (2000) 334-337.

Washburn, S. & Webb, R. A. (1986). Aharonov-Bohm effect in normal metal. Quantum coherence and transport. *Adv. Phys.*, Vol. 35, No. 4, (1986) 375-422.

Van Wees, B. J., de Vries, P., Magnic, P., et al. (1992). Excess conductance of superconductor-semiconductor interfaces due to phase conjunction between electrons and holes. *Phys. Rev. Lett.*, Vol. 69, No. 3, (1992) 510-513.

Zhou, F., Spivak, B., & Zyuzin, A. (1995). Coherence effects in a normal-metal-insulator-superconductor junction. *Phys. Rev. B*, Vol. 52, No. 6, (1995) 4467-4472.

7

Effects of Impurities on a Noncentrosymmetric Superconductor - Application to $CePt_3Si$

Heshmatollah Yavari
University of Isfahan
Iran

1. Introduction

In the past two decades, a number of novel superconducting materials have been discovered where order parameter symmetries are different from an *s*-wave spin singlet, predicted by the Bardeen-Cooper-Schrieffer (BCS) theory of electron-phonon mediated pairing. From the initial discoveries of unconventional superconductivity in heavy-fermion compounds, the list of examples has now grown to include the high-T_c cuprate superconductors, ruthenates, ferromagnetic superconductors, and possibly organic materials.

In most of these materials, there are strong indications that the pairing is caused by the electron correlations, in contrast to conventional superconductors such as Pb, Nb, etc. Nonphononic mechanisms of pairing are believed to favor a nontrivial spin structure and orbital symmetry of the Cooper pairs. For example, the order parameter in the high-T_c superconductors, where the pairing is thought to be caused by the antiferromagnetic correlations, has the *d*-wave symmetry with lines of zeroes at the Fermi surface. A powerful tool of studying unconventional superconducting states is symmetry analysis, which works even if the pairing mechanism is not known.

In general, the superconducting BCS ground state is formed by Cooper pairs with zero total angular momentum. The electronic states are four-fold degenerate $|k\downarrow\rangle$, and $|-k\uparrow\rangle$ have the same energy $\varepsilon(k)$. The states with opposite momenta and opposite spins are transformed to one another under time reversal operation $K|k\uparrow\rangle = |-k\downarrow\rangle$ and states with opposite momenta are transformed to one another under inversion operation $I|k\uparrow\rangle = |-k\uparrow\rangle$. The four degenerate states are a consequence of space and time inversion symmetries. Parity symmetry is irrelevant for spin-singlet pairing, but is essential for spin-triplet pairing. Time reversal symmetry is required for spin-singlet configuration, but is unimportant for spin-triplet state (Anderson, 1959, 1984).

If this degeneracy is lifted, for example, by a magnetic field or magnetic impurities coupling to the electron spins, then superconductivity is weakened or even suppressed. For spin-triplet pairing, Anderson noticed that additionally inversion symmetry is required to obtain the necessary degenerate electron states. Consequently, it became a widespread view that a material lacking an inversion center would be an unlikely candidate for spin-triplet pairing. For example, the absence of superconductivity in the paramagnetic phase of MnSi close to the quantum critical point to itinerant ferromagnetism was interpreted from this point of view (Mathur, 1998; Saxena, 2000). Near this quantum critical point the most natural spin fluctuation mediated Cooper pairing would occur in the spin-triplet channel. However, MnSi has the so-called *B*20 structure ($P2_1$), without an inversion center, inhibiting spin-triplet pairing.

Unusual properties are expected in superconductors whose crystal structure does not possess an inversion center (Edelstein, 1995; Frigeri et al., 2004; Gor'kov & Rashba, 2001; Samokhin et al., 2004; Sergienko& Curnoe, 2004).

Recent discovery of heavy fermion superconductor $CePt_3Si$ has opened up a new field of the study of superconductivity (Bauer et al., 2004). This is because this material does not have inversion center, which has stimulated further studies (Akazawa et al., 2004; Yogi et al., 2005). Because of the broken inversion symmetry, Rashba-type spin–orbit coupling (RSOC) is induced (Edelstein, 1995; Rashba, 1960; Rashba & Bychkov, 1984)), and hence different parities, spin-singlet pairing and spin triplet pairing, can be mixed in a superconducting state (Gor'kov & Rashba, 2001).

From a lot of experimental and theoretical studies, it is believed that the most possible candidate of superconducting state in $CePt_3Si$ is *s*+*p*-wave pairing (Frigeri et al., 2004; Hayashi et al., 2006). This mixing of the pairing channels with different parity may result in unusual properties of experimentally observed quantities such as a very high upper critical field H_{c2} which exceeds the paramagnetic limit (Bauer et al., 2004; Bauer et al., 2005a, 2005b; Yasuda et al., 2004), and the simultaneous appearance of a coherence peak feature in the NMR relaxation rate T_1^{-1} and low-temperature power-law behavior suggesting line nodes in the quasiparticle gap (Bauer et al., 2005a, 2005b; Yogi et al., 2004). The presence of line nodes in the gap of $CePt_3Si$ is also indicated by measurements of the thermal conductivity (Izawa et al., 2005) and the London penetration depth (Bauer et al., 2005; Bonalde et al., 2005).

It is known that the nonmagnetic as well as the magnetic impurities in the conventional and unconventional superconductors already have been proven to be a useful tool in distinguishing between various symmetries of the superconducting state (Blatsky et al., 2006). For example, in the conventional isotropic s-wave superconductor, the single magnetic impurity induced resonance state is located at the gap edge, which is known as Yu-Shiba-Rusinov state (Shiba, 1968). In the case of unconventional superconductor with $d_{x^2-y^2}$ -wave symmetry of the superconducting state, the nonmagnetic impurity-induced bound state appears near the Fermi energy as a hallmark of $d_{x^2-y^2}$ -wave pairing symmetry (Salkalo et al., 1996). The origin of this difference is understood as being due to the nodal structure of two kinds of SC order: in the $d_{x^2-y^2}$ -wave case, the phase of Cooper pairing wave function changes sign across the nodal line, which yields finite density of states (DOS) below the superconducting gap, while in the isotropic s-wave case, the density of states is gapped up to energies of about Δ_0 and thus the bound state can appear only at the gap edge. In principle the formation of the impurity resonance states can also occur in unconventional superconductors if the nodal line or point does not exist at the Fermi surface of a superconductor, as it occurs for isotropic nodeless *p*-wave and/or $d_x + id_y$ -wave superconductors for the large value of the potential strength (Wang Q.H. & Wang,Z.D, 2004).

In unconventional superconductors non-magnetic impurities act as pair-breakers, similar to magnetic impurities in s-wave superconductors. A bound state appears near an isolated non-magnetic strong (scattering phase shift $\frac{\pi}{2}$, or unitarity) scatterer, at the energy close to the Fermi level. The broadening of this bound state to an impurity band at finite disorder leads to a finite density of states at zero energy, $N(0)$, that increases with increasing impurity concentration (Borokowski & Hirschfeld, 1994). The impurity scattering changes the temperature dependence of the physical quantities below T corresponding to the impurity bandwidth: $\Delta\lambda$ changes the behavior from T to T^2 the NMR relaxation rate changes from T^3 to T, and specific heat $C(T)$ changes from T^2 to T. In other words, the impurities modify the power laws, especially at low temperatures.

The problem of a magnetic impurity in a superconductor has been extensively studied, but is not completely solved because of the difficulty of treating the dynamical correlations of the coupled impurity-conduction electron system together with pair correlations. Generally, the behavior of the system can be characterized by the ratio of the Kondo energy scale in the normal metal to the superconducting transition temperature $\frac{T_K}{T_c}$. For $\frac{T_K}{T_c} \ll 1$, conduction electrons scatter from classical spins and physics in this regime can be described by the Abrikosov-Gor'kov theory (Abrikosov & Gor'kov, 1961). In the opposite limit, $\frac{T_K}{T_c} \gg 1$, the impurity spin is screened and conduction electrons undergo only potential scattering. In this regime s-wave superconductors are largely unaffected by the presence of Kondo impurities due to Anderson's theorem. Superconductors with an anisotropic order parameter, e.g. p-wave, d-wave etc., are strongly affected, however and the potential scattering is pair-breaking. The effect of pair breaking is maximal in s-wave superconductors in the intermediate region, $T_K \sim T_c$, while in the anisotropic case it is largest for $\frac{T_K}{T_c} \to \infty$ (Borkowski & Hirschfeld, 1992).

In the noncentrosymmetric superconductor with the possible coexistence of *s*-wave and *p*-wave pairing symmetries, it is very interesting to see what the nature of the impurity state is and whether a low energy resonance state can still occur around the impurity through changing the dominant role played by each of the pairing components. Previously, the effect of nonmagnetic impurity scattering has been studied in the noncentrosymmetric superconductors with respect to the suppression of T_c and the behavior of the upper critical field (Frigeri et al., 2004; Mineev& Samokhin, 2007).

This in turn stimulates me to continue studying more properties. My main goal in this chapter is to find how the superconducting critical temperature, magnetic penetration depth, and spin–lattice relaxation rate of a noncentrosymmetric superconductor depend on the magnetic and nonmagnetic impurity concentration and also discuss the application of our results to a model of superconductivity in $CePt_3Si$. I do these by using the Green's function method when both s-wave and p-wave Cooper pairings coexist.

The chapter is organized as follows. In Sect. 2, the disorder averaged Green's functions in the superconducting states are calculated and the effect of impurity is treated via the self-

energies of the system. In Sect. 3, the equations for the superconducting gap functions renormalized by impurities are used to find the critical temperature T_c.
In Sect. 4, by using linear response theory I calculate the appropriate correlation function to evaluate the magnetic penetration depth. In this system the low temperature behavior of the magnetic penetration depth is consistence with the presence of line nodes in the energy gap.
In Sect. 5, the spin–lattice relaxation rate of nuclear magnetic resonance (NMR) in a superconductor without inversion symmetry in the presence of impurity effect is investigated.
In the last two cases I assume that the superconductivity in $CePt_3Si$ is most likely unconventional and our aim is to show how the low temperature power law is affected by nonmagnetic impurities.
Finally sect. 6 contains the discussion and conclusion remarks of my results.

2. Impurity scattering in normal and superconducting state

By using a single band model with electron band energy ξ_k measured from the Fermi energy where electrons with momentum k and spin s are created (annihilated) by operators $C^{\dagger}_{k,s}$ $(C_{k,s})$, the Hamiltonian including the pairing interaction can be written as

$$H = \sum_{k,s} \xi_k C^{\dagger}_{k,s} C_{k,s} + \frac{1}{2} \sum_{k,k'} \sum_{s,s'} V_{k,k'} C^{\dagger}_{k,s} C^{\dagger}_{-k,s'} C_{-k',s'} C_{k',s} \tag{1}$$

This system possesses time reversal and inversion symmetry $(\xi_k = \xi_{-k})$ and the pairing interaction does not depend on the spin and favors either even parity (spin-singlet) or odd parity (spin-triplet) pairing as required. The absence of inversion symmetry is incorporated through the antisymmetric Rashba-type spin-orbit coupling

$$H_{so} = \sum_{k,s,s'} \alpha \vec{g}_k . \vec{\sigma}_{s,s'} C^{\dagger}_{ks} C_{ks'} \tag{2}$$

which removes parity but conserves time-reversal symmetry, i.e., $IH_{so}I^{-1} = -H_{so}$ and $TH_{so}T^{-1} = H_{so}$. In Eq. (2), σ denotes the Pauli matrices (this satisfies the above condition $I\sigma I^{-1} = -\sigma$ and $T\sigma T^{-1} = \sigma$), g_k is a dimensionless vector [$g_k = -g_{-k}$ to preserve time reversal symmetry], and $\alpha(>0)$ denotes the strength of the spin-orbit coupling. The antisymmetric spin-orbit coupling (ASOC) term $\alpha \vec{g}_k . \vec{\sigma}$ is different from zero only for crystals without an inversion center and can be derived microscopically by considering the relativistic corrections to the interaction of the electrons with the ionic potential (Frigeri et al., 2004; Dresselhaus, 1995). For qualitative studies, it is sufficient to deduce the structure of the g-vector from symmetry arguments (Frigeri et al., 2004) and to treat α as a parameter. I set $\langle g_k^2 \rangle_k = 1$, where $\langle ... \rangle$ denotes the average over the Fermi surface. The ASOC term lifts the spin degeneracy by generating two bands with different spin structure.
In the normal state the eigenvalues of the total Hamiltonan $(H + H_{so})$ are

$$\xi_{k\pm} = \varepsilon_k - \mu \mp \alpha |g_k| \tag{3}$$

where $\varepsilon_k = \dfrac{k^2}{2m}$ and and μ is the chemical potential.

It is obvious from here that the time reversal symmetry is lost and the shape of the Fermi surfaces does not obey the mirror symmetry.

Due to the big difference between the Fermi momenta we neglected the pairing of electronic states from different bands. The structure of theory is now very similar to the theory of ferromagnetic superconductors with triplet pairing (Mineev, 2004).

Effects of disorder are described by potential scattering of the quasiparticles, which in real-space representation is given by

$$H_{imp} = \sum_i \int \psi_s^\dagger(\vec{r}) U_{imp} \psi_s(\vec{r}) d\vec{r} \tag{4}$$

where $U_{imp} = U_n + U_m$, U_n is the potential of a non-magnetic impurity, which we consider rather short-ranged such that s-wave scattering is dominant and $U_m = J(\vec{r})\vec{S}.\vec{\sigma}$ is the potential interaction between the local spin on the impurity site and conduction electrons, here J is the exchange coupling and S is the spin operator.

2.1 Impurity averaging in superconducting state

Let us calculate the impurity-averaged Green`s functions in the superconducting state. The Gor'kov equations with self-energy contributions are formally analogous to those obtained for system with inversion symmetry (Abrikosov et al., 1975).

$$\left(i\omega_n - \xi_{k\pm} - \tilde{\Sigma}_G(i\omega_n)\right)\tilde{\Im}_\pm(k,i\omega_n) + \left(\Delta_k + \tilde{\Sigma}_F(i\omega_n)\right)\tilde{F}_\pm^\dagger(k,\omega_n) = \hat{\sigma}_0 \tag{5}$$

$$\left(i\omega_n + \xi_{-k\pm} + \tilde{\Sigma}_G^\dagger(i\omega_n)\right)\tilde{F}_\pm^\dagger(k,\omega_n) + \left(\Delta_k^\dagger + \tilde{\Sigma}_F^\dagger(i\omega_n)\right)\tilde{\Im}_\pm(k,\omega_n) = 0 \tag{6}$$

where $\omega_n = (2n+1)\pi T$ are the Matsubara Fermionic frequencies, $\hat{\sigma}_0$ is the unit matrix in the spin state, and the impurity scattering enters the self-energy of the Green`s function of the normal, $\tilde{\Sigma}_G$, and the anomalous type, $\tilde{\Sigma}_F$, their mathematical expressions read

$$\tilde{\Sigma}_G(i\omega_n) = \left(n_n|U_n|^2 + n_m|U_m|^2\right)\int \frac{d\vec{k}'}{(2\pi)^3}\tilde{\Im}\left(\vec{k}', i\omega_n\right) \tag{7}$$

$$\tilde{\Sigma}_F(i\omega_n) = \left(n_n|U_n|^2 + n_m|U_m|^2\right)\int \frac{d\vec{k}'}{(2\pi)^3}\tilde{F}\left(\vec{k}', i\omega_n\right) \tag{8}$$

here n_n and n_m are the concentrations of nonmagnetic and magnetic impurities, respectively.

The equations for each band are only coupled through the order parameters given by the self-consistency equations

$$\Delta_{k\pm} = -T \sum_{k',\upsilon,n} V_{\pm\upsilon}\left(\vec{k},\vec{k}'\right) F_\upsilon\left(\vec{k}',\omega_n\right) \tag{9}$$

where $\upsilon = \pm$.

Solving the Gor'kov equations one obtains the following expressions for the disorder-averaged Green's functions

$$\tilde{\Im}_{\pm}\left(\vec{k},\omega_n\right)=\begin{pmatrix}\Im_{\pm}\left(\vec{k},i\omega_n\right) & F_{\pm}\left(\vec{k},i\omega_n\right)\\ F_{\pm}^{\dagger}\left(\vec{k},i\omega_n\right) & -\Im_{\pm}\left(-\vec{k},-i\omega_n\right)\end{pmatrix} \tag{10}$$

where

$$\Im_{\pm}\left(\vec{k},\omega_n\right)=\frac{i\omega_n+\xi_{-k\pm}}{\left(i\omega_n-\Sigma_{imp}+\vec{v}_s.\vec{k}_F-\xi_{k\pm}\right)\left(i\omega_n-\Sigma_{imp}+\vec{v}_s.\vec{k}_F+\xi_{-k\pm}\right)-\Delta_{k\pm}\Delta_{k\pm}^{\dagger}} \tag{11}$$

$$F_{\pm}\left(\vec{k},\omega_n\right)=\frac{-\Delta_{k\pm}}{\left(i\omega_n-\Sigma_{imp}+\vec{v}_s.\vec{k}_F-\xi_{k\pm}\right)\left(i\omega_n-\Sigma_{imp}+\vec{v}_s.\vec{k}_F+\xi_{-k\pm}\right)-\Delta_{k\pm}\Delta_{k\pm}^{\dagger}} \tag{12}$$

here $\Sigma_{imp}=\Sigma_{imp(n)}+\Sigma_{imp(m)}$ is the self energy due to non magnetic and magnetic impurities. The energies of elementary excitations are given by

$$E_{k\pm}=\frac{\xi_{k\pm}-\xi_{-k\pm}}{2}\pm\sqrt{\left(\frac{\xi_{k\pm}+\xi_{-k\pm}}{2}\right)^2+\Delta_{k\pm}\Delta_{k\pm}^{\dagger}} \tag{13}$$

The presence of the antisymmetric spin-orbit coupling would suppress spin-triplet pairing. However, it has been shown by Frigeri et al., (Frigeri et al., 2004) that the antisymmetric spin-orbit coupling is not destructive to the special spin-triplet state with the d vector parallel to $\vec{g}_k$ $\left(\vec{d}_k \parallel \vec{g}_k\right)$. Therefore, by choosing $\vec{g}_k=\sqrt{\frac{3}{2}}\frac{1}{k_F}\left(-k_y,k_x,0\right)$, one adopts the p-wave pairing state with parallel $\vec{d}$ vector $d_k=\Delta\left(-\tilde{k}_y,\tilde{k}_x,0\right)$. Here the unit vector $\tilde{k}=\left(\tilde{k}_x,\tilde{k}_y,\tilde{k}_z\right)=\left(\cos\varphi\sin\theta,\sin\varphi\sin\theta,\cos\varphi\right)$.

By considering this parity-mixed pairing state the order parameter defined in (5) and (6) can be expressed as

$$\Delta\left(\vec{r},\vec{k}\right)=\left[\Delta_0\left(\vec{r}\right)\hat{\sigma}_0+\vec{d}\left(\vec{k}\right).\vec{\sigma}\right]i\hat{\sigma}_y=\left[\Delta_0\left(\vec{r}\right)\hat{\sigma}_0+\Delta\left(\vec{r}\right)\left(-k_y\hat{\sigma}_x+k_x\hat{\sigma}_y\right)\right] \tag{14}$$

with the spin-singlet s-wave component $\Delta_0\left(\vec{r}\right)$ and the $\vec{d}$ vector $\vec{d}_k\left(\vec{r}\right)=\Delta\left(\vec{r}\right)\left(-\vec{k}_y,\vec{k}_x,0\right)$, here, the vector $\vec{r}$ indicates the real-space coordinates. While this spin-triplet part alone has point nodes (axial state with two point nodes), the pairing state of Eq. (14) can possess line nodes in a gap as a result of the combination with the s-wave component (Hayashi et al., 2006; Sergienko 2004). In the presence of uniform supercurrent the gap function has the $\vec{r}$ dependence as

$$\Delta\left(\vec{r},\vec{k}\right)=\Delta_k e^{i2m\vec{v}_s.\vec{r}} \tag{15}$$

where m is the bare electron mass.

The particular form of order parameter prevents the existence of interband terms in the Gor'kov equations

$$\left(i\omega_n - \xi_{k,\pm} - \Sigma_G(i\omega_n)\right)\Im_{\pm}(k,i\omega_n) + \left(\Delta_{k,\pm} + \Sigma_F(i\omega_n)\right)F_{\pm}^{\dagger}(k,\omega_n) = 1 \quad (16)$$

$$\left(i\omega_n + \xi_{-k,\pm} + \Sigma_G^{\dagger}(i\omega_n)\right)F_{\pm}^{\dagger}(k,\omega_n) + \left(\Delta_{k,\pm}^{\dagger} + \Sigma_F^{\dagger}(i\omega_n)\right)\Im_{\pm}(k,\omega_n) = 0 \quad (17)$$

where in this case

$$\tilde{\Sigma}_G(i\omega_n) = \left(n_n|U_n|^2 + n_m|U_m|^2\right)\int\frac{d\vec{k}'}{(2\pi)^3}\left[\Im_+(\vec{k}',i\omega_n) + \Im_-(\vec{k}',i\omega_n)\right] \quad (18)$$

$$\tilde{\Sigma}_F(i\omega_n) = \left(n_n|U_n|^2 + n_m|U_m|^2\right)\int\frac{d\vec{k}'}{(2\pi)^3}\left[F_+(\vec{k}',i\omega_n) + F_-(\vec{k}',i\omega_n)\right] \quad (19)$$

and

$$\Delta_{\pm} = \Delta_0 \pm d|g_k| \quad (20)$$

I consider the superconducting gaps $|\Delta_0 + \Delta\sin\theta|$ and $|\Delta_0 - \Delta\sin\theta|$ on the Fermi surfaces I and II, respectively (such as superconductor $CePt_3Si$). Such a gap structure can lead to line nodes on either Fermi surface I or II (Hayashi et al., 2006). These nodes are the result of the superposition of spin-singlet and spin-triplet contributions (each separately would not produce line nodes). On the Fermi surface I, the gap is $|\Delta_0 + \Delta\sin\theta|$ and is nodeless, (not that we choose $\Delta_0 > 0$ and $\Delta > 0$). On the other hand, the form of the gap on the Fermi surface II is $|\Delta_0 - \Delta\sin\theta|$, where line nodes can appear for $\Delta_0 < \Delta$ (Hayashi et al., 2006).

3. Effects of impurities on the transition temperature of a noncentrosymmetrical superconductor

In the case of large SO band splitting, the order parameter has only intraband components and the gap equation (Eq. (9)) becomes

$$\Delta_{k\pm} = -T\sum_{n,\nu}\int\frac{d^3k'}{(2\pi)^3}V_{\pm\upsilon}(\vec{k},\vec{k}')\frac{\Delta_{k\pm}}{\left(i\omega_n - \Sigma_{imp} - \xi_{k\pm}\right)\left(i\omega_n - \Sigma_{imp} + \xi_{-k\pm}\right) - \Delta_{k\pm}\Delta_{k\pm}^{\dagger}} \quad (21)$$

The coupling constants $V_{\lambda\lambda'}(\vec{k},\vec{k}')$' I have used in previous considerations can be expressed through the real physical interactions between the electrons naturally introduced in the initial spinor basis where BCS type Hamiltonian has the following form

$$H_{\text{int}} = \frac{1}{4\Omega}\sum_{kk'q}\sum_{\alpha\beta\mu\delta}\left[V_{\alpha\beta\mu\delta}^{s}(\vec{k},\vec{k}') + V_{\alpha\beta\mu\delta}^{t}(\vec{k},\vec{k}') + V_{\alpha\beta\mu\delta}^{m}(\vec{k},\vec{k}')\right] \times c_{\vec{k}+\vec{q},\lambda}^{\dagger}c_{-\vec{k},\lambda}^{\dagger}c_{-\vec{k}',\lambda}c_{\vec{k}+\vec{q},\lambda'} \quad (22)$$

where the pairing interaction is represented as a sum of the k-even, k-odd, and mixed-parity terms: $V = V^s + V^t + V^m$. The even contribution is

$$V^{s}_{\alpha\beta\mu\delta}\left(\vec{k},\vec{k}'\right) = V^{s}\left(\vec{k},\vec{k}'\right)\left(i\sigma_2\right)_{\alpha\beta}\left(i\sigma_2\right)^{\dagger}_{\mu\delta} \tag{23}$$

The odd contribution is

$$V^{t}_{\alpha\beta\mu\delta}\left(\vec{k},\vec{k}'\right) = V^{t}_{ij}\left(\vec{k},\vec{k}'\right)\left(i\sigma_i\sigma_2\right)_{\alpha\beta}\left(i\sigma_j\sigma_2\right)^{\dagger}_{\mu\delta} \tag{24}$$

here the amplitudes $V^s\left(\vec{k},\vec{k}'\right)$ and $V^t_{ij}\left(\vec{k},\vec{k}'\right)$ are even and odd with respect to their arguments correspondingly.

Finally, the mixed-parity contribution is

$$\begin{aligned} V^{m}_{\alpha\beta\mu\delta}\left(\vec{k},\vec{k}'\right) &= V^{m}_{i}\left(\vec{k},\vec{k}'\right)\left(i\sigma_i\sigma_2\right)_{\alpha\beta}\left(i\sigma_2\right)^{\dagger}_{\mu\delta} \\ &+ V^{m}_{i}\left(\vec{k}',\vec{k}\right)\left(i\sigma_2\right)_{\alpha\beta}\left(i\sigma_i\sigma_2\right)^{\dagger}_{\mu\delta} \end{aligned} \tag{25}$$

The first term on the right-hand side of Eq. (25) is odd in k and even in k', while the second term is even in k and odd in k'.

The pairing interaction leading to the gap function [Eq. (14)] is characterized by three coupling constants, V_s, V_t, and V_m. Here, V_s, and V_t result from the pairing interaction within each spin channel (s: singlet, t: triplet). V_m is the scattering of Cooper pairs between those two parity channels, present in systems without inversion symmetry. The linearized gap equations acquire simple algebraic form

$$\Delta_0 = V_s \pi T \sum_n \left\langle \mathrm{E}_+ \right\rangle + V_m \pi T \sum_n \left\langle \sin\theta \mathrm{E}_- \right\rangle \tag{26}$$

$$\Delta = V_t \pi T \sum_n \left\langle \sin\theta \mathrm{E}_- \right\rangle + V_m \pi T \sum_n \left\langle \mathrm{E}_+ \right\rangle \tag{27}$$

where the angular brackets denote the average over the Fermi surface, assuming the spherical Fermi surface for simplicity, $\mathrm{E}_{\pm} = \dfrac{\mathrm{E}_I \pm \mathrm{E}_{II}}{2}$, $\mathrm{E}_{I,II} = \dfrac{\Delta_0 \pm \Delta\sin\theta}{\mathrm{B}_{I,II}}$, and

$$\mathrm{B}_{I,II} = \left[\left(\omega_n + i\Sigma_{imp}\right)^2 + \left|\Delta_0 \pm \Delta\sin\theta\right|^2\right]^{1/2} \tag{28}$$

From Eqs. (26) and (27) one obtains then the following expression for the critical temperature

$$\ln\frac{T_{c0}}{T_c} = (1-\mathrm{X})\left[\Psi\left(\frac{1}{2}+\frac{1}{4\pi T\tau_m}\right)-\Psi\left(\frac{1}{2}\right)\right] + \mathrm{X}\left\{\Psi\left[\frac{1}{2}+\frac{1}{4\pi T_c}\left(\frac{1}{\tau_n}+\frac{1}{\tau_m}\right)-\Psi\left(\frac{1}{2}\right)\right]\right\} \tag{29}$$

where

$$\frac{1}{\tau_n} = 2\pi n_n N_0 \left|U_n\right|^2 \qquad \frac{1}{\tau_m} = 2\pi n_m N_0 \left|U_m\right|^2 \tag{30}$$

$\Psi(x)$ is the digamma function, $N_0 = (N_+ + N_-)/2$, $N_\pm$ are the densities of state (DOS) of the two bands at the Fermi level, and T_{c0} is the critical temperature of the clean superconductor.

The coefficient $X = 1 - \frac{\langle \Delta(p) \rangle_{FS}^2}{\langle \Delta^2(p) \rangle_{FS}}$ quantifies the degree of anisotropy of the order parameter on the Fermi surface (FS), where the angular brackets $\langle ... \rangle_{FS}$ stand for a FS average.

For isotropic *s*-wave pairing $\langle \Delta(p) \rangle_{FS}^2 = \langle \Delta^2(p) \rangle_{FS}$ $(X = 0)$ and for any pairing state with angular momentum $l > 1$, e.g., for *p*-wave and d-wave states $(l = 1,2)$, $\left(X = 1, \frac{1}{\tau_m} = 0 \right)$ Eq. (29) reduces to the well-known expressions (Abrikosov, 1993; Abrikosov, A. A. & Gor'kov, 1959).

$$\ln \frac{T_{c0}}{T_c} = \Psi\left(\frac{1}{2} + \frac{1}{4\pi T_c \tau_m} \right) - \Psi\left(\frac{1}{2} \right) \tag{31}$$

$$\ln \frac{T_{c0}}{T_c} = \Psi\left(\frac{1}{2} + \frac{1}{4\pi T_c \tau_n} \right) - \Psi\left(\frac{1}{2} \right) \tag{32}$$

For mixing of *s*-wave state with some higher angular harmonic state , e.g., for example $s + p$ and $s + d$, $\left(0 < X < 1, \frac{1}{\tau_m} = 0 \right)$, Eq. (29) becomes

$$\ln \frac{T_{c0}}{T_c} = X \left[\Psi\left(\frac{1}{2} + \frac{1}{4\pi T_c \tau_n} \right) - \Psi\left(\frac{1}{2} \right) \right] \tag{33}$$

At $\tau_n T_{c0} \gg 1$ and $\tau_m T_{c0} \gg 1$ (weak scattering) one has from Eq. (29):

$$T_{c0} - T_c \approx \frac{\pi}{4} \left[\frac{X}{2} \frac{1}{\tau_n} + \frac{1 - X/2}{\tau_m} \right] \tag{34}$$

In two particular cases of (i) both nonmagnetic and magnetic scattering in an isotropic s-wave superconductor ($X = 0$) and (ii) nonmagnetic scattering only in a superconductor with arbitrary anisotropy of $\Delta(p)$ ($\frac{1}{\tau_m} = 0, 0 < X < 1$), the Eq. (34) reduces to well-known expressions

$$T_{c0} - T_c \approx \frac{\pi}{4\tau_m} \tag{35}$$

$$T_{c0} - T_c \approx \frac{\pi X}{8\tau_n} \tag{36}$$

In the strong scattering limit ($\tau_n T_c \ll 1$, $\tau_m T_c \ll 1$), by using

$$\Psi\left(\frac{1}{2}+\frac{1}{4\pi T_c\tau}\right)-\Psi\left(\frac{1}{2}\right)\approx \ln\left(\frac{\gamma}{\pi\tau T_c}\right)+\frac{2\pi^2}{3}(\tau T)^2+O(\tau T)^3 \quad (37)$$

From Eq. (29) one finds

$$\left(\frac{1}{\tau_m}\right)^{1-X}\left(\frac{1}{\tau_n}+\frac{1}{\tau_m}\right)^{X}=\frac{\pi}{\gamma}T_{c0}2^{X-1} \quad (38)$$

One can see that the left hand side of Eq. (38) increases monotonically with both $\frac{1}{\tau_n}$ and $\frac{1}{\tau_m}$ for any value of X , with the exception of the case $X=0$ which does not depend on magnetic impurities.

For strongly anisotropic gap parameter $(X\sim 1)$, Eq. (38) reduces to

$$\frac{1}{\tau_n}+\frac{1}{\tau_m}=\frac{\pi}{\gamma}T_{c0} \quad (39)$$

i.e., the contribution of magnetic and nonmagnetic impurities to pairing breaking is about the same.

For strongly isotropic case $(X\ll 1)$, one has

$$\frac{1}{\tau_m}=\frac{\pi}{2\gamma}T_{c0} \quad (40)$$

and T_c is determined primarily by magnetic impurities.

For the case of $s+p$ wave pairing in the absence of magnetic impurities, one has

$$\left(\frac{1}{\tau_n}\right)^{X}=\frac{\pi}{\gamma}2^{X-1}T_{c0} \quad (41)$$

In this case the value of T_c asymptotically goes to zero as τ_n^{-1} increase, whereas T_c of a d or p wave superconductor with $X=1$ vanishes at a critical value $\frac{1}{\tau_n^c}=\frac{\pi T_{c0}}{\gamma}$.

In the absence of nonmagnetic impurities one obtains

$$\left(\frac{1}{\tau_m}\right)=\frac{\pi}{\gamma}2^{X-1}T_{c0} \quad (42)$$

And for the s-wave superconductor with $X=0$ one has $\frac{1}{\tau_m^c}=\frac{\pi}{2\gamma}T_{c0}$.

Application of these results to real noncentrosymmetric materials is complicated by the lack of definite information about the superconducting gap symmetry and the distribution of the pairing strength between the bands.

As far as the pairing symmetry is concerned, there is strong experimental evidence that the superconducting order parameter in $CePt_3Si$ has lines of gap nodes (Yasuda et al., 2004; Izawa et al., 2005; Bonalde et al., 2005). The lines of nodes are required by symmetry for all nontrivial one-dimensional representations of C_{4v} (A_2, B_1, and B_2), so that the superconductivity in $CePt_3Si$ is most likely unconventional. This can be verified using the measurements of the dependence of T_c on the impurity concentration: For all types of unconventional pairing, the suppression of the critical temperature is described by the universal Abrikosov-Gor'kov function, see Eq. (32).

It should be mentioned that the lines of gap nodes can exist also for conventional pairing (A_1 representation), in which case they are purely accidental. While the accidental nodes would be consistent with the power-law behavior of physical properties observed experimentally, the impurity effect on T_c in this case is qualitatively different from the unconventional case. In this case in the absence of magnetic impurities one obtains the following equation for the critical temperature:

$$\ln\frac{T_{c0}}{T_c} = \mathrm{X}\left[\Psi\left(\frac{1}{2}+\frac{1}{4\pi T_c\tau_n}\right)-\Psi\left(\frac{1}{2}\right)\right] \tag{43}$$

In the low $(\tau_n T_c \gg 1)$ and dirty $(\tau_n T_{c0} \ll 1)$ limit of impurity concentration one has

$$T_{c0} - T_c \approx \frac{\mathrm{X}\pi}{8\tau_n} \qquad \tau T_{c0} \gg 1 \tag{44}$$

$$T_c = T_{c0}\left(\frac{\pi\tau_n T_{c0}}{e^C}\right)^{\frac{\mathrm{X}}{1-\mathrm{X}}} \qquad \tau T_{c0} \ll 1 \tag{45}$$

This means that anisotropy of the conventional order parameter increases the rate at which T_c is suppressed by impurities. Unlike the unconventional case, however, the superconductivity is never completely destroyed, even at strong disorder.

4. Low temperature magnetic penetration depth of a superconductor without inversion symmetry

To determine the penetration depth or superfluid density in asuperconductor without inversion symmetry one calculates the electromagnetic response tensor $K(\vec{q},\vec{v}_s,T)$, relating the current density $\vec{J}$ to an applied vector potential $\vec{A}$

$$\vec{J}(q) = -K(\vec{q},\vec{v}_s,T)\vec{A}(\vec{q}) \tag{46}$$

The expression for the response function can be obtained as

$$K(\vec{q},\vec{v}_s,T,\omega_m) = \frac{ne^2}{mc}\left(1+\frac{2\pi}{m}\left\langle T\sum_{n,k}\hat{k}_{\parallel}^2\Im_{\pm}\left(\vec{k}_+,\omega_n\right)\Im_{\pm}\left(\vec{k}_-,\omega_n,\omega_m\right)\right\rangle\right) \tag{47}$$

where $k_{\pm} = k \pm \frac{q}{2}$, $\hat{k}_{\|}^{2}$ is the direction of the supercurrent and $\langle .... \rangle$ represents a Fermi surface average.
By using the expression of Green`s function into Eq. (47) one obtains

$$K(\vec{q},\vec{v}_s,T,\omega_n) = \frac{ne^2}{mc} \left(1 + \frac{2\pi T}{m} \sum_n \int \frac{d^2k}{(2\pi)^2} \hat{k}_{\|}^{2} \frac{\left(i\omega_n - \Sigma_{imp} + \vec{v}_s.\vec{k}_F\right)^2 + \xi_{k_+,\pm}\xi_{k_-,\pm} + \Delta_{k_+}\Delta_{k_-}}{\left[\left(i\omega_n - \Sigma_{imp} + \vec{v}_s.\left(\vec{k} + \frac{\vec{q}}{2}\right)\right)^2 - E^2_{k_+,\pm}\right]\left[\left(i\omega_n - \Sigma_{imp} + \vec{v}_s.\left(\vec{k} - \frac{\vec{q}}{2}\right)\right)^2 - E^2_{k_-,\pm}\right]} \right) \tag{48}$$

Now we separate out the response function as

$$K(\vec{q},\vec{v}_s,T) = K(0,0,0) + \delta K(\vec{q},\vec{v}_s,T) \tag{49}$$

where $K(0,0,0) = \frac{c}{4\pi\lambda^2(0)}$ ($\lambda(0) = \left(\frac{mc^2}{4\pi ne^2}\right)^{1/2}$ is the zero temperature London penetration depth).
Doing the summation over Matsubara frequencies for each band one gets

$$\begin{aligned}
\delta K(\vec{q},\vec{v}_s,T) = & -\frac{2n_+e^2}{mc} \left\langle \hat{k}_{\|}^{2} \int_0^{\infty} d\omega \, \mathrm{Re} \frac{\left[f\left(\omega - \vec{v}_s.\vec{k}_F\right) - f\left(-\omega - \vec{v}_s.\vec{k}_F\right)\Delta^2_{k,+}\right]}{\sqrt{\left(\omega - i\Sigma_{imp}\right)^2 - \Delta^2_{k,+}}\left[\Delta^2_{k,+} + \left(\frac{\vec{q}.\vec{k}_F}{2m}\right)^2 - (2\alpha g_k)^2 - \left(\omega - i\Sigma_{imp}\right)^2\right]} \right\rangle \\
& -\frac{2n_e^2}{mc} \left\langle \hat{k}_{\|}^{2} \int_0^{\infty} d\omega \, \mathrm{Re} \frac{\left[f\left(\omega - \vec{v}_s.\vec{k}_F\right) - f\left(-\omega - \vec{v}_s.\vec{k}_F\right)\Delta^2_{k,-}\right]}{\sqrt{\left(\omega - i\Sigma_{imp}\right)^2 - \Delta^2_{k,-}}\left[\Delta^2_{k,-} + \left(\frac{\vec{q}.\vec{k}_F}{2m}\right)^2 - (2\alpha g_k)^2 - \left(\omega - i\Sigma_{imp}\right)^2\right]} \right\rangle \\
= & \frac{-n_+e^2}{mc} \left\{ 1 - 2\left\langle \hat{k}_{\|}^{2} \frac{\sinh^{-1}\left(\frac{\vec{q}.\vec{k}_F}{2m\Delta_{k,+}}\right)}{\left(\frac{\vec{q}.\vec{k}_F}{2m\Delta_{k,+}}\right)\sqrt{1 + \left(\frac{\vec{q}.\vec{k}_F}{2m\Delta_{k,+}}\right)^2 + \left(\frac{2\alpha g_k}{\Delta_{k,+}}\right)^2}} \right\rangle \right. \\
& + \left\langle \hat{k}_{\|}^{2} \ln\left[\frac{\sqrt{\Delta^2_{k,+} + \left(\frac{\vec{q}.\vec{k}_F}{2m}\right)^2} - \left(1 + \left(\frac{\vec{q}.\vec{k}_F}{2m\Delta_{k,+}}\right)^2 - \left(\frac{2\alpha g_k}{\Delta_{k,+}}\right)^2\right)\Sigma_{imp} - \left(\frac{\vec{q}.\vec{k}_F}{2m\Delta_{k,+}}\right)\sqrt{1 + \left(\frac{\vec{q}.\vec{k}_F}{2m\Delta_{k,+}}\right)^2 - \left(\frac{4m\Delta_{k,+}\alpha g_k}{\vec{q}.\vec{k}_F}\right)^2}\sqrt{\Sigma^2_{imp} - \Delta^2_{k,+}}}{\sqrt{\Delta^2_{k,+} + \left(\frac{\vec{q}.\vec{k}_F}{2m}\right)^2} + \left(1 + \left(\frac{\vec{q}.\vec{k}_F}{2m\Delta_{k,+}}\right)^2 - \left(\frac{2\alpha g_k}{\Delta_{k,+}}\right)^2\right)\Sigma_{imp} - \left(\frac{\vec{q}.\vec{k}_F}{2m\Delta_{k,+}}\right)\sqrt{1 + \left(\frac{\vec{q}.\vec{k}_F}{2m\Delta_{k,+}}\right)^2 - \left(\frac{4m\Delta_{k,+}\alpha g_k}{\vec{q}.\vec{k}_F}\right)^2}\sqrt{\Sigma^2_{imp} - \Delta^2_{k,+}}} \right] \right\rangle \\
& \left. + 2\left\langle \hat{k}_{\|}^{2} \int d\omega \, \mathrm{Re} \frac{\left[f\left(\omega - \vec{v}_s.\vec{k}_F\right) + f\left(\omega + \vec{v}_s.\vec{k}_F\right)\right]}{\sqrt{\left(\omega - i\Sigma_{imp}\right)^2 - \Delta^2_{k,+}}\left[\Delta^2_{k,+} + \left(\frac{\vec{q}.\vec{k}_F}{2m}\right)^2 - (2\alpha g_k)^2 - \left(\omega - i\Sigma_{imp}\right)^2\right]} \right\rangle \right\}
\end{aligned} \tag{50}$$

$$
\begin{aligned}
&\frac{-n\_e^2}{mc}\left\{1-2\left\langle \hat{k}_{\|}^2 \frac{\sinh^{-1}\left(\frac{\vec{q}.\vec{k}_F}{2m\Delta_{k,-}}\right)}{\left(\frac{\vec{q}.\vec{k}_F}{2m\Delta_{k,-}}\right)\sqrt{1+\left(\frac{\vec{q}.\vec{k}_F}{2m\Delta_{k,-}}\right)^2+\left(\frac{2\alpha g_k}{\Delta_{k,-}}\right)^2}}\right\rangle\right. \\
&+\left\langle \hat{k}_{\|}^2 \ln\left[\frac{\sqrt{\Delta_{k,-}^2+\left(\frac{\vec{q}.\vec{k}_F}{2m}\right)^2}-\left(1+\left(\frac{\vec{q}.\vec{k}_F}{2m\Delta_{k,-}}\right)^2-\left(\frac{2\alpha g_k}{\Delta_{k,-}}\right)^2\right)\Sigma_{imp}-\left(\frac{\vec{q}.\vec{k}_F}{2m\Delta_{k,-}}\right)\sqrt{1+\left(\frac{\vec{q}.\vec{k}_F}{2m\Delta_{k,-}}\right)^2-\left(\frac{4m\Delta_{k,-}\alpha g_k}{\vec{q}.\vec{k}_F}\right)^2}\sqrt{\Sigma_{imp}^2-\Delta_{k,-}^2}}{\sqrt{\Delta_{k,-}^2+\left(\frac{\vec{q}.\vec{k}_F}{2m}\right)^2}+\left(1+\left(\frac{\vec{q}.\vec{k}_F}{2m\Delta_{k,-}}\right)^2-\left(\frac{2\alpha g_k}{\Delta_{k,-}}\right)^2\right)\Sigma_{imp}-\left(\frac{\vec{q}.\vec{k}_F}{2m\Delta_{k,-}}\right)\sqrt{1+\left(\frac{\vec{q}.\vec{k}_F}{2m\Delta_{k,-}}\right)^2-\left(\frac{4m\Delta_{k,-}\alpha g_k}{\vec{q}.\vec{k}_F}\right)^2}\sqrt{\Sigma_{imp}^2-\Delta_{k,-}^2}}\right]\right\rangle \\
&\left.+2\left\langle \hat{k}_{\|}^2 \int d\omega \,\mathrm{Re}\frac{\left[f\left(\omega-\vec{v}_s.\vec{k}_F\right)+f\left(\omega+\vec{v}_s.\vec{k}_F\right)\right]}{\sqrt{\left(\omega-i\Sigma_{imp}\right)^2-\Delta_{k,-}^2}\left[\Delta_{k,-}^2+\left(\frac{\vec{q}.\vec{k}_F}{2m}\right)^2-\left(2\alpha g_k\right)^2-\left(\omega-i\Sigma_{imp}\right)^2\right]}\right\rangle\right\}
\end{aligned}
\tag{50}
$$

The factor αg_k characterizes and quantifies the absence of an inversion center in a crystal lattice. This is the main result of my work i.e. nonlocality, nonlineary, impurity and nonsentrosymmetry are involved in the response function. The first two terms in Eq. (50) represent the nonlocal correction to the London penetration depth and the third represents the nonlocal and impure renormalization of the response while the forth combined nonlocal, nonlinear, and impure corrections to the temperature dependence.

I consider a system in which a uniform supercurrent flows with the velocity $\vec{v}_s$, so all quasiparticles Matsubara energies modified by the semiclassical Doppler shift $\vec{v}_s.\vec{k}_F$.

The specular boundary scattering in terms of response function can be written as (Kosztin & Leggett, 1997)

$$\frac{\Delta\lambda_{spec}(T)}{\lambda_0}=\frac{2}{\pi}\int_0^{\infty} d\tilde{q}\frac{\delta K(\tilde{q},\vec{v}_s,T)}{\left(\tilde{q}^2+1\right)^2} \tag{51}$$

In the pure case there are four relevant energy scales in the low energy sector in the Meissner state: T, E_{nonlin}, E_{nonloc}, and αg_k. The first two are experimentally controlled parameters while the last two are intrinsic one.

In low temperatures limit the contribution of the fully gap ($|\Delta_0+\Delta\sin\theta|$) Fermi surface I decrease and the effect of the gap $|\Delta_0-\Delta\sin\theta|$ Fermi surface II is enhanced. I consider geometry where the magnetic field is parallel to c axis and thus $\vec{v}_s$ and the penetration direction $\tilde{q}$ are in the ab plane, and in general, $\vec{v}_s$ makes an angle φ with the axis. There are two effective nonlinear energy scales $E^{+}_{nonlin}=v_s k_F u_{\varphi l_1}$ and $E^{-}_{nonlin}=v_s k_F u_{\varphi l_2}$ .where $u_{\varphi l}=|\cos\varphi+l\sin\varphi|$ and $l_1,l_2=\pm 1$.

In the nonlocal $(q\neq 0)$, linear $(v_s\to 0)$ limit, i.e., in the range of temperature where $E_{nonlin}\ll T\ll E_{nonloc}$ one gets

$$\delta K(q,0,T)=\begin{cases}\dfrac{-c}{4\pi\lambda_0^2}(2\ln 2)\dfrac{T}{\Delta_0} & \alpha' w_{\theta l}\ll T\\[2ex] \dfrac{-c}{4\pi\lambda_0^2}\sum_l u_{\theta l}^2\left(\dfrac{\pi}{4}\dfrac{\alpha' w_{\theta l}}{\Delta_0}+\dfrac{3}{2}\xi(3)\dfrac{T^3}{\Delta_0\alpha'^2 w_{\theta l}^2}\right) & \alpha' w_{\theta l}\gg T\end{cases} \tag{52}$$

where $w_{\theta l} = |\sin\theta - l\cos\theta|$, $u_{\theta l} = |cos\theta + l\sin\theta|$, and $\alpha' = \frac{qv_F}{2\sqrt{2}} - \sqrt{2}\alpha g_k$.

Depending on the effective nonlocal energy scales $\left(E_{nonloc}^{+} = {}^{v_F u_{\theta l_1}}\!/_{\lambda_0}, E_{nonloc}^{-} = {}^{v_F u_{\theta l_2}}\!/_{\lambda_0}, l_1, l_2 = \pm 1 \right)$ one obtains

$$\frac{\Delta\lambda_{spec}(T)}{\lambda_0} \propto \begin{cases} T & E_{nonloc}^{+}, E_{nonloc}^{-} \ll T \\ T & E_{nonloc}^{-} \ll T \ll E_{nonloc}^{+} \\ T^2 & E_{nonloc}^{+}, E_{nonloc}^{-} \gg T \end{cases} \tag{53}$$

For CePt$_3$Si superconductor with $T_c = 0.75K$, the linear temperature dependence would crossover to a quadratic dependence below $T_{nonloc}^{*} \sim 0.015K$.

Magnetic penetration depth measurements in CePt3Si did not find a T^2 law as expected for line nodes. I argue that it may be due to the fact that such measurements were performed above 0.015K. On the other hand, it is note that CePt3Si is an extreme type-II superconductor with the Ginzburg-Landau parameter, $K = 140$, and the nonlocal effect can be safely neglected, and because this system is a clean superconductor, neglect the impurity effect can be neglected (Bauer et al., 2004; Bauer et al., 2005).

In the local, clean, and nonlinear limit $(q \to 0, v_s \neq 0)$ the penetration depth is given by

$$\lambda_{spec}^{(loc)} = \left(\frac{c}{4\pi\delta K(q \to 0, \vec{v}_s, T)} \right)^{1/2} \tag{54}$$

Where

$$\delta K(\vec{q}, \vec{v}_s, T) =$$

$$= \frac{-n_+ e^2}{mc} \left\{ 1 - 2\left\langle \hat{k}_{\parallel}^2 \frac{\sinh^{-1}\left(\frac{\vec{q}.\vec{k}_F}{2m\Delta_{k,+}}\right)}{\left(\frac{\vec{q}.\vec{k}_F}{2m\Delta_{k,+}}\right)\sqrt{1+\left(\frac{2\alpha k_F}{\Delta_{k,+}}\right)^2}} \right\rangle + 2\left\langle \hat{k}_{\parallel}^2 \int d\omega \,\mathrm{Re}\frac{\left[f\left(\omega - \vec{v}_s.\vec{k}_F\right) + f\left(\omega + \vec{v}_s.\vec{k}_F\right)\right]}{\sqrt{\omega^2 - \Delta_{k,+}^2\left[\Delta_{k,+}^{2\ 2} - (2\alpha g_k)^2 - \omega^2\right]}} \right\rangle \right\}$$
$$\frac{-n_- e^2}{mc} \left\{ 1 - 2\left\langle \hat{k}_{\parallel}^2 \frac{\sinh^{-1}\left(\frac{\vec{q}.\vec{k}_F}{2m\Delta_{k,-}}\right)}{\left(\frac{\vec{q}.\vec{k}_F}{2m\Delta_{k,-}}\right)\sqrt{1+\left(\frac{2\alpha k_F}{\Delta_{k,-}}\right)^2}} \right\rangle + 2\left\langle \hat{k}_{\parallel}^2 \int d\omega \,\mathrm{Re}\frac{\left[f\left(\omega - \vec{v}_s.\vec{k}_F\right) + f\left(\omega + \vec{v}_s.\vec{k}_F\right)\right]}{\sqrt{\omega^2 - \Delta_{k,-}^2\left[\Delta_{k,-}^2 - (2\alpha g_k)^2 - \omega^2\right]}} \right\rangle \right\} \tag{55}$$

Thus by considering only the second term in the right hand side of Eq. (55) into Eq. (51) one gets

$$\frac{\delta\lambda_{spec}}{\lambda_0} \approx \begin{cases} \frac{1}{2}\ln 2\sum_{l=\pm 1} u_{\theta l}^2 \frac{T}{\Delta_0} & E_{nonlin}^+, E_{nonlin}^- \ll T \\ \frac{1}{2}\ln 2 u_{\theta l_2}^2 \frac{T}{\Delta_0} + \frac{u_{\theta l_1}^3}{2\sqrt{2}}\left[\frac{v_s k_F}{2\Delta_0} + \frac{2\alpha g_k}{\Delta_0}\right] & E_{nonlin}^- \ll T \ll E_{nonlin}^+ \\ \frac{\frac{1}{2}\sum_{l=\pm} u_{\theta l}^3 \left[\frac{v_s k_F}{2\Delta_0} + \frac{2\alpha g_k}{\Delta_0}\right]}{\sqrt{2}} + o\left(Te^{\frac{-(v_s k_F + 4\alpha g_k)}{\sqrt{2}T}}\right) & E_{nonlin}^+, E_{nonlin}^- \gg T \end{cases} \tag{56}$$

The linear temperature dependence of penetration depth is in agreement with Bonalde et al's result (Bonalde et al., 2005).

Thus the T behavior at low temperatures of the penetration depth in Eq. (56) is due to nonlineary indicating the existence of line nodes in the gap parameter in $CePt_3Si$ compound. A T linear dependence of the penetration depth in the low temperature region is expected for clean, local and nonlinear superconductors with line nodes in the gap function.

Now the effect of impurities when both *s*-wave and *p*-wave Cooper pairings coexist is considered.

I assume that the superconductivity in $CePt_3Si$ is unconventional and is affected only by nonmagnetic impurities. The equation of motion for self-energy can be written as

$$\Sigma_{imp(n)}(\vec{p}, i\omega_n) = n_n T(\vec{p}, \vec{p}', i\omega_n) \tag{57}$$

where the T matrix is given by

$$T(\vec{p}, \vec{p}', i\omega_n) = \frac{u\sigma_3}{1 - u\sigma_3 \Im(\vec{p}, \vec{p}', i\omega_n)} \tag{58}$$

here σ_3 is the third Pauli-spin operator.

By using the expression of the Green's function in Eq. (58) one can write

$$T(\vec{p}, \vec{p}', i\omega_n) = \frac{\pi N_0 u_0^2 I}{1 + (\pi N_0 u_0 I)^2} \tag{59}$$

where

$$I = \int_0^{2\pi} \frac{d\Omega}{4\pi} \frac{-\left(\omega + i\Sigma_{imp(n)}\right)}{\left[\left(\omega + i\Sigma_{imp(n)}\right)^2 + \Delta_+\Delta_-\right]^{1/2}} \tag{60}$$

and u_0 is a single s-wave matrix element of scattering potential u .Small u_0 puts us in the limit where the Born approximation is valid, where large $u_0\,(u_0 \to \infty)$, puts us in the unitarity limit.

Theoretically it is known that the nodal gap structure is very sensitive to the impurities. If the spin-singlet and triplet components are mixed, the latter might be suppressed by the impurity scattering and the system would behave like a BCS superconductor. For p-wave gap function the polar and axial states have angular structures, $\Delta_k(T)=\Delta_0(T)\cos\theta_k$ and $\Delta_k(T)=\Delta_0(T)\sin\theta_k$ respectively. The electromagnetic response now depends on the mutual orientation of the vector potential **A** and $\hat{l}$ (unit vector of gap symmetry), which itself may be oriented by surfaces, fields and superflow. A detailed experimental and theoretical study for the axial and polar states was presented in Ref. (Einzel, 1986). In the clean limit and in the absence of Fermi-Liquid effects the following low-temperature asymptotic were obtained for axial and polar states

$$\frac{\Delta\lambda(T)^{\|,\perp}}{\lambda(0)}=a_{\|,\perp}\left(\frac{k_BT}{\Delta_0}\right)^{n_{\|,\perp}} \tag{61}$$

where in the axial state $n=2(4)$ and $a=\pi^2\left(\frac{7\pi^4}{15}\right)$, and in the polar state $n=3(1)$ and $a=\frac{27\pi\xi(3)}{4}\left(\frac{3\pi\ln 2}{2}\right)$, for the orientations $\|(\perp)$.

The influence of nonmagnetic impurities on the penetration depth of a p-wave superconductor was discussed in detail in Ref (Gross et al., 1986). At very low temperatures, the main contribution will originated from the eigenvalue with the lower temperature exponent n, i.e., for the axial state (point nodes) with T^2 low, and for the polar state (line nodes) the dominating contribution with a linear T. The quadratic dependence in axial state may arise from nonlocality.

The low temperature dependence of penetration depth in polar and axial states used by Einzel et al., (Einzel et al. 1986) to analyze the $\lambda(T)\sim T^2$ behavior of Ube$_{13}$ at low temperatures. The axial $\vec{A}\,\|\,\hat{l}$ case seems to be the proper state to analyze the experiment because it was favored by orientation effects and was the only one with T^2 dependence. Meanwhile, it has turned out that T^2 behavior is introduced immediately by T-matrix impurity scattering and also by weak scattering in the polar case. The axial sate., and according to the Andersons theorem the s-wave value of the London penetration depth are not at all affected by small concentration of nonmagnetic impurities.

Thus, for the polar state, Eq. (60) can be written as

$$I=\int_0^{2\pi}\frac{d\theta}{2\pi}\frac{-\left(\omega+i\Sigma_{imp(n)}\right)}{\left[\left(\omega+i\Sigma_{imp(n)}\right)^2+\Delta_0^2 cos^2\theta\right]^{1/2}} \tag{62}$$

Doing the angular integration in Eq. (62) and using Eqs. (57) and (59) one obtains

$$\Sigma_{imp(n)} = \frac{-\left[2\tilde{\omega}N(0)n_n u_0^2 / \sqrt{\tilde{\omega}^2+\Delta_0^2}\right]K\left(\frac{\Delta_0^2}{\sqrt{\tilde{\omega}^2+\Delta_0^2}}\right)}{1+\left[4N(0)^2 u_0^2\tilde{\omega}^2 / \left(\tilde{\omega}^2+\Delta_0^2\right)\right]K^2\left(\frac{\Delta_0^2}{\sqrt{\tilde{\omega}^2+\Delta_0^2}}\right)} \tag{63}$$

here K is the elliptic integral and $\tilde{\omega}=\omega+i\Sigma_{imp(n)}$. We note that in the impurity dominated gapless regime, the normalized frequency $\tilde{\omega}$ takes the limiting form $\tilde{\omega}\to\omega+i\gamma$, where γ is a constant depending on impurity concentration and scattering strength.
In the low temperature limit we can replace the normalized frequency $\tilde{\omega}$ everywhere by its low frequency limiting form and after integration over frequency one gets

$$\delta K(\vec{q},\vec{v}_s,T)=\frac{-N_1 e^2}{mc}\left\{\frac{4\gamma\pi^2T^2}{3}\left\langle \hat{k}_{\parallel}^2\frac{\Delta_k^2}{\left(\Delta_k^2+\gamma^2\right)^{5/2}}\right\rangle\right\} \tag{64}$$

As in the case of d-wave order parameter, from Eqs. (64) and (51) one finds

$$\frac{\delta\lambda(T)}{\lambda(0)}=\frac{\gamma}{4\pi\Delta_0}\ln\left(\frac{4\Delta_0}{\gamma}\right)+\frac{\pi}{24\gamma\Delta_0}T^2 \tag{65}$$

In *p*-wave cuprates, scattering fills in electronic states at the gap nodes, thereby suppressing the penetration depth at low temperatures and changing T-linear to T^2 behavior.

5. Effect of impurities on the low temperature NMR relaxation rate of a noncentrosymmetric superconductor

I consider the NMR spin-lattice relaxation due to the interaction between the nuclear spin magnetic moment $\gamma_n I$ (γ_n is the nuclear gyro magnetic ratio) and the hyperfine field h, created at the nucleus by the conduction electrons. Thus the system Hamiltonian is

$$H=H_0+H_{so}+H_n+H_{\text{int}} \tag{66}$$

where H_0 and H_{so} are defined by Eqs. (1) and (2), $H_n=-\gamma_n IH$ is the Zeeman coupling of the nuclear spin with the external field $\vec{H}$, and $H_{\text{int}}=-\gamma_n Ih$ is the hyperfine interaction.
The spin-lattice relaxation rate due to the hyperfine contact interaction of the nucleus with the band electron is given by

$$R=\frac{1}{T_1T}=-\frac{J^2}{2\pi}\lim_{\omega\to 0}\frac{\mathrm{Im}\,\mathrm{K}_{+-}^R(\omega)}{\omega} \tag{67}$$

where ω is the NMR frequency, $J=\frac{8\pi}{3}\gamma_n\gamma_e$ (γ_e is the electron geomagnetic ratio) is the hyperfine coupling constant, and $\mathrm{K}_{+-}^R(\tau)$, the Fourier transform of the retarded correlation function of the electron spin densities at the nuclear site, in the Matsubara formalism is given by (in our units $k_B = \hbar = 1$)

$$\mathrm{K}_{+-}^{R}(\tau) = -\langle T_\tau S_+(\tau) S_-(0)\rangle \tag{68}$$

here T_τ is the time order operator, τ is the imaginary time, $S_\pm(\tau) = e^{H\tau} S_\pm e^{-H\tau}$, and

$$S_+(r) = \psi_\uparrow^\dagger(\vec{r})\psi_\downarrow(\vec{r}) \qquad S_-(r) = \psi_\downarrow^\dagger(\vec{r})\psi_\uparrow(\vec{r}) \tag{69}$$

with $\psi_\sigma^\dagger(\vec{r})$ and $\psi_\sigma(\vec{r})$ being the electron field operators.
The Fourier transform of the correlation function is given by

$$\mathrm{K}_{+-}^{R}(i\omega_n) = \int_0^\beta d\tau e^{i\omega_n \tau} \mathrm{K}_{+-}^{R}(\tau) \tag{70}$$

The retarded correlation function is obtained by analytical continuation of the Matsubara correlation function $\mathrm{K}_{+-}^{R}(\omega) = \mathrm{K}_{+-}^{R}(i\omega_n)\big|_{i\omega_n \to \omega + i\delta}$.

From Eqs. (66)- (70), one gets

$$\begin{aligned}\frac{1}{T_1 T} &= -\frac{J^2}{2\pi}\lim_{\omega\to 0}\frac{1}{\omega} \\ &\times \mathrm{Im}\left\{ T\sum_{p,p',\omega_n}\left[Tr\left(\Im_\pm(\vec{p}, i\omega_n + i\Omega_m)\Im_\pm(\vec{p}', i\omega_n)\right) - Tr\left(F_\pm(\vec{p}, i\omega_n + i\Omega_m)F_\pm(\vec{p}', i\omega_n)\right)\right]_{i\Omega_n \to \omega + i\delta}\right\}\end{aligned} \tag{71}$$

where $\Omega_m = 2m\pi T$ are the bosonic Matsubara frequencies. By using Eqs. (11) and (12) into Eq. (71), the final result for the relaxation rate is

$$\frac{1}{T_1 T} = J^2 \int_0^\infty d\omega \left(-\frac{\partial f}{\partial \omega}\right)\left\{N_+(\omega)N_-(\omega) + M_+(\omega)M_-(\omega)\right\} \tag{72}$$

where $f(\omega) = \left(e^{\omega/T} + 1\right)^{-1}$ is the Fermi Function., $N_\sigma(\omega)$ and $M_\sigma(\omega)$ defined by the retarded Green's factions as

$$N_\sigma(\omega) = -\sum_{p,\nu=\pm} \mathrm{Im}\,\Im_\nu^R(\vec{p}, \omega) \tag{73}$$

$$M_\sigma(\omega) = -\sum_{p,\nu=\pm} \mathrm{Im}\,F_\nu^R(\vec{p}, \omega) \tag{74}$$

In low temperatures limit the contribution of the fully gap ($|\Delta_0 + \Delta\sin\theta|$) Fermi surface I decrease and the effect of the gap $|\Delta_0 - \Delta\sin\theta|$ Fermi surface II is enhanced.

As I mentioned above, the experimental data for CePt3Si at low temperature seem to point to the presence of lines of the gap nodes in gap parameter (In our gap model for $\Delta_0 < \Delta$, $|\Delta_0 - \Delta\sin\theta|$ has line nodes). Symmetry imposed gap nodes exist only for the order parameters which transform according to one of the nonunity representations of the point group. For all such order parameters $M_\sigma = 0$ .Thus, Eq. (72) can be written as

$$\frac{1}{T_1 T} = \frac{J^2}{4T}\int_0^\infty \frac{d\omega}{\cos h^2\left(\frac{\omega}{2T}\right)}\{N_+(\omega)N_-(\omega)\} \tag{75}$$

In the clean limit the density of state can be calculated from BCS expression

$$N_\sigma(\omega) = N_0 \operatorname{Re}\left\langle \frac{\omega}{\sqrt{\omega^2 - \Delta_\pm^2}} \right\rangle \tag{76}$$

For the gap parameter with line nodes from Eq. (76) one gets

$$N(\omega) = N_0 \frac{\pi}{2}\frac{\omega}{\Delta_0} \tag{77}$$

Thus from Eq. (75) one has

$$\frac{1}{T_1} = \frac{J^2\pi^2 N_0^2 T^3}{2\Delta_0^2} \tag{78}$$

Therefore, line nodes on the Fermi surface II lead to the low-temperature T^3 law in T_1^{-1} which is in qualitative agreement with the experimental results.
In the dirty limit the density of state can be written as

$$N_{imp}(\omega) = \int d\Omega \frac{N_{BCS}(\omega,\theta)}{1 + u_0^2 N_{BCS}^2(\omega,\theta)} \tag{79}$$

In the limit, $\Gamma << \Delta_0$ where $\Gamma = \frac{n_{imp}}{\pi N_0}\left(\frac{N}{V}\right)$ ($\frac{N}{V}$ is the electron density) the density of state is

$$N_{imp}(\omega) \approx N(0) + ac^2\omega \tag{80}$$

where $c = \cot g\delta_0$ (δ_0 is the s-wave scattering phase shift), a is a constant, and $N(0)$ the zero energy $(\omega = 0)$ quasi-particle density of state is given by

$$N(0) = N_o \left(\frac{\varsigma}{\sqrt{1+\frac{1}{4}\varsigma^2}+\frac{1}{2}\varsigma} \right)^{1/2} \tag{81}$$

where $\varsigma = \frac{\Gamma}{\Delta}$.

In the unitary limit $(u_0 \to \infty)$, $c=0$ $(\delta_0 = \pi/2)$, from Eqs. (75) and (80) one obtains

$$\frac{1}{T_1} = J^2 N(0)^2 T \tag{82}$$

Thus the power-low temperature dependence of T_1^{-1} is affected by impurities and it changes to linear temperature dependence characteristic of the normal state Koringa relation again is in agreement with the experimental results.

6. Conclusion

In this chapter I have studied theoretically the effect of both magnetic and nonmagnetic impurities on the superconducting properties of a non-centrosymmetric superconductor and also I have discussed the application of my results to a model of superconductivity in CePt3Si.

First, the critical temperature is obtained for a superconductor with an arbitrary of impurity concentration (magnetic and nonmagnetic) and an arbitrary degree of anisotropy of the superconducting order parameter, ranging from isotropic s wave to p wave and mixed (s+p) wave as particular cases.

The critical temperature is found to be suppressed by disorder, both for conventional and unconventional pairings, in the latter case according to the universal Abrikosov-Gor'kov function.

In the case of nonsentrosymmetrical superconductor CePt3Si with conventional pairing (A_1 representation with purely accidental line nodes), I have found that the anisotropy of the conventional order parameter increases the rate at which T_c is suppressed by impurities. Unlike the unconventional case, however, the superconductivity is never completely destroyed, even at strong disorder.

In section 4, I have calculated the appropriate correlation function to evaluate the magnetic penetration depth. Besides nonlineary and nonlocality, the effect of impurities in the magnetic penetration depth when both *s*-wave and *p*-wave Cooper pairings coexist, has been considered.

For superconductor $CePt_3Si$, I have shown that such a model with different symmetries describes the data rather well. In this system the low temperature behavior of the magnetic penetration depth is consistence with the presence of line nodes in the energy gap and a quadratic dependence due to nonlocality may accrue below $T^*_{nonloc} = 0.015K$. In a dirty superconductor the quadratic temperature dependence of the magnetic penetration depth may come from either impurity scattering or nonlocality, but the nonlocality and nodal behavior may be hidden by the impurity effects.

Finally, I have calculated the nuclear spin-lattice relaxation of $CePt_3Si$ superconductor. In the clean limit the line nodes which can occur due to the superposition of the two spin channels lead to the low temperature T^3 law in T_1^{-1}. In a dirty superconductor the linear temperature dependence of the spin-lattice relaxation rate characteristic of the normal state Koringa relation.

7. Acknowledgment

I wish to thank the Office of Graduate Studies and Research Vice President of the University of Isfahan for their support.

8. References

Abrikosov, A. A. (1993). Influence of the Gap Anisotropy on Superconducting Properties. *Physica C*, Vol. 214, No. 1-2, (1 September 1993), pp.107-110, ISSN 0921-4534

Abrikosov, A. A. & Gor'kov, L. P. (1959).On the theory of superconducting alloys. *Soviet Physics, JETP*, Vol. 8, pp. 1090. ISSN 0038-5654

Abrikosov, A. A. & Gor'kov,L. P. (1961). Contribution to the theory of superconducting alloys with paramagnetic impurities. *Soviet Physics, JETP*, Vol. 12, pp. 1243. ISSN 0038-5654

Abrikosov, A. A; Gor'kov, L. P & Dzyaloshnskii, I. E. (1975). *Methods of Quantum Field Theory in Statistical Physics*, Dover Publications Inc, ISBN 0486632288, New York.

Akazawa, T; Hidaka, H; Fujiwara, T; Kobayashi, T. C; Yamamoto, E; Haga, Y; Settai, R. & Onuki, Y. (2004). Pressure-induced Superconductivity in UIr without inversion symmetry. *Journal of Physics: Condensed Matter*, Vol. 16, No. 4, (16 January 2004), pp. 29-32, ISSN 1361-648X

Anderson, P. W. (1959). Theory of dirty superconductors. *Journal of Physics and Chemistry of Solids*,. Vol. 11, No. 1-2, (September 1959), pp. 26-30, ISSN 0022-3697

Anderson, P. W. (1984). Structure of triplet superconducting energy gap. *Physical Review B*, Vol. 30, No.7, (1 October 1984), pp. 4000-4002, ISSN 1550-235X

Bauer, E; Bonalde, I. & Sigrist, M. (2005). Superconductivity and normal state properties of non-centrosymmetric CePt3Si:a status report. *Low Temp. Phys*, Vol. 31, No. 8, (7 October 2005), pp.748-757,

Bauer, E; Hilscher, G; Michor, H; Paul, Ch; Scheidt, E. W; Gribanov, A; Seropegin, Yu; Noël, H; Sigrist, M. & Rogl, P. (2004). Heavy Fermion Superconductivity and Magnetic Order in Noncentrosymmetric $CePt_3Si$. *Physical Review Letters*, Vol. 92, No. 2, (13 January 2004), pp.027003 [4 pages], ISSN 1079-7114

Bauer, E; Hilscher, G; Michor, H; Sieberer, M; . Scheidt, E. W; Gribanov, A; Seropegin, Yu; Rogl, P; Amato, A; Song, W. Y; Park,J. G; Adroja, D. T; Nicklas, M; Sparn, G; Yogi, M. & Kitaoka, Y. (2005). Unconventional superconductivity and magnetism in $CePt_3Si_{1-x}Ge_x$. *Physica B*, Vol. 359-361, (30 April 2005), Pp. 360-367, ISSN 0921-4526

Blatsky, A.V; Vekhter,I. & Zhu, J.X. (2006). Impurity-induced states in conventional and unconventional superconductors. *Reviews of Modern Physics*, Vol. 78, No. 2, (9 May2006), pp.373–433, ISSN 1539-0756

Bonalde,I; Brämer-Escamilla, W. & Bauer, E. (2005). Evidence for Line Nodes in the Superconducting Energy Gap of Noncentrosymmetric $CePt_3Si$ from Magnetic Penetration Depth Measurements. *Physical Review Letters*, Vol. 94, No. 20, (23 May 2005), pp.207002 [4 pages], ISSN 1079-7114

Borkowski, L. S; & Hirschfeld, P. J. (1992). Kondo effect in gapless superconductors. *Physical Review B,* Vol.46, No.14, (1 October 1992), pp.9274–9277, ISSN 1550-235X

Borkowski, L. S; & Hirschfeld, P. J. (1994). Distinguishing *d*-wave superconductors from highly anisotropic *s*-wave superconductors. *Physical Review B*, Vol. 49, No. 21, (1 June 1994), pp.15404–15407, ISSN 1550-235X

Dresselhaus, G. (1955). Spin-Orbit Coupling Effects in Zinc Blende Structures.Phys. Rev. 100, 580. *Physical Review*, Vol.100, No.2, (15 October1955), pp.580–586,

Einzel, D; Hirschfeld, P. J; Gross, F; Chandrasekhar, B.S; Andres, K; Ott, H R; Beuers, J; Fisk, Z. & Smith, J.L. (1986). Magnetic Field Penetration Depth in the Heavy-Electron Superconductor UBe_{13}. *Physical Review Letters,* Vol. 56, No. 23, (9 June 1986), pp.2513–2516, ISSN 1079-7114

Edelstein, V. M. (1995). Magnetoelectric Effect in Polar Superconductors. *Physical Review Letters,* Vol. 75, No 4, (4 September 1995), pp. 2004–2007, ISSN 1079-7114

Frigeri, P. A; Agterberg, D.F. & Sigrist,M. (2004). Spin susceptibility in superconductors without inversion symmetry. *New Journal of Physics,* Vol.6, No.1, (2 September 2004), pp.115 [9 pages], ISSN 1367-2630

Frigeri, P. A; Agterberg; D. F; Koga, A. & Sigrist, M. (2004). Superconductivity without Inversion Symmetry: MnSi versus $CePt_3Si$. *Physical Review Letters,* Vol.92, No. 9, (3 March 2004), pp.097001 [4 pages], ISSN 1079-7114

Gor'kov, L. P & Rashba, E. I. (2001). Superconducting 2D System with Lifted Spin degeneracy: Mixed Singlet-Triplet State. *Physical Review Letters,* Vol. 87, No. 3, (16July 2001), pp.037004 [4 pages], ISSN 1079-7114

Gross, F; Chandrasekhar, B.S; einzel,D; Andres, K; Herschfeld, P.J; Ott, H.R; Beuers, J; Fisk, Z. & Smith, J.L. (1986). Anomalous temperature dependence of the magnetic field penetration depth in superconducting UBe_{13}. *Zeitschrift für Physik B: Condensed Matter*, Vol. 64, No. 2, pp.175-188, ISSN 0722-3277

Hayashi, N; Wakabayashi, K; Frigeri, P. A. & Sigrist1, M. (2006). Nuclear magnetic relaxation rate in a noncentrosymmetric superconductor. *Physical Review B*, Vol. 73, No. 9, (17 March 2006), pp.092508 [4 pages], ISSN 1550-235x

Hayashi, N; Wakabayashi, K; Frigeri, P. A. & Sigrist, M. (2006). Temperature dependence of the superfluid density in a noncentrosymmetric superconductor. *Physical Review B,* Vol. 73, No. 2, (19 January 2006), pp. 024504[9 pages], ISSN 1550-235X

Izawa,K; Kasahara, Y; Matsuda,Y; Behnia,K; Yasuda,T; Settai,R. & Onuki,Y. (2005) Line Nodes in the Superconducting Gap Function of Noncentrosymmetric $CePt_3Si$. *Physical Review Letters,* Vol. 94, No. 19, (16 May 2005), pp.197002 [4 pages], ISSN 1079-7114

Kosztin, I & Leggett, A.J. (1997) Nonlocal Effects on the Magnetic Penetration Depth in *d*-Wave Superconductors. *Physical Review Letters*, Vol. 79, No. 1, (7 July 1997), pp.135–138, ISSN 1079-7114

Mathur, N.D; Grosche, F. M; Julian S. R; Walker, I. R; Freye, D. M; Haselwimmer R. K. W & Lonzarich G. G. (1998). Magnetically mediated superconductivity in heavy rmion compounds . *Nature,* Vol. 394, (2 July 1998), pp.39-43 ISSN 0028-0836

Mineev, V. P. & Samokhin, K. V. (2007). Effects of impurities on superconductivity in noncentrosymmetric compounds. *Physical Review B,* Vol.75, No. 18, (25 May 2007), pp.184529 [14 pages], ISSN 1550-235X

Mineev, V. P. & Champel, T. (2004) Theory of superconductivity in ferromagnetic superconductors with triplet pairing. Phys. *Physical Review B*, Vol. 69, No. 14, (27 April 2004),pp.144521 [6 pages], ISSN 1550-235X

Rashba,E. I. (1960). Properties of Semiconductors with an Extremum Loop I. Cyclotron and Combinatorial Resonance in a Magnetic Field Perpendicular to the Plane of the Loop. *Soviet Physics – Solid State,* vol. 2, (Jun. 1960), pp. 1109-1122, ISSN 0038-5654.

Rashba,E. I & Bychkov, Yu. A. (1984). Oscillatory effects and the magnetic susceptibility of carriers in inversion layers. *Journal of Physics C: Solid State Physics,* Vol. 17, No. 33, (30 November 1984), pp.6039-6045, ISSN 0022-3719

Salkalo, M.I; Balatsky,A.V. & Scalapino,D.J. (1996) Theory of Scanning Tunneling Microscopy Probe of Impurity States in a *D*-Wave Superconductor. *Physical Review Letters,* Vol. 77, No. 9, (26 August 1996), pp.1841–1844, ISSN 1079-7114

Samokhin, K. V; Zijlstra,E. S; & Bose, S. K. (2004). $CePt_3Si$: An unconventional superconductor without inversion center. *Physical Review B,* Vol. 69, No. 9, (19 March 2004), pp. 094514 [8 pages], ISSN 1550-235X

Saxena, S. S; Agarwal, P; Ahilan, K; Grosche, F. M; Haselwimmer, R. K. W; Steiner, M. J; Pugh, E; Walker, I. R; Julian, S. R; Monthoux, P; Lonzarich, G. G; Huxley, A; Sheikin, I Braithwaite D & Flouquet, J. (2000). Superconductivity on the border of itinerant-electron ferromagnetism in UGe_2. *Nature,* Vol. 406, No. 6796, (10 August 2000), pp.587-592, ISSN 0028-0836

Sergienko, I. A. & Curnoe, S. H. (2004). Order parameter in superconductors with nondegenerate bands. *Physical Review B,* Vol.70, No. 21, (7 December 2004), pp.214510 [8 pages], ISSN 1550-235X

Sergienko, I. A. (2004). Mixed-parity superconductivity in centrosymmetric crystals, *Physical Review B*, Vol. 69, No. 17, (18 May 2004), pp.174502 [5 pages], ISSN 1550-235X

Shiba,H. (1968). Classical Spins in Superconductors. *Progress of Theoretical Physics,* Vol. 40, No. 3 (12 April 1968), pp. 435-451, ISSN 1347-0481

Wang, Q.H. & Wang, Z.D. (2004). Impurity and interface bound states in $d_{x^2-y^2}+id_{xy}$ and p_x+ip_y superconductors. *Physical Review B,* Vol. 69, No. 9, (10 March 2004), pp.092502 [4 pages], ISSN 1550-235X

Yasuda, T; Shishido, H; Ueda,T; Hashimoto, S; Settai, R; Takeuchi,T; Matsuda,T. D; Haga, Y. & Onuki,Y. (2004). Superconducting Property in $CePt_3Si$ under Pressure. *Journal of the Physical Society of Japan,* Vol. **73**, No. 7, (13 April 2004) pp. 1657-1660, ISSN: 1347-4073

Yogi, M; Kitaoka,Y; Hashimoto,S; Yasuda, T; Settai,R; Matsuda, T. D; Haga,Y; Onuki, Y; Rogl, P. & Bauer,E. (2004). Evidence for a Novel State of Superconductivity in

Noncentrosymmetric $CePt_3Si$: A ^{195}Pt-NMR Study. *Physical Review Letters*,Vol. 93, No. 2, (8 July 2004), pp.027003 [4 pages], ISSN 1079-7114

Yogi, M; Kitaoka,Y; Hashimoto, S; Yasuda, T; Settai, R; Matsuda, T. D; . Haga, Y; Onuki, Y; Rogl, P. & Bauer, E. (2005). Novel superconductivity in noncentrosymmetric heavy-fermion compound $CePt_3Si$: a ^{195}Pt-NMR study. *Physica B*,Vol.359-361, (30 April 2005), pp. 389-391. ISSN 0921-4526

8

Foundations of Meissner Superconductor Magnet Mechanisms Engineering

Jose Luis Perez-Diaz and Efren Diez-Jimenez
Dpt. Ingeniería Mecánica – Universidad Carlos III de Madrid
Spain

1. Introduction

It has long been known that a repulsive force arises between a magnetic field (generated, for instance, by a permanent magnet - PM) and a superconductor -Sc (Arkadiev, 1947). This force is due to the repulsion of the magnetic field away from the superconductor - the Meissner effect. Type I superconductors only can be in the Meissner state, which means that a magnetic field will be always expelled from the superconductor, independently of its poles orientation. Nevertheless, type II superconductors may be in two different states: first, provided the magnetic field is low enough, they are at a Meissner state similar to type I superconductors. In this Meissner state they absolutely expel the magnetic field and prevalent repulsive forces appear. Second, for magnetic fields larger than the so-called First Critical Field H_{C1}, the magnetic flux penetrates the superconductor creating a magnetization which contributes to an attractive resulting force. This second state is known as mixed state.

In 1953 Simon first tried to make a superconducting bearing (Simon, 1953) using superconductors in the mixed state.The first engine using a superconducting bearing was made in 1958 (Buchhold, 1960). After the discovery of high critical temperature superconductors (Bednorz & Müller, 1986), the Meissner repulsive force has become a popular way of demonstrating superconducting properties (Early et al., 1988).For calculating forces between a magnet and a superconductor it is necessary to have models that describe both the flux penetration state and the Meissner state repulsion. The first one can be solved by using conventional methods to compute forces between magnetic elements and magnetized volumes. However, for the Meissner state the question has remained open until these last years.

Several models using the method of images to calculate superconducting repulsion forces (Lin, 2006; Yang & Zheng, 2007) have been proposed. However, this method of images is limited to a few geometrical configurations that can be solved exactly, and the physical interpretation of the method is under discussion (Giaro et al., 1990; Perez-Diaz & Garcia-Prada, 2007). Furthermore, some discrepancies within experiments still exist (Hull, 2000).

A general local model based on London's and Maxwell's equations has been developed to describe the mechanics of the superconductor-permanent magnet system (Perez-Diaz et al., 2008). Due to its differential form, this expression can be easily implemented in a finite elements analysis (FEA) and is consequently appliable to any shape of superconductor in pure Meissner state (Diez-Jimenez et al. 2010).

In this chapter, we present the demonstration of the model, the implementation into a finite elements program, the experimental verification and its limit of application. To finish, we show an example of usage.

2. Magneto-mechanics of a superconductor in Meissner state

A superconductor is in a pure Meissner state when it is exposed to an externally applied magnetic field, $\vec{H}^{ap}$, lower than a certain value, H_{C1}. H_{C1} is a characteristic of the material (Alario & Vincent, 1991), which depends on temperature. In this case, it is assumed that both $\vec{H}$ and $\vec{B}$ are equal to zero inside the superconductor. When a magnetic field is then applied -for example by moving a permanent magnet close to the superconductor - a surface current density is generated on the outermost surfaces. According to the London equation, this current is confined only to a depth of λ(T). Type II superconductors, such as the rare earth oxide high temperature superconductors, have the highest values for λ, reaching typical values of thousands of Amstrongs (Umezawa & Crabtree, 1998). Therefore, as this paper deals with macroscopic elements, it can be approximated that current density has an infinitely localized surface current

$$\vec{J} = \vec{j}_s(x,y)\delta(z) \tag{1}$$

where $\vec{j}_s(x,y)$ is a surface current density tangent to the surface vector field and $\delta(z)$ is a Dirac delta function on z. This current density will make $\vec{H}$ discontinuous when passing from the air or vacuum ($z>0$) into the superconductor ($z<0$).The second Maxwell law (Jackson, 1975) relates the magnetic field and the current density in such a way that $\vec{j}_s(x,y)$ is determined by $\vec{H}^{ap}$.Using units from the MKSA system, this second Maxwell law can be written as

$$\vec{\nabla} \times \vec{H} = \vec{j} + \frac{\partial \vec{D}}{\partial t} \tag{2}$$

In the static limit it can be assumed that $\frac{\partial \vec{D}}{\partial t} = 0$. Therefore it may simply be written as:

$$\vec{\nabla} \times \vec{H} = \vec{J} \tag{3}$$

$\vec{H}$ may be decomposed in that externally applied $\vec{H}^{ap}$ and that generated by the superconducting currents $\vec{H}^{sc}$. Furthermore, these three vector fields will be decomposed both in tangent and normal to the surface components:

$$\vec{H}_{//} = \vec{H}^{ap}_{//} + \vec{H}^{sc}_{//} \tag{4}$$

and

$$\vec{H}_{\perp} = \vec{H}^{ap}_{\perp} + \vec{H}^{sc}_{\perp} \tag{5}$$

Note that $\vec{H}^{ap}$ is continuous and fulfils: $\vec{\nabla} \times \vec{H}^{ap} = 0$, provided the permanent magnet does not touch the superconductor surface. On the contrary, both $\vec{H}$ and $\vec{H}^{sc}$ are discontinuous at the superconducting surface. In particular, both $H_{//}$ and $H_{//}^{sc}$ are discontinuous.
By using the divergence theorem (Jackson, 1975) on a small parallelepiped with volume V, a face just above the superconductor surface and another parallel face under it, it can be written that:

$$\int_V \left(\vec{\nabla} \times \vec{H}\right) d^3x = \int_S \left(\vec{n}_s \times \vec{H}\right) dS \tag{6}$$

where S is the surface defining this parallelepiped and n_S its normal vector. By using Maxwell law (3) it can be reduced to:

$$\int_V \vec{J}\ d^3x = \int_S \left(\vec{n}_s \times \vec{H}\right) dS \tag{7}$$

But, taking (1) into account, and considering $\vec{H} = \vec{0}$ under the superconducting surface it can be written that:

$$\int_S \vec{j}_s dS = \int_S \left(\vec{n}_s \times \vec{H}_{//}\right) dS \tag{8}$$

where $\vec{H}_{//}$ is evaluated at z=0+ (limit above the superconductor surface).
As this result is independent of the small parallelepiped previously chosen, the integrands must equal:

$$\vec{j}_s = \vec{n}_s \times \vec{H}_{//}(z \to 0^+) \tag{9}$$

Furthermore, $\vec{H}_{//}(z \to 0^-) = 0$ and $\vec{H}_{//}^{sc}(z \to 0^+) = -\vec{H}_{//}^{sc}(z \to 0^-)$ implies:

$$\vec{H}_{//}^{sc}(z \to 0^+) = \vec{H}_{//}^{ap}(z \to 0^+) \tag{10}$$

Therefore, an expression for the superconducting current as a function of the applied magnetic field may be written:

$$\vec{j}_s = 2\ \vec{n}_s \times \vec{H}_{//}^{ap} = 2\ \vec{n}_s \times \vec{H}^{ap} \tag{11}$$

All expressions shown use the MKS unit system.
Applying the divergence theorem clearly shows that the total charge is always conserved, for whichever surface shape the superconductor has, provided the source of the applied field is outside the superconductor:

$$\int_S \vec{j}_s dS = \int_S 2\ \vec{n}_s \times \vec{H}^{ap} dS = \int_V \vec{\nabla} \times \vec{H}^{ap} d^3x = \vec{0} \tag{12}$$

Furthermore,

$$\vec{H}_\perp = \vec{H}_\perp^{ap} + \vec{H}_\perp^{sc} = \vec{0} \tag{13}$$

which is consistent with the previous expression for the superconducting current.
The total field thus generated for a semi-infinite plane is equivalent to that generated by the method of images (Cansiz & Hull, 2005; Hellman et al. 1988; Hull & Cansiz, 1999; Moon, 1994). As the expression (11) has been derived using only local arguments, it may be used for any shape of superconductor. It does not depend on the curvature of the surface.

2.1 Force calculation

The external force (by unit surface) experienced by the superconductor can be calculated by using Lorentz force.

$$\frac{d\vec{F}}{dS} = \vec{j}_s \times \vec{B}^{ap} \tag{14}$$

Using the previous expression for the superconducting current (1) and the constituent equation of air (15) (the medium in which the field is generated)

$$\vec{B}^{ap} = \mu_0 \vec{H}^{ap} \tag{15}$$

it can be written that:

$$\frac{d\vec{F}}{dS} = 2\mu_0(\vec{n}_s \times \vec{H}^{ap}) \times \vec{H}^{ap} \tag{16}$$

This is a local and exact expression for the "pressure" or more precisely "stress" or force per unit surface on the superconductor, which depends only on the applied magnetic field. It is useful for any shape of superconductor. This differs fundamentally from the general expression used to calculate the magnetic stress between magnetic materials as given by Moon.
According to Newton's law, the force exerted by the superconductor on the magnet is simply the opposite one. Therefore, for any shape of superconductor, the force exerted by the superconductor on the magnet can be written as:

$$\vec{F} = -\iint_{Sc} 2\mu_0(\vec{n}_s \times \vec{H}^{ap}) \times \vec{H}^{ap} dS \tag{17}$$

where the integration extends over the whole surface of the superconductor.

2.2 Torque calculation

The torque suffered by the superconductor can easily be deduced as :

$$\vec{M}_{Sc} = \iint_{Sc} \vec{r} \times (2\mu_0(\vec{n}_s \times \vec{H}^{ap}) \times \vec{H}^{ap}) dS \tag{18}$$

where $\vec{r}$ is the position vector between the differential surface element and the center of mass of the superconductor bulk. Again, the integration extends over the whole surface of the superconductor.
In order to calculate the moment applied over the magnet, the $\vec{r}_{PM}$ must be the position vector between the differential surface element and the center of mass of the permanent magnet. As noted previously, the force exerted by the superconductor on the external

magnetic field (in this case a single permanent magnet) is simply the opposite one, and the same applies for the torque.

$$\vec{M}_{PM} = -\iint_{Sc} \vec{r}_{PM} \times (2\mu_0(\vec{n}_s \times \vec{H}^{ap}) \times \vec{H}^{ap})dS \qquad (19)$$

3. Finite elements implementation

Due to this differential form equation (16) can be easily implemented in a finite element program. A FEM algorithm has been adapted for the commercial software ANSYS. The SURF154 element of ANSYS was used insofar as it has defined a set of useful attributes e.g. the surface normal direction. The algorithm is valid in the context of a common electromechanical simulation. The steps for the simulation were:

- Select Element Type: SOLID98 (with a maximum of one degree of freedom MAG) and SURF154.
- Create the different materials to be used. For the superconductor bulk, air properties were used.
- Generate the geometries of the volumes for the electromagnetic system.
- Assign materials' properties to each volume, selecting air for the superconductor.
- Mesh the whole system with the SOLID98 element (as fine as is considered adequate - discussed further below).
- Mesh the superconductor surface with the SURF154 element.
- Apply the electromechanical loads to the system.
- Solve the electromagnetic equation system.

Once the system has been solved, the algorithm can be applied using a Command List. Fig. 1 shows a flow-diagram of the procedure.

This procedure has to be performed for each piece of superconductor in the system. Should there be more than one piece, a different internal SURF154 element must be created and accordingly, the number of SURF154 elements in the first step (ESEL) must be changed.

These steps provide the three components of the force vector. The torque applied on the superconductor can also be calculated, from which the torques values can then be derived.

3.1 Results provide by the post-processing

The algorithm has been tested using one of the most common experiments found in relevant literature: a permanent magnet oriented vertically over a superconductor pile in any arbitrary position.

Firstly, an electromagnetic system composed of a small magnet suspended over a superconducting cylinder was designed, as shown in Fig. 2.

The dimensions of the superconductor were: 20 mm diameter and 7.5 mm height, and the small magnet was characterized by a coercivity of 875 kA/m and a remanence of 1.18 T, with a 3.5 mm diameter and a 2 mm height. The magnet's centroid was placed 10 mm over the surface of the superconductor. The entire system was placed in surroundings measuring 100x100x100 mm.

The results that can be obtained are the distribution of forces, torque and current densities per surface element. In fig 3, these distributions for an arbitrary position of the magnet over the superconductor are shown.

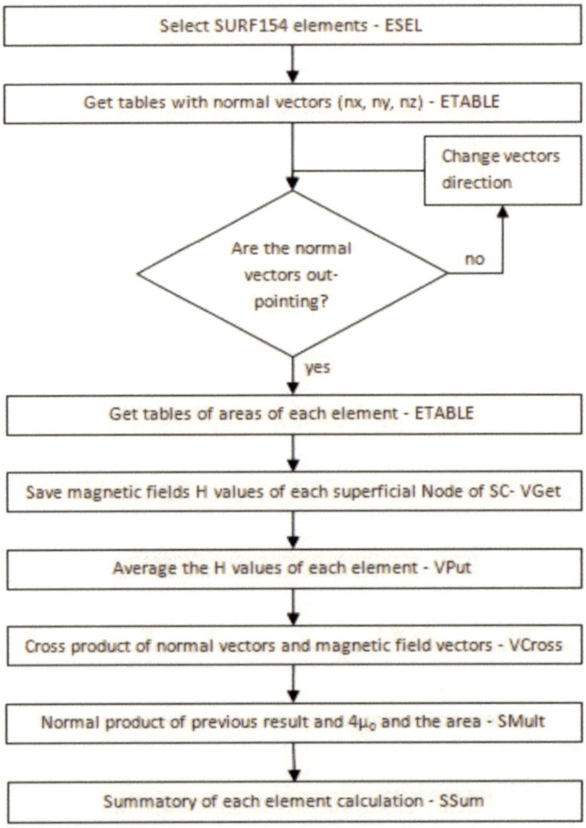

Fig. 1. Flow-diagram of the algorithm.

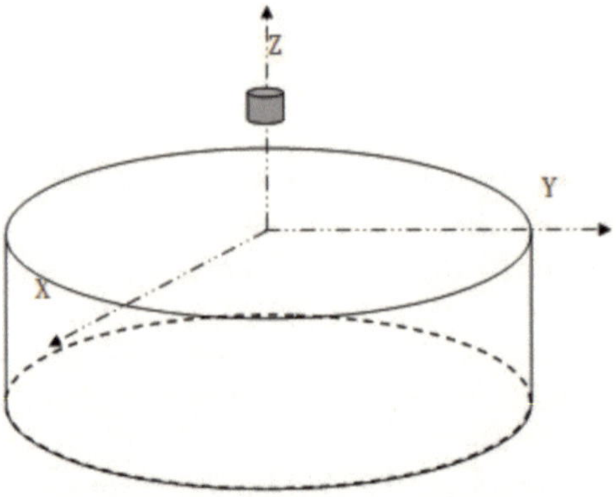

Fig. 2. Small permanent magnet (m=0.016 Am2) over superconductor.

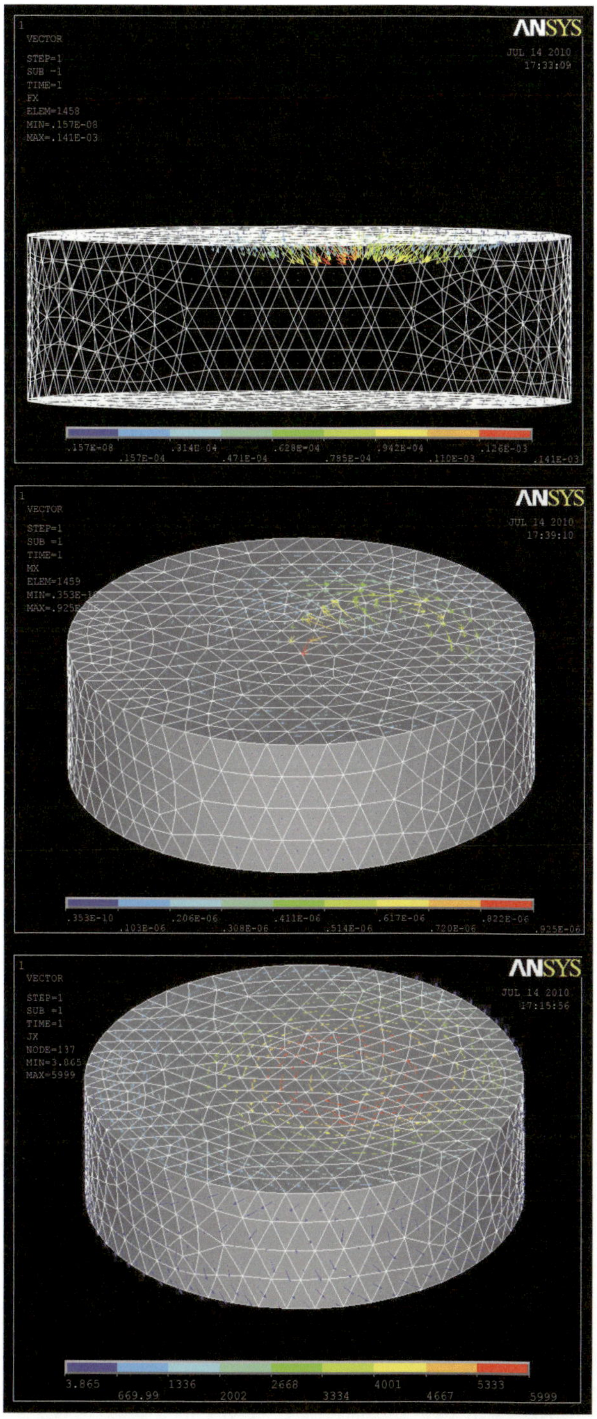

Fig. 3. Force, torque and current density distributions per surface element.

The same simulation was repeated several times with different meshes, increasing the number of elements for the whole simulation. Using the parameter α, the fineness of the mesh can be defined as the ratio between the maximum of the area of the elements and the total area of the superconductor multiplied by 100.

$$\alpha = \frac{\text{max (elements areas)}}{\text{total SC area}} \times 100 \tag{3}$$

Different meshes along with their respective α parameters are shown in fig. 4. The number of SURF154 elements and the values of the solution are also displayed.

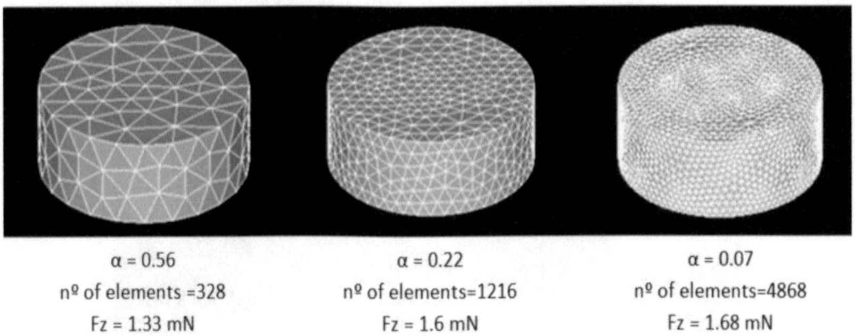

Fig. 4. Different meshes of the superconductor pile.

Fig. 5 shows the relative error of the calculations in relation to the analytical solution (with a magnetic moment of 0.016 A·m²). The higher the number of surface elements, the smaller the relative error of the result. For example, where α is smaller than 0.1 %, the resulting relative error is less than 3 %.

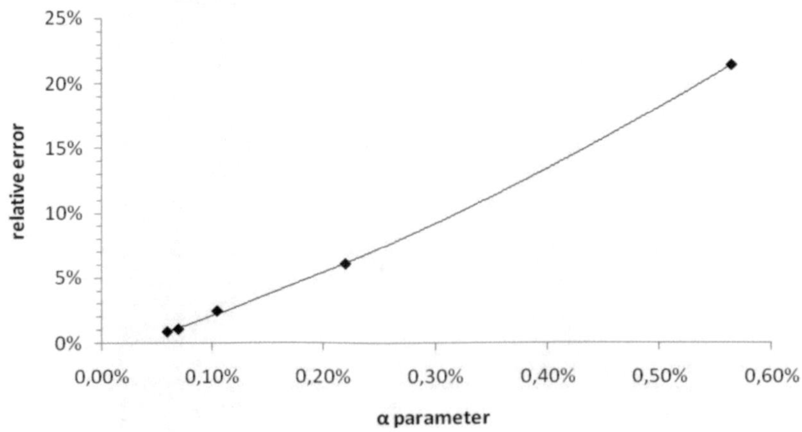

Fig. 5. Relative errors of the results vs. alpha for z=10 mm.

In Fig. force versus z are shown for different values of α. The FEM results tend towards the analytical values as α decreases. It is noted that the magnetic dipole approximation made for the analytical calculation only remains valid where there is a large distance between the permanent magnet and the superconductor.

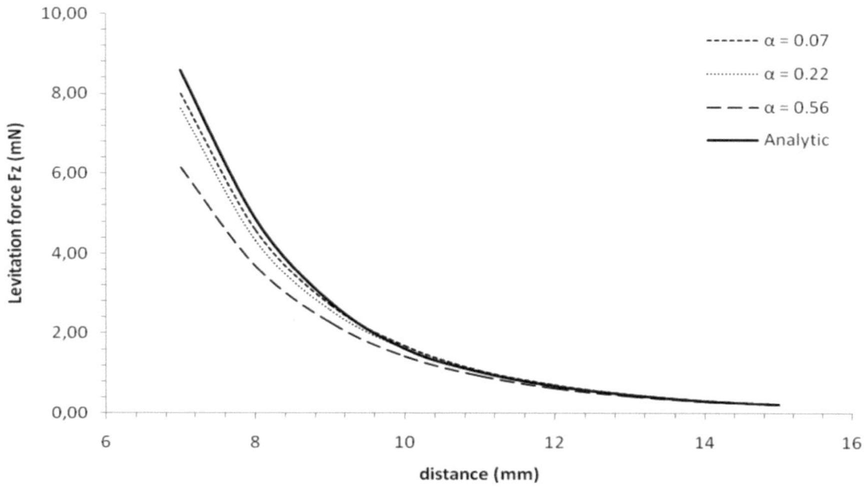

Fig. 6. Levitation force computed by the analytical expression with a point magnetic dipole and by FEM with different α.

The convergence of the algorithm has been checked in relation to the finite elements' size, and compared to analytical solutions for simple geometries. An α parameter has been proposed to assess the relative error in the results. The results showed good accuracy, whilst not requiring high specification computing technology.

4. Experimental verification

Different experiments were carried out in order to check the validity of the model. Some of them will be summarized in the following.

4.1 Force measurement

The following methodology was used to measure the forces: a cylindrical superconductor made of polycrystalline YBa2Cu3O7-x,manufactured by CAN superconductors (Kamenice 25168, Czech Republic) was immersed in a bath of liquid nitrogen N_2 (77 K) at ambient pressure. The cylinder had a diameter of 45 mm and a height of 13 mm. It was fixed to a nitrogen vessel. The vessel, containing the superconductor, was placed on a lab jack stand to adjust the height. A small cylindrical permanent magnet was used, which had a coercivity of 875 kA/m, a remanence of 1.18 T, and had a diameter of 5 mm and a height of 5 mm. All experimental measurements followed the same coordinate system shown in Fig. 7. The origin of the coordinates was set at the center of the upper surface of the superconductor.
The permanent magnet was placed over the superconductor (Z coordinate),and fixed vertically to a PVC cantilever according to its magnetization direction ($\theta = 90^o$). The cantilever had 2 pairs of strain gauges to measure vertical forces at its extremes. This strain gauge configuration is not sensitive to the lateral and axial forces. The torques were neglected due to the size of the magnet. The PVC cantilever was joined to a 3D positioning table. The position of the magnet was then fixed in relation to the superconductor surface with a precision of 0.1 mm. The strain gauges were calibrated using a dynamometer and a set of 12 references forces. The calibration constant was established by least squares fitting in $K = (3.87\pm0.14)\times10^{-4}$ N/με, with a correlation coefficient of $R^2 = 0.997$.

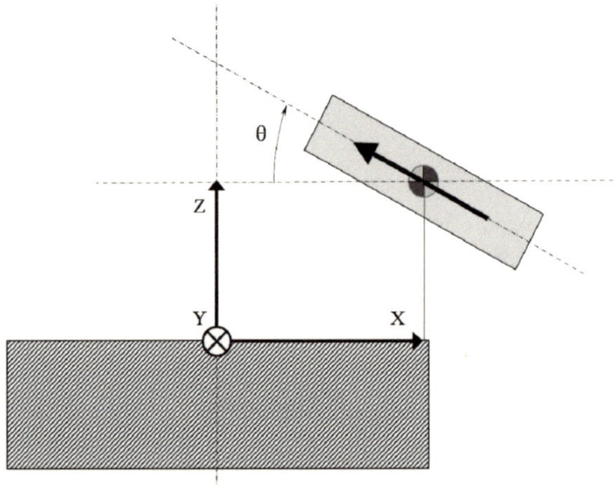

Fig. 7. Coordinate system of the PM-SC configuration. The superconductor is down and the permanent magnet is over it. Figure is not scaled to real sizes.

The measurement for every position was made in zero field cooling conditions (ZFC). The vertical forces were recorded where X = 0.0, 5.0, 10.0, 15.0, 17.5, 20.0, 22.5 and 25.0 ± 0.1 mm; at 3 different heights from the surface of the superconductor: 12.0, 10.0, and 8.0 ± 0.1 mm. Furthermore, measurements were taken in the center of the upper superconductor face, X = 0 mm from Z = 7.0 to 14.0 ± 0.1 mm. The Y position was always fixed at 0 mm.
These positions were chosen in order to avoid exceeding a limit of 3.5 mT of magnetic flux density at any point of the superconductor surface. Using this limit ensures the Meissner state is retained. Regardless of this, after every measurement the remanent magnetization of the superconductor bulk was checked and in most cases no measurable magnetization was found. In order to compare the experimental and theoretical values, expression (16) was implemented in a finite element analysis program. The following figures (Fig 8-11) show the results.

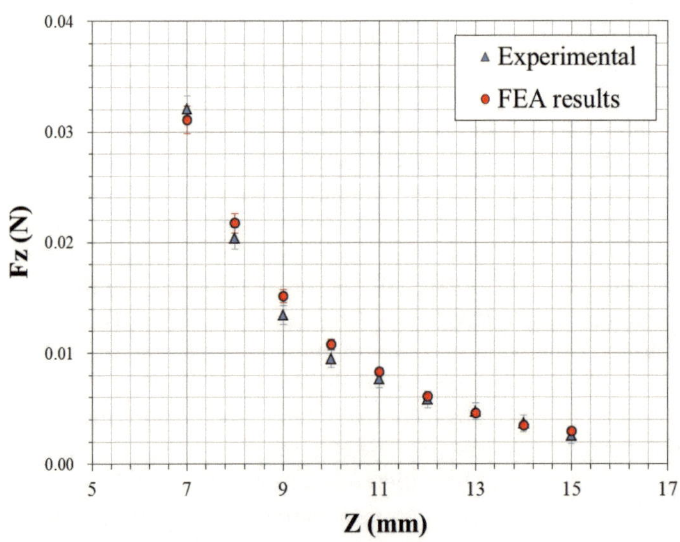

Fig. 8. Z dependence of vertical force for X=0 mm.

The FEA errors are estimated to be less than 3 %.
The figures show positive agreement between experimental and theoretical values. Only Fig. 9 shows an appreciable difference for values approaching X = 20 mm. However, it must be pointed that the radius of the superconductor is 22.5 mm. It is only in these surroundings that a very low remanent magnetization was recorded, which indicates a non complete Meissner state. This explains why some experimental values were lower than those of a complete Meissner state.

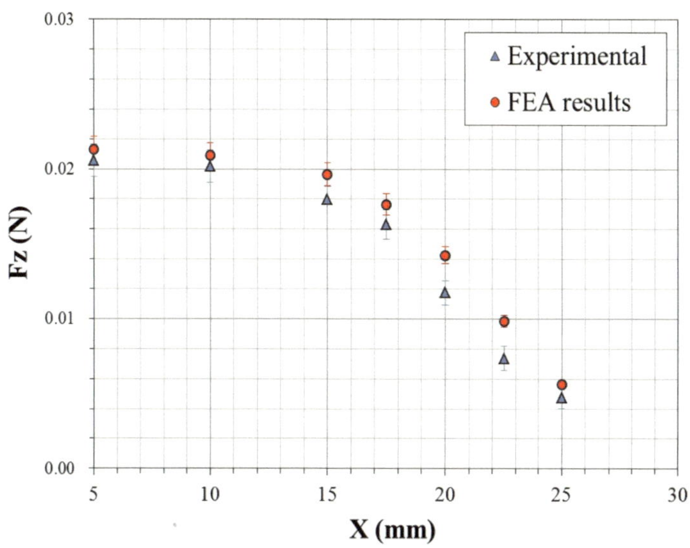

Fig. 9. X dependence of vertical force for Z=8 mm.

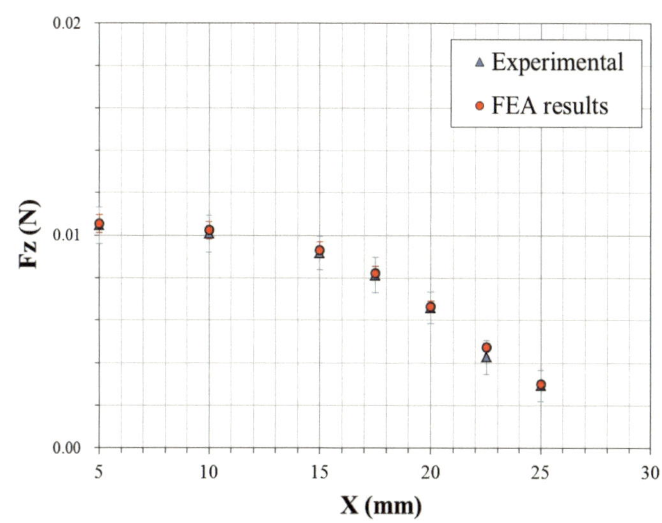

Fig. 10. X dependence of vertical force for Z=10 mm.

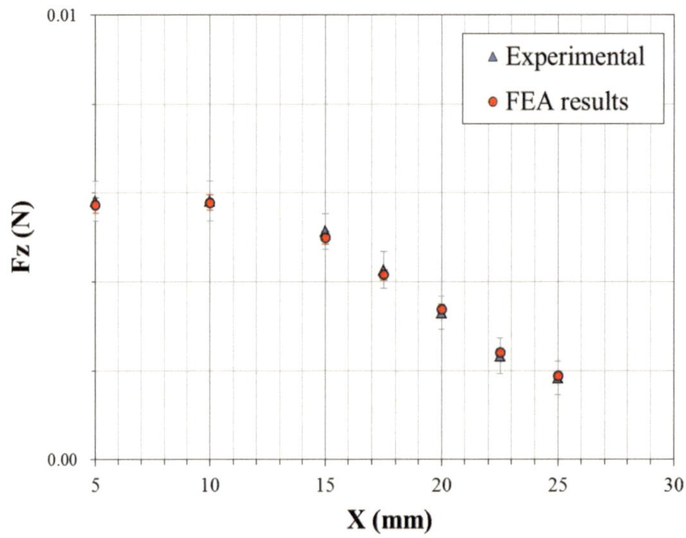

Fig. 11. X dependence of vertical force for Z=12 mm.

4.2 Equilibrium angle measurement

In addition to previous experiments, the mechanical behavior of a magnet which has the ability to tilt over the superconductor in the Meissner state was also studied in this paper. In the present experiment only one degree of freedom was permitted in the tilt angle of the magnet (θ coordinate). The equilibrium angle of the permanent magnet over the cylindrical superconductor was measured for different relative positions. The results can be used to understand not only how the permanent magnet is repelled, but also how it turns when it is released over a superconductor.

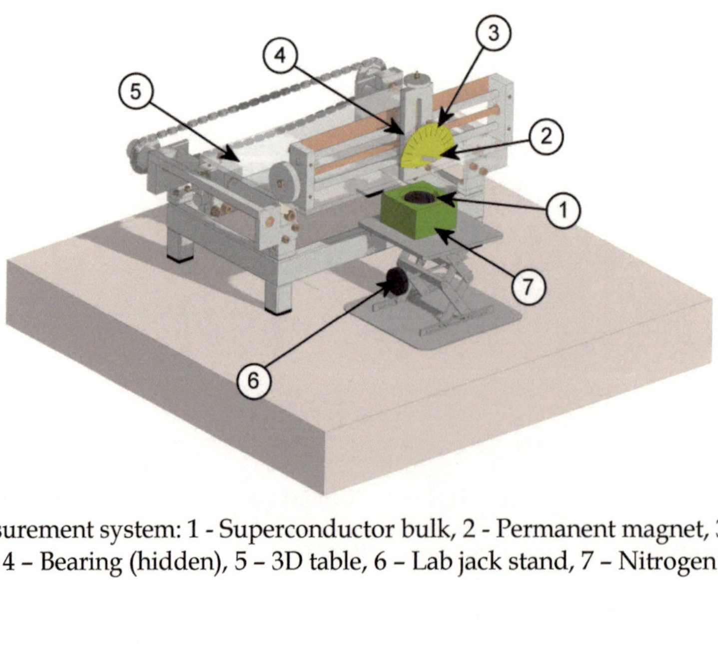

Fig. 12. Measurement system: 1 - Superconductor bulk, 2 - Permanent magnet, 3 - Goniometer, 4 – Bearing (hidden), 5 – 3D table, 6 – Lab jack stand, 7 – Nitrogen vessel.

A cylindrical permanent magnet (made of NdFeB with a coercivity of 875 kA/m and a remanence of 1.29 T) was placed over the superconductor. Their dimensions were 6.3 mm in diameter and 25.4 mm in length and it had a magnetization direction parallel to its axis of revolution. A rigid plastic circular rod was fixed in the center of mass, perpendicular to the axis of revolution. This rod was used as the shaft in a plastic bearing, which was lubricated with oil. The whole bearing system was joined to a 3D displacement table. This arrangement ensured it was possible to control the position of the permanent magnet with an accuracy of 0.1 mm, and the only permitted degree of freedom was the rotation around the Y axis. Concentric to the bearing, a graduate goniometer measured the angle of rotation of the magnet. The whole experiment design is shown in Fig 12.

Fig. 13 shows the comparison between the equilibrium angles measured and those calculated by expression (19).

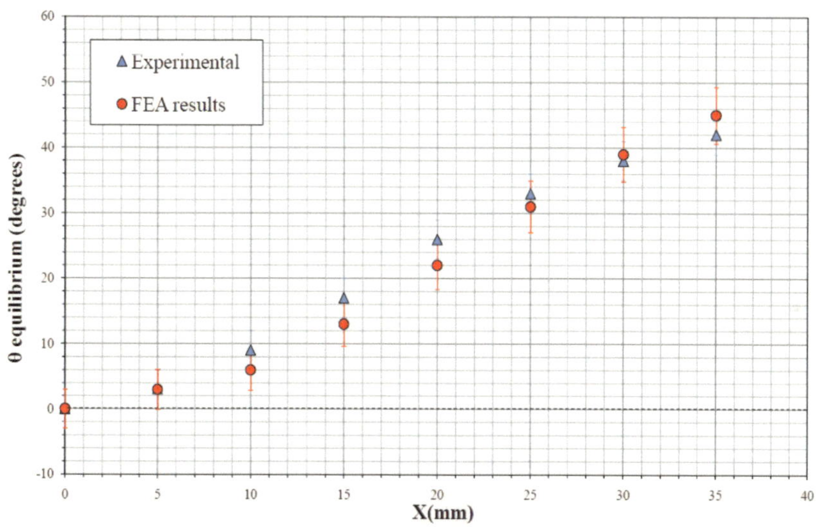

Fig. 13. Comparative graph between experimental and FEA calculus of the equilibrium angle versus x position. Hight z was fixed at + 15 mm.

Again, there was a good agreement between the calculus made according to our model and the experiments. These experiments were carried out in Zero Field cooling condition (ZFC), and consequently there is no remanent magnetization.

5. Limits of application

The lower critical field, H_{c1}, is one of the typical parameters of type II superconductors, which has been experimentally being assessed from the magnetization changes from the Meissner state slope to the reversible mixed-state behavior (Poole, 2007). H_{c1} is directly related to the free energy of a flux line and contains information on essential mixed-state parameters, such as the London penetration depth, λ_L, and the Ginzburg-Landau parameter, κ. Measurements of H_{c1} and, of course, of the upper critical field, H_{c2}, therefore provide a complete characterization of the mixed-state parameters of the superconductor.

Differences between the predicted Meissner forces and the experimentally measured ones indicate that a part of the sample is in the mixed-state. Establishing with precision the instant when the differences begin will permit us to determine the H_{c1} mechanically. Nevertheless, many other experimental techniques have been used to determine the state transition; most of them based on some kind of d.c. or a.c. magnetic measurement, but also on muon spin rotation (μSR) or magneto-optical techniques (Meilikhov & Shapiro, 1992). The basic problem of magnetization measurements introduced by flux pinning lies in the fact that the change of slope at the lower critical field is extremely small, since the first penetrating flux lines are immediately pinned and change the overall magnetization ($M = m / V$) only marginally. Elaborate schemes of subtracting the measured moments from an initial Meissner slope (Vandervoort et al., 1991; Webber et al., 1983) or experiments providing us directly the derivative of magnetization (Hahn & Weber, 1983; Wacenoysky et al., 1989; Weber et al., 1989) have been employed, SQUIDS have also been used to improve the precision of these kind of means (Böhmer et al., 2007).
The method also determines the zone at the sample where transition from Meissner to mixed state occurs.
For a position of the magnet with respect to the superconductor we define the Meissner Efficacy as

$$\eta = \frac{F_{ex}}{F_M} \tag{27}$$

where F_{ex} is the experimentally measured force and F_M is the calculated force according with the Meissner model cited above. For a certain position of the magnet a Meissner Efficacy equal to one ($\eta = 1$) proves that the superconductor is completely in the Meissner state and there is not any flux penetration. On the contrary, values lower than 1 indicate that a part of the superconductor has flux penetration and is in the mixed-state.
The measurement for every position was made in zero field cooling conditions (ZFC). The origin of coordinates was set at the center of the upper surface of the superconductor. The reference point of the magnet was placed in the center of the lower surface of the magnet. Therefore, the Z coordinate is the distance between the faces of the magnet and the superconductor. X is the distance of the center of the magnet to the axis of the superconductor cylinder (radial position). We have recorded the vertical forces for X = 0.0, 5.0, 10.0, 15.0, 17.5, 20.0, 22.5 and 25.0 ± 0.1 mm; at 4 different heights from the surface of the superconductor: 12.0, 10.0, 8.0 and 6.0 ± 0.1 mm.
Fig. 14 shows the Meissner Efficacy versus the maximum of the surface current density distribution J_{surf} for different positions.
We observe that for low values of the maximum surface current density, the Meissner Efficacy is just 1.
From a certain value, the Meissner Efficacy decays linearly. From this data we can derive a weighted mean value of $J_{c1\,surf} = 6452 \pm 353\ A/m$ for a polycrystalline $YBa_2Cu_3O_{7-x}$ sample at 77 K.
In Table 1 H_{c1} values from different authors are shown for comparison. The values are those obtained for the H_{c1} parallel to c-axis in monocrystalline samples. Our value for a polycrystalline sample is of the same order of magnitude than the lowest monocrystalline values.

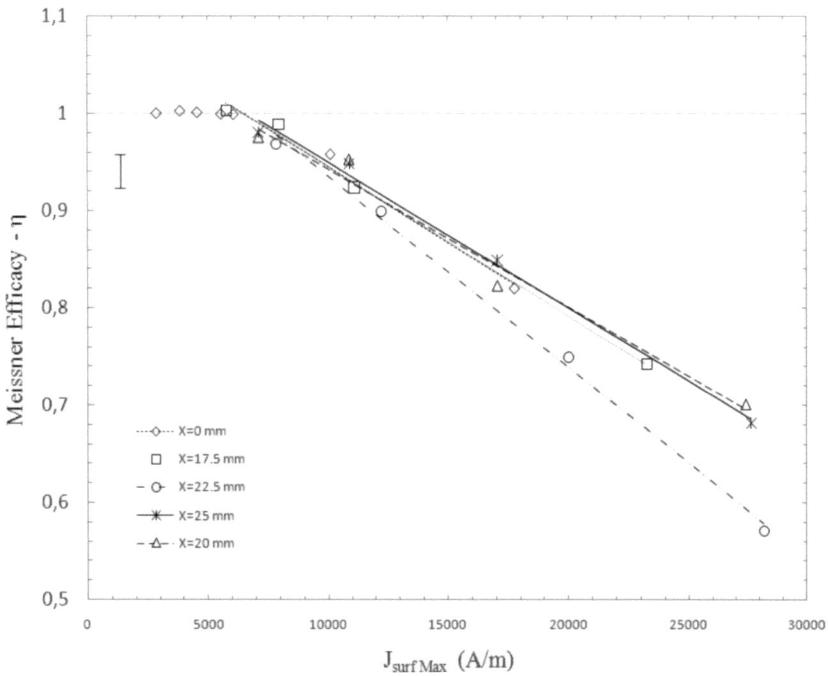

Fig. 14. Meissner Efficacy versus maximum Jsurf for different positions. The values obtained for X=5.0, 10.0, 15.0 mm radial positions are similar to those obtained for the X=0.0 mm values.

Now, if we use a value of λ_L=*4500 Å*, carried out from the literature (Geflbaux & Tazawa, 1998; Mayer & Schuster, 1993) we have a lower critical current density value of Jc_1 = (1.43 ± *0.08)* $\times 10^7 A/cm^2$. By using Eq. 2 we calculate $Hc_1 = 3226 \pm 176\ A/m$.

	C. Bömer et al (2007) (monocr.)	Umewaza et al (2007) (monocr.)	Kaiser et al (1991) (monocr.)	Wu et al (1990) (monocr.)	Mechanical method (polycr.)
Results for H_{c1} (A/m)	2900 ± 250 6000 ± 2300	4500 ± 450	3580 11000	4950 15518	*3226 ± 176*

Table 1. Comparison of the values found in different articles with that measured in this paper. The values and relative errors have been obtained directly from graphs, at 77 K. Available values for H ⊥ (a,b) and H || (c) in monocrystals are shown. H || (c) is always greater than H ⊥ (a,b)

The uncertainty in the determination of $J_{c1\ surf}$ may be reduced by increasing the number of series of measurements (or paths). Therefore, this is a method intrinsically more precise than other common methods.

In fact, the values far from the Meissner state contribute to improve the accuracy of the $J_{c1\ surf}$ determination. The determination of the slopes of straight lines has a propagation of errors

more convenient than that in the case of the measurement of a change in the slope of the tangents to a curve. Other methods, therefore, would require high precision measurements to obtain a reasonable error for Hc1.
This results are in according to the border and thickness effects and border magnetization that have been already described by other authors in an uniform magnetic field (Brandt, 2000; Morozov et al., 1996; Li et al., 2004; Schmidt et al., 1997):

6. Example of application - permanent magnet over a superconducting torus

We calculate the torque exerted between a superconducting torus and a permanent magnet by using this model. We find that there is a flip effect on the stablest direction of the magnet depending on its position. This could be easily used as a digital detector for proximity.
We consider a full superconducting torus and a cylindrical permanent NdFeB magnet over the superconductor axis (Z axis). In figure 15 we can observe the geometrical configuration of both components. Every calculation is referenced with respect of a Cartesian coordinate system placed in the center of mass of the torus which Z axis is coincident with the axis of the torus.

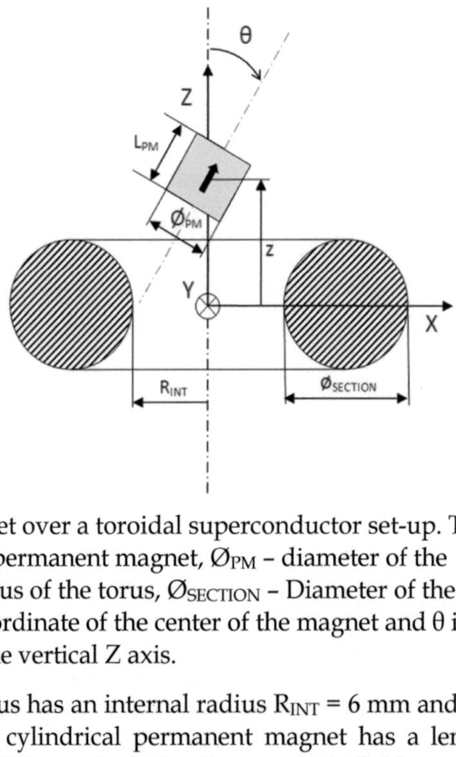

Fig. 15. Permanent magnet over a toroidal superconductor set-up. The dimensions are: L_{PM} - length of the cylindrical permanent magnet, $Ø_{PM}$ – diameter of the cylindrical permanent magnet, R_{INT} – Inner radius of the torus, $Ø_{SECTION}$ – Diameter of the circular section of the torus. z is the vertical coordinate of the center of the magnet and θ is the angle between the axis of the magnet and the vertical Z axis.

The superconducting torus has an internal radius R_{INT} = 6 mm and a diameter of the section $Ø_{SECTION}$ = 10 mm. The cylindrical permanent magnet has a length L_{PM} = 5 mm and a diameter $Ø_{PM}$ = 5 mm. When calculating the magnetic field generated by the magnet we define its magnetic properties as: Coercive magnetic field $H_{COERCIVITY}$ = 875 kA/m and remanent magnetic flux density $B_{REMANENT}$ = 1.18 T. We assume that the direction of magnetization of the permanent magnet coincides with its axis of revolution.
The variables θ and z are the coordinates we modify in order to analyze the mechanical behavior of the magnet over the superconductor. z is the distance along the Z axis between

the center of mass of the torus and the one of the cylindrical permanent magnet. θ is the angle between the axis of the magnet and the vertical Z axis.
The equilibrium angle (θ_{eq}) as a function of z can be determined as follows. For a certain z we calculate the Y component of the torque (M_y) exerted on the magnet by the superconductor as a function of θ and we find the equilibrium angle as the value for which $M_y(\theta_{eq})=0$. The sign of the slope $dM_y/d\theta$ at that point determines the stability or instability of the equilibrium point.

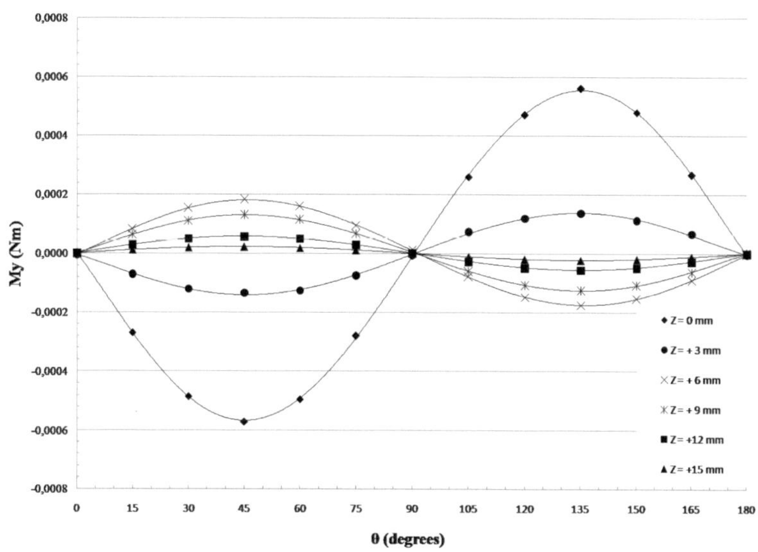

Fig. 16. M_y applied to the permanent magnet by the superconductor as a function of θ for z= 0, 3, 6, 9, 12 and 15 mm.

In figure 16 the torque (M_y) exerted on the magnet by the superconductor as a function of θ is shown for z = 0, 3, 6, 9, 12 and 15 mm. The maximum values for the torque exerted to the permanent magnet appear at θ = 45° and θ = 135° for every z. The remarkable fact is that the sign suddenly changes when moving from z = 3 mm to z = 6 mm. The equilibrium points are always at θ = 0° and θ = 90°, but θ = 0° is a stable equilibrium point for z = 0 mm and z = 3 mm, while it is unstable for the rest of the positions. On the other hand θ = 90° is unstable for z = 0 mm and z = 3 mm, but it is stable for the rest of the positions. That means that if you approach a magnet along the Z axis and it is able to rotate, it will be perpendicular to the Z axis while it is at z ≥ 6 mm, but it will suddenly rotate to be parallel to the Z axis when you pass from z = 6 to z ≤ 3.
In figure 17 the variation of the torque at θ = 45° as a function of z. The torque changes its sign between z =3 mm and z =4 mm.
Finally, figure 18 shows the stable equilibrium angle as a function of z. It is evident that, at a certain position between z =+ 3 and z =+ 4 mm we found that the stable equilibrium angle switches from a vertical orientation of the magnet to an horizontal one describing the flip effect claimed in this work.
Therefore, it can be concluded that if you approach a magnet along the Z axis and it is able to rotate, it will be perpendicular to the Z axis while it is at a certain distance (z ≥ 4 mm in

our example) and it will change to be parallel to the Z axis for closer positions ($z \leq 3$ mm in our example). As the equilibrium angle does not depend on the magnetic moment, the magnet can be much smaller. As a flip in the orientation of a permanent magnet can be easily instrumented, this effect can be easily used as a binary detector for proximity.

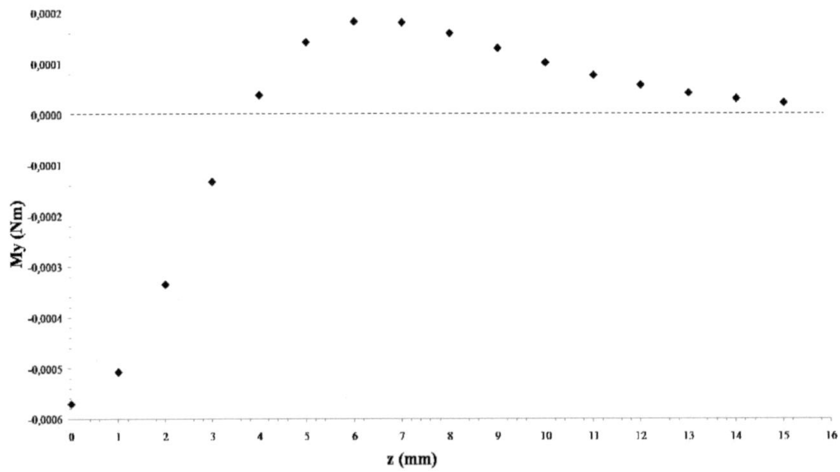

Fig. 17. Torque M_y exerted on the magnet for $\theta = 45^o$ as a function of z.

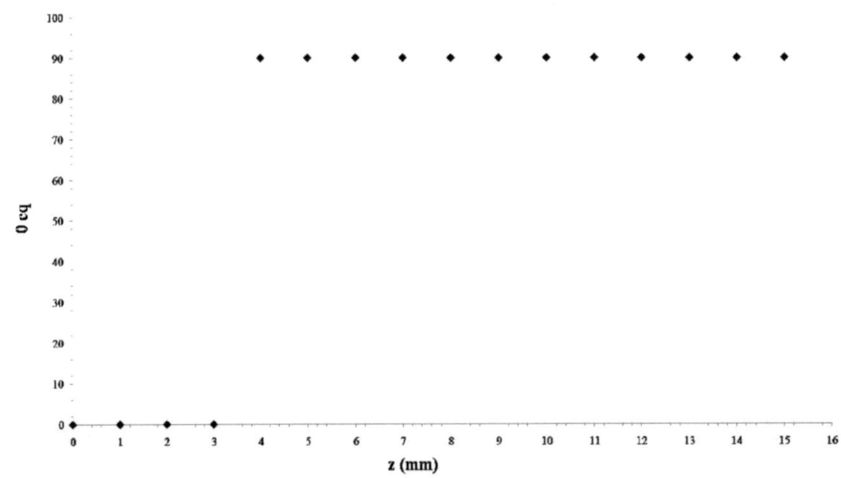

Fig. 18. Stable equilibrium angle (θ_{eq}) as a function of z.

7. Conclusion

Magnet-superconductor forces both in Meissner and mixed states can be calculated with the accuracy required to engineer useful levitating devices.

The implementation of a local differential expression in a finite elements program opens new perspectives to the use of magnet-superconductor devices for engineering. This can be

used to calculate forces whatever the size, shape and geometry of the system, for both permanent magnets and electromagnets.
Accuracy and convergence, in addition to the experimental verification for different cases have been tested. There is a good agreement between experimental results and calculation, even with very low-cost computing resources involved.
Moreover, the expression can be used to determine the point when the mixed state arises in a superconductor piece.

8. References

Alario, M.A. & Vicent, J.L. (1991). *Superconductividad,* 1st ed. (EUDEMA, Madrid, 1991).

Arkadiev, V. (1947). A floating magnet. *Nature* 160, 330 (1947).

Bednorz, J.G. & Müller, K.A. (1986). Possible highT c superconductivity in the Ba- La- Cu- O system. *Zeitschrift Für Physik B Condensed Matte.r* 64, 189–193 (1986).

Böhmer, C.; Brandstatter, G. & Weber, H. (2007). The lower critical field of high-temperature superconductors. *Supercond. Sci. Techn.,* 10 (2007) A1-A10

Brandt, E.H. (2000). Superconductors in realistic geometries: geometric edge barrier versus pinning. *Physica C,* 332 (2000) 99-107

Buchhold, T.A. (1960). Applications of superconductivity. *Scientific American* 3, 74 (1960).

Cansiz, A.; Hull, J.R. & Gundogdu O. (2005). Translational and rotational dynamic analysis of a superconducting levitation system. *Supercond. Sci. Technol.* 18, 990 (2005).

Diez-Jimenez E. et al. (2010). Finite element algorithm for solving supercondcuting Meissner repulsion forces. *Icec 23 - Icmc 2010* (Wroclaw (Poland), 2010).

Early, E.A.; Seaman, C.L.; Yang, K.N. & Maple, *M.B. (*1988) *American Journal Of Physics* 56, 617 (1988).

Geflbaux, X. & Tazawa, *M.* (1998). Étude de la variation de la longueur de London entre 5 et 70 K , dans un film très mince d ' YBaCuOç6 , par spectromérie dans I ' infrarouge lointain ; comparaison avec NbN. *C.R. de l'Academie Sciences.* Fasc. B, 324 (1998) 389-397

Giaro, K.; Gorzkowski, W. & Motylewski, *T.* (1990). A correct description of the interaction between a magnetic moment and its image. *Physica C* 168, 479-481 (1990).

Gijutsucho, K. (2001) *Patent JP2001004652-A,* (2001) Japan

Hahn, P. & Weber, *H.W.* (1983). Automatic device for magnetization measurements on superconductors. *Cryogenics,* 23 (1983) 87-90

Hellman F. et al. (1988). Levitation of a magnet over a flat type II superconductor. *Journal Of applied Physics* 63, 447 (1988).

Hull, J.R. & Cansiz, *A.* (1999). Vertical and lateral forces between a permanent magnet and a high-temperature superconductor. *Journal Of Applied Physics* 86, (1999).

Hull, J.R. (2000). Superconducting bearings. *Supercond. Sci. Techn.,* 13 (2000) R1–R15

Jackson, J.D. (1975) *Classical Electrodynamics* (John Wiley & Sons, New York, 1975).

Kaiser, D.L. & Swartzendruber, *L.J.* (1991). Lower critical field measurements in YBACUO single cristals. *Proc. Adv. in mat. sci. & appl. of HTSC* (1991)

Li, Q.; Suenaga, M. & Ye, Z. (2004). Crossover of thickness dependence of critical current density J(c)(T,H) in YBa2Cu3O7-delta thick films. App. *Physics Lett.* 84 (2004) 3528-3530

Lin, Q. (2006). Theoretical development of the image method for a general magnetic source in the presence of a superconducting sphere or a long superconducting cylinder. *Physical Review B* 74, 24510 (2006).

Mayer, B. & Schuster, S. (1993). Magnetic-field dependence of the critical current in yba2cu3o7-delta bicrystal grain-boundary junctions. *App. Physics Lett.* 63 (1993) 783-785

Meilikhov, E. & Shapiro, V.G. (1992). Critical fields of the htsc superconductors. *Supercond. Sci. Techn.*, 5 (1992) S391-S394

Moon, F.C. (1994). *Superconducting Levitation* (John Wiley & Sons, New York, 1994).

Morozov, N. & Zeldov, E. (1996). Negative local permeability in Bi2Sr2CaCu2O8 crystals. *Phys. Rev. Lett.* 76 (1996) 138-141

Perez-Diaz, J.L. & Garcia-Prada, J.C. (2007). Finite-size-induced stability of a permanent magnet levitating over a superconductor in the Meissner state. *Applied Physics Letters* 91, 142503 (2007).

Perez-Diaz, J.L. & Garcia-Prada, J.C. (2007). Interpretation of the method of images in estimating superconducting levitation. *Physica C* 467, 141-144 (2007).

Perez-Diaz, J.L.; Garcia-Prada, J.C. & Diaz-Garcia, J.A. (2008). Universal Model for Superconductor-Magnet Forces in the Static Limit. *I.R.E.M.E.* 2, (2008).

Poole, C.P. (2007) *Superconductivity*, 2nd ed.,AP, Elsevier, 2007

Schmidt, B.; Morozov, N. & Zeldov, E. (1997). Angular dependence of the first-order vortex-lattice phase transition in Bi2Sr2CaCu2O8. *Phys. Rev. B*, 55 (1997) R8705-R8708

Simon, I. (1953). Forces acting on superconductors in magnetic fields. *Journal Of Applied Physics* 24, 19 (1953).

Umezawa, A. & Crabtree, A. (1998). Anisotropy of the lowr critical field, magnetic penetration depth, and equilibrium shielding current y single-crystal YBACUO. *Physical Review B*. 38, (1998).

Vandervoort, K.G.; Welp, U. & Kessler, J.E. (1991). Magnetic measurements of the upper and lower critical fields of oxygen-deficient yba2cu3o7-delta single-crystals. *Phys. Rev. B*, 43 (1991) 13042-13048

Wacenovsky, M.; Weber, H.; Hyun, O.B. (1989). Magnetization of grain-aligned hoba2cu3o7-delta. *Physica C*, 162 (1989) 1629-1630

Weber, H.W.; Seidl, E. & Laa, C. (1991). Anisotropy effects in superconducting niobium. *Phys. Rev. B*, 44 (1991) 7585-7600

Wu, D.H. & Sridhar, S. (1990). Pinning forces and lower critical fields in YBACUO crystals: temperature dependence and anisotropy. *Phys. Rev. Lett.* 65 (1990) 2074-2077

Yang, Y. & Zheng, X. (2007). Method for solution of the interaction between superconductor and permanent magnet. *Journal Of Applied Physics* 101, 113922 (2007).

9

Properties of Macroscopic Quantum Effects and Dynamic Natures of Electrons in Superconductors

Pang Xiao-feng
Institute of Life Science and Technology, University of Electronic Science and Technology of China, Chengdu, International Centre for Materials Physics, Chinese Academy of Science, Shenyang, China

1. Introduction

So-called macroscopic quantum effects(MQE) refer to a quantum phenomenon that occurs on a macroscopic scale. Such effects are obviously different from the microscopic quantum effects at the microscopic scale as described by quantum mechanics. It has been experimentally demonstrated [1-17] that macroscopic quantum effects are the phenomena that have occurred in superconductors. Superconductivity is a physical phenomenon in which the resistance of a material suddenly vanishes when its temperature is lower than a certain value, Tc, which is referred to as the critical temperature of superconducting materials. Modern theories [18-21] tell us that superconductivity arises from the irresistible motion of superconductive electrons. In such a case we want to know "How the macroscopic quantum effect is formed? What are its essences? What are the properties and rules of motion of superconductive electrons in superconductor?" and, as well, the answers to other key questions. Up to now these problems have not been studied systematically. We will study these problems in this chapter.

2. Experimental observation of property of macroscopic quantum effects in superconductor

(1) Superconductivity of material. As is known, superconductors can be pure elements, compounds or alloys. To date, more than 30 single elements, and up to a hundred alloys and compounds, have been found to possess the characteristics [1-17] of superconductors. When $T \le T_c$, any electric current in a superconductor will flow forever without being damped. Such a phenomenon is referred to as perfect conductivity. Moreover, it was observed through experiments that, when a material is in the superconducting state, any magnetic flux in the material would be completely repelled resulting in zero magnetic fields inside the superconducting material, and similarly, a magnetic flux applied by an external magnetic field can also not penetrate into superconducting materials. Such a phenomenon is

called the perfect anti-magnetism or Maissner effect. Meanwhile, there are also other features associated with superconductivity, which are not present here

How can this phenomenon be explained? After more than 40 years' effort, Bardeen, Cooper and Schreiffier proposed the new idea of Cooper pairs of electrons and established the microscopic theory of superconductivity at low temperatures, the BCS theory [18-21],in 1957, on the basis of the mechanism of the electron-phonon interaction proposed by Frohlich [22-23]. According to this theory, electrons with opposite momenta and antiparallel spins form pairs when their attraction, due to the electron and phonon interaction in these materials, overcomes the Coulomb repulsion between them. The so-called Cooper pairs are condensed to a minimum energy state, resulting in quantum states, which are highly ordered and coherent over the long range, and in which there is essentially no energy exchange between the electron pairs and lattice. Thus, the electron pairs are no longer scattered by the lattice but flow freely resulting in superconductivity. The electron pairs in a superconductive state are somewhat similar to a diatomic molecule but are not as tightly bound as a molecule. The size of an electron pair, which gives the coherent length, is approximately 10^{-4} cm. A simple calculation shows that there can be up to 10^6 electron pairs in a sphere of 10^{-4} cm in diameter. There must be mutual overlap and correlation when so many electron pairs are brought together. Therefore, perturbation to any of the electron pairs would certainly affect all others. Thus, various macroscopic quantum effects can be expected in a material with such coherent and long range ordered states. Magnetic flux quantization, vortex structure in the type-II superconductors, and Josephson effect [24-26] in superconductive junctions are only some examples of the phenomena of macroscopic quantum mechanics.

(2) The magnetic flux structures in superconductor. Consider a superconductive ring. Assume that a magnetic field is applied at T >Tc, then the magnetic flux lines ϕ_0 produced by the external field pass through and penetrate into the body of the ring. We now lower the temperature to a value below Tc, and then remove the external magnetic field. The magnetic induction inside the body of circular ring equals zero ($\vec{B}=0$) because the ring is in the superconductive state and the magnetic field produced by the superconductive current cancels the magnetic field, which was within the ring. However, part of the magnetic fluxes in the hole of the ring remain because the induced current is in the ring vanishes. This residual magnetic flux is referred to as "the frozen magnetic flux". It has been observed experimentally, that the frozen magnetic flux is discrete, or quantized. Using the macroscopic quantum wave function in the theory of superconductivity, it can be shown that the magnetic flux is established by $\Phi'=n\phi_0$ (n=0,1,2,…), where ϕ_0 =hc/2e=2.07×10^{-15} Wb is the flux quantum, representing the flux of one magnetic flux line. This means that the magnetic fluxes passing through the hole of the ring can only be multiples of ϕ_0 [1-12]. In other words, the magnetic field lines are discrete. We ask, "What does this imply?" If the magnetic flux of the applied magnetic field is exactly n, then the magnetic flux through the hole is $n\phi_0$, which is not difficult to understand. However, what is the magnetic flux through the hole if the applied magnetic field is (n+1/4) ϕ_0? According to the above, the magnetic flux cannot be (n+1/4)ϕ_0. In fact, it should be $n\phi_0$. Similarly, if the applied magnetic field is (n+3/4)ϕ_0 and the magnetic flux passing through the hole is not (n+3/4)ϕ_0, but rather (n+1)ϕ_0, therefore the magnetic fluxes passing through the hole of the circular ring are always quantized.

An experiment conducted in 1961 surely proves this to be so, indicating that the magnetic flux does exhibit discrete or quantized characteristics on a macroscopic scale. The above experiment was the first demonstration of the macroscopic quantum effect. Based on quantization of the magnetic flux, we can build a "quantum magnetometer" which can be used to measure weak magnetic fields with a sensitivity of 3×10^{-7} Oersted. A slight modification of this device would allow us to measure electric current as low as 2.5×10^{-9} A.

(3) Quantization of magnetic-flux lines in type-II superconductors. The superconductors discussed above are referred to as type-I superconductors. This type of superconductor exhibits a perfect Maissner effect when the external applied field is higher than a critical magnetic value $\vec{H}_c$. There exists other types of materials such as the NbTi alloy and Nb_3Sn compounds in which the magnetic field partially penetrates inside the material when the external field $\vec{H}$ is greater than the lower critical magnetic field $\vec{H}_{c1}$, but less than the upper critical field $\vec{H}_{c2}$ [1-7]. This kind of superconductor is classified as type-II superconductors and is characterized by a Ginzburg-Landau parameter greater than 1/2. Studies using the Bitter method showed that the penetration of a magnetic field results in some small regions changing from superconductive to normal state. These small regions in normal state are of cylindrical shape and regularly arranged in the superconductor, as shown in Fig.1. Each cylindrical region is called a vortex (or magnetic field line)[1-12]. The vortex lines are similar to the vortex structure formed in a turbulent flow of fluid. Both theoretical analysis and experimental measurements have shown that the magnetic flux associated with one vortex is exactly equal to one magnetic flux quantum ϕ_0, when the applied field $\vec{H} \geq \vec{H}_{c1}$, the magnetic field penetrates into the superconductor in the form of vortex lines, increased one by one. For an ideal type-II superconductor, stable vortices are distributed in triagonal pattern, and the superconducting current and magnetic field distributions are also shown in Fig. 1. For other, non-ideal type-II superconductors, the triagonal pattern of distribution can be also observed in small local regions, even though its overall distribution is disordered. It is evident that the vortex-line structure is quantized and this has been verified by many experiments and can be considered a result of the quantization of magnetic flux. Furthermore, it is possible to determine the energy of each vortex line and the interaction energy between the vortex lines. Parallel magnetic field lines are found to repel each other while anti-parallel magnetic lines attract each other.

(4) The Josephson phenomena in superconductivity junctions [24-26]. As it is known in quantum mechanics, microscopic particles, such as electrons, have a wave property and that can penetrate through a potential barrier. For example, if two pieces of metal are separated by an insulator of width of tens of angstroms, an electron can tunnel through the insulator and travel from one metal to the other. If voltage is applied across the insulator, a tunnel current can be produced. This phenomenon is referred to as a tunneling effect. If two superconductors replace the two pieces of metal in the above experiment, a tunneling current can also occur when the thickness of the dielectric is reduced to about $30\overset{0}{A}$. However, this effect is fundamentally different from the tunneling effect discussed above in quantum mechanics and is referred to as the Josephson effect.

Evidently, this is due to the long-range coherent effect of the superconductive electron pairs. Experimentally, it was demonstrated that such an effect could be produced via many types of junctions involving a superconductor, such as superconductor-metal-superconductor junctions, superconductor-insulator- superconductor junctions, and superconductor bridges. These junctions can be considered as superconductors with a weak link. On the one hand, they have properties of bulk superconductors, for example, they are capable of carrying certain superconducting currents. On the other hand, these junctions possess unique properties, which a bulk superconductor does not. Some of these properties are summarized in the following.

(A) When a direct current (dc) passing through a superconductive junction is smaller than a critical value Ic, the voltage across the junction does not change with the current. The critical current Ic can range from a few tens of μA to a few tens of mA.

(B) If a constant voltage is applied across the junction and the current passing through the junction is greater than Ic, a high frequency sinusoidal superconducting current occurs in the junction. The frequency is given by υ=2eV/h, in the microwave and far-infrared regions of (5-1000)×10⁹Hz. The junction radiates a coherent electromagnetic wave with the same frequency. This phenomenon can be explained as follows: The constant voltage applied across the junction produces an alternating Josephson current that, in turn, generates an electromagnetic wave of frequency, υ. The wave propagates along the planes of the junction. When the wave reaches the surface of the junction (the interface between the junction and its surrounding), part of the electromagnetic wave is reflected from the interface and the rest is radiated, resulting in the radiation of the coherent electromagnetic wave. The power of radiation depends on the compatibility between the junction and its surrounding.

(C) When an external magnetic field is applied over the junction, the maximum dc current, Ice , is reduced due to the effect of the magnetic field. Furthermore, I_c changes periodically as the magnetic field increases. The $I_c - H$ curve resembles the distribution of light intensity in the Fraunhofer diffraction experiment , and the latter is shown in Fig. 2. This phenomenon is called quantum diffraction of the superconductivity junction.

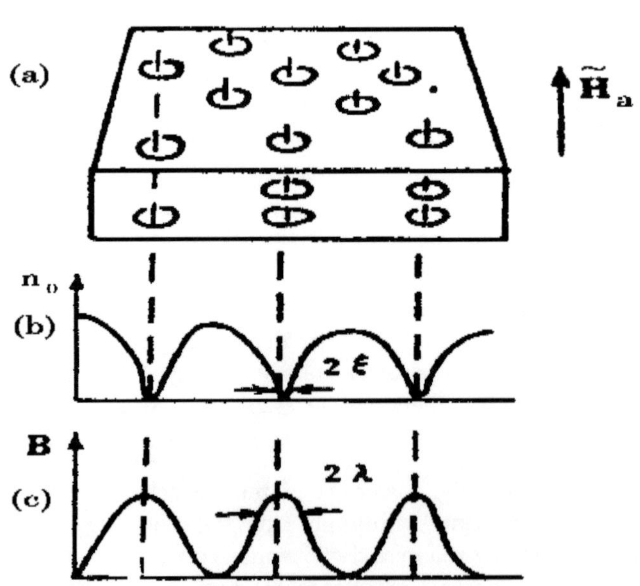

Fig. 1. Current and magnetic field distributionseffect in in a type-II superconductor.

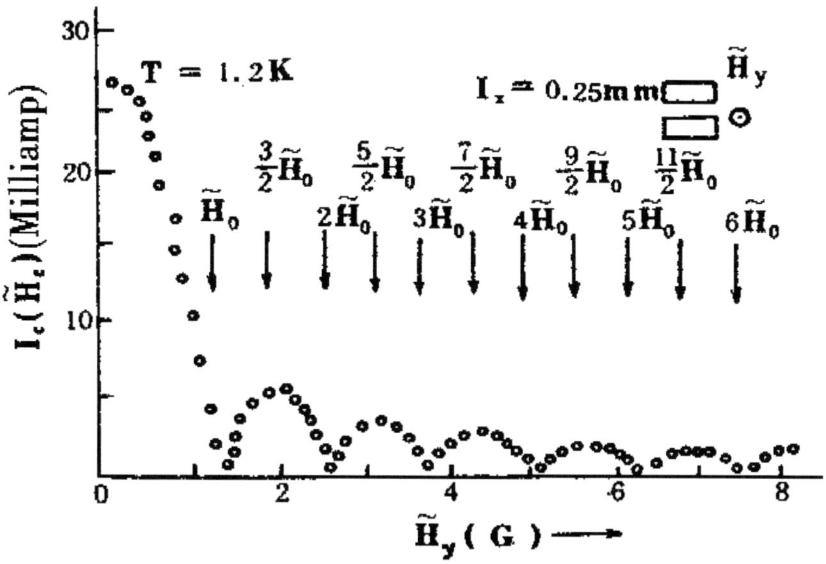

Fig. 2. Quantum diffractionsuperconductor junction

(D) When a junction is exposed to a microwave of frequency, υ, and if the voltage applied across the junction is varied, it can be seen that the dc current passing through the junction increases suddenly at certain discrete values of electric potential. Thus, a series of steps appear on the dc I − V curve, and the voltage at a given step is related to the frequency of the microwave radiation by nυ=2eVn/h(n=0,1,2,3…). More than 500 steps have been observed in experiments.

Josephson first derived these phenomena theoretically and each was experimentally verified subsequently. All these phenomena are, therefore, called Josephson effects [24-26]. In particular, (1) and (3) are referred to as dc Josephson effects while (2) and (4) are referred to as ac Josephson effects. Evidently, Josephson effects are macroscopic quantum effects, which can be explained well by the macroscopic quantum wave function. If we consider a superconducting junction as a weakly linked superconductor, the wave functions of the superconducting electron pairs in the superconductors on both sides of the junction are correlated due to a definite difference in their phase angles. This results in a preferred direction for the drifting of the superconducting electron pairs, and a dc Josephson current is developed in this direction. If a magnetic field is applied in the plane of the junction, the magnetic field produces a gradient of phase difference, which makes the maximum current oscillate along with the magnetic field, and the radiation of the electromagnetic wave occur. If a voltage is applied across the junction, the phase difference will vary with time and results in the Josephson effect. In view of this, the change in the phase difference of the wave functions of superconducting electrons plays an important role in Josephson effect, which will be discussed in more detail in the next section.

The discovery of the Josephson effect opened the door for a wide range of applications of superconductor theory. Properties of superconductors have been explored to produce superconducting quantum interferometer–magnetometer, sensitive ammeter, voltmeter, electromagnetic wave generator, detector, frequency-mixer, and so on.

3. The properties of boson condensation and spontaneous coherence of macroscopic quantum effects

3.1 A nonlinear theoretical model of theoretical description of macroscopic quantum effects

From the above studies we know that the macroscopic quantum effect is obviously different from the microscopic quantum effect, the former having been observed for physical quantities, such as, resistance, magnetic flux, vortex line, and voltage, etc.

In the latter, the physical quantities, depicting microscopic particles, such as energy, momentum, and angular momentum, are quantized. Thus it is reasonable to believe that the fundamental nature and the rules governing these effects are different.

We know that the microscopic quantum effect is described by quantum mechanics. However, the question remains relative to the definition of what are the mechanisms of macroscopic quantum effects? How can these effects be properly described?

What are the states of microscopic particles in the systems occurring related to macroscopic quantum effects? In other words, what are the earth essences and the nature of macroscopic quantum states? These questions apparently need to be addressed.

We know that materials are composed of a great number of microscopic particles, such as atoms, electrons, nuclei, and so on, which exhibit quantum features. We can then infer, or assume, that the macroscopic quantum effect results from the collective motion and excitation of these particles under certain conditions such as, extremely low temperatures, high pressure or high density among others. Under such conditions, a huge number of microscopic particles pair with each other condense in low-energy state, resulting in a highly ordered and long-range coherent. In such a highly ordered state, the collective motion of a large number of particles is the same as the motion of "single particles", and since the latter is quantized, the collective motion of the many particle system gives rise to a macroscopic quantum effect. Thus, the condensation of the particles and their coherent state play an essential role in the macroscopic quantum effect.

What is the concept of condensation? On a macroscopic scale, the process of transforming gas into liquid, as well as that of changing vapor into water, is called condensation. This, however, represents a change in the state of molecular positions, and is referred to as a condensation of positions. The phase transition from a gaseous state to a liquid state is a first order transition in which the volume of the system changes and the latent heat is produced, but the thermodynamic quantities of the systems are continuous and have no singularities. The word condensation, in the context of macroscopic quantum effects has its' special meaning. The condensation concept being discussed here is similar to the phase transition from gas to liquid, in the sense that the pressure depends only on temperature, but not on the volume noted during the process, thus, it is essentially different from the above, first-order phase transition. Therefore, it is fundamentally different from the first-order phase transition such as that from vapor to water. It is not the condensation of particles into a high-density material in normal space. On the contrary, it is the condensation of particles to a single energy state or to a low energy state with a constant or zero momentum. It is thus also called a condensation of momentum. This differs from a first-order phase transition and theoretically it should be classified as a third order phase transition, even though it is really a second order phase transition, because it is related to the discontinuity of the third derivative of a thermodynamic function. Discontinuities can be clearly observed in measured specific heat, magnetic susceptibility of certain systems when condensation occurs. The phenomenon results from a spontaneous breaking of symmetries of the system due to nonlinear interaction within the system under some special conditions such as, extremely low temperatures and high pressures. Different systems have different critical temperatures of condensation. For example, the condensation temperature of a superconductor is its critical temperature T_c, and from previous discussions[27-32].

From the above discussions on the properties of superconductors, and others we know that, even though the microscopic particles involved can be either Bosons or Fermions, those being actually condensed, are either Bosons or quasi-Bosons, since Fermions are bound as pairs. For this reason, the condensation is referred to as Bose-Einstein condensation[33-36] since Bosons obey the Bose-Einstein statistics. Properties of Bosons are different from those of Fermions as they do not follow the Pauli exclusion principle, and there is no limit to the number of particles occupying the same energy levels. At finite temperatures, Bosons can distribute in many energy states and each state can be occupied by one or more particles, and some states may not be occupied at all. Due to the statistical attractions between Bosons in the phase space (consisting of generalized coordinates and momentum), groups of Bosons tend to occupy one quantum energy state under certain conditions. Then when the temperature of the system falls below a critical value, the majority or all Bosons condense to the same energy level (e.g. the ground state), resulting in a Bose condensation and a series of interesting macroscopic quantum effects. Different macroscopic quantum phenomena are observed because of differences in the fundamental properties of the constituting particles and their interactions in different systems.

In the highly ordered state of the phenomena, the behavior of each condensed particle is closely related to the properties of the systems. In this case, the wave function $\phi = fe^{i\theta}$ or $\phi = \sqrt{\rho}e^{i\theta}$ of the macroscopic state[33-35], is also the wave function of an individual condensed particle. The macroscopic wave function is also called the order parameter of the condensed state. This term was used to describe the superconductive states in the study of these macroscopic quantum effects. The essential features and fundamental properties of macroscopic quantum effect are given by the macroscopic wave function ϕ and it can be further shown that the macroscopic quantum states, such as the superconductive states are coherent and are Bose condensed states formed through second-order phase transitions after the symmetry of the system is broken due to nonlinear interaction in the system.

In the absence of any externally applied field, the Hamiltonian of a given macroscopic quantum system can be represented by the macroscopic wave function ϕ and written as

$$H = \int dx \mathrm{H}' = \int dx[-\frac{1}{2}|\nabla\phi|^2 - \alpha|\phi|^2 + \lambda|\phi|^4] \tag{1}$$

Here H'=H presents the Hamiltonian density function of the system, the unit system in which m=h=c=1 is used here for convenience. If an externally applied electromagnetic field does exist, the Hamiltonian given above should be replaced by

$$\mathrm{H} = \int dx \mathrm{H}' = \int dx\left[-\frac{1}{2}|\nabla - ie^*\vec{A}\phi|^2 - \alpha|\phi|^2 + \lambda|\phi|^4 + \frac{\vec{\tilde{H}}^2}{8\pi}\right] \tag{2}$$

or, equivalently

$$H = \int dx H' = \int dx \left[-\frac{1}{2}\left|(\partial_j - ie^* A_j)\phi\right|^2 - \alpha|\phi|^2 + \lambda|\phi|^4 + \frac{1}{4}F_{ji}.F^{ji} \right]$$

where $F_{ji} = \partial_j A_i - \partial_t A_j$ is the covariant field intensity, $\vec{H} = \nabla \times \vec{A}$ is the magnetic field intensity, e is the charge of an electron, and e*=2e, $\vec{A}$ is the vector potential of the electromagnetic field, α and λ can be said to be some of the interaction constants. The above Hamiltonians in Eqs.(1) and (2) have been used in studying superconductivity by many scientists, including Jacobs de Gennes [37], Saint-Jams [38], Kivshar [39-40], Bullough [41-42], Huepe [43], Sonin [44], Davydov [45], et al., and they can be also derived from the free energy expression of a superconductive system given by Landau et al [46-47]. As a matter of fact, the Lagrangian function of a superconducting system can be obtained from the well-known Ginzberg-Landau (GL) equation [47-54] using the Lagrangian method, and the Hamiltonian function of a system can then be derived using the Lagrangian approach. The results, of course, are the same as Eqs. (1) and (2). Evidently, the Hamiltonian operator corresponding to Eqs. (1) and (2) represents a nonlinear function of the wave function of a particle, and the nonlinear interaction is caused by the electron-phonon interaction and due to the vibration of the lattice in BCS theory in the superconductors. Therefore, it truly exists. Evidently, the Hamiltonians of the systems are exactly different from those in quantum mechanics, and a nonlinear interaction related to the state of the particles is involved in Eqs. (1) -(2). Hence, we can expect that the states of particles depicted by the Hamiltonian also differ from those in quantum mechanics, and the Hamiltonian can describe the features of macroscopic quantum states including superconducting states. These problems are to be treated in the following pages. Evidently, the Hamiltonians in Eqs. (1) and (2) possess the U (1) symmetry. That is, they remain unchanged while undergoing the following transformation:

$$\phi(\vec{r},t) \to \phi'(\vec{r},t) = e^{-iQ_j\theta}\phi(\vec{r},t)$$

where Q_j is the charge of the particle, θ is a phase and, in the case of one dimension, each term in the Hamiltonian in Eq. (1) or Eq. (2) contains the product of the $\phi_j(x,t)$s, then we can obtain:

$$\phi_1'(x,t)\phi_2'(x,t)....\phi_n'(x,t) = e^{-i(Q_1+Q_2+\cdots+Q_n)\theta}\phi_1(x,t)\phi_2(x,t)....\phi_n(x,t)$$

Since charge is invariant under the transformation and neutrality is required for the Hamiltonian, there must be $(Q_1 + Q_2 + \cdots + Q_n) = 0$ in such a case. Furthermore, since θ is independent of x, it is necessary that $\nabla\phi_j \to e^{-i\theta Q_j}\nabla\phi_j$. Thus each term in the Hamiltonian in Eqs. (1) is invariant under the above transformation, or it possesses the U(1) symmetry[16-17]. If we rewrite Eq. (1) as the following

$$H' = -\frac{1}{2}(\nabla\phi)^2 + U_{eff}(\phi), U_{eff}(\phi) = -\alpha\phi^2 + \lambda\phi^4 \tag{3}$$

We can see that the effective potential energy, $U_{eff}(\phi)$, in Eq. (3) has two sets of extrema, $\phi_0 = \pm\sqrt{\alpha/2\lambda}$ and ϕ_0 =0, but the minimum is located at

$$\phi_0 = \pm\sqrt{\alpha/2\lambda} = \langle 0|\phi|0\rangle, \tag{4}$$

rather than at $\phi_0 = 0$. This means that the energy at $\phi_0 = \pm\sqrt{\alpha/2\lambda}$ is lower than that at $\phi_0 = 0$. Therefore, $\phi_0 = 0$ corresponds to the normal ground state, while $\phi_0 = \pm\sqrt{\alpha/2\lambda}$ is the ground state of the macroscopic quantum systems .

In this case the macroscopic quantum state is the stable state of the system. This shows that the Hamiltonian of a normal state differs from that of the macroscopic quantum state, in which the two ground states satisfy $\langle 0|\phi|0\rangle \neq -\langle 0|\phi|0\rangle$ under the transformation, $\phi \to -\phi$. That is, they no longer have the U(1) symmetry. In other words，the symmetry of the ground states has been destroyed. The reason for this is evidently due to the nonlinear term $\lambda\phi^4$ in the Hamiltonian of the system. Therefore, this phenomenon is referred to as a spontaneous breakdown of symmetry. According to Landau's theory of phase transition, the system undergoes a second-order phase transition in such a case, and the normal ground state $\phi_0 == 0$ is changed to the macroscopic quantum ground state $\phi_0 = \pm\sqrt{\alpha/2\lambda}$. Proof will be presented in the following example .

In order to make the expectation value in a new ground state zero in the macroscopic quantum state, the following transformation [16-17] is done:

$$\phi' = \phi + \phi_0 \tag{5}$$

so that

$$\langle 0|\phi'|0\rangle = 0 \tag{6}$$

After this transformation, the Hamiltonian density of the system becomes

$$H'(\phi+\phi_0) = \frac{1}{2}|\nabla\phi|^2 + (6\lambda\phi^2 - \alpha)\phi^2 + 4\lambda\phi_0\phi^3 + (4\lambda\phi_0^3 - 2\alpha\phi_0)\phi + \lambda\phi^4 - \alpha\phi_0^2 + \lambda\phi_0^4 \tag{7}$$

Inserting Eq. (4) into Eq. (7), we have $\langle \phi_0 | 4\lambda\phi_0^2 - 2\alpha | \phi_0 \rangle = 0$.

Consider now the expectation value of the variation $\delta H'/\delta\phi$ in the ground state, i.e. $\langle 0|\frac{\delta H'}{\delta\phi}|0\rangle = 0$, then from Eq. (1), we get

$$\langle 0|\frac{\delta H'}{\delta\phi}|0\rangle = \langle 0|-\nabla^2\phi + 2\alpha\phi - 4\lambda\phi^3|0\rangle = 0 \tag{8}$$

After the transformation Eq. (6), it becomes

$$\nabla^2\phi_0 + (4\lambda\phi_0^2 - 2\alpha)\phi_0 + 12\lambda\phi_0\langle 0|\phi^2|0\rangle + 4\lambda\langle 0|\phi^3|0\rangle - (2\alpha - 12\lambda\phi_0^2)\langle 0|\phi|0\rangle = 0 \tag{9}$$

where the terms $\langle 0|\phi^3|0\rangle$ and $\langle 0|\phi|0\rangle$ are both zero, but the fluctuation $12\lambda\phi_0\langle 0|\phi^2|0\rangle$ of the ground state is not zero. However, for a homogeneous system, at T=0K, the term $\langle 0|\phi^2|0\rangle$ is very small and can be neglected.

Then Eq. (9) can be written as

$$-\nabla^2\phi_0 - (4\lambda\phi_0^2 - 2\alpha)\phi_0 = 0 \tag{10}$$

Obviously, two sets of solutions, $\phi_0 = 0$, and $\phi_0 = \pm\sqrt{\alpha / 2\lambda}$, can be obtained from the above equation, but we can demonstrate that the former is unstable, and that the latter is stable.
If the displacement is very small, i.e. $\phi_0 \to \phi_0 + \delta\phi_0 = \phi_0'$, then the equation satisfied by the fluctuation $\delta\phi_0$ is relative to the normal ground state $\phi_0 = 0$ and is

$$\nabla^2\delta\phi_0 - 2\alpha\delta\phi_0 = 0 \tag{11}$$

Its' solution attenuates exponentially indicating that the ground state, $\phi_0 = 0$ is unstable. On the other hand, the equation satisfied by the fluctuation $\delta\phi_0$, relative to the ground state $\phi_0 = \pm\sqrt{\alpha / 2\lambda}$ is $\nabla^2\delta\phi_0 + 2\alpha\delta\phi_0 = 0$. Its' solution, $\delta\phi_0$, is an oscillatory function and thus the macroscopic quantum state ground state $\phi_0 = \pm\sqrt{\alpha / 2\lambda}$ is stable. Further calculations show that the energy of the macroscopic quantum state ground state is lower than that of the normal state by $\varepsilon_0 = -\alpha^2 / 4\lambda < 0$ .Therefore, the ground state of the normal phase and that of the macroscopic quantum phase are separated by an energy gap of $\alpha^2 / (4\lambda)$ so then, at T=0K, all particles can condense to the ground state of the macroscopic quantum phase rather than filling the ground state of the normal phase. Based on this energy gap, we can conclude that the specific heat of the macroscopic quantum systems has an exponential dependence on the temperature, and the critical temperature is given by: $T_c = 1.14\omega_p \exp[-1 / (3\lambda / \alpha)N(0)]$ [16-17]. This is a feature of the second-order phase transition. The results are in agreement with those of the BCS theory of superconductivity.
Therefore, the transition from the state $\phi_0 = 0$ to the state $\phi_0 = \pm\sqrt{\alpha / 2\lambda}$ and the corresponding condensation of particles are second-order phase transitions. This is obviously the results of a spontaneous breakdown of symmetry due to the nonlinear interaction, $\lambda\phi^4$.
In the presence of an electromagnetic field with a vector potential $\vec{A}$, the Hamiltonian of the systems is given by Eq. (2). It still possesses the U (1) symmetry.Since the existence of the nonlinear terms in Eq. (2) has been demonstrated, a spontaneous breakdown of symmetry can be expected. Now consider the following transformation:

$$\phi(x) = \frac{1}{\sqrt{2}}[\phi_1(x) + i\phi_2(x)] \to \frac{1}{\sqrt{2}}[\phi_1(x) + \phi_0 + i\phi_2(x)] \tag{12}$$

Since $\langle 0|\phi_i|0\rangle = 0$ under this transformation, then the equation (2) becomes

$$\begin{aligned} H' = {} & \frac{1}{4}(\partial_i A_j - \partial_j A_i)^2 - \frac{1}{2}(\nabla\phi_2)^2 - \frac{1}{2}(\nabla\phi_1)^2 + \frac{(e^*)^2}{2}[(\phi_1 + \phi_0)^2 + \phi_2^2]A_i^2 - e^*\phi_0 A_i \nabla\phi_2 + \\ & e^*(\phi_2\nabla\phi_1 - \phi_1\nabla\phi_2)A_i - \frac{1}{2}(-12\lambda\phi_0^2 + 2\alpha)\phi_1^2 - \frac{1}{2}(12\lambda\phi_0^2 + 2\alpha)\phi_2^2 + 4\lambda\phi_0\phi_1(\phi_1^2 + \phi_2^2) + \\ & 4\lambda(\phi_1^2 + \phi_2^2)^2 - \phi_0(4\lambda\phi_0^2 + 2\alpha)\phi_1 - \alpha\phi_0^2 + \lambda\phi_0^2 \end{aligned} \tag{13}$$

We can see that the effective interaction energy of ϕ_0 is still given by:

$$U_{eff}(\phi_0) = -\alpha\phi_0^2 + \lambda\phi_0^4 \tag{14}$$

and is in agreement with that given in Eq. (4). Therefore, using the same argument, we can conclude that the spontaneous symmetry breakdown and the second-order phase transition also occur in the system. The system is changed from the ground state of the normal phase, $\phi_0 = 0$ to the ground state $\phi_0 = \pm\sqrt{\alpha / 2\lambda}$ of the condensed phase in such a case. The above result can also be used to explain the Meissner effect and to determine its critical temperature in the superconductor. Thus, we can conclude that, regardless of the existence of any external field macroscopic quantum states, such as the superconducting state, are formed through a second-order phase transition following a spontaneous symmetry breakdown due to nonlinear interaction in the systems.

3.2 The features of the coherent state of macroscopic quantum effects

Proof that the macroscopic quantum state described by Eqs. (1) - (2) is a coherent state, using either the second quantization theory or the solid state quantum field theory is presented in the following paragraphs and pages.

As discussed above, when $\delta H'/\delta\phi$ =0 from Eq. (1), we have

$$\nabla^2\phi - 2\alpha\phi + 4\lambda|\phi|^2\phi = 0 \tag{15}$$

It is a time-independent nonlinear Schrödinger equation (NLSE), which is similar to the GL equation. Expanding ϕ in terms of the creation and annihilation operators, b_p^+ and b_p

$$\phi = \frac{1}{\sqrt{\overline{V}}}\sum_p \frac{1}{\sqrt{2\varepsilon_p}}(b_p e^{-ipx} + b_p^+ e^{ip.x}) \tag{16}$$

where $\overline{V}$ is the volume of the system. After a spontaneous breakdown of symmetry, ϕ_0, the ground-state of ϕ, for the system is no longer zero, but $\phi_0 = \pm\sqrt{\alpha / 2\lambda}$. The operation of the annihilation operator on $|\phi_0\rangle$ no longer gives zero, i.e.

$$b_p|\phi_0\rangle \neq 0 \tag{17}$$

A new field ϕ' can then be defined according to the transformation Eq. (5), where ϕ_0 is a scalar field and satisfies Eq. (10) in such a case. Evidently, ϕ_0 can also be expanded into

$$\phi_0 = -\frac{1}{\sqrt{\overline{V}}}\sum_p \frac{1}{\sqrt{2\varepsilon_p}}(\zeta_p e^{-ip.x} + \zeta_p^+ e^{ip.x}) \tag{18}$$

The transformation between the fields ϕ and ϕ' is obviously a unitary transformation, that is

$$\phi' = U\phi U^{-1} = e^{-s}\phi e^{s} = \phi + \phi_0 \tag{19}$$

where

$$S = i\int dx'[\phi'(x',t)\phi_0(x',t) - \phi_0(x',t)\phi(x',t)] \tag{20}$$

ϕ and ϕ' satisfy the following commutation relation

$$[\phi'(x',t),\phi(x,t)] = i\delta(x' - x) \tag{21}$$

From Eq. (6) we now have $\langle 0|\phi'|0\rangle = \phi_0' = 0$. The ground state $|\phi_0'\rangle$ of the field ϕ' thus satisfies

$$b_p|\phi_0'\rangle = 0 \tag{22}$$

From Eq. (6), we can obtain the following relationship between the annihilation operator a_p of the new field ϕ' and the annihilation operator b_p of the ϕ field

$$a_p = e^{-S}b_p e^{S} = b_p + \zeta_p \tag{23}$$

where

$$\zeta_p = \frac{1}{(2\pi)^{3/2}}\int \frac{dx}{\sqrt{\varepsilon_p}}[\phi_0(x,t)e^{ip.x} + i\phi_0^*(x,t)e^{-ip.x}] \tag{24}$$

Therefore, the new ground state $|\phi_0'\rangle$ and the old ground state $|\phi_0\rangle$ are related through $|\phi_0'\rangle = e^S|\phi_0\rangle$.

Thus we have

$$a_p|\phi_0'\rangle = (b_p + \zeta_p)|\phi_0'\rangle = \zeta_p|\phi_0'\rangle \tag{25}$$

According to the definition of the coherent state, equation (25) we see that the new ground state $|\phi_0'\rangle$ is a coherent state. Because such a coherent state is formed after the spontaneous breakdown of symmetry of the systems, thus, it is referred to as a spontaneous coherent state. But when $\phi_0 = 0$, the new ground state is the same as the old state, which is not a coherent state.The same conclusion can be directly derived from the BCS theory [18-21]. In the BCS theory, the wave function of the ground state of a superconductor is written as

$$|\phi_0'\rangle = \prod_k(\mu_k + \upsilon_k\hat{a}_k^+\hat{a}_{-k}^+)|\phi_0\rangle = \prod_k(\mu_k + \upsilon_k\hat{b}_{k\text{-}k}^+)|\phi_0\rangle \sim \eta'\exp(\sum_k\frac{\upsilon_k}{\mu_k}\hat{b}_{k\text{-}k}^+)|\phi_0\rangle \tag{26}$$

where $\hat{b}_{k\text{-}k}^+ = \hat{a}_k^+\hat{a}_{-k}^+$. This equation shows that the superconducting ground state is a coherent state. Hence, we can conclude that the spontaneous coherent state in superconductors is formed after the spontaneous breakdown of symmetry.

By reconstructing a quasiparticle-operator-free new formulation of the Bogoliubov-Valatin transformation parameter dependence [55], W. S. Lin et al [56] demonstrated that the BCS state is not only a coherent state of single-Cooper-pairs, but also the squeezed state of the double-Cooper- pairs, and reconfirmed thus the coherent feature of BCS superconductive state.

3.3 The Boson condensed features of macroscopic quantum effects

We will now employ the method used by Bogoliubov in the study of superfluid liquid helium 4He to prove that the above state is indeed a Bose condensed state. To do that, we rewrite Eq. (16) in the following form [12-17]

$$\phi(x)=\frac{1}{\sqrt{\overline{V}}}\sum_{\mathrm{p}} q_{\mathrm{p}}e^{\mathrm{ipx}},q_{\mathrm{p}}=\frac{1}{\sqrt{2\varepsilon_{\mathrm{p}}}}\left(b_{\mathrm{p}}+b_{\text{-p}}^{+}\right) \tag{27}$$

Since the field ϕ describes a Boson, such as the Cooper electron pair in a superconductor and the Bose condensation can occur in the system, we will apply the following traditional method in quantum field theory, and consider the following transformation:

$$b_{\mathrm{p}}=\sqrt{N_0}\delta(p)+\gamma_{\mathrm{p}},b_{\mathrm{p}}^{+}=\sqrt{N_0}\delta(p)+\beta_{\mathrm{p}} \tag{28}$$

where N_0 is the number of Bosons in the system and $\delta(p)=\begin{cases}0\text{ ,if p}\neq 0\\ 1\text{ , if p=0}\end{cases}$. Substituting Eqs. (27) and (28) into Eq. (1), we can arrive at the Hamiltonian operator of the system as follows

$$\begin{aligned}\widehat{H}=&\left(\frac{4\lambda N_0^2}{\varepsilon_0^2\overline{V}}-\frac{\alpha}{\varepsilon_0}\right)\sqrt{N_0}\left(\gamma_0+\gamma_0^{+}+\beta_0+\beta_0^{+}\right)+\sum_{\mathrm{P}}\left(\frac{4\lambda}{\varepsilon_0\varepsilon_{\mathrm{P}}}\frac{N_0}{\overline{V}}-\varepsilon_{\mathrm{p}}\right)\left(\gamma_{\mathrm{p}}^{+}\beta_{\text{-p}}^{+}+\gamma_{\mathrm{p}}\beta_{\text{-p}}\right)+\frac{4\lambda N_0^2}{\varepsilon_0^2\overline{V}}-\\&\frac{2N_0\alpha}{\varepsilon_0}+\frac{\lambda N_0}{\varepsilon_0\overline{V}}\sum_{\mathrm{p}}\frac{1}{\varepsilon_{\mathrm{p}}}\begin{pmatrix}\beta_{\mathrm{p}}^{+}\beta_{\text{-p}}^{+}+\beta_{\mathrm{p}}\beta_{\text{-p}}+\gamma_{\mathrm{p}}^{+}\gamma_{\text{-p}}^{+}+\\ \gamma_{\mathrm{p}}\gamma_{\text{-p}}+2\gamma_{\mathrm{p}}^{+}\beta_{\mathrm{p}}^{+}+2\beta_{\mathrm{p}}^{+}\gamma_{\mathrm{p}}\end{pmatrix}+\\&\sum_{\mathrm{P}}\left(\varepsilon_{\mathrm{p}}-\frac{\alpha}{2\varepsilon_0}+\frac{4\lambda N_0}{\varepsilon_0\varepsilon_p\overline{V}}\right)\left(\gamma_{\mathrm{p}}^{+}\gamma_{\mathrm{p}}+\beta_{\mathrm{p}}^{+}\beta_{\mathrm{p}}\right)+\sum_{P}\frac{4\lambda}{\varepsilon_0\varepsilon_{\mathrm{P}}}\frac{N_0}{\overline{V}}+O\left(\frac{\sqrt{N_0}}{\overline{V}}\right)+O\left(\frac{N_0}{\overline{V}^2}\right)\end{aligned} \tag{29}$$

Because the condensed density $N_0/\overline{V}$ must be finite, it is possible that the higher order terms $0\left(\sqrt{N_0}/\overline{V}\right)$ and $0\left(N_0/\overline{V}^2\right)$ may be neglected. Next we perform the following canonical transformation

$$\gamma_{\mathrm{p}}=u_{\mathrm{p}}^{*}c_{\mathrm{p}}+\upsilon_{\mathrm{p}}c_{\text{-p}}^{+},.\beta_{\mathrm{p}}=u_{\mathrm{p}}^{*}d_{\mathrm{p}}+\upsilon_{\mathrm{p}}d_{\text{-p}}^{+} \tag{30}$$

where υ_p and u_p are real and satisfy $\left(u_{\mathrm{p}}^2-\upsilon_{\mathrm{p}}^2\right)=1$. This introduces another transformation

$$\varsigma_{\mathrm{p}}=\frac{1}{\sqrt{2}}\left(u_{\mathrm{p}}\gamma_{\mathrm{p}}^{+}-\upsilon_{\mathrm{p}}\gamma_{\text{-p}}+u_{\mathrm{p}}\beta_{\mathrm{p}}^{+}-\upsilon_{\mathrm{p}}\beta_{\text{-p}}\right),\eta_{\mathrm{p}}^{+}=\frac{1}{\sqrt{2}}\left(u_{\mathrm{p}}\gamma_{\mathrm{p}}^{+}-\upsilon_{\mathrm{p}}\gamma_{\text{-p}}-u_{\mathrm{p}}\beta_{\mathrm{p}}^{+}+\upsilon_{\mathrm{p}}\beta_{\text{-p}}\right) \tag{31}$$

the following relations can be obtained

$$\left[\varsigma_{\mathrm{p}},H\right]=g_{\mathrm{p}}\varsigma_{\mathrm{p}}+M_{\mathrm{p}}\varsigma_{\text{-p}}^{+},\qquad \left[\eta_{\mathrm{p}},H\right]=g_{\mathrm{p}}'\eta_{\mathrm{p}}+M_{\mathrm{p}}'\eta_{\text{-p}}^{+} \tag{32}$$

where

$$\begin{cases}g_{\mathrm{p}}=G_{\mathrm{p}}\left(u_{\mathrm{p}}^2+\upsilon_{\mathrm{p}}^2\right)+F_{\mathrm{p}}2u_{\mathrm{p}}\upsilon_{\mathrm{p}},\ \mathrm{M}_{\mathrm{p}}=F_{\mathrm{p}}\left(u_{\mathrm{p}}^2+\upsilon_{\mathrm{p}}^2\right)+G_{\mathrm{p}}2u_{\mathrm{p}}\upsilon_{\mathrm{p}};\\ g_{\mathrm{p}}'=G_p'\left(u_{\mathrm{p}}^2+\upsilon_{\mathrm{p}}^2\right)+F_{\mathrm{p}}'2u_p\upsilon_{\mathrm{p}},\ \mathrm{M}_{\mathrm{p}}'=F_{\mathrm{p}}'\left(u_{\mathrm{p}}^2+\upsilon_{\mathrm{p}}^2\right)+G_{\mathrm{p}}'2u_{\mathrm{p}}\upsilon_{\mathrm{p}}\end{cases} \tag{33}$$

while

$$G_p = \varepsilon_p - \frac{\alpha}{2\varepsilon_p} + 6\xi'_p,\ F_p = -\frac{\alpha}{2\varepsilon_p} + 6\xi'_p,\ G'_p = \varepsilon_p - \frac{\alpha}{2\varepsilon_p} + 2\xi'_p,\ F'_p = \frac{\alpha}{2\varepsilon_p} - 2\xi'_p \tag{34}$$

where $\xi'_p = \dfrac{\lambda N_0}{\varepsilon_0 \varepsilon_p \bar{V}}$.

We will now study two cases to illustrate the concepts.

(A) Let $M'_p = 0$, then it can be seen from Eq. (32) that η_p^+ is the creation operator of elementary excitation and its energy is given by

$$g'_p = \sqrt{\varepsilon_p^2 + 4\varepsilon_p \xi'_p - 2\alpha} \tag{35}$$

Using this concept, we can obtain the following form from Eqs. (32) and (34)

$$\left(u'_p\right)^2 = \frac{1}{2}\left(1 + \frac{G'_p}{g'_p}\right) \quad \text{and} \quad \left(\upsilon'_p\right)^2 = \frac{1}{2}\left(-1 + \frac{G'_p}{g'_p}\right) \tag{36}$$

From Eq. (32), we know that ξ_p^+ is not a creation operator of the elementary excitation. Thus, another transformation must be made

$$B_p = \chi_p \varsigma'_p + \mu_p \varsigma'^+_p,\ |\chi_p|^2 - |\mu_p|^2 = 1 \tag{37}$$

We can then prove that

$$[B_p, \widehat{H}] = E_p B_p \tag{38}$$

where $E_p = \sqrt{12\varepsilon_p \xi'_p + \varepsilon_p^2 - 2\alpha}$

Now, inserting Eqs. (30), (37)-(38) and $M'_p = 0$ into Eq. (29), and after some reorganization, we have

$$\widehat{H} = U + E_0 + \sum_{p>0}\left[E_p\left(B_p^+ B_p + B_{-p}^+ B_{-p}\right) + g'_p\left(\eta_p^+ \eta_p + \eta_{-p}^+ \eta_{-p}\right)\right] \tag{39}$$

where

$$U = \frac{\lambda N_0^2}{\varepsilon_0^2 \bar{V}} - \frac{2\alpha N_0}{\varepsilon_0} + \sum_p 4\xi'_p + \sum_p\left(\varepsilon_p + \frac{\alpha}{2\varepsilon_p} + 4\xi'\right)4\upsilon_p'^2 + \sum_{p>0} 4\eta \cdot 2u'_p \upsilon'_p$$

$$E_0 = -2\sum_{p>0} E_p |\mu_p|^2 = -\sum_{p>0}\left(g'_p - E_p\right) \tag{40}$$

Both U and E_0 are now independent of the creation and annihilation operators of the Bosons. $U + E_0$ gives the energy of the ground state. N_0 can be determined from the condition, $\dfrac{\delta(U + E_0)}{\delta N_0} = 0$, which gives

$$\frac{N_0}{\bar{V}} = \frac{\alpha\varepsilon_0}{4\lambda} = \frac{1}{2}\varepsilon_0\phi_0^2 \tag{41}$$

This is the condensed density of the ground state ϕ_0. From Eqs. (36), (37) and (40), thus we can arrive at:

$$g'_p = \sqrt{\varepsilon_p^2 - \alpha}, \; E_p = \sqrt{\varepsilon_p^2 - \alpha} \tag{42}$$

These correspond to the energy spectra of η_p^+ and B_p^+,respectively, and they are similar to the energy spectra of the Cooper pair and phonon in the BCS theory. Substituting Eq. (42) into Eq. (36), thus we now have:

$$u_p'^2 = \frac{1}{2}\left(1+\frac{2\varepsilon_p^2\text{-}\alpha}{2\sqrt{\varepsilon_p^2\text{-}\alpha e_p}}\right), \quad \upsilon_p'^2 = \frac{1}{2}\left(\text{-}1+\frac{2\varepsilon_p^2-\alpha}{2\sqrt{\varepsilon_p^2-\alpha\varepsilon_p}}\right) \tag{43}$$

(B) In the case of Mp=0, a similar approach can be used to arrive at the energy spectrum corresponding to ξ_p^+ as $E_p = \sqrt{\varepsilon_p^2 + \alpha}$, while that corresponding to $A_p^+ = \chi_p\eta_p^+ + \mu_p\eta_{\text{-}p}$ is $g'_p = \sqrt{\varepsilon_p^2 + \alpha}$, where

$$u_p^2 = \frac{1}{2}\left(1+\frac{2\varepsilon_p^2 + \alpha}{2\varepsilon_p\sqrt{\varepsilon_p^2+\alpha}}\right), \quad \upsilon_p^2 = \frac{1}{2}\left(\text{-}1+\frac{2\varepsilon_p^2+\alpha}{2\varepsilon_p\sqrt{\varepsilon_p^2+\alpha}}\right) \tag{44}$$

Based on experiments in quantum statistical physics, we know that the occupation number of the level with an energy of ε_p, for a system in thermal equilibrium at temperature $T(\neq 0)$ is shown as:

$$N_p = \left\langle b_p^+ b_p \right\rangle = \frac{1}{e^{\varepsilon_p/K_BT} - 1} \tag{45}$$

where $\langle\cdots\rangle$ denotes Gibbs average, defined as $\langle\cdots\rangle = \dfrac{SP\left[e^{-H/K_BT}\ldots\right]}{SP\left[e^{-H/K_BT}\right]}$, here SP denotes the trace in a Gibbs statistical description. At low temperatures, or $T \to 0\,K$, the majority of the Bosons or Cooper pairs in a superconductor condense to the ground state with $p = 0$. Therefore $\left\langle b_0^+ b_0 \right\rangle \approx N_0$, where N_0 is the total number of Bosons or Cooper pairs in the system and $N_0 >> 1$, i.e. $\left\langle b^+ b \right\rangle = 1 << \left\langle b_0^+ b_0 \right\rangle$.

As can be seen from Eqs. (27) and (28), the number of particles is extremely large when they lie in condensed state, that is to say:

$$\phi_0 = \phi_{p=0} = \frac{1}{\sqrt{2\varepsilon_0\bar{V}}}\left(b_0 + b_0^+\right) \tag{46}$$

Because $\gamma_0|\phi_0\rangle = 0$ and $\beta_0|\phi_0\rangle = 0$, b_0 and b_0^+ can be taken to be $\sqrt{N_0}$. The average value of $\phi^*\phi$ in the ground state then becomes

$$\left\langle \phi_0 \left| \phi^* \phi \right| \phi_0 \right\rangle = \left\langle \phi^* \phi \right\rangle_0 = \frac{1}{2\varepsilon_0 \bar{V}} \cdot 4N_0 = \frac{2N_0}{\varepsilon_0 \bar{V}} \tag{47}$$

Substituting Eq. (41) into Eq. (47), we can see that:

$$\left\langle \phi^* \phi \right\rangle_0 = \frac{\alpha}{2\lambda} \quad \text{or} \quad \left\langle \phi^* \right\rangle_0 = \pm\sqrt{\frac{\alpha}{2\lambda}}$$

which is the ground state of the condensed phase, or the superconducting phase, that we have known. Thus, the density of states, $N_0/\bar{V}$, of the condensed phase or the superconducting phase formed after the Bose condensation coincides with the average value of the Boson's (or Copper pair's) field in the ground state. We can then conclude from the above investigation shown in Eqs. (1) - (2) that the macroscopic quantum state or the superconducting ground state formed after the spontaneous symmetry breakdown is indeed a Bose-Einstein condensed state. This clearly shows the essences of the nonlinear properties of the result of macroscopic quantum effects.

In the last few decades, the Bose-Einstein condensation has been observed in a series of remarkable experiments using weakly interacting atomic gases, such as vapors of rubidium, sodium lithium, or hydrogen. Its' formation and properties have been extensively studied. These studies show that the Bose-Einstein condensation is a nonlinear phenomenon, analogous to nonlinear optics, and that the state is coherent, and can be described by the following NLSE or the Gross-Pitaerskii equation [57-59]:

$$i\frac{\partial \phi}{\partial t'} = -\frac{\partial^2 \phi}{\partial x'^2} - \lambda|\phi|^3 + V(x)\phi \tag{48}$$

where $t'=t/\hbar$, $x'=x\sqrt{2m}/\hbar$. This equation was used to discuss the realization of the Bose-Einstein condensation in the $d+1$ dimensions $(d=1,2,3)$ by H. K. Bullough et.al [41-42]. Too, Elyutin et al [60-61]. gave the corresponding Hamiltonian density of a condensate system as follows:

$$H' = \left|\frac{\partial \phi}{\partial x'}\right|^2 + V(x')|\phi|^2 - \frac{1}{2}\lambda|\phi|^4 \tag{49}$$

where H'=H, the nonlinear parameters of λ are defined as $\lambda = -2Naa_1 / a_0^2$, with N being the number of particles trapped in the condensed state, a is the ground state scattering length, a_0 and a_1 are the transverse (y, z) and the longitudinal (x) condensate sizes (without self-interaction) respectively, (Integrations over y and z have been carried out in obtaining the above equation). λ is positive for condensation with self-attraction (negative scattering length).The coherent regime was observed in Bose-Einstein condensation in lithium. The specific form of the trapping potential V (x') depends on the details of the experimental setup. Work on Bose-Einstein condensation based on the above model Hamiltonian were carried out and are reported by C. F. Barenghi et al [31].

It is not surprising to see that Eq. (48) is exactly the same as Eq. (15), corresponding to the Hamiltonian density in Eq. (49) and, where used in this study is naturally the same as Eq. (1). This prediction confirms the correctness of the above theory for Bose-Einstein condensation. As a matter of fact, immediately after the first experimental observation of this condensation phenomenon, it was realized that the coherent dynamics of the condensed macroscopic wave function could lead to the formation of nonlinear solitary waves. For example, self-localized bright, dark and vortex solitons, formed by increased (bright) or decreased (dark or vortex) probability density respectively, were experimentally observed, particularly for the vortex solution which has the same form as the vortex lines found in type II-superconductors and superfluids. These experimental results were in concordance with the results of the above theory. In the following sections of this text we will study the soliton motions of quasiparticles in macroscopic quantum systems, superconductors. We will see that the dynamic equations in macroscopic quantum systems do have such soliton solutions.

3.4 Differences of macroscopic quantum effects from the microscopic quantum effects

From the above discussion we may clearly understand the nature and characteristics of macroscopic quantum systems. It would be interesting to compare the macroscopic quantum effects and microscopic quantum effects. Here we give a summary of the main differences between them.

1. Concerning the origins of these quantum effects; the microscopic quantum effect is produced when microscopic particles, which have only a wave feature are confined in a finite space, or are constituted as matter, while the macroscopic quantum effect is due to the collective motion of the microscopic particles in systems with nonlinear interaction. It occurs through second-order phase transition following the spontaneous breakdown of symmetry of the systems.
2. From the point-of-view of their characteristics, the microscopic quantum effect is characterized by quantization of physical quantities, such as energy, momentum, angular momentum, etc. wherein the microscopic particles remain constant. On the other hand, the macroscopic quantum effect is represented by discontinuities in macroscopic quantities, such as, the resistance, magnetic flux, vortex lines, voltage, etc. The macroscopic quantum effects can be directly observed in experiments on the macroscopic scale, while the microscopic quantum effects can only be inferred from other effects related to them.
3. The macroscopic quantum state is a condensed and coherent state, but the microscopic quantum effect occurs in determinant quantization conditions, which are different for the Bosons and Fermions. But, so far, only the Bosons or combinations of Fermions are found in macroscopic quantum effects.
4. The microscopic quantum effect is a linear effect, in which the microscopic particles and are in an expanded state, their motions being described by linear differential equations such as the Schrödinger equation, the Dirac equation, and the Klein-Gordon equations.

On the other hand, the macroscopic quantum effect is caused by the nonlinear interactions, and the motions of the particles are described by nonlinear partial differential equations such as the nonlinear Schrödinger equation (17).

Thus, we can conclude that the macroscopic quantum effects are, in essence, a nonlinear quantum phenomenon. Because its' fundamental nature and characteristics are different from those of the microscopic quantum effects, it may be said that the effects should be depicted by a new nonlinear quantum theory, instead of quantum mechanics.

4. The nonlinear dynamic natures of electrons in superconductors

4.1 The dynamic equations of electrons in superconductors

It is quite clear from the above section that the superconductivity of material is a kind of nonlinear quantum effect formed after the breakdown of the symmetry of the system due to the electron-phonon interaction, which is a nonlinear interaction.

In this section we discuss the properties of motion of superconductive electrons in superconductors and the relation of the solutions of dynamic equations in relation to the above macroscopic quantum effects on it. The study presented shows that the superconductive electrons move in the form of a soliton, which can result in a series of macroscopic quantum effects in the superconductors. Therefore, the properties and motions of the quasiparticles are important for understanding the essences and rule of superconductivity and macroscopic quantum effects.

As it is known, in the superconductor the states of the electrons are often represented by a macroscopic wave function,

$$\phi(\vec{r},t) = f(\vec{r},t)\phi_0 e^{i\theta(\vec{r},t)}, \quad \text{or} \quad \phi = \sqrt{\rho}e^{i\theta},$$

as mentioned above, where $\phi_0^2 = \alpha / 2\lambda$. Landau et al [45,46] used the wave function to give the free energy density function, f, of a superconducting system, which is represented by

$$f_s = f_n - \frac{\hbar^2}{2m}|\nabla\phi|^2 - \alpha|\phi|^2 + \lambda|\phi|^4 \tag{50}$$

in the absence of any external field. If the system is subjected to an electromagnetic field specified by a vector potential $\vec{A}$, the free energy density of the system is of the form:

$$f_s = f_n - \frac{\hbar^2}{2m}\left|(\nabla - \frac{ie^*}{c\hbar}\vec{A})\phi\right|^2 - \alpha|\phi|^2 + \lambda|\phi|^4 + \frac{1}{8\pi}\vec{H}^2 \tag{51}$$

where e*=2e , $\vec{H}=\nabla\times\vec{A}$, α and λ are some interactional constants related to the features of superconductor, m is the mass of electron, e* is the charge of superconductive electron, c is the velocity of light, h is Planck constant, $\hbar = h/2\pi$, fn is the free energy of normal state. The free energy of the system is $F_s = \int f_s d^3x$. In terms of the conventional field, $F_{jl} = \partial_j A_l - \partial_l A^j$, (j, l=1, 2, 3), the term $\vec{H}^2/8\pi$ can be written as $F_{jl}F^{jl}/4$. Equations (50) - (51) show the nonlinear features of the free energy of the systems because it is the nonlinear function of the wave function of the particles, $\phi(\vec{r},t)$. Thus we can predict that the superconductive electrons have many new properties relative to the normal electrons. From $\delta F_s / \delta\phi = 0$ we get

$$\frac{\hbar^2}{2m}\nabla^2\phi - \alpha\phi + 2\lambda\phi^3 = 0 \tag{52}$$

and

$$\frac{\hbar^2}{2m}(\nabla - \frac{ie^*}{c\hbar}\vec{A})^2\phi - \alpha\phi + 2\lambda\phi^3 = 0 \tag{53}$$

in the absence and presence of an external fields respectively, and

$$\vec{J} = +\frac{e^*\hbar}{2mi}(\phi^*\nabla\phi - \phi\nabla\phi^*) - \frac{e^*}{mc}|\phi|^2\vec{A} \tag{54}$$

Equations (52) - (54) are just well-known the Ginzburg-Landau (GL) equation [48-54] in a steady state, and only a time-independent Schrödinger equation. Here, Eq. (52) is the GL equation in the absence of external fields. It is the same as Eq. (15), which was obtained from Eq. (1). Equation (54) can also be obtained from Eq. (2). Therefore, Eqs. (1)-(2) are the Hamiltonians corresponding to the free energy in Eqs. (50)- (51).

From equations (52) - (53) we clearly see that superconductors are nonlinear systems. Ginzburg-Landau equations are the fundamental equations of the superconductors describing the motion of the superconductive electrons, in which there is the nonlinear term of $2\lambda\phi^3$. However, the equations contain two unknown functions ϕ and $\vec{A}$ which make them extremely difficult to resolve.

4.2 The dynamic properties of electrons in steady superconductors

We first study the properties of motion of superconductive electrons in the case of no external field. Then, we consider only a one-dimensional pure superconductor [62-63], where

$$\phi = \phi_0\varphi(x,t),\ \xi'^2(T) = \hbar^2 / 2m|\alpha|,\ x' = x/\xi'(T) \tag{55}$$

and where $\xi'(T)$ is the coherent length of the superconductor, which depends on temperature. For a uniform superconductor, $\xi'(T) = 0.94\xi_0[T_c/(T_c - T)]^2$, where T_c is the critical temperature and ξ_0 is the coherent length of superconductive electrons at T=0. In boundary conditions of $\varphi(x'=0)=1$, and $\varphi(x'\to\pm\infty)=0$, from Eqs. (52) and (54) we find easily its solution as:

$$\varphi = \pm\sqrt{2}\,\mathrm{sec}h\left[\frac{x - x_0}{\xi'(T)}\right]$$

or

$$\phi = \pm\sqrt{\frac{\alpha}{\lambda}}\,\mathrm{sec}h[\frac{x - x_0}{\xi'(T)}] = \pm\sqrt{\frac{\alpha}{\lambda}}\,\mathrm{sec}h[\frac{\sqrt{2m\alpha}}{\hbar}(x - x_0)] \tag{56}$$

This is a well-known wave packet-type soliton solution. It can be used to represent the bright soliton occurred in the Bose-Einstein condensate found by Perez-Garcia et. al. [64]. If the signs of α and λ in Eq. (52) are reversed, we then get a kink-soliton solution under the boundary conditions of $\varphi(x'=0)=0$, $\varphi(x'\to\pm\infty)=\pm 1$,

$$\phi = \pm(\alpha/2\lambda)^{1/2}\tanh\{[m\alpha(x - x_0/\hbar^2]^{1/2}\} \tag{57}$$

The energy of the soliton, (56), is given by

$$E_{so1} = \int_{-\infty}^{\infty} \left[\frac{\hbar^2}{2m} (\frac{d\phi}{dx})^2 - \alpha\phi^2 - \lambda\phi^4 \right] dx = \frac{4\hbar\alpha^{3/2}}{3\lambda\sqrt{2m}} \quad (58)$$

We assume here that the lattice constant, r_0=1. The above soliton energy can be compared with the ground state energy of the superconducting state, Eground$=-\alpha^2/4\lambda$. Their difference is $E_{so1} - E_{\text{ground}} = \alpha^{3/2}\left(\sqrt{\alpha} + \frac{16\hbar}{3\sqrt{2m}} \right) / 2\lambda > 0$. This indicates clearly that the soliton is not in the ground state, but in an excited state of the system, therefore, the soliton is a quasiparticle.

From the above discussion, we can see that, in the absence of external fields, the superconductive electrons move in the form of solitons in a uniform system. These solitons are formed by a nonlinear interaction among the superconductive electrons which suppresses the dispersive behavior of electrons. A soliton can carry a certain amount of energy while moving in superconductors. It can be demonstrated that these soliton states are very stable.

4.3 The features of motion of superconductive electrons in an electromagnetic field and its relation to macroscopic quantum effects

We now consider the motion of superconductive electrons in the presence of an electromagnetic field $\vec{A}$; its equation of motion is denoted by Eqs. (53)-(54) · Assuming now that the field $\vec{A}$ satisfies the London gauge $\nabla \cdot \vec{A} = 0$ [65], and that the substitution of $\phi(\vec{r},t) = \varphi(\vec{r},t)\phi_0 e^{i\theta(\vec{r})}$ into Eqs. (53) and (54) yields [66-67]:

$$J = \frac{e^*\phi_0^2}{m}(\hbar\nabla\theta - \frac{e^*}{c}\vec{A})\varphi^2 \quad (59)$$

and

$$\nabla^2\varphi - [(\nabla\theta - \frac{e^*}{\hbar c}\vec{\mathbf{A}})^2\varphi] - \frac{2m}{\hbar^2}(\alpha - 2\lambda\phi_0^2\varphi^2)\varphi = 0 \quad (60)$$

For bulk superconductors, J is a constant (permanent current) for a certain value of $\vec{A}$, and it thus can be taken as a parameter. Let $B^2 = m^2J^2 / \hbar^2(e^*)^2\phi_0^4$, $b = 2m\alpha / \hbar^2 = \xi'^{-2}$, from Eqs. (59) and (60), we can obtain [66-67]:

$$(\hbar\nabla\theta - \frac{e^*}{c}\vec{\mathbf{A}}) = \frac{Jm}{e^*\phi_0^2\varphi^2} \quad (61)$$

$$\frac{d^2\varphi}{dx^2} = -\frac{d}{d\varphi}U_{\text{eff}}(\varphi), U_{\text{eff}}(\varphi) = \frac{B^2}{2\varphi^2} - \frac{1}{2}b\varphi^2 + \frac{1}{4}b\varphi^4 \quad (62)$$

where Ueff is the effective potential of the superconductive electron in this case and it is schematically shown in Fig. 2. Comparing this case with that in the absence of external fields, we found that the equations have the same form and the electromagnetic field changes only the effective potential of the superconductive electron. When $\vec{A} = 0$, the effective potential well is characterized by double wells. In the presence of an electromagnetic field, there are still two minima in the effective potential, corresponding to the two ground states of the superconductor in this condition. This shows that the spontaneous breakdown of symmetry still occurs in the superconductor, thus the superconductive electrons also move in the form of solitons. To obtain the soliton solution, we integrate Eq. (62) and can get:

$$x = \int_{\varphi_1}^{\varphi} \frac{d\varphi}{\sqrt{2[E - U_{\text{eff}}(\varphi)}} \tag{63}$$

Where E is a constant of integration which is equivalent to the energy, the lower limit of the integral, φ_1, is determined by the value of φ at x=0, i.e. $E = U_{\text{eff}}(\varphi_0) = U_{\text{eff}}(\varphi_1)$. Introduce the following dimensionless quantities $\varphi^2 = u,\ \ E = \frac{b}{2}\varepsilon, 2\tilde{d} = \frac{4J^2 m\lambda}{(e^*)^2\alpha^2}$, and equation (63) can be written as the following upon performing the transformation u→–u,

$$-\sqrt{2b}x = \int_{u_1}^{u} \frac{du}{\sqrt{u^3 - 2u^2 - 3\varepsilon u - 2\tilde{d}^2}} \tag{64}$$

It can be seen from Fig. 3 that the denominator in the integrand in Eq. (64) approaches zero linearly when $u=u_1=\varphi_1^2$, but approaches zero gradually when $u=u_2=\varphi_0^2$. Thus we give [66-67]

$$u(x) = \varphi^2(x) = u_0 - g\,\text{sec}h^2\left(\sqrt{\frac{1}{2}}gbx\right) = u_1 + g\tanh^2\left(\sqrt{\frac{1}{2}}gbx\right) \tag{65}$$

where $g= u_0-u_1$ and satisfies

$$(2+g)^2(1-g) = 27\tilde{d}^2,\ 2u_0 + u_1 = 2,\ u_0^2 + 2u_0u_1 = -2\varepsilon,\ u_1u_0^2 = 2\tilde{d}^2 \tag{66}$$

It can be seen from Eq. (65) that for a large part of sample, u_1 is very small and may be neglected; the solution u is very close to u_0. We then get from Eq. (65) that

$$\varphi(x) = \varphi_0 \tanh\left(\sqrt{\frac{1}{2}}gbx\right) \tag{67}$$

Substituting the above into Eq. (61), the electromagnetic field $\vec{A}$ in the superconductors can be obtained

$$\vec{A} = -\frac{\vec{J}mc}{(e^*)^2\phi_0^2}\frac{1}{\varphi^2} - \frac{\hbar c}{e^*}\nabla\theta = \frac{\vec{J}mc}{(e^*)^2\phi_0^2\varphi_0^2}\coth\left(\sqrt{\frac{1}{2}}gbx\right) - \frac{\hbar c}{e^*}\nabla\theta$$

For a large portion of the superconductor, the phase change is very small. Using $\vec{H} = \nabla \times \vec{A}$ the magnetic field can be determined and is given by [66-67]

$$\vec{H} = \frac{\vec{J}mc\sqrt{2gb}}{(e^*)^2\phi_0^2\varphi_0^2}[\coth^3\left(\sqrt{\frac{1}{2}}gbx\right) + \coth\left(\sqrt{\frac{1}{2}}gbx\right)] \tag{68}$$

Equations (67) and (68) are analytical solutions of the GL equation.(63) and (64) in the one-dimensional case, which are shown in Fig. 3. Equation (67) or (65) shows that the superconductive electron in the presence of an electromagnetic field is still a soliton. However, its amplitude, phase and shape are changed, when compared with those in a uniform superconductor and in the absence of external fields, Eq. (66). The soliton here is obviously influenced by the electromagnetic field, as reflected by the change in the form of solitary wave. This is why a permanent superconducting current can be established by the motion of superconductive electrons along certain direction in such a superconductor, because solitons have the ability to maintain their shape and velocity while in motion.

It is clear from Fig.4 that $\tilde{H}(x)$ is larger where $\phi(x)$ is small, and vice versa. When $x \to 0$, $\tilde{H}(x)$ reaches a maximum, while ϕ approaches to zero. On the other hand, when $x \to \infty$, ϕ becomes very large, while $\tilde{H}(x)$ approaches to zero. This shows that the system is still in superconductive state.These are exactly the well-known behaviors of vortex lines-magnetic flux lines in type-II superconductors [66-67]. Thus we explained clearly the macroscopic quantum effect in type-II superconductors using GL equation of motion of superconductive electron under action of an electromagnetic-field.

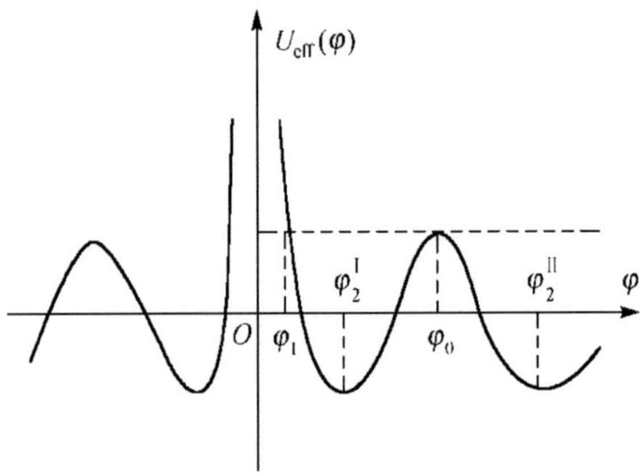

Fig. 3. The effective potential energy in Eq. (67).

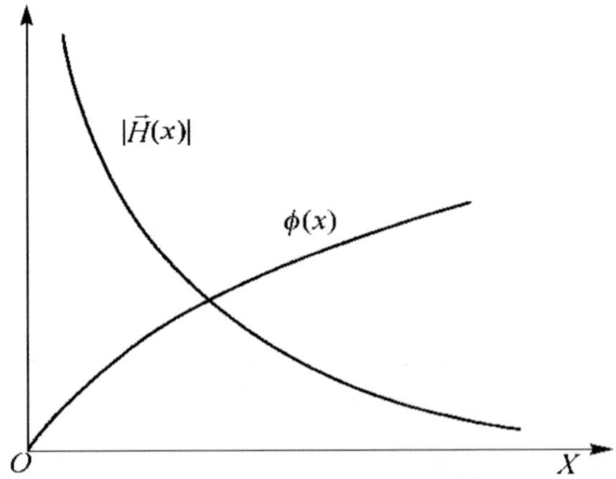

Fig. 4. Changes of $\phi(x)$ and $|\tilde{H}(x)|$ with x in Eqs. (67)-(68)

Recently, Garadoc-Daries et al. [68], Matthews et al. [69] and Madison et al.[70] observed vertex solitons in the Boson-Einstein condensates. Tonomure [71] observed experimentally magnetic vortexes in superconductors. These vortex lines in the type-II-superconductors are quantized. The macroscopic quantum effects are well described by the nonlinear theory discussed above, demonstrating the correctness of the theory.

We now proceed to determine the energy of the soliton given by (67). From the earlier discussion, the energy of the soliton is given by:

$$E=\int_{-\infty}^{+\infty}\left[\frac{1}{2}(\frac{d\varphi}{dx})^2+\frac{b}{2}\varphi^2-\frac{b}{4}\varphi^4-\frac{B^2}{2\varphi^2}\right]dx\approx\varphi_0{}^2\left[\frac{2b\varphi_0^2}{3}-1+\frac{b}{2}(1-\frac{\varphi_0^2}{2})\right]-\frac{B^2}{2\varphi_0^2}$$

which depends on the interaction between superconductive electrons and electromagnetic field.

From the above discussion, we understand that for a bulk superconductor, the superconductive electrons behave as solitons, regardless of the presence of external fields. Thus, the superconductive electrons are a special type of soliton. Obviously, the solitons are formed due to the fact that the nonlinear interaction $\lambda|\phi|^2\phi$ suppresses the dispersive effect of the kinetic energy in Eqs. (52) and (53). They move in the form of solitary wave in the superconducting state. In the presence of external electromagnetic fields, we demonstrate theoretically that a permanent superconductive current is established and that the vortex lines or magnetic flux lines also occur in type-II superconductors.

5. The dynamic properties of electrons in superconductive junctions and its relation to macroscopic quantum effects

5.1 The features of motion of electron in S-N junction and proximity effect

The superconductive junction consists of a superconductor (S) which contacts with a normal conductor (N), in which the latter can be superconductive. This phenomenon refers to a proximity effect. This is obviously the result of long- range coherent property of superconductive electrons. It can be regarded as the penetration of electron pairs from the superconductor into the normal conductor or a result of diffraction and transmission of superconductive electron wave. In this phenomenon superconductive electrons can occur in the normal conductor, but their amplitudes are much small compare to that in the superconductive region, thus the nonlinear term $\lambda|\phi|^2\phi$ in GL equations (53)-(54) can be neglected. Because of these, GL equations in the normal and superconductive regions have different forms. On the S side of the S-N junction, the GL equation is [72]

$$\frac{\hbar^2}{2m}(\nabla-\frac{ie^*}{ch}\vec{A})\phi-\alpha\phi+2\lambda\phi^3=0 \tag{69}$$

while that on the N side of the junction is

$$\frac{\hbar^2}{2m}(\nabla-\frac{ie^*}{ch}\vec{A})\phi-\alpha'\phi=0 \tag{70}$$

Thus, the expression for $\vec{J}$ remains the same on both sides.

$$\vec{J}=\frac{e^{*}\hbar}{2mi}(\phi^{*}\nabla\phi-\phi\nabla\phi^{*})-\frac{(e^{*})^{2}}{mc}|\phi|^{2}\vec{A} \tag{71}$$

In the S region, we have obtained solution of (69) in the previous section, and it is given by (65) or (67) and (68). In the N region, from Eqs. (70)- (71) we can easily obtain

$$\begin{cases}\varphi^{2}=\frac{1}{2}\sqrt{(\varepsilon')^{2}-4\tilde{d}^{2}}\sin(2\sqrt{b'}x)+\frac{\varepsilon'}{2}\\ \phi_{N}^{2}=\varphi^{2}\phi_{0}^{2}e^{-2i\theta}=\frac{1}{2}\sqrt{(\varepsilon')^{2}-4\tilde{d}^{2}}\sin(2\sqrt{b'}x)e^{-i2\theta}+\frac{\varepsilon'}{2}\phi_{0}^{2}e^{-i2\theta}\end{cases} \tag{72}$$

where $b'=\frac{2m\alpha'}{\hbar^{2}}=\frac{1}{\bar{\xi}'^{2}}$, $2\tilde{d}^{2}=\frac{4\vec{J}^{2}m\lambda}{(e^{*})^{2}\alpha'^{2}}$, $E'=\frac{b'}{2}\varepsilon'$. .

here ε' is an integral constant. A graph of ϕ vs. x in both the S and the N regions, as shown in Fig.5, coincides with that obtained by Blackbunu [73]. The solution given in Eq. (72) is the analytical solution in this case. On the other hand, Blackbunu's result was obtained by expressing the solution in terms of elliptic integrals and then integrating numerically. From this, we see that the proximity effect is caused by diffraction or transmission of the superconductive electrons

5.2 The Josephson effect in S-I-S and S-N-S as well as S-I-N-S junctions

A superconductor-normal conductor -superconductor junction (S-N-S) or a superconductor-insulator-superconductor junction (S-I-S) consists of a normal conductor or an insulator sandwiched between two superconductors as is schematically shown in Fig.6a · The thickness of the normal conductor or the insulator layer is assumed to be L and we choose the z coordinate such that the normal conductor or the insulator layer is located at $-L/2\le x\le L/2$. The features of S-I-S junctions were studied by Jacobson et al.[74]. We will treat this problem using the above idea and method [75-76].

The electrons in the superconducting regions ($|x|\ge L/2$) are depicted by GL equation (69). Its' solution was given earlier in Eq.(67). After eliminating u_1 from Eq.(66), we have [73-74]

$J=\frac{1}{2}e^{*}\alpha u_{0}\sqrt{\frac{\alpha}{m\lambda}(1-)u_{0}}$.

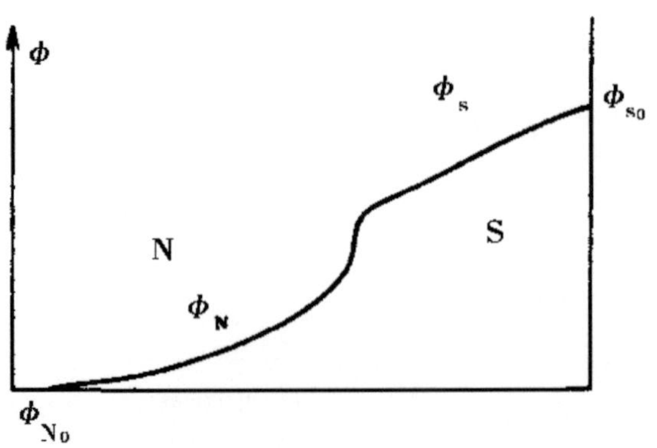

Fig. 5. Proximity effect in S-N junction

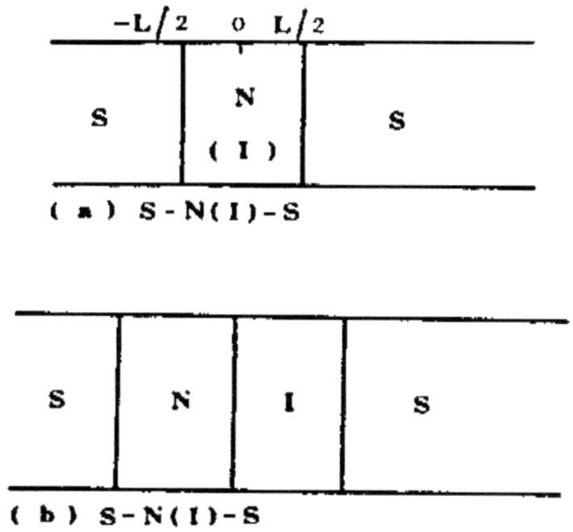

Fig. 6. Superconductive junction of S-N(I)-S and S-N-I-S

The electrons in the superconducting regions ($|x| \geq L/2$) are depicted by GL equation (69). Its' solution was given earlier. Setting $dJ/du_0 = 0$, we get the maximum current $J_c = \frac{e^*\alpha}{3}\sqrt{\frac{\alpha}{3m\lambda}}$. This is the critical current of a perfect superconductor, corresponding to the three-fold degenerate solution of Eq.(66), i.e.,u_1=u_0.

From Eq.(71), we have $\vec{A} = -\frac{m\vec{J}c}{(e^*)^2\phi_0^2\varphi^2} + \frac{hc}{e^*}\nabla\theta$. Using the London gauge, $\nabla.\vec{A} = 0$, we can get[75-76] $\frac{d^2\theta}{dx^2} = \frac{mJ}{e^*\phi_0^2\hbar}\frac{d}{dx}(\frac{1}{\varphi^2})$. Integrating the above equation twice , we get the change of the phase to be

$$\Delta\theta = \frac{mJ}{e^*\phi_0^2\hbar}\int(\frac{1}{\varphi^2} - \frac{1}{\varphi_\infty^2})dx \tag{73}$$

where $\varphi^2 = u$, and $\varphi_\infty^2 = u_0$. Here we have used the following de Gennes boundary conditions in obtaining Eq. (73)

$$\left.\frac{d\phi}{dx}\right|_{|x|\to\infty} = 0, \left.\frac{d\theta}{dx}\right|_{|x|\to\infty} = 0, \ \phi(|x| \to \infty) = \phi_\infty \tag{74}$$

If we substitute Eqs.(64) - (67) into Eq.(73), the phase shift of wave function from an arbitrary point x to infinite can be obtained directly from the above integral, and takes the form of:

$$\Delta\theta_L(x \to \infty) = -\tan^{-1}\sqrt{\frac{u_1}{u_0 - u_1}} + \tan^{-1}\sqrt{\frac{u_1}{u - u_1}} \tag{75}$$

For the S-N-S or S-I-S junction, the superconducting regions are located at $|x| \geq L/2$ and the phase shift in the S region is thus

$$\Delta\theta_s = 2\Delta\theta_L(\frac{L}{2} \to \infty) \approx 2\tan^{-1}\sqrt{\frac{u_1}{u_s - u_1}} \quad (76)$$

According to the results in (70) - (71) and the above similar method, the change of the phase in the I or N region of the S-N-S or S-I-S junction may be expressed as [75-76]

$$\Delta\theta_N = -2\tan^{-1}[\frac{2e^*h'}{\dot{j}}\sqrt{\frac{\alpha^2}{8m\lambda}}\tan(\frac{\sqrt{b'}L}{2})] + \frac{m\dot{j}L}{2e^*h'\mu_0} \quad (77)$$

where $h' = \sqrt{\frac{8m\lambda}{\alpha^2}}\frac{\dot{j}}{2e^*}\frac{\tan(\Delta\theta_N/2)}{\tan(\sqrt{b'}L/2)}$, $\frac{m\dot{j}L}{2e^*h'\mu_0}$ is an additional term to satisfy the boundary conditions (74),and may be neglected in the case being studied.

Near the critical temperature (T<Tc), the current passing through a weakly linked superconductive junction is very small ($\dot{j} \ll 1$), we then have $\mu_1' = \frac{4\dot{j}^2 m\lambda}{(e^*)^2\alpha^2} = 2\bar{A}^{-2}$, and g'=1. Since $\eta\varphi^2$ and $d\varphi^2/dx$ are continuous at the boundary x=L/2, we have

$$\frac{d\mu_s}{dx}\Big|_{x=L/2} = \frac{d\mu_N}{dx}\Big|_{x=L/2}, \quad \eta_s\mu_s\Big|_{x=L/2} = \eta_N\mu_N\Big|_{x=L/2},$$

where η_s and η_N are the constants related to features of superconductive and normal phases in the junction, respectively. These give [75-76]

$$2\sqrt{b'}\bar{A}\sin(2\Delta\theta_N) = \varepsilon_1[1-\cos(2\Delta\theta_s)]\sin(\sqrt{b'}L),$$

$$\cos(\sqrt{b'}L)\sin(2\Delta\theta_s) = \varepsilon\sin(2\Delta\theta_N) + \sin(2\Delta\theta_s + \Delta\theta_N)$$

where $\varepsilon_1 = \eta_N/\eta_S$. From the two equations, we can get

$$\sin(\Delta\theta_s + \Delta\theta_N) = \frac{2\sqrt{2m\lambda}\dot{j}}{e^*\alpha}\sqrt{b'}\sin(\sqrt{b'}L).$$

Thus

$$\dot{j} = \dot{j}_{max}\sin(\Delta\theta_s + \Delta\theta_N) = \dot{j}_{max}\sin(\Delta\theta) \quad (78)$$

where

$$\dot{j}_{max} = \frac{e^*\alpha_s}{2\sqrt{2m\lambda b'}}\cdot\frac{1}{\sin(\sqrt{b'}L)}, \Delta\theta = \Delta\theta_s + \Delta\theta_N \quad (79)$$

Equation (78) is the well-known example of the Josephson current. From Section I we know that the Josephson effect is a macroscopic quantum effect. We have seen now that this effect can be explained based on the nonlinear quantum theory. This again shows that the macroscopic quantum effect is just a nonlinear quantum phenomenon.

From Eq. (79) we can see that the Josephson critical current is inversely proportional to sin ($\sqrt{b'}L$), which means that the current increases suddenly whenever $\sqrt{b'}L$ approaches to $n\pi$, suggesting some resonant phenomena occurs in the system · This has not been observed before. Moreover $\dot{J}_{max}$ is proportional to $e^*\alpha_s/2\sqrt{2m\lambda b'} = (e^*\hbar\alpha_S/4m\sqrt{\lambda\alpha_N})$, which is related to $(T-T_c)^2$ ·

Finally, it is worthwhile to mention that no explicit assumption was made in the above on whether the junction is a potential well (α <0) or a potential barrier (α >0). The results are thus valid and the Josephson effect in Eq. (2.78), occurs for both potential wells and for potential barriers.

We now study Josephson effect in the superconductor -normal conductor-insulator-superconductor junction (SNIS) is shown schematically in Fig. 6b. It can be regarded as a multilayer junction consists of the S-N-S and S-I-S junctions. If appropriate thicknesses for the N and I layers are used (approximately 20 °A- 30 °A), the Josephson effect similar to that discussed above can occur in the SNIS junction. Since the derivations are similar to that in the previous sections, we will skip much of the details and give the results in the following. The Josephson current in the SNIS junction is still given by

$$\dot{J}=\dot{J}_{max}\sin(\Delta\theta)$$

but, where $\Delta\theta = \Delta\theta_{s1} + \Delta\theta_N + \Delta\theta_I + \Delta\theta_{s2}$ and

$$\dot{J}_{max} = \frac{1}{\sqrt{b'_N}}\left\{\frac{\varepsilon_1 \sinh(\sqrt{b'_N}L)}{2[\cosh(\sqrt{b'_N}L) - \cos(2\Delta\theta_N)]}\right\}\times \frac{1}{\sqrt{[1+\cos(2\Delta\theta_N)][1+\cos(2\Delta\theta_I)]} - \sqrt{[1-\cos(2\Delta\theta_N)][1-\cos(2\Delta\theta_I)]}} -$$
$$\frac{1}{\sqrt{b'_N}}\left\{\frac{\varepsilon_1\sqrt{[1-\cos^2(2\Delta\theta_N)]}\sinh(\sqrt{b'_N}L)}{2[\cosh(\sqrt{b'_N}L) - \cos(2\Delta\theta_N)]^2 - 1 + \cos^2(2\Delta\theta_N)}\right\}\times \frac{1}{\sqrt{[1-\cos(2\Delta\theta_N)][1-\cos(2\Delta\theta_I)]} + \sqrt{[1+\cos(2\Delta\theta_N)][1+\cos(2\Delta\theta_I)]}},$$

It can be shown that the temperature dependence of J_{max} is $J_{max} \propto (T_c - T_0)^2$,which is quite similar to the results obtained by Blackburm et al[73] for the SNIS junction and those by Romagnan et al[7] using the Pb-PbO-Sn-Pb junction. Here, we obtained the same results using a complete different approach. This indicates again that we can obtain some results, which agree with the experimental data.

6. The nonlinear dynamic-features of time- dependence of electrons in superconductor

6.1 The soliton solution of motion of the superconductive electron

We studied only the properties of motion of superconductive electrons in steady states in superconductors in section 2.3.2, and which are described by the time-independent GL equation. In such a case, the superconductive electrons move as solitons. We ask, "What are the features of a time-dependent motion in non-equilibrium states of a superconductor?" Naturally, this motion should be described by the time-dependent Ginzburg-Landau (TDGL) equation [48-54,77] in this case. Unfortunately, there are many different forms of the

TDGL equation under different conditions. The one given in the following is commonly used when an electromagnetic field $\vec{A}$ is involved

$$\Gamma\left[\hbar\frac{\partial}{\partial t}-2ie\mu(r)\right]\phi=\frac{-1}{2m}\left(\hbar\nabla-\frac{2ie}{c}\vec{A}\right)^2+\alpha\phi-\lambda|\phi|^2\phi \quad (80)$$

and

$$\vec{J}=\sigma\left[-\frac{1}{c}\frac{\partial\vec{A}}{\partial t}-\nabla\mu(r)\right]+\frac{ie\hbar}{m}(\phi^*\nabla\phi-\phi\nabla\phi^*)-\frac{4e^2}{mc}\vec{A}|\phi|^2 \quad (81)$$

here $i=\sqrt{-1}$, $\nabla\times\nabla\times\vec{A}=\frac{1}{c}\frac{\partial}{\partial t}\left(-\frac{1}{c}\frac{\partial\vec{A}}{\partial t}-\nabla\mu\right)+\frac{4\pi\vec{J}}{c}$ and σ is the conductivity in the normal state, Γ is an arbitrary constant, and μ is the chemical potential of the system. In practice, Eq. (80) is simply a time-dependent Schrödinger equation with a damping effect.
In certain situations, the following forms of the TDGL equation are also used.

$$i\hbar\frac{\partial\phi}{\partial t}=-\frac{\hbar^2}{2m}\left(\nabla-\frac{2ie}{\hbar c}\vec{A}\right)^2\phi+\alpha\phi-\lambda|\phi|^2\phi \quad (82)$$

or

$$i\left(\hbar\frac{\partial}{\partial t}-i2e\mu\right)\phi=\frac{1}{\Gamma}(\alpha-\lambda|\phi|^2)\phi+\frac{\xi'^2}{\Gamma}\left(\nabla-\frac{2ie}{\hbar c}\vec{A}\right)^2\phi \quad (83)$$

here $\xi'=\hbar/\sqrt{2m}$, and equation (82) is a nonlinear Schrödinger equation under an electromagnetic field having soliton solutions. However, these solutions are very difficult to find, and no analytic solutions have been obtained. An approximate solution was obtained by Kusayanage et al [78] by neglecting the ϕ^3 term in Eq. (80) or Eq. (82), in the case of $\vec{A}=(0,\tilde{H}x,0)$, $\mu=-K\tilde{E}x$, $\vec{H}=(0,0,\tilde{H})$ and $\vec{E}=(\tilde{E},0,0)$, where $\vec{H}$ is the magnetic field, while $\vec{E}$ is the electric field .We will solve the TDGL equation in the case of weak fields in the following.
TDGL equation (83) can be written in the following form when $\vec{A}$ is very small[80-81]

$$i\hbar\frac{\partial\phi}{\partial t}+\frac{\hbar^2}{2m\Gamma}\nabla^2\phi+\frac{\lambda}{\Gamma}|\phi|^2\phi=\left(\frac{\alpha}{\Gamma}-2e\mu\right)\phi \quad (84)$$

Where α and Γ are material dependent parameters, λ is the nonlinear coefficient, m is the mass of the superconductive electron. Equation (84) is actually a nonlinear Schrödinger equation in a potential field $\alpha/\Gamma-2e\mu$. Cai, Bhattacharjee et al [79], and Davydov [45] used it in their studies of superconductivity. However, this equation is also difficult to solve · In the following, Pang solves the equation only in the one-dimensional case.
For convenience, let $t'=t/\hbar$, $x'=x\sqrt{2m\Gamma}/\hbar$, then Eq. (84) becomes

$$i\frac{\partial\phi}{\partial t'}+\frac{\partial^2\phi}{\partial x'^2}+\frac{\lambda}{\Gamma}|\phi|^2\phi=\left[\frac{\alpha}{\Gamma}-2e\mu(x')\right]\phi \quad (85)$$

If we let $\frac{\alpha}{\Gamma}-2e\mu=0$, then Eq. (85) is the usual nonlinear Schrödinger equation whose solution is of the form [80-81]

$$\phi_s^0=\varphi_0(x',t')e^{i\theta_0(x',t')}, \tag{86}$$

$$\varphi_0(x',t')=\sqrt{\frac{\Gamma(\upsilon_e^2-2\upsilon_c\upsilon_e)}{2\lambda}}\times\operatorname{sec}h\left[\sqrt{\frac{(\upsilon_e^2-2\upsilon_c\upsilon_e)}{4}}(x'-\upsilon_e t')\right] \tag{87}$$

here $\theta_0(x',t')=\frac{1}{2}\upsilon_e(x'-\upsilon_c t')$. In the case of $\frac{\alpha}{\Gamma}$-$2e\mu\neq 0$, let $\mu=-K\tilde{E}x'$, where K is a constant, and assume that the solution is of the form [80-81]

$$\phi=\varphi'(x',t')e^{i\theta(x',t')} \tag{88}$$

Substituting Eq. (88) into Eq. (86), we get:

$$-\varphi'\frac{\partial\theta}{\partial t'}-\varphi'\left(\frac{\partial\theta}{\partial t'}\right)^2+\frac{\partial^2\varphi'}{\partial(x')^2}+\frac{\lambda}{\Gamma}(\varphi')^3=\left(2Ke\tilde{E}x'+\frac{\alpha}{\Gamma}\right)\varphi' \tag{89}$$

$$\frac{\partial\varphi'}{\partial t'}+2\frac{\partial\varphi'}{\partial x'}\frac{\partial\theta}{\partial x'}+\varphi'\frac{\partial^2\theta}{\partial(x')^2}=0 \tag{90}$$

Now let $\varphi'(x',t')=\varphi(\xi)$, $\xi=x'-u(t')$, $u(t')=-2\tilde{E}Ke(t')^2+\upsilon t'+d$, where $u(t')$ describes the accelerated motion of $\varphi'(x',t')$. The boundary condition at $\xi'\to\infty$ requires $\varphi(\xi)$ to approach zero rapidly. When $2\,\partial\theta/\partial\xi-\dot{u}\neq 0$, equation (90) can be written as: $\varphi^2=\frac{g(t')}{(\partial\theta/\partial\xi-\dot{u}/2)}$, or

$$\frac{\partial\theta}{\partial x'}=\frac{g(t')}{\varphi^2}+\frac{\dot{u}}{2} \tag{91}$$

where $\dot{u}=du/dt'$. Integration of (91) yields:

$$\theta(x',t')=g(t')\int_0^{x'}\frac{dx''}{\varphi^2}+\frac{\dot{u}}{2}x'+h(t') \tag{92}$$

and where $h(t')$ is an undetermined constant of integration. From Eq. (92) we can get:

$$\frac{\partial\theta}{\partial t'}=\dot{g}(t')\int_0^{x'}\frac{dx''}{\varphi^2}-\frac{g\dot{u}}{\varphi^2}+\frac{g\dot{u}}{\varphi^2}\Big|_{x'=0}+\frac{\ddot{u}}{2}x'+\dot{h}(t') \tag{93}$$

Substituting Eqs. (92) and (93) into Eq. (89), we have:

$$\frac{\partial^2\varphi}{\partial(x')^2}=\left[\left(2K\tilde{E}ex'+\frac{\alpha}{\Gamma}\right)+\frac{\ddot{u}}{2}x'+\dot{h}(t')+\frac{\dot{u}^2}{4}+\dot{g}\int_0^{x'}\frac{dx''}{\varphi^2}+\frac{g\dot{u}}{\varphi^2}\Big|_{x'=0}\right]\varphi-\frac{\lambda}{\Gamma}\varphi^3+\frac{g^2}{\varphi^3} \tag{94}$$

Since $\frac{\partial^2\varphi}{\partial(x')^2}=\frac{d^2\varphi}{d\xi^2}$, which is a function of ξ only, the right-hand side of Eq. (94) is also a function of ξ only, so it is necessary that $g(t')=g_0=\text{constant}$, and $(2K\tilde{E}ex'+\frac{\alpha}{\Gamma})+\frac{\ddot{u}}{2}x'+\dot{h}(t')+\frac{\dot{u}^2}{4}+\frac{g\dot{u}}{f^2}|_{x'=0}=\overline{V}(\xi)$. Next, we assume that $V_0(\xi)=\overline{V}(\xi)-\beta$, where β is real and arbitrary, then

$$2K\tilde{E}ex'+\frac{\alpha}{\Gamma}=V_0(\xi)-\frac{\ddot{u}}{2}x' + \left[\beta-\frac{g\dot{u}}{\varphi^2}|_{x'=0}-\dot{h}(t')-\frac{\dot{u}^2}{4}\right] \quad (95)$$

Clearly in the case discussed, $V_0(\xi)=0$, and the function in the brackets in Eq. (95) is a function of t′. Substituting Eq. (95) into Eq. (94), we can get [80-81]:

$$\frac{\partial^2\tilde{\varphi}}{\partial\xi^2}=\beta\tilde{\varphi}-\frac{\lambda}{\Gamma}\tilde{\varphi}^3+g_0^2/\tilde{\varphi}^3 \quad (96)$$

This shows that $\tilde{\varphi}$ is the solution of Eq. (96) when β and g are constant. For large $|\xi|$, we may assume that $|\tilde{\varphi}|\le\beta/|\xi|^{1+\Delta}$, when Δ is a small constant. To ensure that $\tilde{\varphi}$ and $d^2\tilde{\varphi}/d\xi^2$ approach zero when $|\xi|\to\infty$, only the solution corresponding to $g_0=0$ in Eq. (96) is kept, and it can be shown that this soliton solution is stable in such a case. Therefore, we choose $g_0=0$ and obtain the following from Eq. (91):

$$\partial\theta/\partial x'=\dot{u}/2 \quad (97)$$

Thus, we obtain from Eq. (95) that

$$2K\tilde{E}ex'+\frac{\alpha}{\Gamma}=-\frac{\ddot{u}}{2}x'+\beta-\dot{h}(t')-\frac{\dot{u}^2}{4},\qquad h(t')=\left(\beta-\frac{\alpha}{\Gamma}-\frac{1}{4}\upsilon^2\right)t'-\frac{4}{3}(K\tilde{E}e)^2(t')^3+e\upsilon K\tilde{E}(t')^2 \quad (98)$$

Substituting Eq. (98) into Eqs. (92) - (93), we obtain:

$$\theta=\left(-2K\tilde{E}et'+\frac{1}{2}\upsilon\right)x'+\left(\beta-\frac{\alpha}{\Gamma}-\frac{1}{4}\upsilon^2\right)t'-\frac{4}{3}(K\tilde{E}e)^2(t')^3+e\upsilon K\tilde{E}(t')^2 \quad (99)$$

Finally, substituting the Eq. (99) into Eq. (96), we can get

$$\frac{\partial^2\tilde{\varphi}}{\partial\xi^2}-\beta\tilde{\varphi}+\frac{\lambda}{\Gamma}\tilde{\varphi}^3=0 \quad (100)$$

When $\beta>0$, the solution of. Eq. (100) is of the form

$$\tilde{\varphi}=\sqrt{\frac{2\beta\Gamma}{\lambda}}\,\mathrm{sec}h\left(\sqrt{\beta}\xi\right) \quad (101)$$

Thus [80-81]

$$\phi=\sqrt{\frac{2\beta\Gamma}{\lambda}}\operatorname{sech}\left[\sqrt{\beta}\left(\sqrt{\frac{2m\Gamma}{\hbar^2}}x+\frac{2eKEt^2-\upsilon t-d}{\hbar}\right)\right]\times$$
$$\exp\left\{i\left[\left(\frac{-2eKEt}{\hbar}+\frac{\upsilon}{2}\right)\sqrt{\frac{2m\Gamma}{\hbar^2}}x+\left(\beta-\frac{\alpha}{\Gamma}-\frac{1}{4}\upsilon^2\right)\frac{t}{\hbar}-\frac{4(eK\tilde{E})^2t^3}{3\hbar^3}+\frac{\upsilon Ke\tilde{E}t^2}{\hbar}\right]\right\} \tag{102}$$

This is also a soliton solution, but its shape, amplitude and velocity have been changed relatively as compared to that of Eq. (87). It can be shown that Eq. (102) does indeed satisfy Eq. (85). Thus, equation (85) has a soliton solution. It can also be shown that this solition solution is stable.

6.2 The properties of soliton motion of the superconductive electrons

For the solution of Eq. (102), we may define a generalized time-dependent wave number, $k=\frac{\partial\theta}{\partial x'}=\frac{\upsilon}{2}-2K\tilde{E}et'$ and a frequency

$$\begin{aligned}\omega=-\frac{\partial\theta}{\partial t'}&=2K\tilde{E}ex'-\left(\beta-\frac{\alpha}{\Gamma}-\frac{1}{4}\upsilon^2\right)+e(K\tilde{E}e)^2(t')^2-\\ 2K\tilde{E}e\upsilon t'&=2K\tilde{E}ex'-\beta-\frac{\alpha}{\Gamma}+k^2\end{aligned} \tag{103}$$

The usual Hamilton equations for the superconductive electron (soliton) in the macroscopic quantum systems are still valid here and can be written as [80-81] $\frac{dk}{dt'}=-\frac{\partial\omega}{\partial x'}\big|_k=-2K\tilde{E}e$, then the group velocity of the superconductive electron is

$$\upsilon_g=\frac{dx'}{dt'}=\frac{\partial\omega}{\partial k}\Big|_{x'}=2\left(\frac{\upsilon}{2}-2K\tilde{E}et'\right)=\upsilon-4K\tilde{E}et' \tag{104}$$

This means that the frequency ω still represents the meaning of Hamiltonian in the case of nonlinear quantum systems. Hence, $\frac{d\omega}{dt'}=\frac{d\omega}{dk}\big|_{x'}\frac{dk}{dt'}+\frac{\partial\omega}{\partial x'}\big|_k\frac{dx'}{dt'}=0$, as seen in the usual stationary linear medium.

These relations in Eqs. (103)-(104) show that the superconductive electrons move as if they were classical particles moving with a constant acceleration in the invariant electric-field, and that the acceleration is given by $-4K\tilde{E}e$. If $\upsilon>0$, the soliton initially travels toward the overdense region, it then suffers a deceleration and its velocity changes sign. The soliton is then reflected and accelerated toward the underdense region.The penetration distance into the overdense region depends on the initial velocity υ.

From the above studies we see that the time-dependent motion of superconductive electrons still behaves like a soliton in non-equilibrium state of superconductor. Therefore, we can conclude that the electrons in the superconductors are essentially a soliton in both time-independent steady state and time-dependent dynamic state systems. This means that the soliton motion of the superconductive electrons causes the superconductivity of material. Then the superconductors have a complete conductivity and nonresistance property because the solitons can move over a macroscopic distances retaining its amplitude, velocity, energy and other quasi- particle features. In such a case the motions of the electrons in the superconductors are described by a nonlinear Schrödinger equations (52), or (53) or (80) or (82) or (84). According to the soliton theory, the electrons in the superconductors are localized and have a wave-corpuscle duality due to the nonlinear interaction, which is completely different from those in the quantum mechanics. Therefore, the electrons in superconductors should be described in nonlinear quantum mechanics[16-17].

7. The transmission features of magnetic-flux lines in the Josephson junctions

7.1 The transmission equation of magnetic-flux lines

We have learned that in a homogeneous bulk superconductor, the phase $\theta(\vec{r},t)$ of the electron wave function $\phi(\vec{r},t)=f(\vec{r},t)e^{i\theta(\vec{r},t)}$ is constant, independent of position and time. However, in an inhomogeneous superconductor such as a superconductive junction discussed above, θ becomes dependent of $\vec{r}$ and t. In the previous section, we discussed the Josephson effects in the S-N-S or S-I-S, and SNIS junctions starting from the Hamiltonican and the Ginzburg-Landau equations satisfied by $\phi(\vec{r},t)$, and showed that the Josephson current, whether dc or ac, is a function of the phase change, $\varphi=\Delta\theta=\theta_1-\theta_2$. The dependence of the Josephson current on φ is clearly seen in Eq. (78) . This clearly indicates that the Josephson current is caused by the phase change of the superconductive electrons. Josephson himself derived the equations satisfied by the phase difference φ, known as the Josephson relations, through his studies on both the dc and ac Josephson effects. The Josephson relations for the Josephson effects in superconductor junctions can be summarized as the following,

$$J_s=J_m\sin\varphi,\quad \hbar\frac{\partial\varphi}{\partial t}=2eV\ ,\ \hbar\frac{\partial\varphi}{\partial x}=2ed'\overline{H}_y/c\ ,\ \hbar\frac{\partial\varphi}{\partial y}=2ed'\overline{H}_x/c \tag{105}$$

where d′ is the thickness of the junction. Because the voltage V and magnetic field $\vec{H}$ are not determined, equation (105) is not a set of complete equations. Generally, these equations are solved simultaneously with the Maxwell equation $\nabla\times\vec{H}=(4\pi/c)\vec{J}$. Assuming that the magnetic field is applied in the xy plane, i.e. $\vec{H}=(\overline{H}_x,\overline{H}_y,0)$, the above Maxwell equation becomes

$$\frac{\partial}{\partial x}\overline{H}_y(x,y,t)-\frac{\partial}{\partial y}\overline{H}_x(x,y,t)=\frac{4\pi}{c}J(x,y,t) \tag{106}$$

In this case, the total current in the junction is given by $J=J_s(x,y,t)+J_n(x,y,t)+J_d(x,y,t)+J_0$

In the above equation, $J_s(x,y,t)$ is the superconductive current density, $J_n(x,y,t)$ is the normal current density in the junction (J_n =V/R(V) if the resistance in the junction is R(V) and a voltage V is applied at two ends of the junction), $J_d(x,y,t)$ is called a displacement current and it is given by $J_d=CdV(t)/dt$, where C is the capacity of the junction, and J_0 is a constant current density. Solving the equations in Eqs,(102) and (106) simultaneously, we can get

$$\nabla^2\varphi - \frac{1}{v_0^2}\left(\frac{\partial^2\varphi}{\partial t^2} - \gamma_0\frac{\partial\varphi}{\partial t}\right) = \frac{1}{\lambda_J^2}\sin\varphi + I_0 \quad (107)$$

where $v_0 = \sqrt{c^2 / 4\pi Cd'}, \gamma_0 = 1/RC, \lambda_J = \sqrt{c^2\hbar / 4\pi d'e^*}, I_0 = 4e^*\pi J_0 / c^2\hbar, e^* = 2e$.
Equation (107) is the equation satisfied by the phase difference. It is a Sine-Gordon equation (SGE) with a dissipative term. From Eq.(105), we see that the phase difference φ depends on the external magnetic field $\vec{H}$, thus the magnetic flux in the junction $\Phi' = \int \vec{H} ds = \oint \vec{A} \bullet d\vec{l} = \frac{c\hbar}{c^*}\oint \varphi dl$ can be specified in terms of φ, where $\vec{A}$ is vector potential of electromagnetic field, $d\vec{l}$ is line element of vortex lines. Equation (107) represents transmission of superconductive vortex lines. It is a nonlinear equation. Therefore, we know clearly that the Josephson effect and the related transmission of the vortex line, or magnetic flux, along the junctions are also nonlinear problems. The Sine-Gordon equation given above has been extensively studied by many scientists including Kivshar and Malomed[39-40]. We will solve it here using different approaches.

7.2 The transmission features of magnetic-flux lines

Assuming that the resistance R in the junction is very high, so that $J_n \to 0$, or equivivalently $\gamma_0 \to 0$, setting also $I_0 = 0$, equation (107) reduces to

$$\nabla^2\varphi - \frac{1}{v_0^2}\frac{\partial^2\varphi}{\partial t^2} = \frac{1}{\lambda_J^2}\sin\varphi \quad (108)$$

Define $X = x/\lambda_J, T = v_0 t/\lambda_J$, then in one-dimension, the above equation becomes

$$\frac{\partial^2\varphi}{\partial X^2} - \frac{\partial^2\varphi}{\partial T^2} = \sin\varphi$$

which is the 1D Sine-Gordon equation. If we further assume that $\varphi = \varphi(X,T) = \varphi(\theta')$ with

$$\theta' = X' - X_0' - vT', X' = X / \sqrt{hc / 2LI_0 e}, T' = T / \sqrt{2I_0 e / hc}$$

it becomes $(1 - v^2)\varphi_{\theta'}^2(\theta') = 2(A' - \cos\varphi)$, where A′ is a constant of integration. Thus

$$\int_{\varphi_0}^{\varphi(\theta')} [(A' - \cos\varphi)]^{-1/2} d\varphi = \sqrt{2}\delta\nu\theta'$$

where $\nu = 1/\sqrt{1 - v^2}, \delta = \pm 1$. Choosing A′=1, we have

$$\int_{\pi}^{\varphi(\theta')} [\sin(\varphi/2)]^{-1/2} d\varphi = 2\nu\theta'$$

A kink soliton solution can be obtained as follows $\pm\nu\theta' = \ln[\tan(\varphi/2)]$, 'or $\varphi(\theta') = 4\tan^{-1}[\exp(\pm\nu\theta')]$. Thus yields

$$\varphi(X', T') = 4\tan^{-1}\{\exp[\delta\nu(X' - X_0' - vT')]\} \quad (109)$$

From the Josephson relations, the electric potential difference across the junction can be written as

$$V = \frac{\hbar}{2e}\frac{d\varphi}{dT'} = \frac{\varphi_0}{2c\pi}\frac{d\varphi}{dT'} = 2\delta v\nu\sqrt{\frac{2I_o e}{\hbar c^2}}\frac{\varphi_0}{2c\pi}\operatorname{sec}h[\nu(X' - X_0' - vT')]$$

where $\varphi_0 = \pi\hbar c = 2\times10^{-7} Gauss / cm^{-2}$ 'is a quantum fluxon, c is the speed of light. A similar expression can be derived for the magnetic field

$$\overline{H}_z = \frac{\hbar}{2e}\frac{d\varphi}{dX'} = \frac{\varphi_0}{2c\pi}\frac{d\varphi}{dX'} = \pm 2\delta\nu\sqrt{\frac{2I_o e}{\hbar c^2}}\frac{\varphi_0}{2c\pi}\operatorname{sec}h[\nu(X' - X_0' - vT')]$$

We can then determine the magnetic flux through a junction with a length of L and a cross section of 1 cm2. The result is

$$\Phi' = \int_{-\infty}^{\infty} H_x(x,t)dx = B_0'\int_{-\infty}^{\infty} H_x(X',T')dX' = \delta\varphi_0$$

Therefore, the kink ($\delta = +1$) carries a single quantum of magnetic flux in the extended Josephson junction. Such an excitation is often called a fluxon, and the Sine-Gordon equation or Eq.(107) is often referred to as transmission equation of quantum flux or fluxon. The excitation corresponding to $\delta = -1$ is called an antifluxon. Fluxon is an extremely stable formation. However, it can be easily controlled with the help of external effects. It may be used as a basic unit of information.

This result shows clearly that magnetic flux in superconductors is quantized and this is a macroscopic quantum effect as mentioned in Section 1. The transmission of the quantum magnetic flux through the superconductive junctions is described by the above nonlinear dynamic equation (107) or (108).The energy of the soliton can be determined and it is given by $E = 8m^2/\beta$, where $m^2/\beta = 1/\lambda_J^2$.

However, the boundary conditions must be considered for real superconductors. Various boundary conditions have been considered and studied. For example, we can assume the following boundary conditions for a 1D superconductor, $\varphi_x(0,t) = \varphi_x(L,t) = 0$. Lamb[47] obtained the following soliton solution for the SG equation (108)

$$\varphi(x,t) = 4\tan^{-1}[h(x)g(t)] \tag{110}$$

where h and g are the general Jacobian elliptical functions and satisfy the following equations

$$[h(x)]^2 = a'h^4 + (1+b')h^2 - c', \quad [g(x)]^2 = c'h^4 + b'h^2 - a'$$

with a′, b′, and c′ being arbitrary constants. Coustabile et al. also gave the plasma oscillation, breathing oscillation and vortex line oscillation solutions for the SG equation under certain boundary conditions. All of these can be regarded as the soliton solution under the given conditions.

Solutions of Eq.(108) in two and three-dimensional cases can also be found[80-81]. In two-dimensional case, the solution is given by

$$\varphi(X,Y,T)=4\tan^{-1}[\frac{g(X,Y,T)}{f(X,Y,T)}] \tag{111}$$

where $X=x/\lambda_J, Y=y/\lambda_J, T=v_0t/\lambda_J$, $f=1+a(1,2)e^{y_1+y_2}=a(2,3)e^{y_2+y_3}a(3,1)e^{y_1+y_3}$,

$$g=e^{y_1}+e^{y_2}+a(1,2)a(2,3)a(3,1)e^{y_1+y_2+y_3}$$

and

$$y_1=p_iX+q_iY-\Omega_i\tau-y_i^0, p_i^2+q_i^2-\Omega_i^2=1,(i=1,2,3)$$

$$a(i,j)=\frac{(p_i-p_j)^2+(q_i-q_j)^2+(\Omega_i-\Omega_j)^2}{(p_i+p_j)^2+(q_i+q_j)^2+(\Omega_i+\Omega_j)^2},(1\le i\le j\le 3)$$

In addition, P_i, q_i and Ω_i satisfy $\det\begin{vmatrix} p_1 & q_1 & \Omega_1 \\ p_2 & q_2 & \Omega_2 \\ p_3 & q_3 & \Omega 3 \end{vmatrix}=0$. In the three-dimensional case, the solution is given by

$$\varphi(X,Y,Z,T)=4\tan^{-1}[\frac{g(X,Y,Z,T)}{f(X,Y,Z,T)}],, \tag{112}$$

where X, Y , and T are similarly defined as in the 2D case given above, and $Z=z/\lambda_J$. The functions f and g are defined as

$$f=dX_2e^{y_1+y_2}+dY_3e^{y_2+y_3}+dZ_3e^{y_1+y_3}+1,\quad g=e^{y_1}+e^{y_2}+e^{y_3}+dX_2dY_3dZ_3e^{y_1+y_2+y_3},$$

$$y_i=a_{i1}X+a_{i2}Y+a_{i3}+b_iT-C_i, {a_{i1}}^2+{a_{i2}}^2+{a_{i3}}^2-b_i^2=1,(i=X,Y,Z)$$

with

$$d(i,j)=\frac{\sum_{k=1}^{3}[(a_{ik}-a_{jk})^2-(b_i-b_j)^2]}{\sum_{k}^{3}[(a_{ik}+a_{jk})^2-(b_i+b_j)^2]},(1\le j\le 3),$$

here y_3 is a linear combination of y_1 and y_2 , i.e., $y_3=\alpha y_1+\beta y_2$.
We now discuss the SG equation with a dissipative term $\gamma_0\partial\varphi/\partial t$. First we make the following substitutions to simplify the equation

$$X=x/\lambda_J, T=v_0t/\lambda_J=t/\omega_J, a=\gamma_0\lambda_J^2/v_0, B'=I_0\lambda_J^2.$$

In terms of these new parameters, the 1D SG equation (107) can be rewritten as

$$\frac{\partial^2\varphi}{\partial X^2}-\frac{\partial^2\varphi}{\partial T^2}-a\frac{\partial\varphi}{\partial T}=\sin\varphi+B' \tag{113}$$

The analytical solution of Eq.(113) is not easily found. Now let

$$\alpha = \frac{1-v_0^2}{a^2 v_0^2}, \eta = \frac{1}{\sqrt{\alpha}} \frac{X - v_0 T}{a v_0}, q' = \frac{a v_0}{\sqrt{1-v_0^2}}, \varphi = \pi + \varphi' \tag{114}$$

Equation (113) then becomes

$$\frac{\partial^2 \varphi}{\partial \eta^2} + q' \frac{\partial \varphi}{\partial \eta} + \sin \varphi - B' = 0 \tag{115}$$

This equation is the same as that of a pendulum being driven by a constant external moment and a frictional force which is proportional to the angular displacement. The solution of the latter is well known, generally there exists an stable soliton solution[80-81]. Let $Y = d\varphi'/d\eta$, equation (115) can be written as

$$\frac{\partial Y}{\partial \eta} + q' Y + \sin \varphi' - B' = 0 \tag{116}$$

For $0 < B' < 1$, we can let $B' = \sin \varphi_0 (0 < \varphi_0 < \pi/2)$ and $\varphi' = -\pi - \varphi_0 + \varphi_1$, then, equation (116) becomes

$$Y \frac{\partial Y}{\partial \eta} = -q' Y + \sin \varphi_0 + \sin(\varphi_1 - \varphi_0) \tag{117}$$

Expand Y as a power series of φ_1, i.e., $Y = \sum_n c_n \varphi_1^n$, and inserting it into Eq.(117), and comparing coefficients of terms of the same power of φ_1 on both sides, we get

$$c_1 = -\frac{q'}{2} \pm \sqrt{\frac{q'^2}{4} + \cos \varphi_0}, c_2 = -\frac{1}{q' + 3c_1} \frac{\sin \varphi_0}{2}, \; c_3 = \frac{1}{q' + 4c_1}(-2c_2^2 - \frac{\cos \varphi_0}{6}), c_4 = \frac{1}{q' + 5c_1}(-5c_2 c_3 - \frac{\sin \varphi_0}{24}) \tag{118}$$

and so on. Substituting these c_n's into $Y = d\varphi'/d\eta = \sum_n c_n \varphi_1^2$, the solution of φ_1 may be found by integrating $\eta = \int d\varphi_1 / \sum_n c_n \varphi_1^2$. In general, this equation has soliton solution or elliptical wave solution. For example, when $d\varphi'/d\eta = c_1 \varphi_1 + c_2 \varphi_1^2 + c_3 \varphi_1^3$ it can be found that

$$\eta = \frac{2}{\sqrt{A-C}} F(\sqrt{\frac{A-B}{A-C}}, \sin^{-1}(\sqrt{\frac{A-\varphi_1}{A-B}}))$$

where $F(k, \varphi_1)$ is the first Legendre elliptical integral, and A, B and C are constants. The inverse function φ_1 of $F(k, \varphi_1)$ is the Jacobian amplitude ' $\varphi_1 = amF$. Thus,

$$\sin^{-1}(\sqrt{\frac{A-\varphi_1}{A-B}}) = am\sqrt{\frac{A-C}{A-B}}\eta \quad \text{or} \quad \sqrt{\frac{A-\varphi_1}{A-B}}) = sn(\sqrt{\frac{A-C}{A-B}}\eta)$$

where snF is the Jacobian sine function. Introducing the symbol cscF = 1/snF, the solution can be written as

$$\varphi_1 = A - (A-B)[\csc(\sqrt{\frac{A-C}{A-B}}\eta)]^2 \tag{119}$$

This is a elliptic function. It can be shown that the corresponding solution at $|\eta| \to \infty$ is a solitary wave.

It can be seen from the above discussion that the quantum magnetic flux lines (vortex lines) move along a superconductive junction in the form of solitons. The transmission velocity v_0 can be obtained from $h = \alpha v_0 \sqrt{1-v_0^2}$ and c_n in Eq. (118) and it is given by $v_0 = 1/\sqrt{1+[\alpha/h(\varphi_0)]^2}$.

That is, the transmission velocity of the vortex lines depends on the current I_0 injected and the characteristic decaying constant α of the Josephson junction. When α is finite, the greater the injection current I_0 is, the faster the transmission velocity will be; and when I_0 is finite, the greater the α is, the smaller the v_0 will be, which are realistic.

8. Conclusions

We here first reviewed the properties of superconductivity and macroscopic quantum effects, which are different from the microscopic quantum effects, obtained from some experiments. The macroscopic quantum effects occurred on the macroscopic scale are caused by the collective motions of microscopic particles , such as electrons in superconductors, after the symmetry of the system is broken due to nonlinear interactions. Such interactions result in Bose condensation and self-coherence of particles in these systems. Meanwhile, we also studied the properties of motion of superconductive electrons, and arrived at the soliton solutions of time-independent and time-dependent Ginzburg-Landau equation in superconductor, which are, in essence, a kind of nonlinear Schrödinger equation. These solitons, with wave-corpuscle duality, are due to the nonlinear interactions arising from the electron-phonon interaction in superconductors, in which the nonlinear interaction suppresses the dispersive effect of the kinetic energy in these dynamic equations, thus a soliton states of the superconductive electrons, which can move over a macroscopic distances retaining the energy, momuntum and other quasiparticle properties in the systems, are formed. Meanwhile, we used these dynamic equations and their soliton solutions to obtain, and explain, these macroscopic quantum effects and superconductivity of the systems. Effects such as quantization of magnetic flux in superconductors and the Josephson effect of superconductivity junctions,thus we concluded that the superconductivity and macroscopic quantum effects are a kind of nonlinear quantum effects and arise from the soliton motions of superconductive electrons. This shows clearly that studying the essences of macroscopic quantum effects and properties of motion of microscopic particles in the superconductors has important significance of physics.

9. References

[1] Parks, R. D., Superconductivity, Marcel. Dekker, 1969.
[2] Rogovin, D. and M. Scully, Superconductiviand macroscopic quantum phenomena, Phys. Rep. 25(1976) 178.
[3] Rogovin, D., Electrodynamics of Josephson junctions, Phys. Rev. B11 (1975) 1906-108
[4] Abrikosov, A. A. and L. P. Gorkov, I.V. Dzyaloshinkii, Quantum field theoretical mothods in statistic phyics, Pregamon Press, Oxfordea, 1965
[5] Rogovin, D., Josephson tunneling: An example of steady-state superradiance,Phys. Rev. B12 (1975) 130-133.

[6] Ginzburg V L, Superconductivity, Superdiamagnetism, Superfluidity , Moscow: MIR Publ., 1987
[7] Ginzburg V L, Superconductivity, Moscow-Leningrad: Izd. Moscoew, AN SSSR, 1946
[8] Ginzburg, V. L.. Superconductivity and superfluidity *Phys.-Usp.* 40 (1997) 407
[9] Leggett, A. J., Macroscopic Effect of *P*- and *T*-Nonconserving Interactions in Ferroelectrics: A Possible Experiment? Phys. Rev. Lett. 41 (1978) 586
[10] Leggett, A. J., in Percolation, Localization and Superconductivity, eds. by A. M. Goldlinan,S. A. Bvilf, Plenum Press, New York, 1984. pp. 1- 41.
[11] Leggett, A. J., Low temperature physics, Springer, Berlin, 1991, pp. 1-93;
[12] Pang Xiao-feng, Investigations of properties of motion of superconductive electrons in superconductors by nonlinear quantum mechanics, J. Electronic Science and Technology of China, 6(2)(2008)205-211
[13] Pang, Xiao-feng, macroscopic quantum effects, Chinese J. Nature, 5 (1982) 254,
[14] Pang Xiao-feng, Investigations of properties and essences of macroscopic quantum effects in superconductors by nonlinear quantum mechanics, Nature Sciences, 2(1)(2007) 42
[15] Pang, Xiao-feng, Investigation of solutions of a time-dependent Ginzburg-Landau equation in superconductor by nonlinear quantumtheory, IEEE Compendex, 2009, 274-277, DOI: 10.1109/ ASEMD.2009.5306641(EI)
[16] Pang, Xiao-feng, Theory of Nonlinear Quantum Mechanics, Chongqing Press, Chongqing, 1994.p35-97
[17] Pang Xiao-feng Nonlinear Quantum Mechanics, Beijing, Chinese Electronic Industry Press, Beijing, 2009,p20-63
[18] Bardeen, L. N., L. N. Cooper and J. R. Schrieffer, Superconductivity theory, Phys. Rev. 108 (1957) 1175;
[19] Cooper, L. N., The bound electronic pairs in degenerated Fermi gas Phys. Rev. 104 (1956) 1189.
[20] Schrieffer, J. R., Superconductivity, Benjamin, New York, 1969.
[21] Schrieffer, J. R., Theory of Superconductivity, Benjamin, New York, 1964.
[22] Frohlich.H., Theory of superconductive states, Phys. Rev.79(1950)845;
[23] Frohlich.H., On superconductivity theory : one-dimensional case, Proc. Roy.Soc.A, 223(1954)296
[24] Josephson, B.D, Possible New Effects in Superconducting Tunnelling , Phys. Lett. 1 (1962) 251
[25] Josephson, B.D, Supercurrents through barriers, Adv. Phys. 14 (1965) 419.
[26] Josephson, B.D., Thesis, unpublised, Cambridge University (1964)
[27] Pang, Xiao-feng, The relation between the physical parameters and effective spectrum of phonon, Southwest Inst. For Nationalities, . 17 (1991) 1.
[28] Pang, Xiao-feng, On the solutions of the time-dependent Ginzburg-Landau equations for a superconductor in a weak field, J. Low Temp. Physics, 58(1985)333
[29] Pang, Xiao-feng, The isotope effects of superconductor, Chinese. J. Low Temp. Supercond.,No. 3 (1982) 62
[30] Pang, Xiao-feng, The properties of soliton motion for superconductivity electrons. Proc. ICNP, Shanghai, 1989, p139.
[31] Rayfield, G. W. and F. Reif, Evidence for the creation and motion of quantized vortex rings in superfluid helium, Phys. Rev. Lett. 11 (1963) 305.
[32] Perring, J. K., and T.H.R.Skyrme, A model unified field equation,Nucl. Phys. 31 (1962) 550.

[33] Barenghi, C. F., R. J. Donnerlly and W. F.Vinen, Quantized Vortex Dynamics and Superfluid Turbulence, Springer, Berlin, 2001.
[34] Bogoliubov, N. N., Quantum statistics, Nauka, Moscow, 1949
[35] Bogoliubov, N. N., V. V. Toimachev and D. V. Shirkov, A New Method in the Theory of Superconductivity, AN SSSR, Moscow, 1958.
[36] London, F., superfluids Vol.1, Weley, New York 1950.
[37] de Gennes, P. G., Superconductivity of Metals and Alloys, W. A. Benjamin, New York, 1966.
[38] Suint-James, D., et al., Type-II Superconductivity, Pergamon, Oxford, 1966.
[39] Kivshar, Yu. S. and B. A. Malomed, Dynamics of solitons in nearly integrable systems，Rev. Mod. Phys. 61 (1989) 763.
[40] Kivshar, Yu. S., T. J. Alexander and S. K. Turitsy, Nonlinear modes of a macroscopic quantum oscillator， Phys. Lett. A 278(2001) 225.
[41] Bullough, R. K., N. M. Bogolyubov, V. S. Kapitonov, C. Malyshev, J. Timonen, A. V. J.Rybin, A.V., Vazugin, G.G. and Lindberg, M, Quantum integrable and nonintegrable nonlinear Schrodinger models for realizable Bose-Einstein condensation in d+1 dimensions (d=1,2,3)， Theor. Math.Phys. 134(2003)47
[42] Bullough, R. K. and P. T. Caudeey, Solitons, Plenum Press, New York, 1980.
[43] Huepe, C. and M. E. Brachet, *Scaling laws for vortical nucleation solutions in a model of superflow*, Physica D 140 (2000) 126.
[44] Sonin, E. B., Nucleation and creep of vortices in superfluids and clean superconductors,Physica，B210（1995）234-250
[45] Davydov.A.S. and V.N.Ermakov, Stability of a Superconducting Condensate of Bisolitons, Phys. Stat.Sol.B148(1988)305
[46] Landau, L. D. and E. M. Lifshitz, Quantum mechanics, Pergamon Press, Oxford, 1987
[47] Lamb, G. L., Analytical descriptions of ultrashort optical pulse propagation in a resonant medium，Rev. Mod. Phys. 43 (1971) 99.
[48] Ginzberg, V. L. and L. D. Landau, On the theory of superconductivity，Zh. Eksp. Theor. Fiz. 20 (1950) 1064;
[49] Ginzburg V.L., On superconductivity and superfluidity, Physics -Usp. 47 (2004) 1155 - 1170
[50] Ginzberg, V. L. and D. A. Kirahnits, Problems in High-Temperature Superconductivity, Nauka, Moscow, 1977.
[51] Gorkov, L. P., On the energy spectrum of superconductors, Sov. Phys. JETP 7(1958) 505.
[52] Gorkov, L. P., Microscopic derivation of the Ginzburg-Landau equation in the theory of superconductivity, Sov. Phys. JETP 9 (1959) 1364.
[53] Abrikosov, A. A., On the magnetic properties of superconductors of the second group，Zh. Eksp. Theor. Fiz. 32 (1957) 1442.
[54] Abrikosov, A. A. and L. P. Gorkov, Zh. Eksp. Theor. Phys. 39 (1960) 781;
[55] Valatin. J.G. , *Comments on the theory of superconductivity,* Nuovo Cimento7(1958)843
[56] Liu, W. S. and X. P. Li, BCS states as squeezed fermion-pair states, European Phys. J. D2 (1998) 1.
[57] Gross, E. F.,structure of a quantized vertex in boson systems, II Nuovo Cimento, 20 (1961) 454.
[58] Pitaevskii, L. P.. Vortex lines in an imperfect Bose gas. Sov. Phys. JETP-USSR, 13 (1961)451
[59] Pitaevskii, L. P. and Stringari, S. *Bose–Einstein Condensation*. Oxford, Clarendon Press, 2003

[60] Elyutin P. V. and A. N. Rogovenko, Stimulated transitions between the self-trapped states of the nonlinear Schrödinger equation, Phys. Rev. E63 (2001) 026610.
[61] Elyutin, P. V., Buryak A V, Gubernov V V, Sammut R A and Towers I N, Interaction of two one-dimensional Bose-Einstein solitons: Chaos and energy exchange, Phys. Rev. E64 (2001) 016607.
[62] Pang, Xiao-feng, The properties of motion of superconductive electrons in superconductor, J. XinJiang Univ. (nature) 5(1988)33
[63] Pang Xiao-feng, The G-L theory of superconductivity bin magnetic-superconductor, Investigations of metal materials, 12(1986) 31
[64] Perez-Garcia, V. M., M. Michinel and H. Herrero, Bose-Einstein solitons in highly asymmetric traps, Phys. Rev. A57 (1998) 3837.
[65] London, F. and H. London, The electromagnetic equations of the superconductor, Proc. Roy. Soc. (London) A 149 (1935) 71
[66] Pang, Xiao-feng, Interpretation of proximity effect in superconductive junctions by G-L theory, J. Kunming Tech. Sci. Univ., 14. (1989) 78
[67] Pang, Xiao-feng, Properties of transmission of vortex lines along the superconductive junctions. J. Kunming Tech. Sci. Univ., 14 (1989) 83.
[68] Caradoc-Davies, B. M., R. J. Ballagh and K. Bumett, Coherent Dynamics of Vortex Formation in Trapped Bose-Einstein Condensates, Phys. Rev. Lett. 83 (1999) 895.
[69] Matthews, M. R., B. P. Anderson*, P. C. Haljan, D. S. Hall†, C. E. Wieman, and E. A. Cornell, Vortices in a Bose-Einstein Condensate, Phys. Rev. Lett. 83 (1999) 2498.
[70] Madison, K. W., F. Chevy, W. Wohlleben, and J. Dalibard , Vortex formation in a stirred Bose-Einstein condensate, Phys. Rev. Lett. 84 (2000) 806.
[71] Tonomura A, Kasai H, Kamimura O, Matsuda T, Harada K, Yoshida T, Akashi T, Shimoyama J, Kishio K, Hanaguri T, Kitazawa K, Masui T, Tajima S, Koshizuka N, Gammel PL, Bishop D, Sasase M, Okayasu S, Observation of structures of chain vortices inside anisotropic high- Tc superconductors, Phys. Rev. Lett. 88(2002) 237001
[72] Pang, Xiao-feng, Investigation of solutions of Josephson equation in three dimension superconductive junctions, J. Chinghai Normal Univ. Sin. No. 1 (1989) 37.
[73] Blackbunu, J. A., H.J.T. Smith and N.L. Rowell, Proximity effects and the generalized Ginzburg-Landau equation, Phys. Rev. B11 (1975) 1053.
[74] Jacobson, D. A., Ginzburg-Landau equations and the Josephson effect, Phys. Rev. B8 (1965) 1066;
[75] Pang, Xiao-feng, The Bose condensation properties of superconductive states, J. Science Exploration Sin. 1(4)(1986) 70.
[76] Pang, Xiao-feng, The features of coherent state of superconductive states, J. Southwest Inst. For Nationalities Sin. 17 (1991) 18.
[77] Dewitt, B. S., Superconductors and Gravitational Drag, Phys. Rev. Lett. 16 (1966) 1092
[78] Kusayanage.E, T. Kawashima and K.Yamafuji,Flux flow in nonideal type-II superconductor, J. Phys. Soc. Japan 33 (1972) 551.
[79] Cai, S. Y. and A. Bhattacharjee, Ginzburg-Landau equation: A nonlinear model for the radiation field of a free-electron laser, Phys. Rev. A43 (1991) 6934.
[80] Pang, Xiao-feng, Soliton Physics, Press of Sichuan Sci. and Tech., Chengdu, 2003
[81] Guo Bai-lin and Pang Xiao-feng, solitons, Chinese Science Press, Beijing, 1987

10

FFLO and Vortex States in Superconductors With Strong Paramagnetic Effect

M. Ichioka, K.M. Suzuki, Y. Tsutsumi and K. Machida
Department of Physics, Okayama University
Japan

1. Introduction

In type-II superconductors (Fetter & Hohenberg, 1969), magnetic fields penetrate into superconductors as quantized flux lines with flux quanta $\phi_0 = hc/2e$, where h is Planck constant, c velocity of light, e electron's charge. Around a flux line, pair potential $\Delta(\mathbf{r})$ of Cooper pair has a vortex structure, where $\Delta(\mathbf{r})$ has phase winding 2π reflecting screening super-current around a flux line. At the vortex core, amplitude $|\Delta(\mathbf{r})|$ is suppressed, and low energy excitations appear within the superconducting gap of the electronic states.
Vortex physics makes important roles in the study of unconventional superconductors, because unconventional characters hidden in the uniform superconducting state at a zero field appear around vortices. For example, in superconductors with anisotropic superconducting gap on Fermi surface in momentum space, electronic states around a vortex show real-space anisotropy in the local density of states (LDOS) $N(E,\mathbf{r})$ around vortex core. The vortex core image is observed by scanning tunneling microscopy (STM) (Hess et al., 1990; Nishimori et al., 2004). Around vortex cores, local zero-energy electronic states at Fermi level are seen as star shape with tails extending toward node or weak-gap directions (Hayashi et al., 1996; 1997; Ichioka et al., 1996; Schopohl & Maki, 1995). From the spatial average of the zero-energy states, we can estimate the zero-energy density of states (DOS) $N(E=0)$, which determines low temperature (T) behaviors of physical quantities. Due to the differences of electronic states around the vortex core, $N(E=0)$ shows different magnetic field (H) dependences. These H-dependences are studied to identify the pairing symmetry in the experiments for vortex states, such as, electronic specific heat (Moler et al., 1994; Nohara et al., 1999), electronic thermal conductivity, and paramagnetic susceptibility (Zheng et al., 2002). For example, the H-dependence of low temperature specific heat $C(H)$ is often used to distinguish the presence of nodes in the pairing potential. As for Sommerfeld coefficient $\gamma(H) \equiv \lim_{T\to 0} C(H)/T$, $\gamma(H) \propto H$ in s-wave pairing with full gap, and $\gamma(H) \propto \sqrt{H}$ by the Volovik effect in d-wave pairing with line nodes (Ichioka et al., 1999a;b; Miranović et al., 2003; Nakai et al., 2004; Volovik, 1993). The curves of $\gamma(H)$ are expected to smoothly recover to the normal state value towards the upper critical field H_{c2}. However, in some heavy fermion superconductors, $C(H)$ deviates from these curves. In $CeCoIn_5$, $C(H)$ shows convex curves, i.e., $C(H) \propto H^{\alpha}$ ($\alpha > 1$) at higher fields(Ikeda et al., 2001). This behavior is not understood only by effects of the pairing symmetry. A similar $C(H)$ behavior is observed also in UBe_{13} (Ramirez et al., 1999). The experimental data of magnetization curve $M_{\rm total}(H)$ in $CeCoIn_5$ show a convex curve at higher fields, instead of a conventional concave curve (Tayama et al., 2002). As an

unconventional behavior of $CeCoIn_5$, the small angle neutron scattering (SANS) experiment reported anomalous H-dependence of flux line lattice (FLL) form factor determined from the Bragg intensity (Bianchi et al., 2008; DeBeer-Schmitt et al., 2006; White et al., 2010). While the form factor shows exponential decay as a function of H in many superconductors, it increases until near H_{c2} for $H \parallel c$ in $CeCoIn_5$. In some heavy fermion superconductors, the paramagnetic effects due to Zeeman shift are important to understand the properties of the vortex states, because the superconductivity survives until under high magnetic fields due to the effective mass enhancement. A heavy fermion compound $CeCoIn_5$ is a prime candidate of a superconductor with strong Pauli-paramagnetic effect (Matsuda & Shimahara, 2007). There at higher fields H_{c2} changes to the first order phase transition (Bianchi et al., 2002; Izawa et al., 2001; Tayama et al., 2002) and new phase, considered as Fulde-Ferrell-Larkin-Ovchinnikov (FFLO) state, appears (Bianchi, Movshovich, Capan, Pagliuso & Sarrao, 2003; Radovan et al., 2003). As for properties of $CeCoIn_5$, the contribution of antiferromagnetic fluctuation and quantum critical point (QCP) is also proposed in addition to the strong paramagnetic effect (Bianchi, Movshovich, Vekhter, Pagliuso & Sarrao, 2003; Paglione et al., 2003). Therefore, it is expected to study whether properties of vortex states in $CeCoIn_5$ are theoretically explained only by the paramagnetic effect. Theoretical studies of the H-dependences also help us to estimate strength of the paramagnetic effect, in addition to pairing symmetry, from experimental data of the H-dependences in various superconductors.

In this chapter, we concentrate to discuss the paramagnetic effect in the vortex states, to see how the paramagnetic effect changes structures and properties of vortex states. The BCS Hamiltonian in magnetic field is given by

$$\mathcal{H} - \mu_0 \mathcal{N} = \sum_{\sigma=\uparrow,\downarrow} \int \mathrm{d}^3\mathbf{r}\, \psi_\sigma^\dagger(\mathbf{r}) K_\sigma(\mathbf{r}) \psi_\sigma(\mathbf{r}) - \int \mathrm{d}^3\mathbf{r}_1 \int \mathrm{d}^3\mathbf{r}_2 \left\{ \Delta(\mathbf{r}_1,\mathbf{r}_2)\psi_\uparrow^\dagger(\mathbf{r}_1)\psi_\downarrow^\dagger(\mathbf{r}_2) + \Delta^*(\mathbf{r}_1,\mathbf{r}_2)\psi_\downarrow(\mathbf{r}_2)\psi_\uparrow(\mathbf{r}_1) \right\} \tag{1}$$

for superconductors of spin-singlet pairing, with

$$K_\sigma(\mathbf{r}) = \frac{\hbar^2}{2m}\left(\frac{\nabla}{\mathrm{i}} + \frac{\pi}{\phi_0}\mathbf{A}\right)^2 + \sigma\mu_{\mathrm{B}} B(\mathbf{r}) - \mu_0, \tag{2}$$

$\sigma = \pm 1$ for up/down spin electrons. Suppression of superconductivity by magnetic field occurs by two contributions. One is diamagnetic pair-breaking from vector potential $\mathbf{A}$ in Hamiltonian inducing screening current of vortex structure. And the other is paramagnetic pair-breaking from Zeeman term, which induces splitting of up-spin and down-spin Fermi surfaces as schematically presented in Fig. 1. Due to the Zeeman shift, in normal states, numbers of occupied electron states are imbalance between up-spin and down-spin electrons. The imbalance induces paramagnetic moment. In superconducting state with spin-singlet pairing, formations of Cooper pair between up-spin and down-spin electrons reduce the imbalance, and suppress the paramagnetic moment. However, the paramagnetic moment may appear at place where superconductivity is locally suppressed, such as around vortex core. Therefore, it is important to quantitatively estimate the spatial structure of paramagnetic moment and the contributions to properties of superconductors in vortex states.

One of other paramagnetic effect is paramagnetic pair breaking. When the Zeeman effect is negligible, as in Fig. 1(a), for Cooper pair of up-spin and down-spin electrons at Fermi level, total momentum Q of the pair is zero, i.e., $Q = k + (-k) = 0$. However, in the presence of

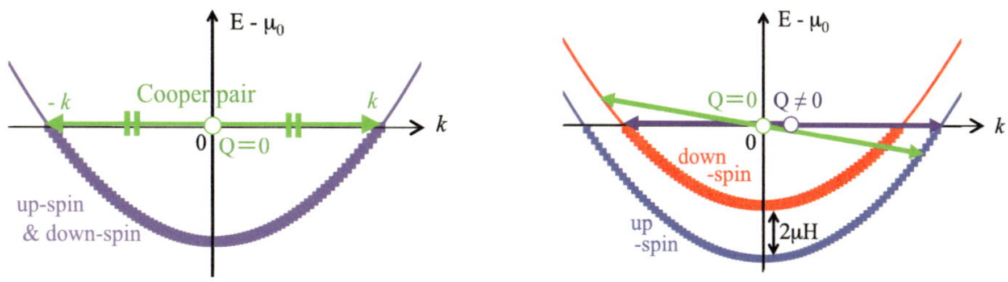

(a) (b)

Fig. 1. Paramagnetic effect by Zeeman shift of energy dispersion is schematically presented. Bold lines indicate occupied states. (a) The case when Zeeman shift is negligible. For Cooper pairs at Fermi level, total momentum $Q = k + (-k) = 0$. (b) When Zeeman shift is significant, the energy dispersions of up-spin and down-spin electrons are separated. When $Q = 0$, the electrons of Cooper pair are not at Fermi level. In FFLO states, $Q \neq 0$ so that electrons of Cooper pair are located at Fermi level.

Zeeman splitting, in order to keep $Q = 0$, Cooper pair is formed between electrons far from Fermi level, as shown in Fig. 1(b). Since the energy gain by this pairing is smaller than that of negligible paramagnetic case, the Zeeman splitting induces paramagnetic pair-breaking of superconductivity. In addition to H_{c2} suppressed by the paramagnetic pair-breaking, it is important to quantitatively estimate the contribution of paramagnetic pair-breaking on properties of vortex states at $H < H_{c2}$.

When paramagnetic effect by Zeeman shift is further significant, transition to FFLO state occurs at high magnetic fields near H_{c2}. In FFLO state, as shown in Fig. 1(b), electrons at Fermi level form Cooper pair with non-zero total momentum ($Q \neq 0$), which indicates periodic modulation of pair potential (Fulde & Ferrell, 1964; Larkin & Ovchinnikov, 1965; Machida & Nakanishi, 1984). When FFLO state appears in vortex state, we have to estimate properties of the FFLO state, considering both of vortex and FFLO modulation (Adachi & Ikeda, 2003; Houzet & Buzdin, 2001; Ichioka et al., 2007; Ikeda & Adachi, 2004; Mizushima et al., 2005a;b; Tachiki et al., 1996). Another system for significant paramagnetic effect is superfluidity of neutral ^{6}Li atom gases under the population imbalance of two species for pairing (Machida et al., 2006; Partridge et al., 2006; Takahashi et al., 2006; Zwierlein et al., 2006). There, we can study vortex state by rotating fermion superfluids, under control of paramagnetic effect by loaded population imbalance.

For theoretical studies of vortex states including electronic structure, we have to use formulation of microscopic theory, such as Bogoliubov-de Gennes (BdG) theory (Mizushima et al., 2005a;b; Takahashi et al., 2006) or quasi-classical Eilenberger theory (Eilenberger, 1968; Klein, 1987). In this chapter, based on the selfconsistent Eilenberger theory (Ichioka et al., 1999a;b; 1997; Miranović et al., 2003), we discuss interesting phenomena of vortex states in superconductors with strong paramagnetic effect, i.e., (i) anomalous magnetic field dependence of physical quantities, and (ii) FFLO vortex states. We study the spatial structure of the vortex states with and without FFLO modulation, in the presence of the paramagnetic effect due to Zeeman-shift (Hiragi et al., 2010; Ichioka et al., 2007; Ichioka & Machida, 2007; Watanabe et al., 2005). Since we calculate the vortex structure in vortex lattice states, self-consistently with local electronic states, we can quantitatively estimate the field dependence of some physical quantities. We will clarify the paramagnetic effect on the vortex core structure, calculating the pair potential, paramagnetic moment, internal magnetic field, and local electronic states. We also study the paramagnetic effect by quantitatively estimating the H-dependence of low temperature specific heat, Knight shift, magnetization and FLL form factors. For quantitative estimate, it is important to appropriately determine vortex core structure by selfconsistent calculation in vortex lattice states. These theoretical studies of the magnetic field dependences help us to evaluate the strength of the paramagnetic effect from the experimental data of the H-dependences in various superconductors.

After giving our formulation of selfconsistent Eilenberger theory in Sec. 2, we study the paramagnetic effect in vortex states without FFLO modulation in Sec. 3, where we discuss the H-dependence of paramagnetic susceptibility, low temperature specific heat, magnetization curve, FLL form factor, and their comparison with experimental data in $CeCoIn_5$. We also show the paramagnetic contributions on the vortex core structure, and the local electronic state in the presence of Zeeman shift. Section 4 is for the study of FFLO vortex state, in order to theoretically estimate properties of the FFLO vortex states, and to show how the properties appear in experimental data. We study the spatial structure of pair potential, paramagnetic moment, internal field, and local electronic state, including estimate of magnetic field range for stable FFLO vortex state. As possible methods to directly observe the FFLO vortex state, we discuss the NMR spectrum and FLL form factors, reflecting FFLO vortex structure. Last section is devoted to summary and discussions.

2. Quasiclassical theory including paramagnetic effect

One of the methods to study properties of superconductors by microscopic theory is a formulation of Green's functions. With field operators $\psi_\uparrow$, $\psi_\downarrow$, Green's functions are defined as

$$
\begin{aligned}
&G(\mathbf{r},\tau;\mathbf{r}',\tau') = -\langle T_\tau[\psi_\uparrow(\mathbf{r},\tau)\psi_\uparrow^\dagger(\mathbf{r}',\tau')]\rangle, \\
&F(\mathbf{r},\tau;\mathbf{r}',\tau') = -\langle T_\tau[\psi_\uparrow(\mathbf{r},\tau)\psi_\downarrow(\mathbf{r}',\tau')]\rangle, \quad F^\dagger(\mathbf{r},\tau;\mathbf{r}',\tau') = -\langle T_\tau[\psi_\downarrow^\dagger(\mathbf{r},\tau)\psi_\uparrow^\dagger(\mathbf{r}',\tau')]\rangle
\end{aligned} \tag{3}
$$

in imaginary time formulation, where T_τ indicates time-ordering operator of τ, and $\langle\cdots\rangle$ is statistical ensemble average. The Green's functions obey Gor'kov equation derived from the BCS Hamiltonian of Eq. (1). Behaviors of Green's functions include rapid oscillation of atomic short scale at the Fermi energy. Thus, in order to solve Gor'kov equation or BdG equation for vortex structure, we need heavy calculation treating all atomic sites within a unit cell of vortex lattice. To reduce the task of the calculation, we adopt quasiclassical approximation to integrate out the rapid oscillation of the atomic scale $\sim 1/k_{\rm F}$ ($k_{\rm F}$ is Fermi wave number), and consider only the spatial variation in the length scale of the superconducting coherence length ξ_0. This is appropriate when $\xi_0 \gg 1/k_{\rm F}$, which is satisfied in most of superconductors in solid state physics. The quasiclassical Green's functions are defined as

$$
\begin{aligned}
&g(\omega_n,\mathbf{k}_{\rm F},\mathbf{r}) = \int \frac{{\rm d}\xi}{{\rm i}\pi} G(\omega_n,\mathbf{k},\mathbf{r}), \\
&f(\omega_n,\mathbf{k}_{\rm F},\mathbf{r}) = \int \frac{{\rm d}\xi}{\pi} F(\omega_n,\mathbf{k},\mathbf{r}), \quad f^\dagger(\omega_n,\mathbf{k}_{\rm F},\mathbf{r}) = \int \frac{{\rm d}\xi}{\pi} F^\dagger(\omega_n,\mathbf{k},\mathbf{r}),
\end{aligned} \tag{4}
$$

where we consider the Fourier transformation of the Green's functions; from $\tau - \tau$ to Matsubara frequency ω_n, and from $\mathbf{r}-\mathbf{r}'$ to relative momentum $\mathbf{k}$, and integral about $\xi \equiv k^2/2m - \mu_0$, i.e., momentum directions perpendicular to the Fermi surface. Thus, the quasiclassical Green's functions depends on the momentum $\mathbf{k}_{\rm F}$ on the Fermi surface, and the center-of-mass coordinate $(\mathbf{r}+\mathbf{r}')/2 \rightarrow \mathbf{r}$.

From the Gor'kov equation, Eilenberger equations for quasiclassical Green's functions are derived as

$$\{\omega_n + i\mu B + \mathbf{v}\cdot(\nabla + i\mathbf{A})\} f = \Delta(\mathbf{r},\mathbf{k}_F) g,$$
$$\{\omega_n + i\mu B - \mathbf{v}\cdot(\nabla - i\mathbf{A})\} f^{\dagger} = \Delta^*(\mathbf{r},\mathbf{k}_F) g, \tag{5}$$

with $\mathbf{v}\cdot\nabla g = \Delta^*(\mathbf{r},\mathbf{k}_F) f - \Delta(\mathbf{r},\mathbf{k}_F) f^{\dagger}$, $g = (1 - f f^{\dagger})^{1/2}$, $\mathrm{Re} g > 0$, $\Delta(\mathbf{r},\mathbf{k}_F) = \Delta(\mathbf{r})\phi(\mathbf{k}_F)$, and $\mu = \mu_B B_0/\pi k_B T_c$. In this chapter, length, temperature, Fermi velocity, magnetic field and vector potential are, respectively, in units of R_0, T_c, $\bar{v}_F$, B_0 and $B_0 R_0$. Here, $R_0 = \hbar \bar{v}_F / 2\pi k_B T_c$ is in the order of coherence length, $B_0 = \hbar c/2|e|R_0^2$, and $\bar{v}_F = \langle v_F^2 \rangle_{\mathbf{k}_F}^{1/2}$ is an averaged Fermi velocity on the Fermi surface. $\langle \cdots \rangle_{\mathbf{k}_F}$ indicates the Fermi surface average. Energy E, pair potential Δ and Matsubara frequency ω_n are in unit of $\pi k_B T_c$. We set the pairing function $\phi(\mathbf{k}_F) = 1$ in the s-wave pairing, and $\phi(\mathbf{k}_F) = \sqrt{2}(k_a^2 - k_b^2)/(k_a^2 + k_b^2)$ in the d-wave pairing. The vector potential is given by $\mathbf{A} = \frac{1}{2}\bar{\mathbf{B}} \times \mathbf{r} + \mathbf{a}$ in the symmetric gauge, with an average flux density $\bar{\mathbf{B}} = (0, 0, \bar{B})$. The internal field is obtained as $\mathbf{B}(\mathbf{r}) = \bar{\mathbf{B}} + \nabla \times \mathbf{a}$.
The pair potential is selfconsistently calculated by

$$\Delta(\mathbf{r}) = g_0 N_0 T \sum_{0 \le \omega_n \le \omega_{\mathrm{cut}}} \left\langle \phi^*(\mathbf{k}_F) \left(f + f^{\dagger *} \right) \right\rangle_{\mathbf{k}_F} \tag{6}$$

with $(g_0 N_0)^{-1} = \ln T + 2T \sum_{0 \le \omega_n \le \omega_{\mathrm{cut}}} \omega_n^{-1}$. We set high-energy cutoff of the pairing interaction as $\omega_{\mathrm{cut}} = 20 k_B T_c$. The vector potential is selfconsistently determined by the paramagnetic moment $\mathbf{M}_{\mathrm{para}} = (0, 0, M_{\mathrm{para}})$ and the supercurrent $\mathbf{j}_s$ as

$$\nabla \times \nabla \times \mathbf{a}(\mathbf{r}) = \mathbf{j}_s(\mathbf{r}) + \nabla \times \mathbf{M}_{\mathrm{para}}(\mathbf{r}) \equiv \mathbf{j}(\mathbf{r}), \tag{7}$$

with

$$\mathbf{j}_s(\mathbf{r}) = -\frac{2T}{\kappa^2} \sum_{0 \le \omega_n} \langle \mathbf{v}_F \mathrm{Im}\{g\} \rangle_{\mathbf{k}_F}, \tag{8}$$

$$M_{\mathrm{para}}(\mathbf{r}) = M_0 \left(\frac{B(\mathbf{r})}{\bar{B}} - \frac{2T}{\mu \bar{B}} \sum_{0 \le \omega_n} \langle \mathrm{Im}\{g\} \rangle_{\mathbf{k}_F} \right). \tag{9}$$

Here, the normal state paramagnetic moment $M_0 = (\mu/\kappa)^2 \bar{B}$, $\kappa = B_0/\pi k_B T_c \sqrt{8\pi N_0}$, N_0 is DOS at the Fermi energy in the normal state.
The unit cell of the vortex lattice is given by $\mathbf{r} = w_1(\mathbf{u}_1 - \mathbf{u}_2) + w_2 \mathbf{u}_2 + w_3 \mathbf{u}_3$ with $-0.5 \le w_i \le 0.5$ (i=1, 2, 3), $\mathbf{u}_1 = (a, 0, 0)$, $\mathbf{u}_2 = (\zeta a, a_y, 0)$ with $\zeta = 1/2$, and $\mathbf{u}_3 = (0, 0, L)$. For triangular vortex lattice $a_y/a = \sqrt{3}/2$, and $a_y/a = 1/2$ for square vortex lattice. For the FFLO modulation, we assume $\Delta(x,y,z) = \Delta(x,y,z+L)$ and $\Delta(x,y,z) = -\Delta(x,y,-z)$. Then, $\Delta(\mathbf{r}) = 0$ at the FFLO nodal planes $z = 0$, and $\pm 0.5L$. These configurations of the FFLO vortex structure are schematically shown in Fig. 2, which show the unit cell in the xz plane including vortex lines, and in the xy plane. We divide w_i to N_i-mesh points in our numerical studies, and calculate the quasiclassical Green's functions, $\Delta(\mathbf{r})$, $M_{\mathrm{para}}(\mathbf{r})$ and $\mathbf{j}(\mathbf{r})$ at each mesh point in the three dimensional (3D) space. Typically we set $N_1 = N_2 = N_3 = 31$ for the calculation of vortex states with FFLO modulation. For the vortex states without FFLO modulation, we assume uniform structure along the magnetic field direction, and set $N_1 = N_2 = 41$.
We solve Eq. (5) for g, f, $f^{\dagger}$, and Eqs. (6)-(9) for $\Delta(\mathbf{r})$, $M_{\mathrm{para}}(\mathbf{r})$, $\mathbf{A}(\mathbf{r})$, alternately, and obtain selfconsistent solutions, by fixing a unit cell of the vortex lattice and a period L of the FFLO

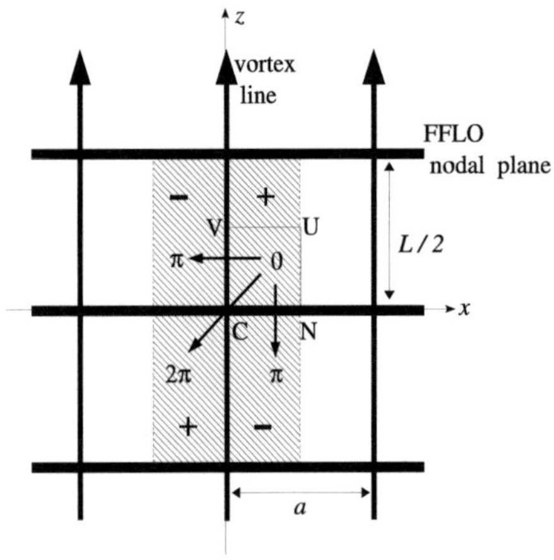

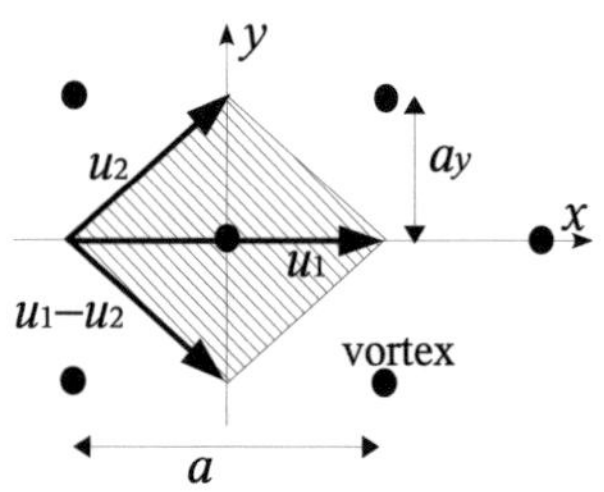

Fig. 2. Configurations of the vortex lines and the FFLO nodal planes are schematically presented in the xz plane including vortex lines (a) and in the xy plane (b). The inter-vortex distance is a in the x direction, and the distance between the FFLO nodal planes is $L/2$. The hatched region indicates the unit cell. In (a), along the trajectories presented by "$0 \longrightarrow \pi$", the pair potential changes the sign ($+ \rightarrow -$) across the vortex line or across the FFLO nodal plane, due to the π-phase shift of the pair potential. Along the trajectory presented by "$0 \longrightarrow 2\pi$", the sign of the the pair potential does not change ($+ \rightarrow +$) across the intersection point of the vortex line and the FFLO nodal plane, since the phase shift is 2π. In (b), • indicates the vortex center. $\mathbf{u}_1 - \mathbf{u}_2$ and $\mathbf{u}_2$ are unit vectors of the vortex lattice.

modulation. When we solve Eq. (5), we estimate $\Delta(\mathbf{r})$ and $\mathbf{A}(\mathbf{r})$ at arbitrary positions by the interpolation from their values at the mesh points, and by the periodic boundary condition of the unit cell including the phase factor due to the magnetic field. The boundary condition is given by

$$\Delta(\mathbf{r}+\mathbf{R}) = \Delta(\mathbf{r})e^{i\chi(\mathbf{r},\mathbf{R})} \tag{10}$$

$$\chi(\mathbf{r},\mathbf{R}) = 2\pi\{\frac{1}{2}((m+n\zeta)\frac{y}{a_y} - n\frac{x}{a_x}) + \frac{mn}{2} + (m+n\zeta)\frac{y_0}{a_y} - n\frac{x_0}{a_x}\} \tag{11}$$

for $\mathbf{R} = m\mathbf{u}_1 + n\mathbf{u}_2$ (m, n : integer), when the vortex center is located at $(x_0, y_0) - \frac{1}{2}(\mathbf{u}_1 + \mathbf{u}_2)$. In the selfconsistent calculation of $\mathbf{a}$, we solve Eq. (7) in the Fourier space $\mathbf{q}_{m_1,m_2,m_3}$, taking account of the current conservation $\nabla \cdot \mathbf{j}(\mathbf{r}) = 0$, so that the average flux density per unit cell of the vortex lattice is kept constant. The wave number $\mathbf{q}$ is discretized as

$$\mathbf{q}_{m_1,m_2,m_3} = m_1\mathbf{q}_1 + m_2\mathbf{q}_2 + m_3\mathbf{q}_3 \tag{12}$$

with integers m_i ($i = 1,2,3$), where $\mathbf{q}_1 = (2\pi/a, -\pi/a_y, 0)$, $\mathbf{q}_2 = (2\pi/a, \pi/a_y, 0)$, and $\mathbf{q}_3 = (0, 0, 2\pi/L)$. The lattice momentum is defined as $\mathbf{G}(\mathbf{q}_{m_1,m_2,m_3}) = (G_x, G_y, G_z)$ with $G_x = [N_1 \sin(2\pi m_1/N_1) + N_2 \sin(2\pi m_2/N_2)]/a$, $G_y = [-N_1 \sin(2\pi m_1/N_1) + N_2 \sin(2\pi m_2/N_2)]/2a_y$, and $G_z = N_3 \sin(2\pi m_3/N_3)/L$. We obtain the Fourier component of $\mathbf{a}(\mathbf{r})$ as $\mathbf{a}(\mathbf{q}) = \mathbf{j}'(\mathbf{q})/|\mathbf{G}|^2$, where $\mathbf{j}'(\mathbf{q}) = \mathbf{j}(\mathbf{q}) - \mathbf{G}(\mathbf{G}\cdot\mathbf{j}(\mathbf{q}))/|\mathbf{G}|^2$ ensuring the current conservation $\nabla \cdot \mathbf{j}'(\mathbf{r}) = 0$, and $\mathbf{j}(\mathbf{q})$ is the Fourier component of $\mathbf{j}(\mathbf{r})$ in Eq. (7) (Klein, 1987). The final selfconsistent solution satisfies $\nabla \cdot \mathbf{j}(\mathbf{r}) = 0$.

Using selfconsistent solutions, we calculate free energy, external field, and LDOS. In Eilenberger theory, free energy is given by

$$F = \int_{\text{unitcell}} \mathrm{d}\mathbf{r} \Big\{ \kappa^2 |\mathbf{B}(\mathbf{r}) - \mathbf{H}|^2 - \mu^2 |B(\mathbf{r})|^2 + |\Delta(\mathbf{r})|^2 (\ln T + 2T \sum_{0<\omega_n<\omega_{\text{cut}}} \omega_n^{-1}) - T \sum_{|\omega_n|<\omega_{\text{cut}}} \langle I(\mathbf{r},\mathbf{k},\omega_n) \rangle_{\mathbf{k}_\mathrm{F}} \Big\} \tag{13}$$

with

$$I(\mathbf{r},\mathbf{k},\omega_n) = \Delta\phi f^\dagger + \Delta^* \phi^* f + (g - \frac{\omega_n}{|\omega_n|}) \Big\{ \frac{1}{f} (\omega_n + \mathrm{i}\mu B + \mathbf{v} \cdot (\nabla + \mathrm{i}\mathbf{A})) f + \frac{1}{f^\dagger} (\omega_n + \mathrm{i}\mu B + \mathbf{v} \cdot (\nabla - \mathrm{i}\mathbf{A})) f^\dagger \Big\}. \tag{14}$$

Using Eqs. (5) and (6), we obtain

$$F = \int_{\text{unitcell}} \mathrm{d}\mathbf{r} \Big\{ \kappa^2 |\mathbf{B}(\mathbf{r}) - \mathbf{H}|^2 - \mu^2 |B(\mathbf{r})|^2 + T \sum_{|\omega_n|<\omega_{\text{cut}}} \mathrm{Re} \left\langle \frac{g-1}{g+1} (\Delta\phi f^\dagger + \Delta^*\phi^* f) \right\rangle_{\mathbf{k}_\mathrm{F}} \Big\}. \tag{15}$$

Using Doria-Gubernatis-Rainer scaling (Doria et al., 1990; Watanabe et al., 2005), we obtain the relation of $\bar{B}$ and the external field H as

$$H = \left(1 - \frac{\mu^2}{\kappa^2}\right) \left(\bar{B} + \left\langle (B(\mathbf{r}) - \bar{B})^2 \right\rangle_{\mathbf{r}} / \bar{B}\right) + \frac{T}{\kappa^2 \bar{B}} \left\langle \sum_{0<\omega_n} \left\langle \mu B(\mathbf{r}) \mathrm{Im}\,\{g\} + \frac{1}{2} \mathrm{Re} \left\{ \frac{(f^\dagger \Delta + f \Delta^*) g}{g+1} \right\} + \omega_n \mathrm{Re}\{g-1\} \right\rangle_{\mathbf{k}_\mathrm{F}} \right\rangle_{\mathbf{r}}, \tag{16}$$

where $\langle \cdots \rangle_{\mathbf{r}}$ indicates the spatial average. We consider the case when $\kappa = 89$ and low temperature $T/T_c = 0.1$. For two-dimensional (2D) Fermi surface, $\kappa = (7\zeta(3)/8)^{1/2} \kappa_{\text{GL}} \sim \kappa_{\text{GL}}$ (Miranović & Machida, 2003). Therefore we consider the case of typical type-II superconductors with large Ginzburg-Landau (GL) parameter. In these parameters, $|\bar{B} - H| < 10^{-4} B_0$.

When we calculate the electronic states, we solve Eq. (5) with $\mathrm{i}\omega_n \to E + \mathrm{i}\eta$. The LDOS is given by $N(\mathbf{r}, E) = N_\uparrow(\mathbf{r}, E) + N_\downarrow(\mathbf{r}, E)$, where

$$N_\sigma(\mathbf{r}, E) = N_0 \langle \mathrm{Re}\{ g(\omega_n + \mathrm{i}\sigma\mu B, \mathbf{k}_\mathrm{F}, \mathbf{r})|_{i\omega_n \to E + i\eta} \} \rangle_{\mathbf{k}_\mathrm{F}} \tag{17}$$

with $\sigma = 1$ (-1) for up (down) spin component. We typically use $\eta = 0.01$, which is small smearing effect of energy by scatterings. The DOS is obtained by the spatial average of the LDOS as $N(E) = N_\uparrow(E) + N_\downarrow(E) = \langle N(\mathbf{r}, E) \rangle_{\mathbf{r}}$.

3. Vortex states in superconductors with strong paramagnetic effect

In this section, we study the paramagnetic effect in vortex state without FFLO modulation. For simplicity, we consider fundamental case of isotropic Fermi surface, that is, 2D cylindrical Fermi surface with $\mathbf{k}_\mathrm{F} = k_\mathrm{F}(\cos\theta, \sin\theta)$ and Fermi velocity $\mathbf{v}_\mathrm{F} = v_{\mathrm{F}0}(\cos\theta, \sin\theta)$. Magnetic field is applied along the z direction. Even before the FFLO transition, the strong paramagnetic effect induces anomalous field dependence of some physical quantities by paramagnetic vortex core and paramagnetic pair-breaking. There are some theoretical approaches to the study of paramagnetic effect, such as by BdG theory (Takahashi et al., 2006), or by Landau level expansion in Eilenberger theory(Adachi et al., 2005). Here, we report results of quantitative estimate by selfconsistent Eilenberger theory given in previous section (Ichioka & Machida, 2007).

3.1 Field dependence of paramagnetic susceptibility and zero-energy DOS

First, we discuss the field dependence of zero-energy DOS $\gamma(H) = N(E = 0)/N_0$ and paramagnetic susceptibility $\chi(H) = \langle M_{\rm para}(\mathbf{r})\rangle_{\mathbf{r}}/M_0$, which are normalized by the normal state values. From low temperature specific heats C, we obtain $\gamma(H) \propto C/T$ experimentally. And $\chi(H)$ is observed by the Knight shift in NMR experiments, which measure the paramagnetic component via the hyperfine coupling between a nuclear spin and conduction electrons. As shown in Fig. 3, γ (dashed lines) and χ (solid lines) show almost the same behavior at low temperatures. First, we see the case of d-wave pairing with line nodes in Fig. 3(a). There $\gamma(H)$ and $\chi(H)$ describe $\sqrt{H}$-like recovery smoothly to the normal state value ($\gamma = \chi = 1$ at H_{c2}) in the case of negligible paramagnetic effect ($\mu = 0.02$). With increasing the paramagnetic parameter μ, H_{c2} is suppressed and the Volovik curve $\gamma(H) \propto \sqrt{H}$ gradually changes into curves with a convex curvature. For large μ, H_{c2} changes to first order phase transition. We note that at lower fields all curves exhibit a $\sqrt{H}$ behavior because the paramagnetic effect ($\propto H$) is not effective. Further increasing H, $\gamma(H)$ behaves quite differently. There we find a turning point field which separates a concave curve at lower H and a convex curve at higher H. H/H_{c2} at the inflection point increases as μ decreases. From these behaviors, we can estimate the strength of the paramagnetic effect, μ.

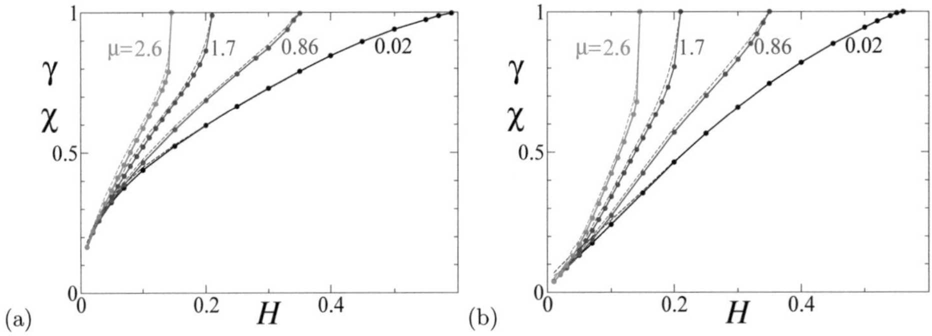

Fig. 3. The magnetic field dependence of paramagnetic susceptibility $\chi(H)$ (solid lines) and zero-energy DOS $\gamma(H)$ (dashed lines) at $T = 0.1T_c$ for various paramagnetic parameters $\mu = 0.02$, 0.86, 1.7, and 2.6 in the d-wave (a) and s-wave (b) pairing cases.

To examine effects of the pairing symmetry, we show $\gamma(H)$ and $\chi(H)$ also for s-wave pairing in Fig. 3(b). In the H-dependence of $\gamma(H)$ and $\chi(H)$, differences by the vortex lattice configuration of triangular or square are negligibly small. The difference in the H-dependences of Figs. 3(a) and 3(b) at low fields comes from the gap structure of the pairing function. In the full gap case of s-wave pairing, $\gamma(H)$ and $\chi(H)$ show H-linear-like behavior at low fields. With increasing the paramagnetic effect, H-linear behaviors gradually change into curves with a convex curvature. As seen in Figs. 3(a) and 3(b), paramagnetic effects appear similarly at high fields both for s-wave and d-wave pairings.

The H-dependence of $\gamma(H)$ for $H \parallel c$ and $H \parallel ab$ was used to identify the pairing symmetry and paramagnetic effect in URu_2Si_2 (Yano et al., 2008).

3.2 Field dependence of magnetization

We discuss the paramagnetic effect on the magnetization curves. The magnetization $M_{\text{total}} = \bar{B} - H$ includes both the diamagnetic and the paramagnetic contributions. In Fig. 4, magnetization curves are presented as a function of H for various μ at $T = 0.1T_c$ for s-wave and d-wave pairings. When the paramagnetic effect is negligible, we see typical magnetization curve of type-II superconductors. There, $|M_{\text{total}}|$ in s-wave pairing is larger, compared with that in d-wave pairing. Dashed lines in Fig. 4 indicate the magnetization in normal states, which shows linear increase of paramagnetic moments as a function of magnetic fields. When paramagnetic effect is strong for large μ, $M_{\text{total}}(H)$ exhibits a sharp rise near H_{c2} by the paramagnetic pair breaking effect, and that $M_{\text{total}}(H)$ has convex curvature at higher fields, instead of a conventional concave curvature. These behaviors are qualitatively seen in experimental data of $CeCoIn_5$ (Tayama et al., 2002).

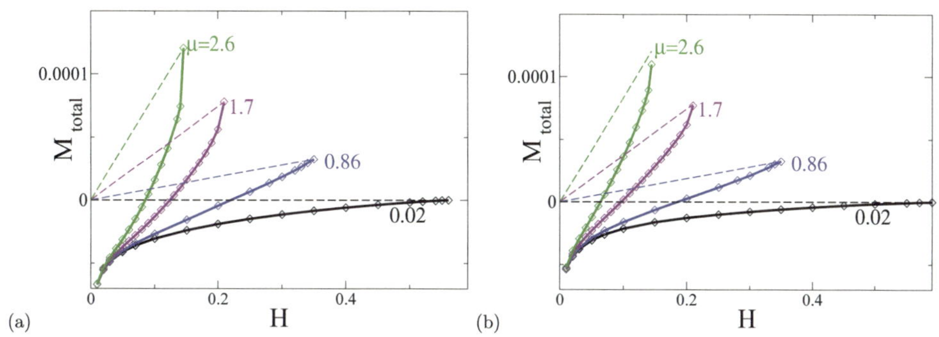

Fig. 4. Magnetization curve M_{total} as a function of H at $T/T_c = 0.1$ for $\mu = 0.02, 0.86, 1.7$ and 2.6 in s-wave (a) and d-wave (b) pairings. Dashed lines are normal state magnetization.

In Fig. 5(a), magnetization curves are presented as a function of H for various T at $\mu = 1.7$. With increasing T, the rapid increase of $M_{\text{total}}(H)$ near H_{c2} is smeared. In Fig. 5(b), M_{total} is plotted as a function of T^2 for various $\bar{B}$. We fit these curves as $M_{\text{total}}(T, H) = M_0 + \frac{1}{2}\beta(H)T^2 + O(T^3)$ at low T. The slope $\beta(H) = \lim_{T\to 0} \partial^2 M_{\text{total}}/\partial T^2$ decreases on raising H at lower fields. However, at higher fields approaching H_{c2}, the slope $\beta(H)$ sharply increases. Thus, as shown in Fig. 5(c), $\beta(H)$ as a function of H exhibits a minimum at intermediate H and rapid increase near H_{c2} by the paramagnetic effect when $\mu = 1.7$. This is contrasted with the case of negligible paramagnetic effect ($\mu = 0.02$), where $\beta(H)$ is a decreasing function of H until H_{c2}. The behavior of $\beta(H)$ is consistent with that of $\gamma(H)$, since there is a relation $\beta(H) \propto \partial\gamma(H)/\partial H$ obtained from a thermodynamic Maxwell's relation $\partial^2 M_{\text{total}}/\partial T^2 = \partial(C/T)/\partial B$ and $B \sim H$ (Adachi et al., 2005). In Fig. 3, we see that for $\mu = 1.7$ the slope of $\gamma(H)$ is decreasing function of H at low H, but changes to increasing function near H_{c2}. This behavior correctly reflects the H-dependence of $\beta(H)$.

3.3 Paramagnetic contribution on vortex core structure

In order to understand contributions of the paramagnetic effect on the vortex structure, we illustrate the local structures of the pair potential $|\Delta(\mathbf{r})|$, paramagnetic moment $M_{\text{para}}(\mathbf{r})$, and internal magnetic field $B(\mathbf{r})$ within a unit cell of the vortex lattice in Fig. 6. Since we assume d-wave pairing with the line node gap here, the vortex core structure is deformed to fourfold symmetric shape around a vortex core (Ichioka et al., 1999a;b; 1996). It is noted that the paramagnetic moment is enhanced exclusively around the vortex core, as shown in Fig. 6(b). Since the contribution of the paramagnetic vortex core is enhanced with increasing

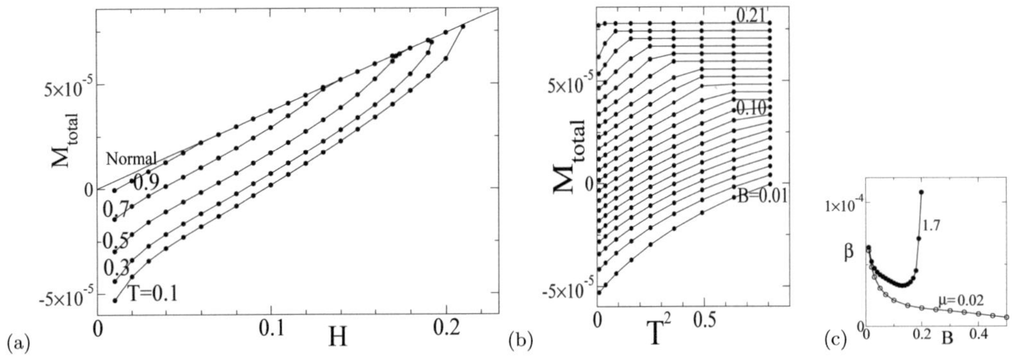

Fig. 5. (a) Magnetization curve $M_{\rm total}$ as a function of H for $\mu = 1.7$ at $T/T_{\rm c} = 0.1, 0.3, 0.5,$ $0.7, 0.9$ and 1.0 (normal state) in d-wave pairing. (b) $M_{\rm total}$ as a function of T^2 at $H = 0.01,$ $0.02, 0.03, \cdots, 0.21$. (c) H-dependence of factor $\beta(H)$ at $\mu = 0.02$ and 1.7.

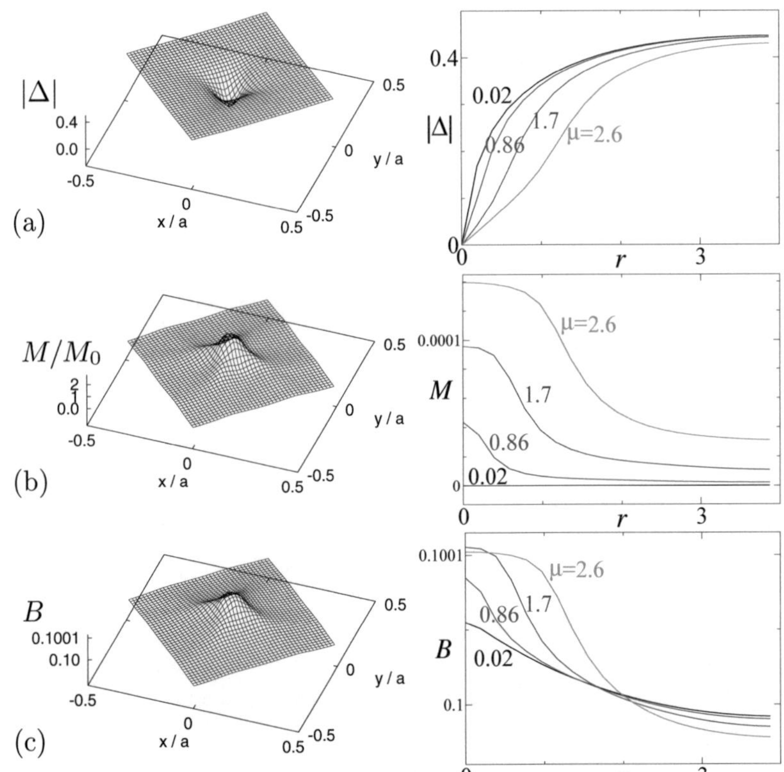

Fig. 6. Spatial structure of the pair potential (a), paramagnetic moment (b) and internal magnetic field (c) at $T = 0.1T_{\rm c}$ and $H \sim \bar{B} = 0.1B_0$, where $a = 11.2R_0$, in d-wave pairing. The left panels show $|\Delta(\mathbf{r})|$, $M_{\rm para}(\mathbf{r})$, and $B(\mathbf{r})$ within a unit cell of the square vortex lattice at $\mu = 1.7$. The right panels show the profiles along the trajectory r from the vortex center to a midpoint between nearest neighbor vortices. $\mu = 0.02, 0.86, 1.7,$ and 2.6.

μ, internal field $B(\mathbf{r})$ consisting of diamagnetic and paramagnetic contributions is further enhanced around the vortex core by the paramagnetic effect, as shown in Fig. 6(c). When μ is large, the pair potential $|\Delta(\mathbf{r})|$ is slightly suppressed around the paramagnetic vortex core, and the vortex core radius is enlarged, as shown in Fig. 6(a).

The enhancement of $M_{\rm para}(\mathbf{r})$ around vortex core is related to spatial structure of the LDOS $N_\sigma(\mathbf{r},E)$. As shown in Fig. 7(a), the LDOS spectrum shows *zero-energy peak* at the vortex center, but the spectrum is shifted to $E = \pm\mu H$ due to Zeeman shift. There is a relation between the LDOS spectrum and local paramagnetic moment, as

$$M_{\rm para}(\mathbf{r}) = -\mu_{\rm B}\int_{-\infty}^{0}\{N_\uparrow(E,\mathbf{r}) - N_\downarrow(E,\mathbf{r})\}{\rm d}E. \tag{18}$$

In Fig. 7(a), the peak states at $E > 0$ is empty for $N_\uparrow(E,\mathbf{r})$, and the peak at $E < 0$ is occupied for $N_\downarrow(E,\mathbf{r})$. Therefore, because of Zeeman shift of the zero-energy peak at the vortex core, large $M_{\rm para}(\mathbf{r})$ appears due to the local imbalance of up- and down-spin occupation around the vortex core. As shown in Figs. 7(b) and 7(c), moving from the vortex center to outside, the peak of the spectrum is split into two peaks, which are shifted to higher and lower energies, respectively. When one of split peaks crosses $E = 0$, the imbalance of up- and down-spin occupation is decreased. Thus, $M_{\rm para}(\mathbf{r})$ is suppressed outside of vortex cores. This corresponds to the behavior of Knight shift, i.e., the paramagnetic moment is suppressed in uniform states of spin-singlet pairing superconductors by the formation of Cooper pair between spin-up and spin-down electrons.

In Figs. 7(d) and 7(e), we present the spectrum of spatially-averaged DOS. In the DOS spectrum, peaks of the LDOS are smeared by the spatial average. Because of the flat spectrum at low energies, paramagnetic susceptibility $\chi(H)$ shows almost the same H-behavior as the zero-energy DOS $\gamma(H) \sim N(E=0)$ even for large μ, as shown in Fig. 3, while $\chi(H)$ counts the DOS contribution in the energy range $|E| < \mu H$, i.e., from Eq. (18),

$$\chi(H) \sim \int_0^{\mu H} N_\uparrow(E){\rm d}E/\mu H. \tag{19}$$

3.4 Field dependence of flux line lattice form factor

One of the best ways to directly see the accumulation of the paramagnetic moment around the vortex core is to observe the Bragg scattering intensity of the FLL in SANS experiment. The intensity of the (h,k)-diffraction peak is given by $I_{h,k} = |F_{h,k}|^2/|\mathbf{q}_{h,k}|$ with the wave vector $\mathbf{q}_{h,k} = h\mathbf{q}_1 + k\mathbf{q}_2$, $\mathbf{q}_1 = (2\pi/a, -\pi/a_y, 0)$ and $\mathbf{q}_2 = (2\pi/a, \pi/a_y, 0)$. The Fourier component $F_{h,k}$ is given by $B(\mathbf{r}) = \sum_{h,k} F_{h,k}\exp({\rm i}\mathbf{q}_{h,k}\cdot\mathbf{r})$. In the SANS for FLL observation, the intensity of the main peak at $(h,k) = (1,0)$ probes the magnetic field contrast between the vortex cores and the surrounding.

The field dependence of $|F_{1,0}|^2$ in our calculations is shown in Fig. 8(a). In the case of negligible paramagnetic effect ($\mu = 0.02$), $|F_{1,0}|^2$ decreases exponentially as a function of H. This exponential decay is typical behavior of conventional superconductors. With increasing paramagnetic effect, however, the decreasing slope of $|F_{1,0}|^2$ becomes gradual, and changes to increasing functions of H at lower fields in strong paramagnetic case ($\mu = 2.6$).

The reason of anomalous enhancement of $|F_{1,0}|$ at high fields is because $|F_{1,0}|$ reflects the enhanced internal field around the vortex core, shown in Fig. 6(c), by the induced paramagnetic moment at the core. We present H-dependence of $|F_{1,0}|$ with the paramagnetic contribution $|M_{1,0}|$ in Fig. 8(b). Fourier component $M_{1,0}$ is calculated from paramagnetic

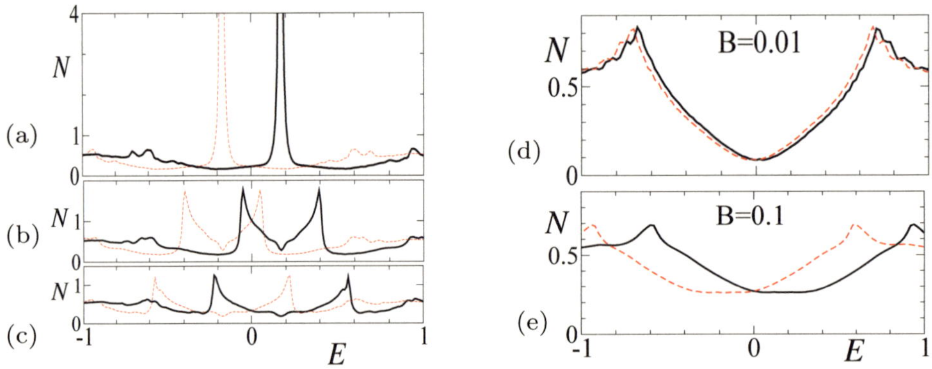

Fig. 7. Local density of states at $r/R_0 = 0$ (a), 0.8 (b) and 1.6 (c) from the vortex center towards the nearest neighbor vortex direction in *d*-wave pairing. Solid lines show $N_\uparrow(\mathbf{r}, E)/N_0$ for up-spin electrons, and dashed lines show $N_\downarrow(\mathbf{r}, E)/N_0$ at $H = 0.1B_0$. $\mu = 1.7$ and $T = 0.1T_c$. Spatial-averaged DOS at $H/B_0 = 0.01$ (d) and 0.1 (e) in *d*-wave pairing. Solid lines show $N_\uparrow(E)/N_0$ for up-spin electrons, and dashed lines show $N_\downarrow(E)/N_0$.

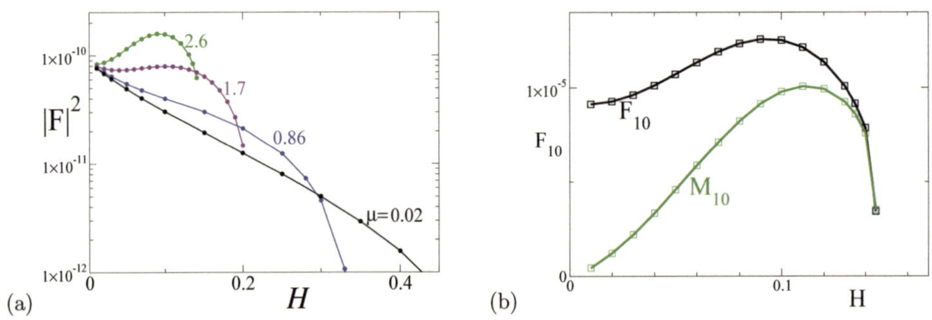

Fig. 8. Field dependence of FLL form factor $F_{1,0}$ for $\mu = 0.02$, 0.86, 1.7, and 2.6 at $T = 0.1T_c$ in *d*-wave pairing. (a) $|F_{1,0}|^2$ is plotted as a function of H. The vertical axis is in logarithmic scale. (b). Field dependence of $|F_{1,0}|$ and the paramagnetic contribution $|M_{1,0}|$ for $\mu = 2.6$. The vertical axis is in linear scale.

moment $M_{\text{para}}(\mathbf{r})$. From Fig. 8(b), we see that the increasing behavior of $|F_{1,0}|$ is due to the paramagnetic contribution $M_{1,0}$ proportional to μH. In Fig. 9, we present how profiles of $M_{\text{para}}(\mathbf{r})$ and $B(\mathbf{r})$ change, depending on magnetic fields. The form factors $|F_{1,0}|$ and $|M_{1,0}|$ reflect the contrast of the variable range in the figures. Increasing magnetic field at low fields ($H = 0.02, 0.06$), $M_{\text{para}}(\mathbf{r})$ is enhanced at vortex core. Reflecting this, $B(\mathbf{r})$ is also enhanced at the core, and the form factor $|F_{1,0}|$ increases as a function of a magnetic field. At higher fields ($H = 0.10, 0.12, 0.14$), inter-vortex distance becomes short. Because of overlap of the regions around vortex core with those of neighbor vortices, the contrasts of enhanced $M_{\text{para}}(\mathbf{r})$ and $B(\mathbf{r})$ around vortex core are smeared. Therefore, form factors $|F_{1,0}|$ and $|M_{1,0}|$ decrease at high fields near H_{c2} in Fig. 8(b).

The SANS experiment in $CeCoIn_5$ for $H \parallel c$ reported that $|F_{1,0}|^2$ increases until near H_{c2} instead of exponential decay (Bianchi et al., 2008; DeBeer-Schmitt et al., 2006; White et al., 2010). The anomalous increasing H-dependence of the SANS intensity in $CeCoIn_5$ can be explained qualitatively by the strong paramagnetic effect, as shown by our calculation. The detailed comparison with the experimental data will be discussed later. Anomalous

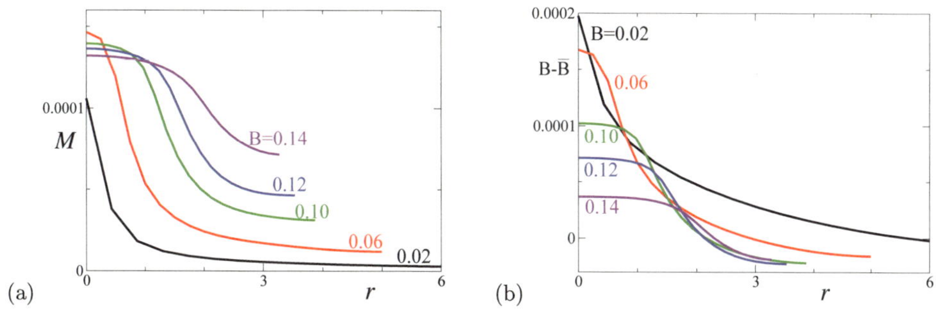

Fig. 9. Profile of paramagnetic moment $M_{\text{para}}(\mathbf{r})$ (a) and internal field $B(\mathbf{r}) - \bar{B}$ (b) as a function of radius r until a midpoint between vortices along nearest neighbor vortex directions. $\mu = 2.6$ and $H = 0.02, 0.06, 0.10, 0.12$ and 0.14.

enhancement of FLL form factor was also observed in $TmNi_2B_2C$, and explained by effective strong paramagnetic effect (DeBeer-Schmitt et al., 2007).

3.5 Comparison with experimental data in $CeCoIn_5$

Here, we discuss anomalous field dependence of low T specific heat, magnetization curve, and FFL form factor in $CeCoIn_5$, based on the comparison with theoretical estimates of strong paramagnetic effect by Eilenberger theory. In Fig. 10(a), we present H-dependence of zero-energy DOS $N(E=0)$ and low-T specific heat (Ikeda et al., 2001). Both H-dependences show rapid increase at higher H. However, we see quantitative differences between theory (line A) and experimental data (circles). Compared to the theoretical estimates, C/T by experiments is smaller at low H and increase more rapidly at higher H. In order to quantitatively reproduce the H-dependence of C/T, we phenomenologically introduce factor $N_0(H)$ coming from the H-dependence of normal state DOS. So far, N_0 was assumed to be a constant in theoretical calculation. Thus, in calculation of Fermi surface average, we modify $\langle \cdots \rangle_{\mathbf{k}_\text{F}} \to \langle \cdots \rangle_{\mathbf{k}_\text{F}} N_0(H)/N_0(H_{c2})$. As shown in Fig. 10(a), the H-dependence of C/T can be reproduced, if we set $N_0(H)/N_0(H_{c2}) = 1 - 0.53\{\tanh 4(1 - H/H_{c2})\}^3$. This expression of $N_0(H)$ is phenomenological one to reproduce the experimental behavior, without microscopic theoretical consideration. This H-dependence of $N_0(H)$ indicates that normal states DOS is enhanced near H_{c2}, and may be related to the effective mass enhancement near QCP (Bianchi, Movshovich, Vekhter, Pagliuso & Sarrao, 2003; Paglione et al., 2003), which is suggested to exist at $H_{c2}(T=0)$ in $CeCoIn_5$.

Theoretical and experimental (Tayama et al., 2002) magnetization curve is presented in Fig. 10(b). There we see rapid increase at high fields and jump at H_{c2} by strong paramagnetic effects. The differences between experimental data (average of magnetization curves for increasing and decreasing H) and theoretical estimate with constant N_0 (line A) are improved by considering the H-dependence of $N_0(H)$ (line B). There, by $N_0(H)$, slope of $M_{\text{total}}(H)$ becomes similar to that of experimental curve.

The H-dependence of FLL form factors using $N_0(H)$ is presented in Fig. 11. There, $|F_{1,0}|^2$ shows further increases until higher H. This sharp peak at high fields resembles to the anomalous increasing behavior observed by SANS experiment in $CeCoIn_5$ (Bianchi et al., 2008; DeBeer-Schmitt et al., 2006). For higher T, the peak is smeared and the peak position is shifted to lower fields. This T-dependence is consistent to those in experimental observation in $CeCoIn_5$ (White et al., 2010).

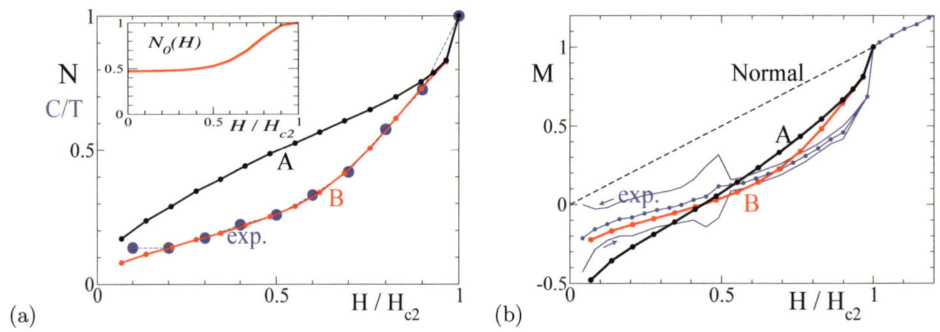

Fig. 10. (a) H-dependence of theoretical estimate (lines) for $N(E = 0)$, and experimental data (Ikeda et al., 2001) of low-T specific heat C/T (circles). Line A is an original estimate for constant N_0. For line B, we assume $N_0(H)/N_0(H_{c2}) = 1 - 0.53\{\tanh 4(1 - H/H_{c2})\}^3$. Inset shows $N_0(H)/N_0(H_{c2})$ as a function of H/H_{c2}. (b) Magnetization curve $M_{\rm total}(H)$ for constant N_0 (line A), and for $N_0(H)$ (line B). Experimental magnetization curves for increasing H(line with right arrow), decreasing H (line with left arrow), and their average (line with dots) are presented (Tayama et al., 2002). We compare the average line with lines A and B.

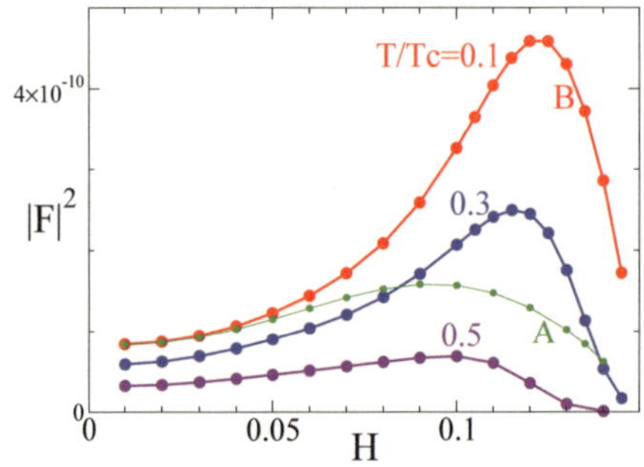

Fig. 11. H-dependence of FLL form factor $|F_{10}|^2$ at $T/T_c = 0.1$ (line B), 0.3, and 0.5 for $N_0(H)$. The line A is for constant N_0 and $T = 0.1T_c$. $\mu = 2.6$.

The above phenomenological discussion by $N_0(H)$ indicates that anomalous H-dependences observed in $CeCoIn_5$ is qualitatively reproduced by theoretical estimate considering strong paramagnetic effect, but they still show systematic quantitative deviations from theoretical estimate. These indicate that we need to consider additional effect, such as effective mass enhancement near QCP, in addition to strong paramagnetic effect, in order to understand anomalous H-dependence in $CeCoIn_5$.

4. Fulde-Ferrell-Larkin-Ovchinnikov (FFLO) vortex state

The FFLO state (Fulde & Ferrell, 1964; Larkin & Ovchinnikov, 1965) is an exotic superconducting state expected to appear at low temperatures and high fields, when the paramagnetic effect due to the Zeeman shift is significant. In the FFLO state, since the Fermi surfaces for up-spin and down-spin electron bands are split by the Zeeman shift, Cooper pairs of up- and down-spins acquire non-zero momentum for the center of mass coordinate of the Cooper pair, inducing the spatial modulation of the pair potential. The possible FFLO state is widely discussed in various research fields, ranging from superconductors in condensed matters, neutral Fermion superfluids in an atomic cloud (Machida et al., 2006; Partridge et al., 2006; Zwierlein et al., 2006), to color superconductivity in high energy physics (Casalbuoni & Nardulli, 2004).

Experimentally, the FFLO state is suggested in a high field phase of a quasi-two dimensional (Q2D) heavy Fermion superconductor $CeCoIn_5$ for $H \parallel ab$ and $H \parallel c$ (Bianchi, Movshovich, Capan, Pagliuso & Sarrao, 2003; Radovan et al., 2003), as reviewed by Matsuda & Shimahara (2007). There, it is supposed that nodal planes of the pair potential run perpendicular to the vortex lines. For $H \parallel ab$, since spin density wave (SDW) appears in the high-field phase (Kenzelmann et al., 2010; 2008; Koutroulakis et al., 2010; Young et al., 2007), we are interested in the relation of FFLO and SDW.

In theoretical studies, many calculations for the FFLO states have been done by neglecting vortex structure. However, we have to consider the vortex structure in addition to the FFLO modulation, because the FFLO state appears at high fields in the mixed states. Among the FFLO states, there are two possible spatial modulation of the pair potential Δ. One is the Fulde-Ferrell (FF) state (Fulde & Ferrell, 1964) with phase modulation such as $\Delta \propto e^{iqz}$, where q is the modulation vector of the FFLO states. The other is the Larkin-Ovchinnikov (LO) state (Larkin & Ovchinnikov, 1965) with the amplitude modulation such as $\Delta \propto \sin qz$, where the pair potential shows periodic sign change, and $\Delta = 0$ at the nodal planes. We discuss the case of the LO states in this section, since some experimental (Matsuda & Shimahara, 2007) and theoretical (Houzet & Buzdin, 2001; Ikeda & Adachi, 2004) works support the LO state for the FFLO states in $CeCoIn_5$. In the FFLO vortex state, it is instructive to clarify the role of the FFLO nodal plane in order to obtain clear evidence of the FFLO states among the experimental data.

When we consider vortex structure in the LO state, there are two possible choices of the configuration for the vortex lines and the FFLO modulation. That is, the modulation vector of the FFLO state is parallel (Tachiki et al., 1996) or perpendicular (Klein et al., 2000; Shimahara, 1994) to the applied magnetic field. In our study, 3D structure of the former case is investigated by the selfconsistent Eilenberger theory. We calculate the spatial structures of pair potentials, paramagnetic moments, internal magnetic fields and electronic states in the vortex lattice state with the FFLO modulation. In our study, fully 3D structures of the vortex and the FFLO modulation are determined by the selfconsistent calculation with local electronic states. Since we can consider the system of vortex lattice and periodic FFLO modulation by the periodic boundary condition, we can discuss the overlaps between tails of the neighbor vortex cores or FFLO nodal planes. These calculations for the periodic systems make us possible to estimate the resonance line shapes in the NMR experiments and FLL form factors in SANS experiments.

On the other hand, the vortex and FFLO nodal plane structures in the FFLO state were calculated by the BdG theory for a single vortex in a superconductor under a cylindrical symmetry situation (Mizushima et al., 2005b). This study clarifies that the topological structure of the pair potential plays important roles to determine the electronic structures in the FFLO vortex state. The pair potential has 2π-phase winding around the vortex line, and π-phase shift at the nodal plane of the FFLO modulation. These topologies of the pair potential structure affect the distribution of paramagnetic moment and low energy electronic states inside the superconducting gap. For example, the paramagnetic moment is enhanced at the vortex core and the FFLO nodal plane. These structures are related to the bound states due to the π-phase shift of the pair potential.

In this section, we report our study of FFLO vortex states for a fundamental case of s-wave pairing and 3D spherical Fermi surface, where $\mathbf{k}_{\rm F} = k_{\rm F}(\sin\theta\cos\phi, \sin\theta\sin\phi, \cos\theta)$ and Fermi velocity $\mathbf{v}_{\rm F} = v_{\rm F0}(\sin\theta\cos\phi, \sin\theta\sin\phi, \cos\theta)$. The calculations of FFLO vortex states for Q2D Fermi surface with rippled cylinder-shape and $H \parallel ab$ both for s-wave and d-wave pairings were reported elsewhere (Ichioka et al., 2007). Main characteristic properties of FFLO vortex state do not seriously depend on the pairing symmetry.

4.1 Spatial structure of FFLO vortex states

In the left panels of Fig. 12, we show the spatial structure of the FFLO vortex state within a unit cell in the slice of the xz plane, i.e., the hatched region shown in Fig. 2(a). Right panels of Fig. 12 are for profiles of the spatial structure along the path UNCVU shown in Fig. 2(a). The point C ($x = y = z = 0$) is the intersection point of a vortex and a nodal plane. The point V ($x = y = 0, z = L/4$) is at the vortex center and far from the FFLO nodal plane. The point N ($x = a/2, y = z = 0$) is at the FFLO nodal plane and outside of the vortex. The point U ($x = a/2, y = 0, z = L/4$) is far from both the vortex and the FFLO nodal plane.

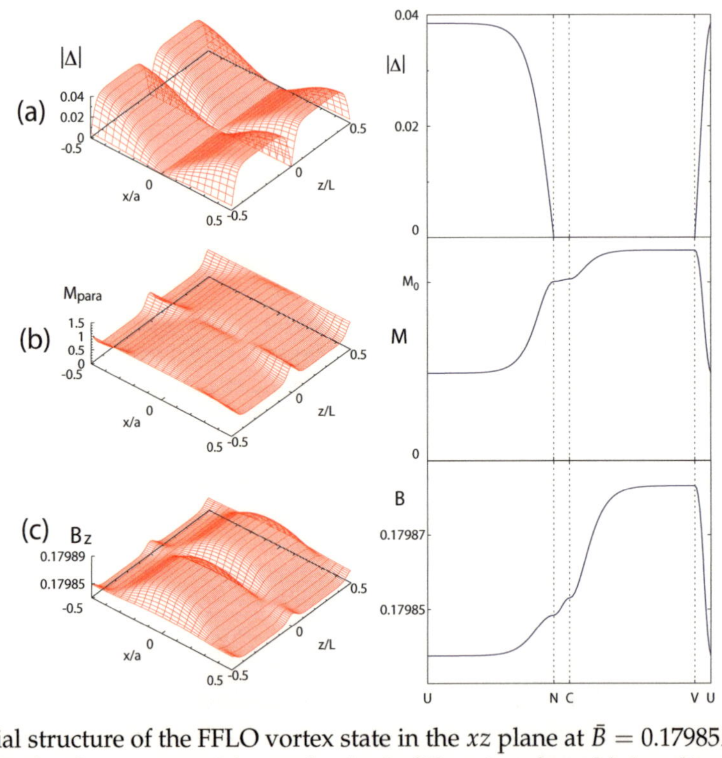

Fig. 12. Spatial structure of the FFLO vortex state in the xz plane at $\bar{B} = 0.17985B_0$, $T = 0.2T_c$ and $L = 100R_0$ for the s-wave pairing and spherical Fermi surface. (a) Amplitude of the pair potential $|\Delta(\mathbf{r})|$. (b) Paramagnetic moment $M_{\rm para}(\mathbf{r})$. (c) Internal magnetic field $B_z(\mathbf{r})$. The left panels show the spatial variation within a unit cell, i.e., hatched region in Fig. 2(a). The right panels present the profiles along the path UNCVU shown in Fig. 2(a).

In the left panel of Fig. 12(a), we show the amplitude of the order parameter, $|\Delta(\mathbf{r})|$, which is suppressed near the vortex center at $x = y = 0$ and the FFLO nodal plane at $z = 0, \pm 0.5L$. Far from the FFLO nodal plane such as $z = 0.25L$ [along path VU], $|\Delta(\mathbf{r})|$ shows a typical profile of the conventional vortex. When we cross the vortex line or the FFLO nodal plane, the sign of $\Delta(\mathbf{r})$ changes due to the π-phase shift of the pair potential as schematically shown in Fig. 2(a). In the profile of $|\Delta(\mathbf{r})|$ presented in the right panels of Fig. 12(a), $|\Delta(\mathbf{r})| = 0$ along the FFLO nodal plane NC and along the vortex line CV.

Correspondingly, paramagnetic moment $M_{\text{para}}(\mathbf{r})/M_0$ is presented in Fig. 12(b). The paramagnetic moment is suppressed, as Knight shift, at uniform-Δ region in the spin-singlet pairing superconductors. In the figures, we see that $M_{\text{para}}(\mathbf{r})$ is suppressed outside of vortex core and far from the FFLO nodal plane, as expected. However, $M_{\text{para}}(\mathbf{r})$ is enhanced at the vortex core or at the FFLO nodal plane. The reason for these structures of $M_{\text{para}}(\mathbf{r})$ is discussed later in connection with the LDOS. At the FFLO nodal plane $M_{\text{para}}(\mathbf{r}) \sim M_0$ [path NC in Fig. 12(b)]. Along the vortex line, $M_{\text{para}}(\mathbf{r})$ is enhanced more than M_0 far from the FFLO nodal planes [position V in Fig. 12(b)].

Figure 12(c) presents the z-component of the internal field, $B_z(\mathbf{r})$. Due to the contribution of the enhanced $M_{\text{para}}(\mathbf{r})$, $B_z(\mathbf{r})$ is enhanced at the FFLO nodal plane even outside of the vortex. A part of the contributions by $M_{\text{para}}(\mathbf{r})$ is compensated by the diamagnetic contribution, because the average flux density per unit cell of the vortex lattice in the xy plane should conserve along the magnetic field direction. Therefore, due to the conservation, the enhancement of $B_z(\mathbf{r})$ at the FFLO nodal plane [path NC in Fig. 12(c)] is smaller, compared with the enhancement of $M_{\text{para}}(\mathbf{r})$ at the FFLO nodal plane [path NC in Fig. 12(b)]. While $B_z(\mathbf{r})$ is largely enhanced than $\bar{B}$ at the vortex core far from the FFLO nodal plane [position V in Fig. 12(c)], $B_z(\mathbf{r})$ is not largely enhanced at the vortex core in the FFLO nodal plane [position C]. Therefore $B_z(\mathbf{r}) \sim \bar{B}$ at the FFLO nodal plane [path NC].

To estimate magnetic field range where the FFLO vortex state is stable, and the FFLO wave number $q = 2\pi/L$, we present the field dependence of the free energy F for some L in Fig. 13(a). At $H < 0.9987H_{c2}$ conventional Abrikosov vortex state with $q = 0$ is stable, but $H > 0.9987H_{c2}$ FFLO vortex state with finite q becomes stable. This is an estimate in the presence of vortices in addition to FFLO modulation. At higher H, q increases for stable FFLO state, as shown in Fig. 13(b), which indicates that the FFLO period L becomes shorter at higher H. In Figs. 13(c) and 13(d), respectively, we present profiles of $\Delta(\mathbf{r})$ and $M_{\text{para}}(\mathbf{r})$ along the z-direction at a midpoint between vortices, i.e., along a line thorough UN in Fig. 2(a). When L is longer at lower H, the FFLO vortex states have wide region of constant $|\Delta(\mathbf{r})|$ and $M_{\text{para}}(\mathbf{r})$. They change only near the FFLO nodal plane, where $\Delta(\mathbf{r})$ has sign change and $M_{\text{para}}(\mathbf{r})$ locally accumulates as in soliton structure. On the other hand, when L becomes shorter at higher H, the region near FFLO nodal plane overlaps with that of neighbor nodal planes. Thus, both $|\Delta(\mathbf{r})|$ and $M_{\text{para}}(\mathbf{r})$ become spatial structure of sinusoidal wave along z-directions.

Due to the presence of FFLO vortex states at high fields, instead of conventional Abrikosov vortex state, H_{c2} to normal state [$F = 0$ in Fig. 13(a)] is enhanced. We note that the FFLO vortex state is stable only in narrow H range near H_{c2} at $T = 0.2T_{c2}$ and $\mu = 2$ for spherical Fermi surface. At lower T or for larger paramagnetic parameter μ, the FFLO vortex states becomes stable in wider H-range.

4.2 Electronic structure in the FFLO vortex state

The LDOS spectrum for up- and down-spin electrons are presented at some positions in Fig. 14. In the quasiclassical theory, $N_\sigma(E,\mathbf{r})$ are symmetric by $E \leftrightarrow -E$ in the absence of the paramagnetic effect ($\mu = 0$). In the presence of the paramagnetic effect, the LDOS spectrum for up- (down-) spin electrons is shifted to positive (negative) energy by μH due to the Zeeman shift. In this case, we have a relation $N_\uparrow(E,\mathbf{r}) = N_\downarrow(-E,\mathbf{r})$ within the quasiclassical theory.

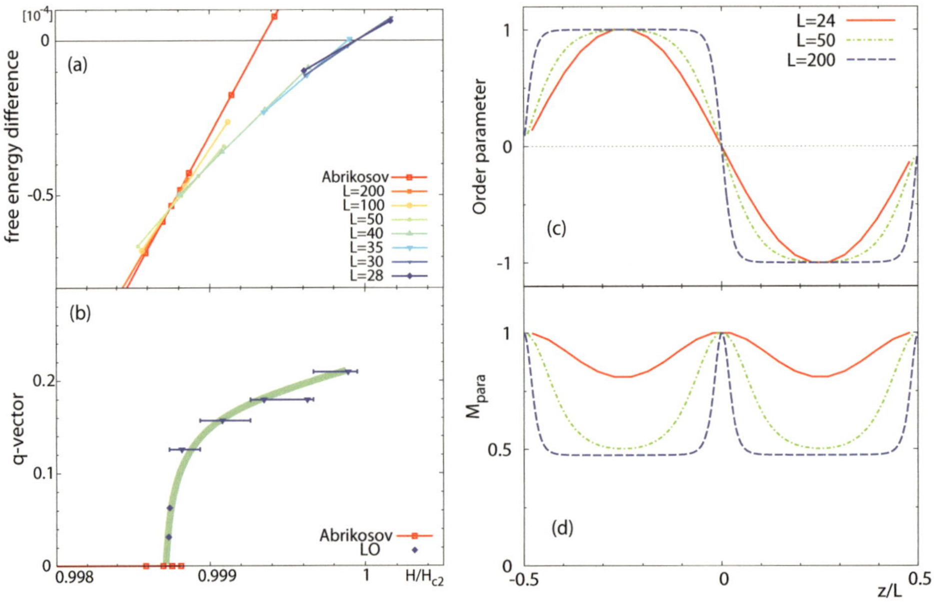

Fig. 13. (a) Field dependence of free energy F for conventional Abrikosov vortex state ($q = 0$) and FFLO vortex state for some $q = 2\pi/L(\neq 0)$. (b) Field dependence of FFLO wave number q estimated from (a). (c) Profile of pair potential $\Delta(\mathbf{r})$ along z-direction at midpoints between vortices for L =24, 50, 200. Normalized value $\Delta(\mathbf{r})/\Delta(z = -0.25L)$ is presented. (d) The same as (c) but for $M_{\mathrm{para}}(\mathbf{r})$. We present normalized value $M_{\mathrm{para}}(\mathbf{r})/M_{\mathrm{para}}(z = 0)$. $\mu = 2$ and $T = 0.2T_{\mathrm{c}}$.

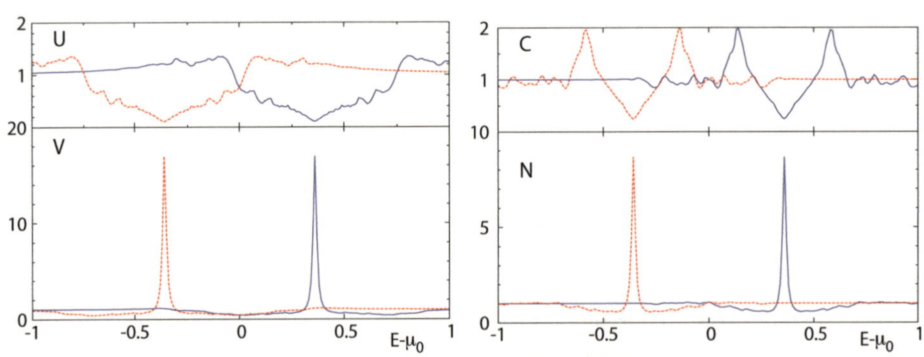

Fig. 14. Spectrum of the LDOS for up-spin electrons $N_{\uparrow}(E,\mathbf{r})/(0.5N_0)$ (solid lines) and for down-spin electrons $N_{\downarrow}(E,\mathbf{r})/(0.5N_0)$ (dashed lines) at positions U, V, N, and C, whose locations are shown in Fig. 2(a). $T = 0.2T_{\mathrm{c}}$, $\bar{B} = 0.17985B_0$, and $L = 100R_0$ in the s-wave pairing.

Far from the FFLO nodal plane and outside of vortex, as shown in the spectrum at position U in Fig. 14, we see Zeeman shift of full-gap structure in s-wave superconductors. There, small LDOS also appears at low energies inside the gap due to the low energy excitations extending from the vortex cores and the FFLO nodal planes at finite magnetic fields. Since the LDOS are occupied at $E < 0$, and empty at $E > 0$, there is a relation of Eq. (18) between the

LDOS spectrum and local paramagnetic moment. Because of superconducting gap structure, the LDOS within superconducting gap is suppressed. Thus, difference of occupation number between up- and down-spin electrons is small, since the LDOS at $E < 0$ are occupied similarly in $N_\uparrow(E,\mathbf{r})$ and $N_\downarrow(E,\mathbf{r})$, except for small LDOS within the gap. This is the reason why $M_{\rm para}(\mathbf{r})$ is suppressed at the position U. Small but finite $M_{\rm para}(\mathbf{r})$ comes from the small LDOS weight of low energy states inside the gap at U in FFLO vortex states.

In the LDOS spectra at the position V on the vortex center and at the position N on the FFLO nodal plane presented in Fig. 14, $N_\uparrow(E,\mathbf{r})$ and $N_\downarrow(E,\mathbf{r})$, respectively, have a sharp peak at $E = \mu_+$ and $E = \mu_-$, with $\mu_\pm \equiv \mu_0 \pm \mu H$. These peaks are related to the topological structure of the pair potential, as schematically shown in Fig. 2. Since a vortex has phase winding 2π, along the trajectory through the vortex center, $\Delta(\mathbf{r})$ changes the sign by the π-phase shift across the vortex center. Also at the trajectory through the FFLO nodal plane, $\Delta(\mathbf{r})$ changes the sign across the nodal plane. The bound states appear as *zero-energy peak*, when the pair potential has the π-phase shift. This peak is shifted to $E = \mu_+$ or $E = \mu_-$ due to the Zeeman effect. Since the peak of the LDOS spectrum for up-spin electrons is an empty state ($E > 0$) and the peak of the LDOS for down-spin electrons is an occupied state ($E < 0$), $M_{\rm para}(\mathbf{r})$ becomes large at these positions, from the relation in Eq. (18).

On the other hand, along the trajectory through the intersection point of a vortex and a nodal plane, $\Delta(\mathbf{r})$ does not change the sign, because the phase shift is 2π by summing π due to vortex and π due to the nodal plane, as schematically shown in Fig. 2. Thus, the sharp peaks do not appear at $E = \mu_\pm$ as seen from the LDOS spectrum at the position C in Fig. 14. Instead, $N_\uparrow(E,\mathbf{r})$ has two broad peaks at finite energies shifted upper or lower from μ_+. In this situation, $M_{\rm para}(\mathbf{r})$ is still large at position C, as in positions V and N, since the LDOS in both peaks are empty ($E > 0$) in $N_\uparrow(E,\mathbf{r})$, and occupied ($E < 0$) in $N_\downarrow(E,\mathbf{r})$.

4.3 NMR spectrum in FFLO vortex states

In the NMR experiment, resonance frequency spectrum of the nuclear spin resonance is determined by the internal magnetic field and the hyperfine coupling to the spin of the conduction electrons. Therefore, in a simple consideration, the effective field for the nuclear spin is given by $B_{\rm eff}(\mathbf{r}) = B_z(\mathbf{r}) + A_{\rm hf} M_{\rm para}(\mathbf{r})$, where $A_{\rm hf}$ is a hyperfine coupling constant depending on species of the nuclear spins. The resonance line shape of NMR is given by

$$P(\omega) = \int \delta(\omega - B_{\rm eff}(\mathbf{r}))\mathrm{d}\mathbf{r}, \tag{20}$$

i.e., the intensity at each resonance frequency ω comes from the volume satisfying $\omega = B_{\rm eff}(\mathbf{r})$ in a unit cell. When the contribution of the hyperfine coupling is dominant, the NMR signal selectively detects $M_{\rm para}(\mathbf{r})$. This is the experiment observing the Knight shift in superconductors. As the NMR spectrum of the Knight shift, we calculate the distribution function $P(M) = \int \delta(M - M_{\rm para}(\mathbf{r}))\mathrm{d}\mathbf{r}$ from the spatial structure of the paramagnetic moment $M_{\rm para}(\mathbf{r})$ shown in Fig. 12(b). On the other hand, in the case of negligible hyperfine coupling, the NMR signal is determined by the internal magnetic field distribution. This resonance line shape is called *Redfield pattern* of the vortex lattice. The distribution function $P(B) = \int \delta(B - B_z(\mathbf{r}))\mathrm{d}\mathbf{r}$ is calculated from the internal field $B_z(\mathbf{r})$.

First we discuss the line shape of the distribution function $P(M)$, shown in Fig. 15(a). The spectrum of $P(M)$ in the conventional vortex state without FFLO modulation is shown by the lowest line in Fig. 15(a). There, the peak of $P(M)$ comes from the signal from the outside of the vortex core. Shift of the peak position from M_0 gives Knight shift in superconductors. The spectrum of $P(M)$ has a tail toward larger M by the vortex core contribution of large $M_{\mathrm{para}}(\mathbf{r})$. The vortex core contribution is a one-dimensional (1D) structure, their volume contribution is small in the spectrum, compared with the peak intensity due to the large volume contribution from outside of the vortex core. After the FFLO transition, the line shape $P(M)$ becomes double peak structure in the FFLO vortex states, as presented by upper lines in Fig. 15(a). The height of the main peak decreases, and there appears a new peak coming from the FFLO nodal plane near $M_{\mathrm{para}} \sim M_0$. The contribution from 2D structure of the FFLO nodal plane appears in $P(M)$ more clearly than that of the 1D structure of the vortex line. When the period L becomes shorter at higher H, new peak at M_0 is enhanced, because the relative volume ratio of region near FFLO nodal plane becomes larger.

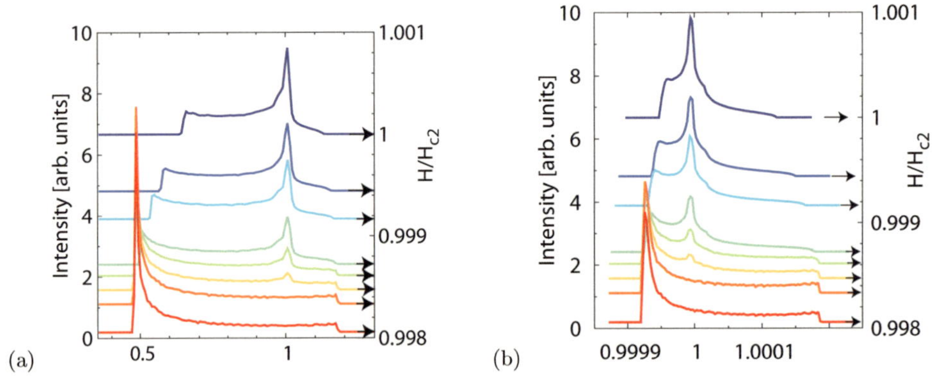

Fig. 15. (a) Distribution function of the paramagnetic moment. We show $P(M)$ as a function of M_{para}/M_0. (b) Distribution function of the internal magnetic field. We show $P(B)$ as a function of $B_z/\overline{B}$. $T = 0.2T_c$ and $\mu = 2$. Right-side axis pointed by arrows from each line indicates applied field H/H_{c2} for each NMR spectrum. Lowest line is for conventional Abrikosov vortex state of $q = 0$. Other upper lines are for FFLO vortex states. The heights of $P(M)$ and $P(B)$ are scaled so that $\int P(M)\mathrm{d}M = \int P(B)\mathrm{d}B = 1$.

Second, we discuss the distribution function $P(B)$ of the internal magnetic field, presented in Fig. 15(b). There, in the absence of the FFLO modulation (the lowest line), the Redfield pattern $P(B)$ has sharp peak corresponding to saddle points of the internal field distribution. The tail to higher B comes from the vortex core region of larger $B_z(\mathbf{r})$. In the presence of the FFLO modulation (other upper lines), the height of the original peak is decreased, and a new peak appears at $B \sim \bar{B}$ as the contribution of the FFLO nodal plane. In the line shape of $P(B)$, new peak by FFLO nodal plane is located near original saddle-point peak, compared with the line shape of $P(M)$. When the period L becomes shorter at higher H, new peak at $\overline{B}$ is enhanced.
The experimental observation of the NMR resonance line shape is a method to identify the FFLO vortex state in high-field phase of $CeCoIn_5$ (Kakuyanagi et al., 2005; Kumagai et al., 2006; 2011). For $H \parallel c$, the NMR spectrum shows the double peak structure in the FFLO phase, appearing new peak in addition to the main peak in the vortex state. For $H \parallel ab$, we see double peak structure in NMR spectrum, but it reflects magnetic moments of SDW state (Koutroulakis et al., 2010; Young et al., 2007). The SDW structure in high field phase was observed also by neutron scattering (Kenzelmann et al., 2010; 2008). However, in NMR experiments at some species of nuclear spin, we can observe $P(M)$ or $P(B)$, excluding the signal by SDW (Kumagai et al., 2011). Thus, we expect that the relation of the SDW and FFLO for $H \parallel ab$ will be clarified in future studies.

4.4 Small Angle Neutron Scattering (SANS) in FFLO vortex states

The modulation of the internal magnetic field $B_z(\mathbf{r})$ may be observed by SANS experiment. If the periodic modulation along the z-direction is observed, it can be direct evidence of the FFLO modulation. Therefore, we discuss the neutron scattering in the FFLO vortex state. The intensity of the (h,k,l)-diffraction peak is given by $I_{h,k,l} = |F_{h,k,l}|^2/|q_{h,k,l}|$ with the wave vector $\mathbf{q}_{h,k,l}$ given in Eq. (12). Here we write $(m_1, m_2, m_3) = (h,k,l)$ following notations of the neutron scattering. The Fourier component $F_{h,k,l}$ is given by $B_z(\mathbf{r}) = \sum_{h,k,l} F_{h,k,l} \exp(\mathrm{i}\mathbf{q}_{h,k,l} \cdot \mathbf{r})$. The spots at $(h,k,l) = (1,0,0)$ and $(0,1,0)$ are used to determine the configuration and the orientation of the vortex lattice in SANS experiments (Bianchi et al., 2008; DeBeer-Schmitt et al., 2006), and the higher component $F_{h,k,0}$ is used to estimate the detailed structure of the internal magnetic field $B_z(\mathbf{r})$ (Kealey et al., 2000; White et al., 2010). It is noted that $F_{0,0,0} = \bar{B}$ and $F_{0,0,l} = 0$ for $l \neq 0$, because average flux density $\bar{B}$ within the unit cell of the vortex lattice is constant along the z-direction. Therefore, to detect the FFLO modulation, we have to use the spot $(1,0,2)$. The spot $(1,0,2)$ is near the spot $(1,0,0)$, which is used in the conventional SANS experiment to observe the stable vortex lattice configuration.

Change of intensity $|F_{1,0,0}|^2$ in the FFLO vortex state is presented in Fig. 16(a). This shows narrow H-range near H_{c2} among the H-dependence of $|F_{1,0}|^2$ in Fig. 8. After the transition from conventional Abrikosov vortex state (q=0) to FFLO vortex state ($q \neq 0$), $|F_{1,0,0}|^2$ shows rapid decrease. This is because $B_z(\mathbf{r})$ of vortex core expands at FFLO nodal plane, and the contrast of $B_z(\mathbf{r})$ between vortex core and outside is smeared after the average along the z-direction. When L becomes shorter at higher H, the relative volume ratio of the FFLO nodal plane increases, and $|F_{1,0,0}|^2$ decreases. As presented in Fig. 16(b), intensity $|F_{1,0,2}|^2$ for the signal of the FFLO vortex state appears at the FFLO transition. When L becomes shorter at higher H, $|F_{1,0,2}|^2$ decreases, due to the overlap between neighbor FFLO nodal regions.

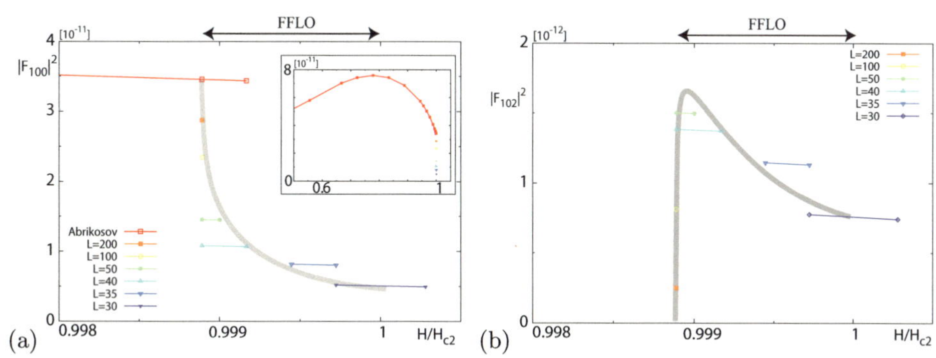

Fig. 16. Magnetic field dependence of FLL form factor $|F_{1,0,0}|^2$ (a) and $|F_{1,0,2}|^2$ (b) in FFLO vortex states. FFLO wave number $q = 2\pi/L$ at each H is given in Fig. 13. $T = 0.2T_c$ and $\mu = 2$. Lines are guide for the eye. Inset in (a) presents wider H-range.

5. Summary and discussion

We discussed interesting phenomena of vortex states in superconductors with strong paramagnetic effect, based on quasi-classical Eilenberger theory. The paramagnetic effect comes from splitting of up-spin and down-spin Fermi surfaces due to the Zeeman effect. In our calculations, since spatial structures of the order parameter and the internal field are calculated in vortex lattice states self-consistently with local electronic states, we can quantitatively estimate the field dependence of physical quantities from obtained quasi-classical Green's functions in Eilenberger theory. These theoretical calculations give helpful information to evaluate contributions of pairing symmetries and paramagnetic effects etc. in experimental data observing physical properties of vortex states in unconventional superconductors.

First, we discussed anomalous field dependence of physical quantities by strong paramagnetic effect in vortex states at lower fields than the FFLO transition. Calculating the spatial structure of the vortex states and local electronic states, we clarified the paramagnetic effects in the vortex core structure. There, the core radius is enlarged and the internal field around the core is further enhanced, due to the enhanced paramagnetic moments at the vortex core. This occurs as a result of Zeeman splitting of bound electronic states at the vortex core. We estimated the magnetic field dependence of low temperature specific heat, Knight shift, magnetization, and flux line lattice form factor. There we found anomalous field dependence when the paramagnetic effect is strong. The specific heat, Knight shift, and magnetization show rapid increase near H_{c2}, due to the paramagnetic pair breaking which is eminent at higher fields. Anomalous enhancement of the FLL form factor as a function of magnetic field observed in $CeCoIn_5$ may reflect the paramagnetic vortex core structure by the strong paramagnetic effect. We quantitatively compared the anomalous magnetic field dependence of specific heat, magnetization curve, and the FLL form factor observed in $CeCoIn_5$ with results of our theoretical calculations. The paramagnetic effect can explain the anomalous field dependences qualitatively. However we found systematic quantitative deviation between the theory and the experimental data. Therefore, we showed that the deviation can be improved by considering phenomenological field dependence of normal state density of states, which reflects mass enhancement near quantum critical point at H_{c2}.

Next, we studied the FFLO states coexisting with vortices. When the paramagnetic effect is very strong, at high magnetic fields we can expect a transition to the FFLO phase where the order parameter has periodic oscillation originating from the Zeeman splitting of the Fermi surface. To discuss the FFLO states suggested in high field phase of $CeCoIn_5$, we have to consider vortices in addition to the FFLO modulation. By Eilenberger theory, we selfconsistently calculated fully 3D spatial structure of the pair potential, the internal magnetic field, the paramagnetic moment, and local electronic states in the vortex lattice state with FFLO nodal planes perpendicular to vortex lines. In the FFLO vortex states, topological structures of the pair potential determine their qualitative properties. At the FFLO nodal plane or at the vortex line, π-phase shift of the pair potential gives rise to sharp peaks in the LDOS at Fermi level of electronic states, and the Zeeman shift of the peaks enhances the local paramagnetic moment. Based on these spatial structures, we discussed NMR spectrum and neutron scattering, to identify characteristic behaviours in the FFLO states. We estimated the period of FFLO modulation and the phase diagram as a function of magnetic field H, and discussed the field dependence of NMR spectrum and FLL form factors in the FFLO vortex states. We hope that these features will be used to identify the FFLO vortex structure in the high-field phase of $CeCoIn_5$ for $H \parallel c$ and for $H \parallel ab$. For the latter, the FLLO modulation may coexist with SDW states.

6. Acknowledgments

The authors are grateful for useful discussions and communications with T. Mizushima, H. Adachi, N. Nakai, K. Kumagai, Y. Matsuda, and M.R. Eskildsen.

7. References

Adachi, H., Ichioka, M. & Machida, K. (2005). Mixed-state thermodynamics of a superconductor with moderately large paramagnetic effects, *J. Phys. Soc. Jpn.* 74: 2181–2184.

Adachi, H. & Ikeda, R. (2003). Effects of Pauli paramagnetism on the superconducting vortex phase diagram in strong fields, *Phys. Rev. B* 68: 184510, 1–17.

Bianchi, A. D., Kenzelmann, M., DeBeer-Schmitt, L., White, J. S., Forgan, E. M., Mesot, J., Zolliker, M., Kohlbrecher, J., Movshovich, R., Bauer, E. D., Sarrao, J. L., Fisk, Z., Petrović, C. & Eskildsen, M. R. (2008). Superconducting vortices in $CeCoIn_5$: Toward the Pauli-limiting field, *Science* 319: 177–180.

Bianchi, A., Movshovich, R., Capan, C., Pagliuso, P. G. & Sarrao, J. (2003). Possible Fulde-Ferrell-Larkin-Ovchinnikov superconducting state in $CeCoIn_5$, *Phys. Rev. Lett.* 91: 187004, 1–4.

Bianchi, A., Movshovich, R., Oeschler, N., Gegenwart, P., Steglich, F., Thompson, J. D., Pagliuso, P. G. & Sarrao, J. (2002). First-order superconducting phase transition in $CeCoIn_5$, *Phys. Rev. Lett.* 89: 137002, 1–4.

Bianchi, A., Movshovich, R., Vekhter, I., Pagliuso, P. & Sarrao, J. (2003). Field dependent coherence length in the superclean, high-κ superconductor $CeCoIn_5$, *Phys. Rev. Lett.* 91: 257001, 1–4.

Casalbuoni, R. & Nardulli, G. (2004). Inhomogeneous superconductivity in condensed matter and QCD, *Rev. Mod. Phys.* 76: 263–320.

DeBeer-Schmitt, L., Dewhurst, C. D., Hoogenboom, B. W., Petrović, C. & Eskildsen, M. R. (2006). Field dependent coherence length in the superclean, high-κ superconductor $CeCoIn_5$, *Phys. Rev. Lett.* 97: 127001, 1–4.

DeBeer-Schmitt, L., Eskildsen, M., Ichioka, M., Machida, K., Jenkins, N., Dewhurst, C. D., Abrahamsen, A. B., Bud'ko, S. L. & Canfield, P. (2007). Pauli paramagnetic effects on vortices in superconducting $TmNi_2B_2C$, *Phys. Rev. Lett.* 99: 167001, 1–4.

Doria, M. M., Gubernatis, J. E. & Rainer, D. (1990). Solving the Ginzburg-Landau equations by simulated annealing, *Phys. Rev. B* 41: 6335–6340.

Eilenberger, G. (1968). Transformation of Gorkov's equation for type II superconductors into transport-like equations, *Z. Phys.* 214: 195–213.

Fetter, A. L. & Hohenberg, P. C. (1969). Theory of type II superconductors, *in* R. D. Parks (ed.), *Superconductivity*, Dekker, New York, pp. 817–923(Chap.14).

Fulde, P. & Ferrell, R. (1964). Superconductivity in a strong spin-exchange field, *Phys. Rev.* 135: A550–A563.

Hayashi, N., Ichioka, M. & Machida, K. (1996). Star-shaped local density of states around vortices in a type-II superconductor, *Phys. Rev. Lett.* 77: 4074–4077.

Hayashi, N., Ichioka, M. & Machida, K. (1997). Effects of gap anisotropy upon the electronic structure around a superconducting vortex, *Phys. Rev. B* 56: 9052–9063.

Hess, H. F., Robinson, R. B. & Waszczak, J. V. (1990). Vortex-core structure observed with a scanning tunneling microscope, *Phys. Rev. Lett.* 64: 2711–2714.

Hiragi, M., Suzuki, K., Ichioka, M. & Machida, K. (2010). Vortex state and field-angle resolved specific heat oscillation for H//ab in d-wave superconductors, *J. Phys. Soc. Jpn.* 79: 094709, 1–9.

Houzet, M. & Buzdin, A. (2001). Structure of the vortex lattice in the Fulde-Ferrell-Larkin-Ovchinnikov state, *Phys. Rev. B* 63: 184521, 1–5.

Ichioka, M., Adachi, H., Mizushima, T. & Machida, K. (2007). Vortex state in a Fulde-Ferrell-Larkin-Ovchinnikov superconductor based on quasiclassical theory, *Phys. Rev. B* 76: 014503, 1–10.

Ichioka, M., Hasegawa, A. & Machida, K. (1999a). Field dependence of the vortex structure in d-wave and s-wave superconductors, *Phys. Rev. B* 59: 8902–8916.

Ichioka, M., Hasegawa, A. & Machida, K. (1999b). Vortex lattice effects on low-energy excitations in d-wave and s-wave superconductors, *Phys. Rev. B* 59: 184–187.

Ichioka, M., Hayashi, N., Enomoto, N. & Machida, K. (1996). Vortex structure in d-wave superconductors, *Phys. Rev. B* 53: 15316–15326.

Ichioka, M., Hayashi, N. & Machida, K. (1997). Local density of states in the vortex lattice in a type-II superconductor, *Phys. Rev. B* 55: 6565–6576.

Ichioka, M. & Machida, K. (2007). Vortex states in superconductors with strong Pauli-paramagnetic effect, *Phys. Rev. B* 76: 064502, 1–7.

Ikeda, R. & Adachi, H. (2004). Modulated vortex lattice in high fields and gap nodes, *Phys. Rev. B* 69: 212506, 1–4.

Ikeda, S., Shishido, H., Nakashima, M., Settai, R., Aoki, D., Haga, Y., Harima, H., Aoki, Y., Namiki, T., Sato, H. & Onuki, Y. (2001). Unconventional superconductivity in $CeCoIn_5$ studied by the specific heat and magnetization measurements, *J. Phys. Soc. Jpn.* 70: 2248–2251.

Izawa, K., Yamaguchi, H., Matsuda, Y., Shishido, H., Settai, R. & Onuki, Y. (2001). Angular position of nodes in the superconducting gap of quasi-2D heavy-fermion superconductor $CeCoIn_5$, *Phys. Rev. Lett.* 87: 057002, 1–4.

Kakuyanagi, K., Saitoh, M., Kumagai, K., Takashima, S., Nohara, M., Takagi, H. & Matsuda, Y. (2005). Texture in the superconducting order parameter of $CeCoIn_5$ revealed by nuclear magnetic resonance, *Phys. Rev. Lett.* 94: 047602, 1–4.

Kealey, P. G., Riseman, T. M., Forgan, E. M., Galvin, L. M., Mackenzie, A. P., Lee, S. L., Paul, D. M., Cubitt, R., Agterberg, D. F., Heeb, R., Mao, Z. Q. & Maeno, Y. (2000). Reconstruction from small-angle neutron scattering measurements of the real space magnetic field distribution in the mixed state of Sr_2RuO_4, *Phys. Rev. Lett.* 84: 6094–6097.

Kenzelmann, M., Gerber, S., Egetenmeyer, N., Gavilano, J. L., Strässle, T., Bianchi, A. D., Ressouche, E., Movshovich, R., Bauer, E. D., Sarrao, J. L. & Thompson, J. D. (2010). Evidence for a magnetically driven superconducting Q phase of $CeCoIn_5$, *Phys. Rev. Lett.* 104: 127001, 1–4.

Kenzelmann, M., Strässle, T., Niedermayer, C., Sigrist, M., Padmanabhan, B., Zolliker, M., Bianchi, A. D., Movshovich, R., Bauer, E. D., Sarrao, J. L. & Thompson, J. D. (2008). Coupled superconducting and magnetic order in $CeCoIn_5$, *Science* 321: 1652–1654.

Klein, U. (1987). Microscopic calculations on the vortex state of type II superconductors, *J. Low Temp. Phys.* 69: 1–37.

Klein, U., Rainer, D. & Shimahara, H. (2000). Interplay of Fulde-Ferrell-Larkin-Ovchinnikov and vortex states in two-dimensional superconductors, *J. Low Temp. Phys.* 118: 91–104.

Koutroulakis, G., Stewart, Jr. M. D., Mitrović, V. F., Horvatić, M., Berthier, C., Lapertot, G. & Flouquet, J. (2010). Field evolution of coexisting superconducting and magnetic orders in $CeCoIn_5$, *Phys. Rev. Lett.* 104: 087001, 1–4.

Kumagai, K., Saitoh, M., Oyaizu, T., Furukawa, Y., Takashima, S., Nohara, M., Takagi, H. & Matsuda, Y. (2006). Fulde-Ferrell-Larkin-Ovchinnikov state in a perpendicular field of quasi-two-dimensional $CeCoIn_5$, *Phys. Rev. Lett.* 97: 227002, 1–4.

Kumagai, K., Shishido, H., Shibauchi, T. & Matsuda, Y. (2011). Evolution of paramagnetic quasiparticle excitations emerged in the high-field superconducting phase of $CeCoIn_5$. Preprint, arXiv:1103.1440 (to appear in *Phys. Rev. Lett.*).

Larkin, A. I. & Ovchinnikov, Y. N. (1965). Inhomogeneous state of superconductors, *Sov. Phys. JETP* 20: 762–769.

Machida, K., Mizushima, T. & Ichioka, M. (2006). Generic phase diagram of fermion superfluids with population imbalance, *Phys. Rev. Lett.* 97: 120407, 1–4.

Machida, K. & Nakanishi, H. (1984). Superconductivity under a ferromagnetic molecular field, *Phys. Rev. B* 30: 122–133.

Matsuda, Y. & Shimahara, H. (2007). Fulde-Ferrell-Larkin-Ovchinnikov state in heavy fermion superconductors, *J. Phys. Soc. Jpn.* 76: 051005, 1–16.

Miranović, P. & Machida, K. (2003). Thermodynamics and magnetic field profiles in low-κ type-II superconductors, *Phys. Rev. B* 67: 092506, 1–4.

Miranović, P., Nakai, N., Ichioka, M. & Machida, K. (2003). Orientational field dependence of low-lying excitations in the mixed state of unconventional superconductors, *Phys. Rev. B* 68: 052501, 1–4.

Mizushima, T., Machida, K. & Ichioka, M. (2005a). Direct imaging of spatially modulated superfluid phases in atomic fermion systems, *Phys. Rev. Lett.* 94: 060404, 1–4.

Mizushima, T., Machida, K. & Ichioka, M. (2005b). Topological structure of a vortex in the Fulde-Ferrell-Larkin-Ovchinnikov state, *Phys. Rev. Lett.* 95: 117003, 1–4.

Moler, K. A., Baar, D. J., Urbach, J. S., Liang, R., Hardy, W. N. & Kapitulnik, A. (1994). Magnetic field dependence of the density of states of $YBa_2Cu_3O_{6.95}$ as determined from the specific heat, *Phys. Rev. B* 73: 2744–2747.

Nakai, N., Miranović, P., Ichioka, M. & Machida, K. (2004). Field dependence of the zero-energy density of states around vortices in an anisotropic-gap superconductor, *Phys. Rev. B* 70: 100503(R), 1–4.

Nishimori, H., Uchiyama, K., Kaneko, S., Tokura, A., Takeya, H., Hirata, K. & Nishida, N. (2004). First observation of the fourfold-symmetric and quantum regime vortex core in YNi_2B_2C by scanning tunneling microscopy and spectroscopy, *J. Phys. Soc. Jpn.* 73: 3247–3250.

Nohara, M., Isshiki, M., Sakai, F. & Takagi, H. (1999). Quasiparticle density of states of clean and dirty s-wave superconductors in the vortex state, *J. Phys. Soc. Jpn.* 68: 1078–1081.

Paglione, J., Tanatar, M. A., Hawthorn, D. G., Boaknin, E., Hill, R. W., Ronning, F., Sutherland, M., Taillefer, L., Petrovic, C. & Canfield, P. (2003). Field-induced quantum critical point in $CeCoIn_5$, *Phys.Rev. Lett.* 91: 246405, 1–4.

Partridge, G. B., Li, W., Kamar, R. I., Liao, Y. & Hulet, R. (2006). Pairing and phase separation in a polarized fermi gas, *Science* 311: 503–505.

Radovan, H. A., Fortune, N. A., Murphy, T. P., Hannahs, S. T., Palm, E. C., Tozer, S. W. & Hall, D. (2003). Magnetic enhancement of superconductivity from electron spin domains, *Nature (London)* 425: 51–55.

Ramirez, A. P., Varma, C. M., Fisk, Z. & Smith, J. L. (1999). Fermi-liquid renormalization in the superconducting state of UBe_{13}, *Philos. Mag. B* 79: 111–117.

Schopohl, N. & Maki, K. (1995). Quasiparticle spectrum around a vortex line in a d-wave superconductor, *Phys. Rev. B* 52: 490–493.

Shimahara, H. (1994). Fulde-Ferrell state in quasi-two-dimensional superconductors, *Phys. Rev. B* 50: 12760–12765.

Tachiki, M., Takahashi, S., Gegenwart, P., Weiden, M., Lang, M., Geibel, C., Steglich, F., Modler, R., Paulsen, C. & Onuki, Y. (1996). Generalized Fulde-Ferrell-Larkin-Ovchinnikov state in heavy-fermion and intermediate-valence systems, *Z. Physik B* 100: 369–380.

Takahashi, M., Mizushima, T., Ichioka, M. & Machida, K. (2006). Vortex-core structure in neutral fermion superfluids with population imbalance, *Phys. Rev. Lett.* 97: 180407, 1–4.

Tayama, T., Harita, A., Sakakibara, T., Haga, Y., Shishido, H., Settai, R. & Onuki, Y. (2002). Unconventional heavy-fermion superconductor $CeCoIn_5$: dc magnetization study at temperatures down to 50 mK, *Phys. Rev. B* 65: 180504(R), 1–4.

Volovik, G. E. (1993). Superconductivity with lines of gap nodes: density of states in the vortex, *JETP Lett.* 58: 469–473.

Watanabe, K., Kita, T. & Arai, M. (2005). Magnetic-field dependence of thermodynamic quantities in the vortex state of type-II superconductors, *Phys. Rev. B* 71: 144515, 1–8.

White, J. S., Das, P., Eskildsen, M. R., DeBeer-Schmitt, L., Forgan, E. M., Bianchi, A. D., Kenzelmann, M., Zolliker, M., Gerber, S., Gavilano, J. L., Mesot, J., Movshovich, R., Bauer, E. D., Sarrao, J. L. & Petrović, C. (2010). Observations of Pauli paramagnetic effects on the flux line lattice in $CeCoIn_5$, *New J. Phys.* 12: 023026, 1–11.

Yano, K., Sakakibara, T., Tayama, T., Yokoyama, M., Amitsuka, H., Homma, Y., Miranović, P., Ichioka, M., Tsutsumi, Y. & Machida, K. (2008). Field-angle-dependent specific heat measurements and gap determination of a heavy fermion superconductor URu_2Si_2, *Phys. Rev. Lett.* 100: 017004, 1–4.

Young, B.-L., Urbano, R., Curro, N., Thompson, J., Sarrao, J. L., Vorontsov, A. B. & Graf, M. (2007). Microscopic evidence for field-induced magnetism in $CeCoIn_5$, *Phys. Rev. Lett.* 98: 036402, 1–4.

Zheng, G. -q., Ozaki, H., Kitaoka, Y., Kuhns, P., Reyes, A. & Moulton, W. (2002). Delocalized quasiparticles in the vortex state of an overdoped high- T_c superconductor probed by ^{63}Cu NMR, *Phys. Rev. Lett.* 88: 077003, 1–4.

Zwierlein, M. W., Schirotzek, A., Schunck, C. H. & Ketterle, W. (2006). Fermionic superfluidity with imbalanced spin populations, *Science* 311: 492–496.

11

Development of Josephson Voltage Standards

Johannes Kohlmann and Ralf Behr
Physikalisch-Technische Bundesanstalt (PTB)
Germany

1. Introduction

Exciting applications of superconductivity are based on the macroscopic quantum state which exists in a superconductor. In this chapter we investigate the behaviour of junctions consisting of two weakly coupled superconductors. These junctions are nowadays called Josephson junctions[1] (Josephson, 1962). The macroscopic quantum state results in an exceptional behaviour of these Josephson junctions. They are the basis for various applications in superconductive electronics (cf. Anders et al, 2010), e.g. in the field of metrology for high-precision measurements. The most significant representative of a metrological application is the Josephson voltage standard. This quantum standard enables the reference of the unit of voltage, the volt, just to physical constants. It is nowadays used in many laboratories world-wide for high-precision voltage measurements. The main component of each modern Josephson voltage standard is the highly integrated series array consisting of tens of thousands of Josephson junctions fabricated in thin-film technology.

While Josephson junctions are conceptually simple, nearly 50 years of developments were needed to progress from single junctions delivering a few millivolt at most to highly integrated series arrays containing more than 10,000 or even 100,000 junctions. These large series arrays enable the generation of dc and ac voltages at the 10 V level, which is relevant for most applications. Conventional Josephson voltage standards based on underdamped Josephson junctions are used for dc applications. The increasing interest in highly precise ac voltages has stimulated different attempts to develop measurement tools on the basis of Josephson arrays for ac applications, namely programmable Josephson voltage standards containing binary-divided arrays and pulse-driven Josephson voltage standards both based on overdamped Josephson junctions. This chapter describes the development of these modern dc and ac Josephson voltage standards as well as their fundamentals and applications. The development and use of Josephson voltage standards have also been described recently in several review papers (amongst others: Niemeyer, 1998; Hamilton, 2000; Yoshida, 2000; Behr et al., 2002; Kohlmann et al., 2003; Benz & Hamilton, 2004; Jeanneret & Benz, 2009).

---

[1] When Brian D. Josephson was a 22-year-old graduate student at Trinity College in Cambridge, UK, he theoretically derived equations for the current and voltage across a junction consisting of two weakly coupled superconductors in 1962. His discovery won him a share of the 1973 Nobel Prize in Physics.

2. Fundamentals - the Josephson effects

A superconductor as a macroscopic object is quantum mechanically described by a macroscopic wavefunction. This macroscopic wavefunction is an important aspect of the BCS theory of superconductivity named after the authors Bardeen, Cooper, and Schrieffer[2] (1957). Brian Josephson investigated the behaviour of two weakly coupled superconductors on the basis of the BCS theory a few years after its publication (Josephson, 1962). He predicted two effects due to the tunnelling of Cooper pairs across the connection, i.e. a coupling of the macroscopic wavefunction of the two superconductors: (1) a dc supercurrent $I = I_c \sin \phi$ can flow across this junction (I_c denotes the critical current and ϕ the phase between the macroscopic wavefunction of the two superconductors); (2) an ac supercurrent of frequency $f_J = (2e/h)\cdot V$ occurs if the junction is operated at a non-zero voltage V, i.e. a Josephson junction is an oscillator (e is the elementary charge and h is Planck's constant). Irradiation of the junction by external microwaves of frequency f vice versa produces constant-voltage steps due to the phase locking of the Josephson oscillator by the external oscillator: $V_n = n\cdot(h/2e)\cdot f$ (n = 1, 2, 3, … denotes the integer step number). As an illustration, the generation of constant-voltage steps can also be described as a specific transfer of flux quanta $\Phi_0 = h/2e$ through the Josephson junction. The irradiation of the Josephson junctions with external microwaves of frequency f effects this specific transfer and produces constant-voltage steps V_n:

$$V_n = n \cdot \Phi_0 \cdot f \quad (1)$$

The Josephson effect thus reduces the reproduction of voltages to the determination of a frequency, which can be finely controlled with high precision and accurately referenced to atomic clocks. The constant-voltage steps were observed soon after by Shapiro (1963). A single Josephson junction operated at the first-order constant-voltage step generates about 145 μV, when irradiated by 70 GHz microwaves. Highly integrated junction series arrays are therefore needed to achieve practical output voltages up to 1 V or 10 V.

The frequency range for the best operation of Josephson junctions is determined by their dynamic characteristics. The most important parameter is the characteristic voltage $V_c = I_c \cdot R_n$ (R_n denotes the normal state resistance of the junctions). The characteristic voltage is related to the characteristic frequency by equation (1): $f_c = (2e/h)\cdot V_c = (2e/h)\cdot I_c\cdot R_n$.

The dynamics of a Josephson junction is often investigated using the resistively-capacitively-shunted-junction (RCSJ) model (Stewart, 1968; McCumber, 1968). Within this model, the real Josephson junction is described as a parallel shunting of an ohmic resistance R, a capacitance C, and an ideal Josephson element. In the linear approximation, the resonance frequency is given by the plasma frequency $f_p = (e\cdot j_c/\pi\cdot h\cdot C_s)^{1/2}$ (j_c denotes the critical current density, $C_s = C/A$ the specific junction capacitance, and A the junction area). Details of the behaviour depend on the kind of junction, which can be characterized by the dimensionless McCumber parameter $\beta_c = Q^2$ being equal to the square of the quality factor $Q = 2\pi\cdot f_p\cdot R\cdot C$ of the junction. Underdamped junctions with $\beta_c > 1$ show a hysteretic current-voltage characteristic, overdamped junctions with $\beta_c \leq 1$ a non-hysteretic one as schematically shown in Fig. 1. Detailed descriptions of the Josephson effects and Josephson junctions have been

[2] Bardeen, Cooper, and Schrieffer were awarded the 1972 Nobel Prize in Physics for their theory of superconductivity.

given in several reviews (e.g. Josephson, 1965; Kautz, 1992; Rogalla, 1998) and textbooks (e.g. Barone & Paternò, 1982; Likharev, 1986; Kadin, 1999).

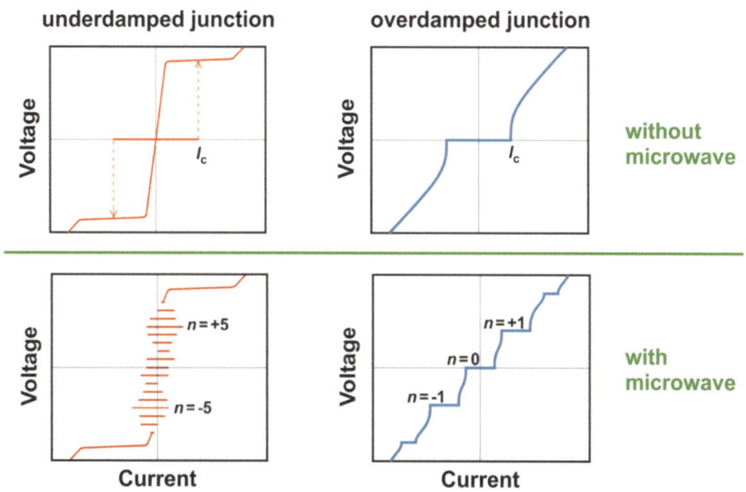

Fig. 1. Schematic current-voltage characteristic of underdamped (left) and overdamped (right) Josephson junctions without (top) and with (bottom) microwave irradiation. Some constant-voltage steps are marked.

3. Realization of Josephson junctions and series arrays

A Josephson junction is composed of two weakly coupled superconductors. While Josephson (1962) originally investigated the tunnelling of Cooper pairs through a barrier, i.e. an insulator, he also mentioned that similar effects should occur when two superconductors are separated by a thin normal region. These two junction types are nowadays indeed the most important ones for Josephson junctions, namely the so-called SIS junctions and SNS junctions, respectively (S: Superconductor, I: Insulator, N: Normal metal). SIS junctions are typically underdamped junctions, while SNS junctions are overdamped ones. Moreover, further possibilities for the realization of Josephson junctions exist such as e.g. SINIS junctions, grain boundary junctions (especially for high-temperature superconductors), and junctions consisting of two superconductors connected by a narrow constriction. As junctions for Josephson voltage standards are mainly based on SIS, SNS, or SINIS junctions, these types will be described in more detail in the following. The fabrication of the integrated circuits containing these junctions is based on the same main steps; the fabrication processes differ only in detail.

3.1 Fabrication process

The development of Josephson voltage standards is intimately connected with improvements of the fabrication technology for series arrays. The fabrication process should be as simple and reliable as possible, and must be realized in thin-film technology, in order to enable the fabrication of highly integrated circuits containing thousands of junctions in a similar way to in the semiconductor industry. Josephson junctions and the first series arrays in the 1980s were fabricated in lead/lead alloy technology (cf. Niemeyer et al, 1984); but the main problem was the susceptibility to damage of the lead alloy circuits by humidity and thermal cycling. The main important breakthrough in the development of a more robust fabrication process was the invention of the Nb/Al-Al_2O_3 technology by Gurvitch et al (1983). This technology combines the use of the durable and chemically stable metal Nb with the high critical temperature of about 9.2 K, the outstanding covering of thin Al layers on Nb, and the formation of a very homogeneous and stable oxide of Al by thermal oxidation. The adaptation of this process and several improvements made possible the fabrication of voltage standard arrays consisting of Nb/Al-Al_2O_3/Nb Josephson junctions in 1986 (Niemeyer et al, 1986). Nowadays, all Josephson arrays for voltage standard applications are fabricated in processes fundamentally based on this invention.

Sputtered Nb is typically used at present for the superconducting layers and NbN in case of operation at 10 K, respectively. Dielectric layers are realized by SiO_2. Lithography is made optically or by electron-beam depending on the dimensions of the structure and its complexity. The different layers are patterned by adapted fluorine-based dry etching processes. For a reliable process, the trilayer or multilayer defining the junctions are deposited as a sandwich structure without breaking the vacuum. This process requires an additional wiring layer for connecting neighbouring junctions by a window technology. The barrier material is also sputtered; if the barrier includes an oxide, a metallic layer is thermally oxidized. SIS junctions contain an Al_2O_3 barrier realized by thermal oxidation of the Al layer. SINIS junctions consist of a multilayer of Nb/Al_2O_3/Al/Al_2O_3/Nb. SIS junctions are typically operated at around 70 GHz. The characteristic voltage of SINIS junctions can be tuned over a wide range enabling operation either at frequencies around 15 GHz or around 70 GHz.

Different materials have been investigated and used for the N layer of SNS junctions. As the specific resistance of most metals is rather low, high-resistive materials are preferred in order to increase the characteristic voltage. Most SNS junctions are therefore operated at frequencies between 10 GHz and 20 GHz. The high resistivity for the N layer is reached by binary alloys as PdAu (Benz et al, 1997), HfTi (Hagedorn et al, 2006), or $MoSi_2$ (Chong et al, 2005). Junctions containing an N layer of Ti (Schubert et al, 2001a) or TiN (Yamamori et al, 2008) have also been realized. Recently, a new type of junction has increasingly gained in importance: its barrier consists of a semiconductor such as Si doped with a metal and being near a metal insulator transition (Baek et al, 2006). Although these junctions behave like SNS junctions, they are more their own class of junctions and sometimes called SI'S junctions. A promising version of these SI'S junctions is realized by an amorphous Si barrier doped by Nb. Nb and Si are co-sputtered from two sputter targets; the Nb content is varied by adjusting the power for sputtering.

The thickness of the superconducting layers is typically above about 150 nm and therefore roughly twice the superconducting penetration depth at least. The superconducting layers are consequently both thick enough, to ensure appropriate microwave behaviour, and thin enough, to allow reliable thin-film processes. The barrier is between 10 nm and 30 nm thick depending on the details of the material. Stacked junctions have also been investigated in order to increase the integration density of junctions. They contain multilayers of superconducting Nb and barrier material. Adapted etching processes guarantee vertical edges and thus an identical size of each individual junction in order to yield homogeneous electrical parameters of the junction stacks. Arrays of double- and triple-stacked junctions have successfully been fabricated delivering output voltages between a few volts and even 10 V (Chong et al, 2005; Yamamori et al, 2008).

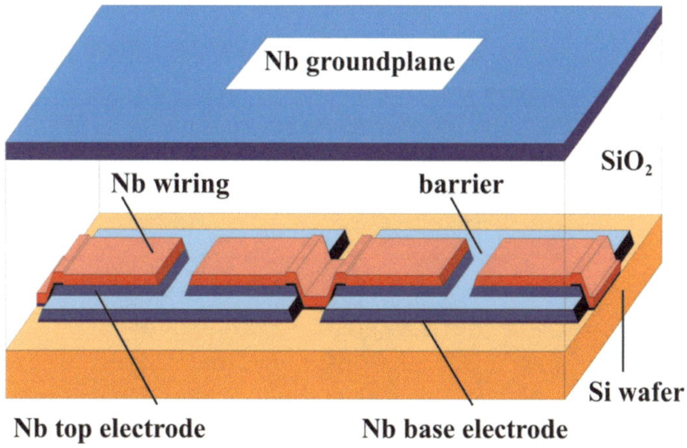

Fig. 2. Cross section of a microstripline.

3.2 Designs - a brief survey

An important requirement for the design of the circuits is the uniform microwave power distribution over all Josephson junctions in order to generate wide and stable constant-voltage steps. The step width of the constant-voltage steps depends on the applied microwave power; in some cases, the dependence is given by a Bessel function (Kautz, 1992 & 1995). A uniform power distribution is achieved by the integration of the Josephson junctions into adapted microwave transmission lines. Most modern Josephson voltage standards are based on one of three different microwave lines: a low-impedance microstrip line (cf. Fig. 2), a 50 Ω coplanar waveguide transmission line (CPW) (cf. Fig. 9), and a 50 Ω coplanar stripline (CPS). The microstrip line caused the breakthrough for the first version of modern voltage standards, i.e. the conventional Josephson voltage standard (cf. Niemeyer et al, 1984), and is mainly used to date for circuits operated in the frequency range around 73 GHz. Circuits based on CPWs have been introduced for programmable Josephson voltage standards operated in the frequency range from 10 GHz to 20 GHz (cf. Benz, 1995). Coplanar striplines were first used for conventional voltage standards operated at 75 GHz (Schubert et al, 2001b). CPW and CPS offer the advantage of a rather simple required fabrication technology compared to the microstrip line that needs an additional ground plane and a dielectric layer. An advantage of the microstrip line is that it enables a rather simple possibility of splitting a single high-frequency line in two parallel ones; this splitting can be performed several times. Each microwave branch is terminated by a matched lossy microwave line that serves as a load. Microwave reflections are therefore suppressed, which consequently provides a uniform microwave distribution by avoiding standing waves.

Most conventional dc Josephson voltage standards are based on microstrip line designs. The design of programmable Josephson voltage standards depends on the frequency range for their operation. Most programmable standards operated around 73 GHz are also based on microstrip line designs. Circuits for operation between 10 GHz and 20 GHz use CPWs (cf. Benz et al, 1997; Dresselhaus et al, 2009). The design is determined in detail by the high-frequency behaviour of the Josephson junctions.

Fig. 3 shows, as an example, the PTB design of a 10 V SNS array for operation at 70 GHz and this is briefly described in the following. An antipodal finline taper serves as an antenna. It connects the microstrip line, containing the Josephson junctions, to the E-band rectangular

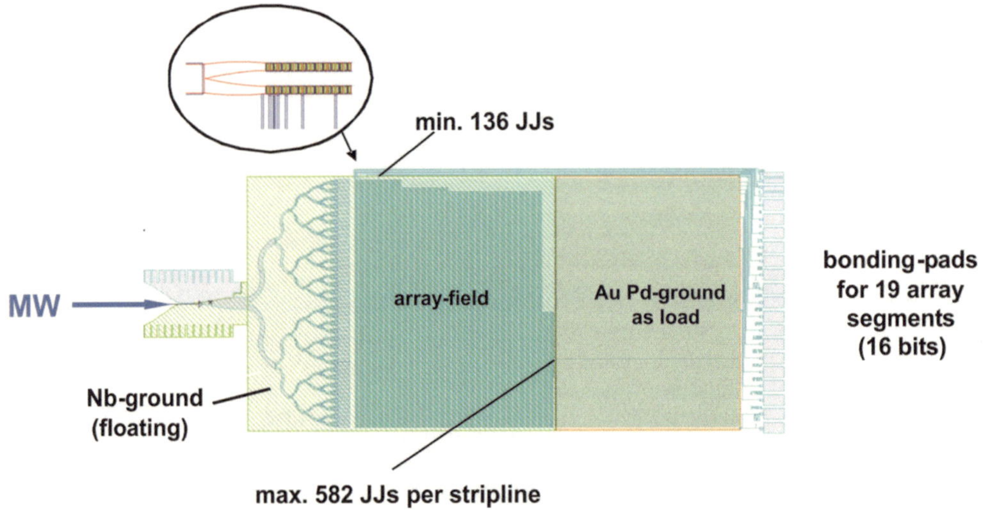

Fig. 3. Design of a 10 V SNS Josephson series array developed at PTB. The array consist of 69,632 junctions embedded into 128 parallel low-impedance microstriplines. The length and width of a single junctions is 6 µm x 20 µm. The size of the total chip is 24 mm x 10 mm.

waveguide while simultaneously matching the impedance of the waveguide (about 520 Ω) to that of the microstrip line (about 5 Ω). The microstrip line is split in several stages forming parallel branches. The design of conventional 10 V circuits contains two stages resulting in four parallel branches. The design of programmable 1 V (10 V) circuits consists of 6 (7) stages forming 64 (128) parallel branches. The reason for these differences can be understood by using the RCSJ model (cf. section 2). For SIS junctions, the ohmic resistance R_n is of the order of 50 Ω, while the impedance of the capacitive branch $Z_d = 1/(2\pi \cdot f \cdot C)$ is of the order of 50 mΩ for a junction capacitance of 50 pF. High-frequency currents therefore flow mainly capacitively resulting in a very low attenuation of the microwave power from about 1 dB/1,000 junctions to 2 dB/1,000 junctions. Each branch can therefore contain a lot of junctions (about 3,500 junctions in the real design) without loosing a uniform microwave power distribution to each junction. The conditions are completely different for over-damped SINIS junctions. Now, R_n and Z_d are comparable (about 50 mΩ each) leading to the significant dissipation of the microwave current and thus to a significant attenuation of the microwave power of about 50 dB/1,000 junctions (Schulze et al, 1999). The high attenuation is, however, compensated in part by an active contribution of the junctions; the junctions act as oscillators. The single branches of programmable series arrays consist therefore of 128 junctions (1 V design) and up to 582 junctions (10 V design), respectively. Overdamped SNS junctions integrated into a low-ohmic microstrip line show similar behaviour, as a significant part of the microwave is dissipated resistively.

Another situation is found for overdamped SNS junctions embedded into the centre line of a CPW. The ratio of the low junction impedance to the 50 Ω impedance of the CPW leads to a situation which is similar to that of the microstrip line for conventional SIS arrays: Attenuation of the microwave power is low, because the junctions are loosely linked to the CPW. Each branch can therefore contain more junctions than in the microstrip line designs. Typical numbers for 1 V (10 V) arrays are 8 (32) branches with 4096 (8400) junctions each (Benz et al, 1997; Burroughs et al, 2009a).

4. DC measurements - conventional Josephson voltage standards

While at the beginning of Josephson voltage standards the voltage of a single junction in the millivolt range was used as a reference (cf. Niemeyer, 1998; Hamilton, 2000), the chapter of modern Josephson voltage standards was opened by two new ideas: First, Levinson et al (1977) suggested the use of highly underdamped junctions with hysteretic current-voltage characteristics producing constant-voltage steps whose current ranges overlap one another for small bias currents. A single bias current source can consequently be used to bias all junctions of a series array on the quantized constant-voltage steps. Secondly, the Josephson junctions are embedded into an adapted microwave transmission line resulting in first 1 V arrays realized by Niemeyer et al (1984). Because of this arrangement, the Josephson junction series array is connected in series for the dc bias and acts as a microstrip line at rf frequencies. As the microwave power is mainly capacitively coupled to the underdamped junctions, the rf attenuation of the series array is very low, therefore, enabling uniform rf bias of all junctions.

Since the mid 1980s Josephson voltage standards based on these concepts have been available. Underdamped Josephson junctions are typically realized by SIS junctions (S: Superconductor, I: Insulator). Large series arrays of Josephson junctions are needed to reach the voltage level essential for real applications, namely 1 V or especially 10 V. A 10 V series array typically contains between about 14,000 and 20,000 Josephson junctions depending on the details of the specific design. The circuits developed and fabricated at PTB consist of about 14,000 junctions distributed to four parallel low-impedance microstrip-lines. Typical arrays show under 70 GHz microwave irradiation a step width above 20 μA, best arrays up to 50 μA. This kind of so-called conventional Josephson voltage standard has been successfully operated to date for dc applications in many national, industrial, and military standards labs around the world. They are now commercially offered by two companies.[3]

In spite of their very successful use for dc applications, conventional Josephson voltage standards have two important drawbacks due to the ambiguity of the constant-voltage steps: First, they do not enable switching rapidly and reliably between different specific steps. Secondly, the constant-voltage steps are only metastable so that electromagnetic interference can cause spontaneous switching between steps.

5. From DC to AC - programmable Josephson voltage standards

As described in the previous section, conventional Josephson voltage standards are operated very successfully for dc applications. The increasing interest in rapidly switching arrays and in highly precise ac voltages stimulated research activities in the mid 1990s to develop measurement tools based on Josephson junctions to meet these requirements. Different attempts have been suggested and partly realized. The main important ones are programmable voltage standards based on binary-divided arrays (cf. 5.1), pulse-driven arrays (cf. 5.3), and a d/a converter based on the dynamic logic of processing single flux quanta (SFQ) (cf. Semenov & Polyakov, 2001). In the following, the first two versions are described in more detail, as most research activities are presently focused on these two, and promising results have meanwhile been demonstrated. Both are intended to extend the use of high-precision Josephson voltage standards from dc to ac.

---

[3] Hypres Inc., USA: www.hypres.com and Supracon AG, Germany: www.supracon.com.

5.1 Programmable voltage standards based on binary-divided arrays

The limitations of conventional Josephson voltage standards are mainly due to the overlapping steps resulting from the hysteretic current-voltage characteristic of underdamped Josephson junctions. Therefore, Josephson junctions showing a non-hysteretic current-voltage characteristic have been investigated. Such behaviour is shown by an overdamped Josephson junction. The current voltage-characteristic is non-hysteretic and remains single-valued under microwave irradiation (cf. Fig. 1). The constant-voltage steps are consequently inherently stable and can rapidly be selected by external biasing. All junctions are operated on the same constant-voltage step (typically the first one) in contrast to those of conventional standards, which are operated at the fourth to fifth step as average. The number of junctions necessary to attain a given voltage must be increased correspondingly. The series array of junctions must additionally be divided into segments in order to enable the generation of different voltage levels. The Josephson array is hence operated as a multi-bit digital-to-analogue (d/a) converter based on a series array of overdamped Josephson junctions divided into segments containing numbers of junctions belonging e.g. to a binary sequence of independently biased smaller arrays (cf. Fig. 4). Any integral number of constant-voltage steps permitted by that sequence can consequently be generated by these arrays, often called programmable Josephson voltage standards.

A programmable Josephson voltage standard was suggested and demonstrated for the first time by Hamilton et al (1995). In that case 2,048 junctions of an array containing 8,192 externally shunted SIS junctions were operated at 75 GHz and delivered an output voltage of about 300 mV. As the critical current and consequently the step width are limited to a few hundred microamperes due to design restrictions of externally shunted SIS arrays, and a design for these junctions is rather complex and challenging, other junction types have subsequently been investigated. The final breakthrough of programmable voltage standards was enabled by the implementation of SNS junctions (Benz, 1995), whereupon calculations by Kautz (1995) had given important hints for their realization (S: Superconductor, N: Normal metal).

The first practical 1 V arrays were realized by Benz et al (1997). A total of 32,768 SNS junctions containing PdAu as the normal metal were embedded into the middle of a coplanar waveguide transmission line (CPW) with an impedance of 50 Ω. The width of the constant-voltage steps exceeds 1 mA under microwave operation around 16 GHz. This low microwave frequency gives rise to a drawback of SNS junctions, namely the large number of junctions needed to reach the 1 V (32,000 junctions) or the 10 V level (300,000 junctions).

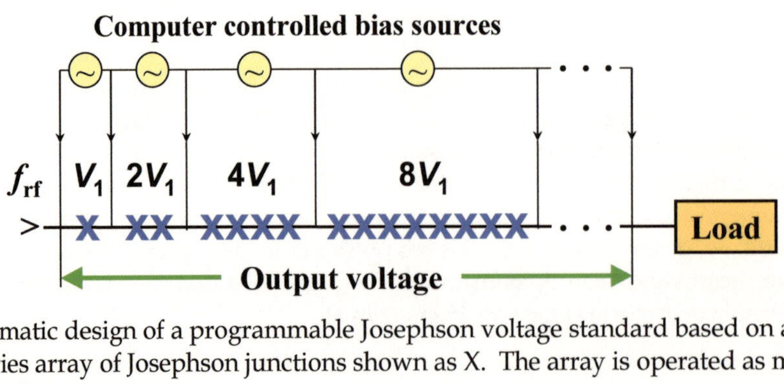

Fig. 4. Schematic design of a programmable Josephson voltage standard based on a binary-divided series array of Josephson junctions shown as X. The array is operated as multi-bit digital-to analogue converter.

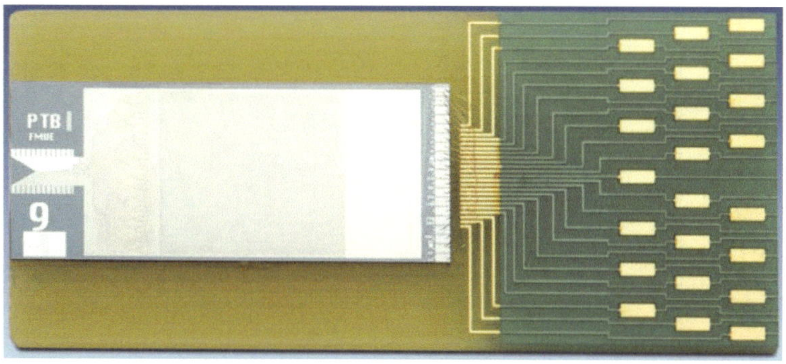

Fig. 5. Photo of a 10 V programmable Josephson junction series array.

This huge number of junctions causes enormous challenges for the microwave design and for the fabrication technology. The use of stacked junctions was subsequently investigated in order to handle this huge number of junctions. For example, arrays of double- and triple-stacked junctions containing $MoSi_2$ barriers were developed generating voltages up to 3.9 V (Chong et al, 2005).

Other kinds of junctions have therefore been investigated, in order to reach characteristic voltages of about 150 μV which allows operation at 70 GHz. A successful development has been SINIS junctions consisting of a multilayer superconductor-insulator-normal metal-insulator-superconductor originally investigated for electronic applications (Maezawa & Shoji, 1997; Sugiyama et al, 1997). The first small series arrays and 1 V arrays were subsequently fabricated (Schulze et al, 1998; Behr et al, 1999). The 1 V arrays contain 8,192 junctions. The first 10 V arrays consisting of 69,120 junctions were also developed shortly afterwards (Schulze et al, 2000) and later significantly improved (Mueller et al, 2007).

In spite of their successful use, a serious drawback of SINIS junctions is their sensitivity to particular steps during fabrication often resulting in a few shorted junctions of a SINIS series array (typically between 0 and 10 of 10,000 junctions) probably due to the very thin insulating oxide barriers (cf. Mueller et al, 2009). The search for more robust barrier materials led to an amorphous silicon layer doped with a metal such as niobium (Baek et al, 2006). The niobium content is tuned to a value near a metal-insulator transition observed at a niobium concentration of about 11.5% (Hertel et al, 1983). This region combining a high resistivity and a sufficient conductivity allows the fabrication of 1 V and 10 V arrays for operation at 70 GHz (Mueller et al, 2009). Fig. 5 shows a photo of a 10 V programmable Josephson junction series array. Measurements showed that a few 10 V arrays consisting of 69,632 junctions had been realized without any shorted junction, which was never achieved using SINIS junctions. Step widths above 1 mA have meanwhile been reached (cf. Fig. 6). This junction type currently enables the most reliable fabrication process.

Series arrays of junctions with an amorphous Nb_xSi_{1-x} barrier were originally used for circuits operated around 15 GHz. Burroughs et al (2009a) developed 10 V arrays containing three-junction stacks with 268,800 junctions arranged in 32 parallel branches. Constant-voltage steps at 10 V were generated under microwave irradiation between about 18 GHz and 20 GHz. Tapered CPWs have been used in order to assure a homogeneous microwave power distribution along 8,400 junctions in each branch (Dresselhaus et al, 2009).

Some other kinds of junctions have also been investigated. While most Josephson arrays are operated in liquid helium at 4.2 K, Yamamori et al (2006) developed arrays for operation at

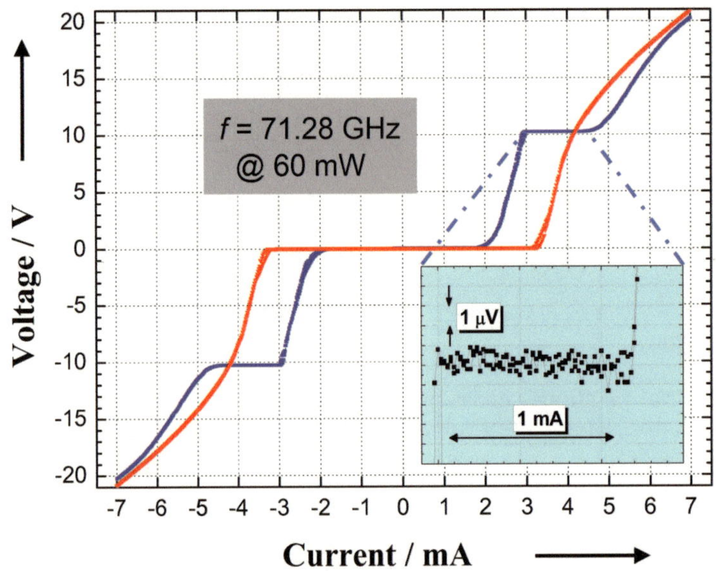

Fig. 6. Current-voltage characteristic of a 10 V programmable Josephson junction series array without (red) and with (blue) 70 GHz microwave irradiation. The inset shows the constant-voltage step at the 10 V level with high resolution.

temperatures around 10 K by using NbN for the superconducting layers and TiN for the barrier. The arrays consisting of more than 500,000 junctions for operation at 16 GHz generate voltages up to 17 V (Yamamori et al, 2008). Another version for 70 GHz operation is based on an improved design of 3315 externally shunted SIS junctions operated on the third-order constant-voltage step (Hassel et al, 2005). Recently 1 V SNIS arrays were developed by Lacquaniti et al (2011) using a slightly oxidized thick Al layer (up to 100 nm) as a barrier.

5.2 Applications using binary-divided programmable Josephson voltage standards

Conventional Josephson voltage standards are used for dc applications, namely to calibrate voltage references e.g. Weston elements or Zener references, and to measure the linearity of voltmeters. The Josephson voltage standards in many countries around the world have been verified by international comparisons. The Bureau International des Poids et Mesures (BIPM) developed a travelling Josephson voltage standard for performing direct comparisons, typically achieving uncertainties of 1 part in 10^{10} (Wood & Solve, 2009). The advantage of programmable Josephson voltage standards over conventional ones is given in the speed required to adjust a precise voltage. In direct comparisons using a null-detector at room temperature, the main uncertainty source is the type-A uncertainty from the null-detector's noise. In speeding up a comparison the uncertainty can be reduced by a factor $\sqrt{n}$ where n is the number of polarity reversals. Using two programmable 10 V Josephson voltage standards, the polarity reversing procedure can be easily automated. This has been demonstrated (Palafox et al, 2009) with a type-A uncertainty of 3 parts in 10^{12}.

Binary-divided Josephson arrays were originally developed aiming at d/a converters with fundamental accuracy as a source for ac calibrations. Fig. 7 shows a step-wise approximated sine wave. It was tested to calibrate thermal transfer standards (Hamilton et al, 1995). The synthesized waveforms contain small parts of undefined voltages during transients between well-defined quantized voltage levels. To improve achievable uncertainties, the transients have been made faster and faster, from 1 μs (Hamilton et al, 1997) to below 100 ns (Williams et al, 2007). Measurements on thermal transfer standards have shown possible uncertainties better than 1 μV/V for frequencies below 200 Hz (Behr et al, 2005) but for higher frequencies transients dominate uncertainties. Different error analyses (Lee et al, 2009; Burroughs et al, 2009b) confirm that transients will make it very difficult to further improve the predictability of these quantized voltage sources as the transients depend on too many parameters like applied bias current, microwave power or helium levels in the dewar. The only way for further improvements seems to require specific assumptions for the device under test (Séron et al, 2011).

Due to this fundamental limitation from transients the idea came up of combining the stepwise approximated Josephson waveforms with sampling methods. In a first experiment, a sampling voltmeter was calibrated by sampling the quantized voltage levels (Ihlenfeld et al, 2005). Later stepwise approximated waveforms and sampling were used to demonstrate an ac quantum voltmeter measuring ac voltage differentially (Behr et al, 2007). Both methods are used nowadays to link a power standard directly to a quantum basis (Palafox et al, 2007 & 2009; Rüfenacht et al, 2009). By introducing faster sampling systems and pre-amplifiers for a wide range of ac applications like ac-dc transfer calibrations, this idea has been further improved. As here the Josephson system is acting as a voltage reference, it also allows combining it with an external ac source traced back or locked to the Josephson voltage (Rüfenacht et al, 2011). For certain applications this is favourable as ac sources can drive a current to low-impedance devices. Driving a current from a Josephson voltage standard is very limited as typically step widths are not much larger than 1 mA, accordingly the impedance must be larger than 10 kΩ for 10 V Josephson arrays.

Towards higher frequencies sampling methods are limited due to the bandwidth of a/d converters which are affected by fast voltage edges in stepwise approximated waveforms and a decreasing aperture time for raising frequencies. The frequency limit is determined by the number of samples taken for a period. When using rectangular waveforms, i.e. the

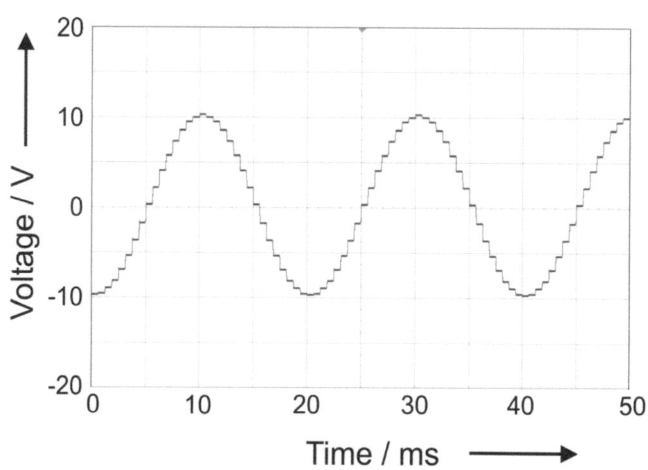

Fig. 7. Synthesis of a step-wise approximated 50 Hz sine wave using a 10 V Josephson junction series array.

minimum number of samples, frequencies up to 6 kHz have been used to calibrate impedance ratios (Lee et al, 2011), while typically 16 to 256 samples reduce the bandwidth to clearly below 1 kHz (Kim et al, 2010).

Another way to minimize the effect of transients is to use the rectangular waveforms and to just look at the fundamental tone of the waveform. Practically this is easy when a lock-in amplifier is used as a null-detector. Internally the lock-in amplifier multiplies the rectangular waveform with a sine wave heavily weighting the quantized plateaus and almost neglecting the transients (Jeanneret et al, 2010). The influence of the transients is suppressed to below parts in 10^8 which is being utilized fully for impedance ratio measurements (Lee et al, 2010).

However, the only way to completely avoid transients at all is to use the so-called pulse-driven Josephson arbitrary waveform synthesizer. This method is described in detail in the next paragraph.

5.3 Pulse-driven arrays

The interest in quantum-accurate ac waveform synthesis led to the development of another version of Josephson voltage standards for ac applications (Benz & Hamilton, 1996). Those Josephson voltage standards described so far are operated by sinusoidal microwaves in order to effect the transfer of flux quanta through Josephson junctions. This works well, if the operating frequency is close to the characteristic frequency of the junctions (cf. chapter 2 and equation (1); Kautz, 1992 & 1995). A modulation of the output voltage by changing the frequency of the irradiated microwaves over a wide frequency range is therefore not possible. Nevertheless, a direct time-dependent manipulation of the flux quanta transfer seems to be very promising for an ac voltage standard, in order to enable the synthesis of spectrally pure waveforms and to avoid those drawbacks related to the multi-bit d/a converter operation of binary-divided arrays.

Indeed, the limitations of sinusoidal operation do not appear, if Josephson junctions are operated by a train of short current pulses as shown first by calculations (Monaco, 1990). The width of the constant-voltage steps is nearly independent of the pulse repetition frequency between zero and the characteristic frequency, if rise and fall time of the pulses are short compared to the characteristic frequency (10 GHz corresponds to 100 ps). The train of pulses then determines the number of flux quanta transferred through the Josephson junctions at any time. The waveform to be generated is encoded in the pulse train. A high pulse repetition rate generates high voltages; the voltage decreases with decreasing pulse repetition rate. Fig. 8 schematically shows the principle of operation. Arbitrary output waveforms can be synthesized by modulating the pulse train using a pulse pattern generator; sometimes this version of pulse-driven Josephson arrays is therefore also called Josephson Arbitrary Waveform Synthesizer (JAWS).

The pulse train is typically created by the use of a second-order sigma-delta (SD) modulation (cf. Benz et al, 1998; Kieler et al, 2009). This procedure shifts the quantization noise to high frequencies; noise contributions are then removed by appropriate filtering. The Josephson junctions act as a quantizer due to the transfer of flux quanta. Spectrally pure waveforms are synthesized that way with higher harmonics suppressed by more than 100 dB (cf. Benz et al, 2009a; Kieler et al, 2009). The easiest way to prove perfect quantization of a synthesized signal is to generate and measure a sine wave, whose spectrum should show a single tone without any additional harmonics.

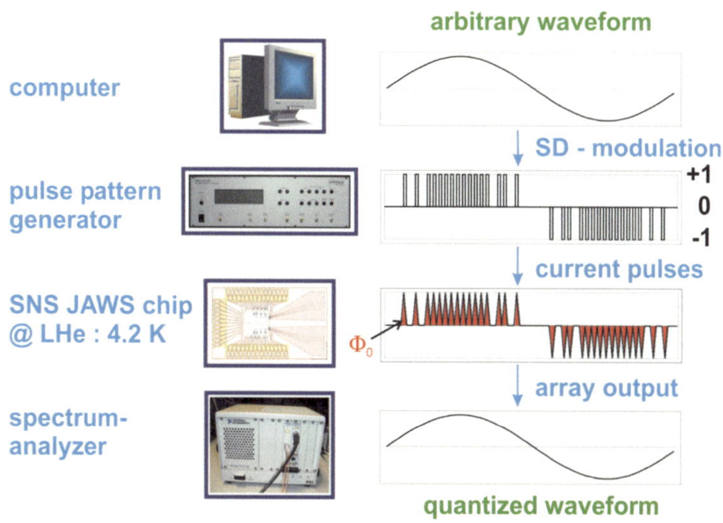

Fig. 8. Schematic of operation for pulse-driven arrays.

Pulse-driven arrays need overdamped Josephson junctions, which have predominantly been realized by SNS junctions. Different materials have been used for the barrier such as e.g. PdAu (Benz et al, 2001), HfTi (Hagedorn et al, 2006) or Nb_xSi_{1-x} (Benz et al, 2007). SINIS junctions have also been investigated (Kohlmann et al, 2006).

Pulse-driven arrays were suggested and first demonstrated by Benz and Hamilton (1996). An array of 512 junctions generated constant-voltage steps up to 265 μV under operation by unipolar pulses with a repetition frequency up to 250 MHz. Continuous enhancements gradually improved the spectra of the synthesized signals and increased the output voltages. The first important steps ahead have been, amongst others: a code generator allowing a pulse repetition frequency of about 10 GHz (Benz et al, 1998) and the use of a bipolar drive signal (Benz et al, 1999). The overdamped Josephson junctions are embedded into the middle of a coplanar waveguide transmission line (CPW). As the pulses consist of broadband frequency components ranging from dc to about 30 GHz, a complicated microwave assembly is required in order to enable the transmission of these broadband signals.

The broadband pulse drive including dc and low-frequency components causes additional requirements in operation compared to sinusoidal driven arrays. The dc component must be delivered to the array, e.g. by a direct connection to the code generator. A resistive microwave termination at the end of the CPW would produce an unwanted common mode voltage; in order to avoid this common mode voltage, the initially used arrays were designed as lumped elements, whose junction series array are directly grounded. Finally, a simple splitting of the array in parallel microwave paths is not possible.

The configuration as lumped arrays, however, limits the length of the series array, which must be short compared to the wavelength λ of the highest significant frequency. A length of typically $\lambda/8$ ensures a uniform distribution of the high-frequency power comprised in the pulses to all junctions ($\lambda \approx 12$ mm for a frequency of 10 GHz within a CPW on a Si wafer). The number of junctions is therefore restricted to about 2,000 at most using sub-μm junction technology (Hagedorn et al, 2006). A promising suggestion for increasing the number of junctions is their arrangement within a meander-like structure as shown in Fig. 9 (Kieler et al, 2007a). Arrays containing more than 10,000 junctions were realized; the

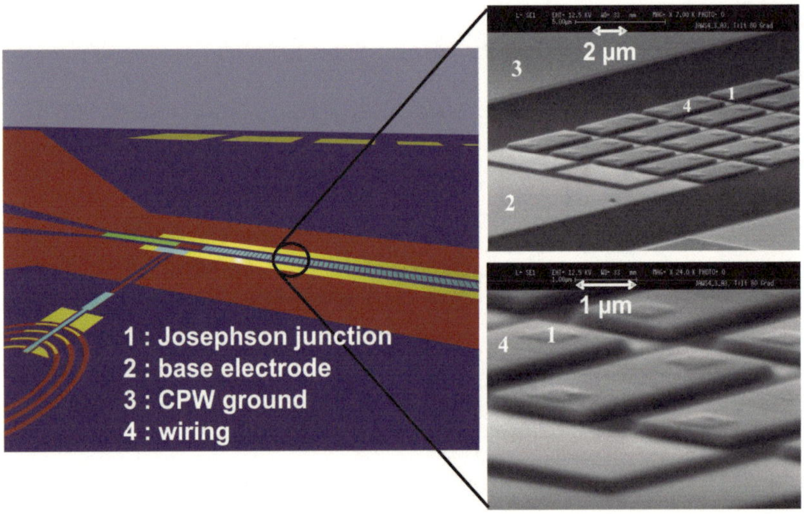

Fig. 9. Design of a Josephson junction series array for pulse drive (left). The scanning electron microscope pictures (right) show a part of the middle of the CPW containing Josephson junctions arranged in a meander-like structure.

synthesis of spectrally pure waveforms with low distortion has, however, been successful only in part so far (Kieler et al, 2007b).

A way of avoiding the limitations related to lumped arrays and of solving the common mode problem is the ac-coupling technique for the operation of Josephson arrays (Benz et al, 2001). Here, the broadband pulse drive is split into high-frequency and low-frequency signals (split around 10 MHz). While the high-frequency signal is capacitively coupled to the series array, the low-frequency part is separately applied by an additional compensation bias. A resistive microwave termination can now be placed at the end of the array without causing common-mode voltages. Therefore, extended series arrays can be used, which consequently enables a significant increase in the number of junctions. Further improvements resulted in output voltages up to 275 mV rms (Benz et al, 2009a). Two arrays containing 6,400 junctions each were simultaneously operated by using the data output and the complementary data output of the code generator, respectively. Higher harmonics are suppressed by more than 110 dB (Benz et al, 2007 & 2009a).

In spite of these very encouraging results the synthesis of voltages at 1 V or more remains very challenging. It will probably require a parallel operation of several arrays using adapted electronics (Benz et al, 1999 & 2009a) or the approach for the operation of multiple arrays that has been suggested by Kohlmann et al (2006). It is based on balanced photodiodes arranged at each array and operated by short optical pulses (Williams et al, 2004). The operation of Josephson arrays by optical pulses has also been investigated by Urano et al (2010).

The pulse train is typically provided by a commercial pulse pattern generator (bitstream generator). Fifteen years ago these generators just delivered unipolar pulses. As bipolar signals are preferred for metrological applications, and the peak-to-peak voltage is simply doubled, ways and means have been investigated to generate bipolar pulse trains even with unipolar pulses. The initially used procedure for this purpose is the suitable superposition of a high-frequency sine wave and a two-level digital signal as first proposed by Benz et al

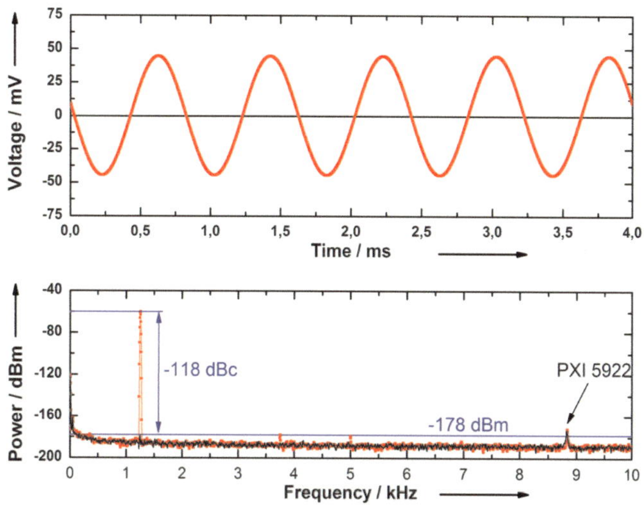

Fig. 10. Synthesized 1.25 kHz sine wave (top) and its frequency spectrum (bottom). Higher harmonics are suppressed by 118 dBc. The small signal at about 8.8 kHz is not related to the synthesized signal, as it is also present at the noise level when no waveform is synthesized.

(1999). Today the direct generation of bipolar pulses using a three-level code generator is easy as corresponding instruments have recently been made available (van den Brom et al, 2008). Now the measurement setup is less complex (cf. Fig. 8) and more temporally stable when this three-level code generator is used (van den Brom et al, 2007 & 2008). Different waveforms were synthesized over a wide frequency range from about 150 Hz to above 100 kHz using arrays containing nearly 4,800 junctions; higher harmonics are suppressed up to 118 dBc (Kieler et al, 2010). In addition, the operation margins of the arrays were significantly improved, and 200 mV (rms) signals at 1 kHz were synthesized by simultaneously operating two arrays containing 5120 junctions each (Houtzager et al, 2009).

A comparison between the output voltages of a pulse-driven and a binary divided Josephson voltage standard at 8 mV showed an excellent agreement of both systems within a relative deviation of $5 \cdot 10^{-7}$ (Kohlmann et al, 2009).

The arbitrary perfect waveforms synthesized by pulse-driven arrays are useful for different metrological applications. First of all, pulse-driven arrays were used as synthesizers for arbitrary waveforms up to 100 kHz with very pure frequency spectra and quantum-accurate voltages (cf. Benz et al, 2009a; Houtzager et al, 2009; Kieler et al, 2009). Then, pulse-driven arrays were utilized for calibrations of thermal converters and transfer standards, which are well-established devices in ac metrology (cf. Lipe et al, 2008; Benz et al, 2009a). Single- or multi-tone signals were, in addition, used for the characterization of electronic components like filters or a/d converters (cf. Toonen and Benz, 2009). The use of pulse-driven arrays was also suggested in combination with a binary-divided array; the spectrum of the pulse-driven array is adjusted to modify the spectrum of the 1 V or 10 V signal generated by the binary divided array (Kohlmann et al, 2007). In addition, pulse-driven arrays provide the opportunity for synthesizing a calculable pseudo-noise waveform consisting of a comb of random-phase harmonics each having identical voltage amplitude. A low-voltage version of this noise source is used in a quantum-based Johnson noise thermometry system to measure the voltage noise of the resistor, and thus its temperature (Benz et al, 2009b).

6. Conclusions

100 years after the discovery of superconductivity and nearly 50 years after the discovery of the Josephson effect, Josephson voltage standards play an essential role in electrical metrology and high-precision voltage measurements. The significant progress of the fabrication technology has been a major prerequisite for the development of large series arrays for Josephson voltage standards containing tens of thousands Josephson junctions. Conventional 10 V Josephson voltage standards are well established for dc measurements and commercially available. Programmable voltage standards opened up the world of ac applications and have, hence, been the next step in the exciting story of the applications of the Josephson effect in metrology. While 1 V arrays are meanwhile fabricated routinely, the first 10 V arrays containing tens or even hundreds of thousands of Josephson junctions are now available. Conventional Josephson voltage standards will be replaced in the future more and more by these programmable Josephson voltage standards, as they are easier to operate and provide exciting additional possibilities and applications. The synthesis of real quantum-based ac voltages is enabled by pulse-driven arrays. Very promising results have been achieved; output voltages of about 275 mV were synthesized with higher harmonics suppressed by about 120 dBc. However, the aim to generate 1 V ac voltages is very challenging due to the complex operation by short current pulses. The value of ac Josephson voltage standards has successfully been demonstrated in initial experiments. Further developments will establish these Josephson voltage standards as a quantum basis for ac metrology.

7. Acknowledgment

The authors would like to thank the Josephson voltage standard team at PTB.
The work within this EURAMET joint research project leading to these results was supported in part by the European Community's Seventh Framework Programme, ERA-NET Plus, under Grant Agreement 217257 (JoSy project).

8. References

Anders, S.; Blamire, M.G.; Buchholz, F.-Im. et al (2010). European roadmap on superconductive electronics - status and perspectives. *Physica C,* Vol.470, No.23-24, (December 2010) pp. 2079-2126

Baek, B.; Dresselhaus, P.D. & Benz, S.P. (2006). Co-sputtered amorphous Nb_xSi_{1-x} barriers for Josephson-junction circuits. *IEEE Transactions on Applied Superconductivity,* Vol.16, No.4 (December 2006) pp. 1966-1970

Bardeen, J.; Cooper, L.N. & Schrieffer, J.R. (1957). Theory of Superconductivity. *Physical Review,* Vol. 108, No. 5, (December 1957) pp. 1175-1204

Barone A. & Paterno G. (1982). Physics and applications of Josephson effect, John Wiley & Sons, ISBN 0-471-01469-9, New York, USA

Behr, R.; Schulze, H.; Müller, F.; Kohlmann, J. & Niemeyer J. (1999). Josephson arrays at 70 GHz for conventional and programmable voltage standards. *IEEE Transactions on Instrumentation and Measurement,* Vol.48, No.2, (April 1999) pp. 270-273

Behr, R.; Müller, F. & Kohlmann, J. (2002). Josephson junction arrays for voltage standards, In: *Studies of Josephson junction arrays II: Studies of high temperature superconductors,*

Vol.40, A.V. Narlikar (Ed.), 155-184, Nova Science Publishers, ISBN 1-59033-204-0, Hauppauge, NY, USA

Behr, R.; Williams, J.M..; Patel, P.; Janssen, T.J.B.M.; Funck, T. & Klonz, M. (2005). Synthesis of precision AC waveforms using a SINIS Josephson junction array. *IEEE Transactions on Instrumentation and Measurement*, Vol.54, No.2, (April 2005) pp. 612-615

Behr, R.; Palafox L.; Ramm G.; Moser H. & Melcher, J. (2007). Direct comparison of Josephson waveforms using an AC quantum voltmeter. *IEEE Transactions on Instrumentation and Measurement,* Vol.56, No.2, (April 2007) pp. 235-238

Benz, S.P. (1995). Superconductor-normal-superconductor junctions for programmable voltage standards. *Applied Physics Letters,* Vol.67, No.18, (October 1995) pp. 2714-2716

Benz, S.P. & Hamilton, C.A. (1996). A pulse-driven programmable Josephson voltage standard. *Applied Physics Letters,* Vol.68, No.22, (May 1996) pp. 3171-3173

Benz, S.P.; Hamilton, C.A.; Burroughs, C.J.; Harvey, T.E. & Christian, L.A. (1997). Stable 1-volt programmable voltage standard. *Applied Physics Letters,* Vol.71, No.13, (September 1997) pp. 1866-1868

Benz, S.P.; Hamilton, C.A.; Burroughs, C.J.; Harvey, T.E.; Christian, L.A. & Przybysz, J.X. (1998). Pulse-driven Josephson digital/analog converter. *IEEE Transactions on Applied Superconductivity,* Vol.8, No.2, (June 1998) pp. 42-47

Benz, S.P.; Hamilton, C.A.; Burroughs, C.J. & Harvey, T.E. (1999). AC and DC bipolar voltage source using quantized pulses. *IEEE Transactions on Instrumentation and Measurement,* Vol.48, No.2, (April 1999) pp. 266-269

Benz, S.P.; Burroughs, C.J. & Dresselhaus, P.D. (2001). AC coupling technique for Josephson waveform synthesis. *IEEE Transactions on Applied Superconductivity,* Vol.11, No.1, (March 2001) pp. 612-616

Benz, S.P. & Hamilton, C.A. (2004). Application of the Josephson effect to voltage metrology. *Proceedings of the IEEE,* Vol.92, No.10, (October 2004) pp. 1617-1629

Benz, S.P.; Dresselhaus, P.D.; Burroughs, C.J. & Bergren, N.F. (2007). Precision measurements using a 300 mV Josephson arbitrary waveform synthesizer. *IEEE Transactions on Applied Superconductivity,* Vol.17, No.2, (June 2007) pp. 864-869

Benz, S.P.; Dresselhaus, P.D.; Rüfenacht, A.; Bergren, N.F.; Kinard, J.R. & Landim, R.P. (2009a). Progress toward a 1 V pulse-driven AC Josephson voltage standard. *IEEE Transactions on Instrumentation and Measurement,* Vol.58, No.4, (April 2009) pp. 838-843

Benz, S.P.; Jifeng Qu; Rogalla, H.; White, D.R.; Dresselhaus, P.D.; Tew, W.L. & Sae Woo Nam (2009b). Improvements in the NIST Johnson noise thermometry system. *IEEE Transactions on Instrumentation and Measurement,* Vol.58, No.4, (April 2009) pp. 884-890

van den Brom, H.E.; Houtzager, E.; Chevtchenko, O.; Wende, G.; Schubert, M.; May, T.; Meyer, H.-G.; Kieler, O. & Kohlmann, J. (2007). Synthesis of sinusoidal signals with a Josephson arbitrary waveform synthesizer. *Superconductor Science and Technology,* Vol.20, No.5 (May 2007) pp. 413-417

van den Brom, H.E.; Houtzager, E.; Brinkmeier, B.E.R. & Chevtchenko, O.A. (2008). Bipolar pulse-drive electronics for a Josephson arbitrary waveform synthesizer. *IEEE Transactions on Instrumentation and Measurement,* Vol.57, No.2, (February 2008) pp. 428-431

Burroughs, C.J.; Rüfenacht, A.; Dresselhaus, P.D.; Benz, S.P. & Elsbury, M.M. (2009a). A 10 Volt "turnkey" programmable Josephson voltage standard for DC and stepwise-approximated waveforms. *NCSLI Measure,* Vol.4, No.3, (September 2009) pp. 70-75

Burroughs, C.J.; Rüfenacht, A.; Benz, S.P. & Dresselhaus, P.D. (2009b). Systematic error analysis of stepwise-approximated AC waveforms generated by programmable Josephson voltage standards. *IEEE Transactions on Instrumentation and Measurement*, Vol.58, No.2, (April 2009) pp. 761-767

Chong, Y.; Burroughs, C.J.; Dresselhaus, P.D.; Hadacek, N.; Yamamori, H. & Benz, S.P. (2005). Practical high-resolution programmable Josephson voltage standards using double- and triple-stacked $MoSi_2$-barrier junctions. *IEEE Transactions on Applied Superconductivity,* Vol.15, No.2, (June 2005) pp. 461-464

Dresselhaus, P.D.; Elsbury, M.M. & Benz, S.P. (2009). Tapered transmission lines with dissipative junctions. *IEEE Transactions on Applied Superconductivity,* Vol.19, No.3, (June 2009) pp. 993-998

Gurvitch, M.; Washington, M.A. & Huggins, H.A. (1983). High quality refractory Josephson tunnel junctions utilizing thin aluminum layers. *Applied Physics Letters,* Vol.42, No.5, (March 1983) pp. 472-474

Hagedorn, D.; Kieler, O.; Dolata, R.; Behr, R.; Müller, F.; Kohlmann, J. & Niemeyer, J. (2006). Modified fabrication of planar sub-μm superconductor-normal metal-superconductor Josephson junctions for use in a Josephson Arbitrary Waveform Synthesizer. *Superconductor Science and Technology,* Vol.19, No.4, (April 2006) pp. 294-298

Hamilton, C.A.; Burroughs, C.J. & Kautz, R.L. (1995). Josephson D/A converter with fundamental accuracy. *IEEE Transactions on Instrumentation and Measurement,* Vol.44, No.2, (April 1995) pp. 223-225

Hamilton, C.A.; Burroughs, C.J.; Benz, S.P. & Kinard, J.R. (1996). AC Josephson voltage standard: progress report. *IEEE Transactions on Instrumentation and Measurement*, Vol.46, No.2, (April 1997) pp. 224-228

Hamilton, C.A. (2000). Josephson voltage standards. *Review of Scientific Instruments,* Vol.71, No.10, (October 2000) pp. 3611-2623

Hassel, J.; Helistö, P.; Grönberg, L.; Seppä, H.; Nissilä, J. & Kemppinen, A. (2005). Stimulated power generation in ES-SIS junction arrays. *IEEE Transactions on Instrumentation and Measurement,* Vol.54, No.2, (April 2005) pp. 632-635

Hertel, G.; Bishop, D.J.; Spencer, E.G.; Rowell, J.M. & Dynes, R.C. (1983). Tunneling and transport measurements at the metal-insulator transition of amorphous Nb:Si. *Physical Review Letters,* Vol.50, No.10, (March 1983) pp. 743-746

Houtzager, E.; Benz, S.P. & van den Brom, H.E. (2009). Operating margins for a pulse-driven Josephson arbitrary waveform synthesizer using a ternary bit-stream generator. *IEEE Transactions on Instrumentation and Measurement,* Vol.58, No.4, (April 2009) pp. 775-780

Ihlenfeld, W.G.K.; Mohns, E.; Behr, R; Williams, J.M.; Patel, P.; Ramm G. & Bachmair, H. (2005). Characterization of a high-resolution analogue-to-digital converter with an AC Josephson voltage source. *IEEE Transactions on Instrumentation and Measurement*, Vol.54, No.2, (April 2005) pp. 649-652

Jeanneret, B. & Benz, S.P. (2009). Applications of the Josephson effect in electrical metrology. *The European Physical Journal Special Topics,* Vol.172, (June 2009) pp. 181-206

Jeanneret, B.; Overney, F.; Rüfenacht, A. & Nissilä, J. (2010). Strong attenuation of the transients' effect in square waves synthesized with a programmable Josephson voltage standard. *IEEE Transactions on Instrumentation and Measurement*, Vol.59, No.7, (July 2010) pp. 1894-1899

Josephson, B.D. (1962). Possible new effects in superconducting tunneling. *Physics Letters*, Vol.1, No.7, (July 1962) pp. 251-253

Josephson, B.D. (1965). Supercurrents through barriers. *Advances in Physics*, Vol.14, No.56, (October 1965) pp. 419-451

Kadin, A.M. (1999). Introduction to superconducting circuits, Wiley, ISBN 0-471-31432-3, New York, USA

Kautz, R.L. (1992). Design and operation of series-array Josephson voltage standards, In: *Metrology at the Frontiers of Physics and Technology*, L. Crovini and T.J. Quinn, (Eds.), 259-296, North-Holland, ISBN 0-444-89770-4, Amsterdam, The Netherlands

Kautz, R.L. (1995). Shapiro steps in large-area metallic-barrier Josephson junctions. *Journal of Applied Physics*, Vol.78, No. 9, (November 1995) pp. 5811-5819

Kieler, O.F.; Kohlmann, J.; Behr, R.; Müller, F.; Palafox, L. & Niemeyer, J. (2007a) SNS Josephson junction series arrays for the Josephson arbitrary waveform synthesizer. *IEEE Transactions on Applied Superconductivity*, Vol.17, No.2, (June 2007) pp. 187-190

Kieler, O.F.; Kohlmann, J. & Müller, F. (2007b). Improved design of superconductor/normal conductor/superconductor Josephson junction series arrays for an ac Josephson voltage standard. *Superconductor Science and Technology*, Vol.20, No.11, (November 2007) pp. S318-S322

Kieler, O.F.; Iuzzolino, R. & Kohlmann, J. (2009). Sub-µm SNS Josephson junction arrays for the Josephson arbitrary waveform synthesizer. *IEEE Transactions on Applied Superconductivity*, Vol.19, No.3, (June 2009) pp. 230-233

Kieler, O.F.; Schleussner, D.; Kohlmann, J. & Behr, R. (2010). Josephson arbitrary waveform synthesizer for analysis of AC components, In: *2010 Conference on Precision Electromagnetic Measurements (CPEM 2010)*, Yang Sup Song (Ed.), 157-158, Institute of Electrical and Electronic Engineers, ISBN 978-1-4244-6794-5, Piscataway, NJ, USA

Kim, M.-S.; Kim, K.-T.; Kim, W.S.; Chong, Y. & Kwon, S.-W. (2010). Analog-to-digital conversion for low frequency waveforms based on the Josephson voltage standard. *Measurement Science and Technology*, Vol.21, No.11, (November 2010) 115102 (6 pp.)

Kohlmann, J.; Behr, R. & Funck, T. (2003). Josephson voltage standards. *Measurement Science and Technology*, Vol.14, No.8, (August 2003) pp. 1216-1228

Kohlmann, J.; Müller, F.; Behr, R.; Hagedorn, D.; Kieler, O.; Palafox, L. & Niemeyer, J. (2006). Development of Josephson junction series arrays for synthesis of AC voltages and arbitrary waveforms. *Journal of Physics: Conference Series*, Vol.43, No.1, (2006) pp. 1385-1388

Kohlmann, J.; Müller, F.; Kieler, O.; Behr, R.; Palafox, L.; Kahmann, M. & Niemeyer, J. (2007). Josephson series arrays for programmable 10-V SINIS Josephson voltage standards and for Josephson arbitrary waveform synthesizers based on SNS junctions. *IEEE Transactions on Instrumentation and Measurement*, Vol.56, No.2, (April 2007) pp. 472-475

Kohlmann, J.; Kieler, O.F.; Iuzzolino, R.; Lee, J.; Behr, R.; Egeling, B. & Müller, F. (2009). Development and investigation of SNS Josephson arrays for the Josephson arbitrary waveform synthesizer. *IEEE Transactions on Instrumentation and Measurement*, Vol.58, No.4, (April 2009) pp. 797-802

Lacquaniti, V.; De Leo, N.; Fretto, M.; Sosso, A.; Müller F. & Kohlmann, J. (2011). 1 V programmable voltage standards based on SNIS Josephson junction series arrays. *Superconductor Science and Technology*, Vol.24, No.4, (April 2011) 045004 (4 pp.)

Lee, J.; Behr, R.; Katkov, A. & Palafox, L. (2009). Error contributions in stepwise synthesized Josephson sine waves. *IEEE Transactions on Instrumentation and Measurement*, Vol.58, No.2, (April 2009) pp. 803-808

Lee, J.; Schurr, J.; Nissilä, J.; Palafox, L. & Behr, R. (2010). The Josephson two-terminal-pair impedance bridge. *Metrologia* Vol. 47, No.4, (August 2010) pp. 453-459

Lee, J.; Schurr, J.; Nissilä, J.; Palafox, L.; Behr, R. & Kibble, B. (2011). Impedance measurements with programmable Josephson systems. *IEEE Transactions on Instrumentation and Measurement*, to be published (2011)

Levinson, M.T.; Chiao, R.Y.; Feldman, M.J. & Tucker, B.A. (1977). An inverse ac Josephson effect voltage standard. *Applied Physics Letters*, Vol.31, No.11, (December 1977) pp. 776-778

Likharev, K.K. (1986). Dynamics of Josephson junctions and circuits, Gordon & Breach, ISBN 2-881-24042-9, New York, USA

Lipe, T.E.; Kinard, J.R.; Tang, Y.-H.; Benz, S.P.; Burroughs, C.J. & Dresselhaus, P.D. (2008). Thermal voltage converter calibrations using a quantum ac standard. *Metrologia*, Vol.45, No.3, (June 2008) pp. 275-280

Maezawa, M. & Shoji, A. (1997). Overdamped Josephson junctions with Nb/AlO_x/Al/AlO_x/Nb structure for integrated circuit application. *Applied Physics Letters*, Vol.70, No.26, (June 1997) pp. 3603-3605

McCumber, D.E. (1968). Effect of ac impedance on dc voltage-current characteristics of superconductor weak-link junctions. *Journal of Applied Physics*, Vol.39, No. 7, (June 1968) pp. 3113-3118

Monaco, R. (1990). Enhanced ac Josephson effect. *Journal of Applied Physics*, Vol.68, No.2, (July 1990) pp. 679-687

Mueller, F.; Behr, R.; Palafox, L.; Kohlmann, J.; Wendisch, R. & Krasnopolin, I. (2007). Improved 10 V SINIS series arrays for applications in AC voltage metrology. *IEEE Transactions on Applied Superconductivity*, Vol.17, No.2, (June 2007) pp. 649-652

Mueller, F.; Behr, R.; Weimann, T.; Palafox, L.; Olaya, D.; Dresselhaus, P.D. & Benz, S.P. (2009). 1 V and 10 V SNS programmable voltage standards for 70 GHz. *IEEE Transactions on Applied Superconductivity*, Vol.19, No.3, (June 2009) pp. 981-986

Niemeyer, J.; Hinken, J.H. & Kautz, R.L. (1984). Microwave-induced constant voltage steps at one volt from a series array of Josephson junctions. *Applied Physics Letters*, Vol.45, No.4, (August 1984) pp. 478-480

Niemeyer, J.; Sakamoto, Y.; Vollmer, E.; Hinken, J.H.; Shoji, A.; Nakagawa, H.; Takada, S. & Kosaka, S. (1986). Nb/Al-oxide/Nb and NbN/MgO/NbN tunnel junctions in large series arrays for voltage standards. *Japanese Journal of Applied Physics*, Vol.25, No.5, (May 1986) pp. L343-L345

Niemeyer, J. (1998). Josephson voltage standards, In: *Handbook of Applied Superconductivity*, B. Seeber, (Ed.), 1813-1834, Institute of Physics Publishing, ISBN 0-750-30377-8, Bristol, UK

Palafox, L.; Ramm, G.; Behr, R.; Kürten Ihlenfeld, W.G. & Moser H. (2007). Primary AC power standard based on programmable Josephson arrays. *IEEE Transactions on Instrumentation and Measurement*, Vol.56, No.2, (April 2007) pp. 534-537

Palafox, L.; Behr, R.; Ihlenfeld, W.G.K.; Müller, F.; Mohns, E.; Seckelmann, M. & Ahlers, F. (2009). The Josephson effect based primary power standard at PTB: progress report. *IEEE Transactions on Instrumentation and Measurement*, Vol.58, No.4, (April 2009) pp. 1049-1053

Rogalla, H. (1998). Josephson junctions, In: *Handbook of Applied Superconductivity*, B. Seeber, (Ed.), 1759-1775, Institute of Physics Publishing, ISBN 0-750-30377-8, Bristol, UK

Rüfenacht, A.; Burroughs, C.J.; Benz, S.P.; Dresselhaus, P.D.; Waltrip, B. & Nelson, T.L. (2009). Precision differential sampling measurements of low-frequency synthesized sine waves with an AC programmable Josephson voltage standard. *IEEE Transactions on Instrumentation and Measurement*, Vol.58, No.4, (April 2009) pp. 809-815

Rüfenacht, A.; Overney, F.; Mortara, A. & Jeanneret, B. (2011). Thermal transfer standard validation of the Josephson-voltage-standard-locked sine wave synthesizer. *IEEE Transactions on Instrumentation and Measurement*, to be published (2011)

Schubert, M.; Fritzsch, L.; Wende, G. & Meyer H.-G. (2001a). SNS junction on Nb-Ti base for microwave circuits. *IEEE Transactions on Applied Superconductivity*, Vol.11, No.1, (March 2001) pp. 1066-1069

Schubert, M.; May, T.; Wende, G.; Fritzsch, L. & Meyer, H.-G. (2001b). Coplanar strips for Josephson voltage standard circuits. *Applied Physics Letters*, Vol.79, No.7, (August 2001) pp. 1009-1011

Schulze, H.; Behr, R.; Müller, F. & Niemeyer, J. (1998). Nb/Al/AlO_x/Al/AlO_x/Al/Nb Josephson junctions for programmable voltage standards. *Applied Physics Letters*, Vol.73, No.7, (August 1998) pp. 996-998

Schulze, H.; Müller, F.; Behr, R.; Kohlmann, J.; Niemeyer, J. & Balashov, D. (1999). SINIS Josephson junctions for programmable Josephson voltage standard circuits. *IEEE Transactions on Applied Superconductivity*, Vol.9, No.2, (June 1999) pp. 4241-4244

Schulze, H.; Behr, R.; Kohlmann, J.; Müller, F. & Niemeyer, J. (2000). Design and fabrication of 10 V SINIS Josephson arrays for programmable voltage standards. *Superconductor Science and Technology*, Vol.13, No.9, (September 2000) pp. 1293-1295

Semenov, V.K. & Polyakov, Yu.A. (2001). Circuit improvements for a voltage multiplier. *IEEE Transactions on Applied Superconductivity*, Vol.11, No.1, (March 2001) pp. 550-553

Séron, O.; Budovsky, I.; Djordjevic, S.; Hagen, T.; Behr, R. & Palafox L. (2011). Precision AC-DC transfer measurements with a Josephson waveform synthesizer and a buffer amplifier. *IEEE Transactions on Instrumentation and Measurement*, to be published (2011)

Shapiro, S. (1963). Josephson currents in superconducting tunneling: The effect of microwaves and other observations. *Physical Review Letters*, Vol.11, No.2, (July 1963) pp. 80-82

Stewart, W.C. (1968). Current-voltage characteristics of Josephson junctions. *Applied Physics Letters*, Vol.22, No.8, (April 1968) pp. 277-280

Sugiyama, H.; Yanada A.; Ota, M.; Fujimaki, A. & Hayakawa, H. (1997). Characteristics of Nb/AlO_x/Al/AlO_x/Nb junctions based on the proximity effect. *Japanese Journal of Applied Physics*, Vol.36, No.9A/B (September 1997) pp. L1157-L1160

Toonen, R.C. & Benz, S.P. (2009). Nonlinear behavior of electronic components characterized with precision multitones from a Josephson arbitrary waveform synthesizer. *IEEE Transactions on Applied Superconductivity,* Vol.19, No.3, (June 2009) pp. 715-718

Urano, C.; Maruyama, M.; Kaneko, N.; Yamamori, H.; Shoji, A.; Maezawa, M.; Hashimoto, Y.; Suzuki, H.; Nagasawa, S.; Satoh, T.; Hidaka, M. & Kiryu, S. (2010). A new coding technique in serial data transmission and demodulation with Josephson junctions array. *Journal of Physics: Conference Series,* Vol.234, No. 4, (2010) 042037 (5 pp.)

Williams, J.M.; Janssen, T.J.B.M.; Palafox, L.; Humphreys, D.A.; Behr, R.; Kohlmann, J. & Müller, F. (2004). The simulation and measurement of the response of Josephson junctions to optoelectronically generated short pulses. *Superconductor Science and Technology,* Vol.17, No.6, (June 2004) pp. 815-818

Williams, J.M.; Henderson, D.; Patel, P.; Behr, R. & Palafox L. (2007). Achieving sub-100 ns switching of programmable Josephson arrays. *IEEE Transactions on Instrumentation and Measuremen*t, Vol.56, No.2, (April 2007) pp. 651-654

Wood, B. & Solve, S. (2009). A review of Josephson comparison results. *Metrologia,* Vol.46, No.6, (December 2009) pp. R13-R20

Yamamori, H.; Ishizaki, M.; Shoji, A.; Dresselhaus, P. D. & Benz, S. P. (2006). 10 V programmable Josephson voltage standard circuits using $NbN/TiN_x/NbN/TiN_x/NbN$ double-junction stacks. *Applied Physics Letters,* Vol.88, No.4, (January 2006) 042503 (3pp.)

Yamamori, H.; Yamada, T.; Sasaki, H. & Shoji, A. (2008). A 10 V programmable Josephson voltage standard circuit with a maximum output voltage of 20 V. *Superconductor Science and Technology,* Vol.21, No.10, (October 2008) pp. 105007 (6 pp.)

Yoshida, H. (2000). Application of the ac Josephson effect for precise measurements. *IEICE Transactions on Electronics,* Vol.E83-C, No.1, (January 2000) pp. 20-26

12

Critical State Analysis Using Continuous Reading SQUID Magnetometer

Zdeněk Janu[1], Zdeněk Švindrych[2], Ahmed Youssef[3] and Lucia Baničová[4]
[1,2]*Institute of Physics AS CR, v.v.i., Prague*
[3,4]*Charles University in Prague, Faculty of Mathematics and Physics, Prague*
Czech Republic

1. Introduction

The critical state in type II superconductors determines the maximum current the superconductor can carry without an energy dissipation. The critical state results from a competition between the Lorentz force acting on flux lines (quantized vortices), thermal agitation, pinning force, and repulsive interaction between flux lines. The pinning force localizes the flux lines on crystal lattice defects (dislocations, voids or impurities) and favors glassy state of flux lines, whereas the repulsive interaction between vortices results in a regular flux line lattice. Materials with a strong pinning are called hard superconductors. Such materials are relevant for power application of superconductors: solenoids for high magnetic fields or cables for large transport currents. Recently, high temperature superconductor (HTS) materials with the critical current density j_c of the order of 100 GA m^{-2} at zero temperature and zero applied field were prepared. The second generation of HTS wires (2GHTSC) is constituted from RE-$Ba_2Cu_3O_{6+x}$ (YBCO) films. The critical current density is one or two orders higher than was achieved in $Bi_2Sr_2CaCu_2O_{8+x}$ (BSCCO) round wires or MgB_2, Nb-Ti, Nb_3Sn, and Nb_3Al wires. Unlike BSCCO wires whose performance is lowered by a flux flow at temperature above 35 K the YBCO wires operate even at liquid nitrogen temperature. Another important field of application of superconductors is superconducting electronics. Most of today's superconducting electronics like superconducting quantum interferometer devices (SQUIDs), radiation detectors (SIS mixers), etc. are made of Nb, NbN, or HTS films. The flux lines trapped in the superconducting film may deteriorate sensor sensitivity as the moving flux lines generate noise (Wellstood et al., 1987). The above mentioned elucidates an interest in flux dynamics in thin films, particularly models to a disk and stripe.

The critical state is affected by material properties, the wire or sensor geometry (shape), applied current, field, and temperature. Conventionally the critical state is studied (judged) using contact measurements (four probe resistive method) or magnetic measurements (local magnetization profile or magnetization loops). The latter method eliminates the need for electrical contacts and allows us to study the response of the critical state to an applied magnetic field. Frequency dependent magnetization loops reveal a flux creep or flux flow while nonlinear magnetization loops reveal surface or bulk pinning. In order to analyze these magnetic measurements we need appropriate models. In general, these model represent solution of 3D+t partial differential equations for a magnetic vector potential or flux density.

Numerical methods apply to conductors and superconductors with axial symmetry, but otherwise with an arbitrary cross section like cylinders of finite length, thin and thick disks, cones, spheres, and rotational ellipsoids. The specimen may even be inhomogeneous and anisotropic as long as axial symmetry pertains (Brandt, 1998). Complete analytical solutions are known only for particular geometries and quasistatic behavior of the magnetic flux when the problem may be reduced to 2D. Two such examples are thin disk and strip in Bean critical state in perpendicular magnetic field.

For magnetization loop measurements, one needs a low frequency magnetic field and low frequency sensor of the magnetic moment of the sample. The whole system should be of high linearity, flat frequency and phase dependence - good choice is the superconducting solenoid and SQUID magnetometer. However, commercial SQUID magnetometers are not suitable for such measurements because the solenoid operates in a persistent mode during a measurement and settling time (dead time) affects (slows down) the measurement.[1] Further, a residual field in the high field solenoid causes a nonlinear $H(I)$ dependence. Since the magnetic moment of the sample is measured differentially, reciprocating the sample punctuates the measurement.

2. Continuous reading SQUID magnetometer

An operation of a continuous reading SQUID magnetometer (CRSM) with an immobile sample is based on detection coils in a gradiometer arrangement, which are insensitive to the homogeneous time varying applied magnetic field, but respond to the magnetic sample placed in proximity of one of the coils. A spontaneous or induced magnetic moment of the sample creates a difference in a magnetic flux in the coils and generates a current in an input coil of the SQUID. The SQUID thus measures the variations in the magnetic moment of the sample. Since the sample is immobile no noise or disturbances are generated due to a sample motion and measurement is not interrupted due to a reciprocating sample or sample positioning. The applied field is generated by a superconducting solenoid operating in a nonpersistent mode.

We use SQUID magnetometers in two basic configurations: Standard Sensitivity and High Sensitivity. In a Standard Sensitivity SQUID Magnetometer (SSSM), the superconducting solenoid, gradiometer, and SQUID are immersed in a liquid helium bath, see Fig. 1. The sample holder with a sample temperature sensor is placed inside an anticryostat.

In a High Sensitivity SQUID Magnetometer (HSSM), the superconducting solenoid, gradiometer, SQUID, and a sample holder with a temperature sensor and heater are placed in a copper vacuum chamber with an inset lead can, see Fig. 1. While the solenoid, gradiometer and SQUID are thermally anchored to the vacuum chamber immersed in a cooling liquid helium bath, the sample is mounted on a block suspended on a support with a low thermal conductivity.

2.1 Applied field generation

The applied homogeneous field is generated using a superconducting solenoid operating in the non-persistent mode. The solenoid is wound with a Nb-Ti wire (number of layers) on a coil-former. The solenoid is supplied from a current source driven by a digital to analog converter (DAC) of a data generation/acquisition card.[2] These $\Sigma - \Delta$ DAC

[1] Quantum Design.

[2] National Instruments PC card model PCI-4451.

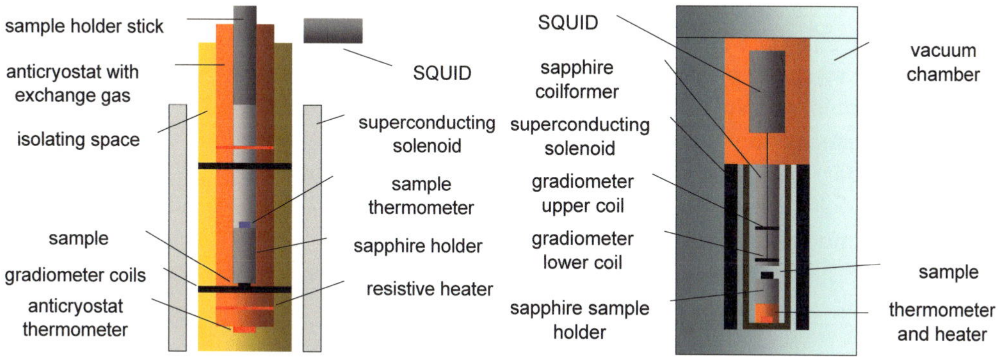

(a) The standard sensitivity magnetometer. (b) The high sensitivity magnetometer.

Fig. 1. Schematic drawing of the SQUID magnetometer

have superior linearity and dynamic range. An applied field $h(t)$ may essentially be of an arbitrary waveform: an AC field superimposed on a DC field for measurement of temperature dependence of susceptibility, with a linear or sinusoidal sweep for measurement of magnetization loops, pulse or step-like for relaxation measurements, frequency sweep, etc. The waveform is designed numerically.

	SSSM	HSSM
Field range (Setting resolution)	$\pm$25 mT ()	$\pm$4 mT ()
Frequency range	DC - 100 Hz	DC - 100 Hz
Temperature range	4.2 - 300 K	4.2 - 150 K
Temperature rate	0.001 - 1 K/min	0.001 - 1 K/min
Sensitivity	7 pA m^2 Hz$^{-1/2}$	5 fA m^2 Hz$^{-1/2}$

Table 1. The parameters of the magnetometers.

Another important property of a SQUID magnetometer is the degree of homogeneity of the applied magnetic field (both in z and r direction). High homogeneity solenoids generating a DC bias field have the homogeneity of the order of 10^{-4} over 4 cm (Vrba, 2001).

2.2 Detection system

The detection system includes superconducting flux transformer and the SQUID. The transformer comprises of coils in a gradiometric arrangement. Two coils with an opposite winding (sense) direction and areas $\mathbf{S}_1$ and $\mathbf{S}_2$ form a first order axial gradiometer which is insensitive to the homogeneous applied field H_0. A balance of the gradiometer, defined as

$$\eta = (\mathbf{S}_1 + \mathbf{S}_2) \cdot \mathbf{S}_1 / |\mathbf{S}_1|^2 \tag{1}$$

is $\eta = 0$ in an ideal case. In practise, any gradiometer is manufactured with a finite mechanical precision and the balance $\eta = 0.0001$ may be achieved with a careful construction (Vrba, 2001).

An inhomogeneous applied field and imperfect gradiometer balance result in a crosstalk of the field to the SQUID and reduce a dynamic range of the CRSM. In SSSM a compensation coil wound on an upper part of the solenoid and supplied with an adjustable current derived from the solenoid supply current minimizes crosstalk. A careful design and construction keeps down deformation of the field affected by a proximity of magnetic or superconducting materials (solder) and frequency dependent eddy currents in metallic (nonsuperconducting) parts.
The magnetic moment of the sample is

$$\mathbf{m} = \frac{1}{2}\int_V (\mathbf{r} \times \mathbf{j})\, d^3r. \tag{2}$$

A vector potential of the induced or spontaneous magnetic moment **m** of the sample is

$$\mathbf{A} = \mu \frac{\mathbf{m} \times \mathbf{r}}{r^3}. \tag{3}$$

The magnetic flux in the pickup coil is

$$\Phi = \oint_\Gamma \mathbf{A} \cdot d\mathbf{l}, \tag{4}$$

where Γ is the coil circumference. The SQUID indicates difference in the flux in an upper and lower coil, $\Delta\Phi = \Phi_{upper} - \Phi_{lower}$, and thus the SQUID output voltage is proportional to a projection of the measured magnetic moment on a gradiometer axis, $m(t) \propto \Delta\Phi(t)$.
Since the detection system is superconducting, the output voltage $m(t)$ is proportional to the magnetic moment of the sample and not to a rate of change of the magnetic moment like in case of induction magnetometers (ac susceptometer (ACS) or vibrating sample magnetometer (VSM)).
Both the SSSM and HSSM use bulk Nb SQUID of the Zimmerman type operating at the rf frequency of about 40 MHz. The Josephson junction is a point contact type in the SSSM and thin film bridge in the HSSM. Both SQUIDs have an equivalent input flux noise density of the order of $10^{-4}\ \Phi_0\ \mathrm{Hz}^{-1/2}$ in a white noise region (> 1 Hz) and range $\pm 500\ \Phi_0$ limited by a slew rate $10^4\ \Phi_0/\mathrm{s}$.[3]
A shielding of an external dc and time varying electromagnetic field originating from an earth magnetic field and man-made sources is necessary to utilize the extraordinary sensitivity of the SQUIDs. The shielding is ensured by a soft magnetic materials (the cryostat is placed inside the shielding) and superconducting shielding (Tsoy et al., 2000).

2.3 Sample mounting and temperature reading and control

In SSSM a sample is glued on a bottom surface of a cylindrical sapphire holder using a varnish or grease. A sample temperature sensor, the Si or GaAlAs diode[4], is mounted on the upper surface. The sapphire holder is connected to a (nonmagnetic, nonconducting) polyethylene straw that extends a thin wall stainless tube suspended in an anticryostat. Another Si diode

[3] iMAG 303 SQUID: The equivalent input noise for the standard LTS SQUID system is less than $10^{-5}\ \Phi_0\ \mathrm{Hz}^{-1/2}$, from 1 Hz to 50 kHz in the $\pm 500\ \Phi_0$ range. The response is flat from DC to the 3 dB points, slow slew mode 500 Hz (- 3 dB), normal slew mode 50 kHz (- 3 dB). The input inductance of the LTS SQUID is 1.8×10^{-6} H.

[4] Lake Shore or CryoCon

temperature sensor measures temperature of the anticryostat to facilitate better closed-loop temperature control. Two section resistance wire (constantan) heater is wound around the top and bottom part of the anticryostat to ensure uniform warming. Heat is removed from the sample by a ^{4}He gas at atmospheric pressure.
In HSSM the sample is mounted on the upper surface of the sapphire holder. The holder is embedded in a copper block whose temperature is measured using the Si diode sensor. The block is heated using a resistance wire heater and suspended on a low thermal conductivity fibreglass support which removes heat to liquid ^{4}He bath. The sample is in vacuum.
In both magnetometers, a temperature controller[5] connected to the computer regulates temperature with relative stability of 10 ppm and 1 ppm in SSSM and HSSM, respectively, and controls cooling or warming with rate from 1 mK/min to 10 K/min.

2.4 Measurement modes

The magnetometers are designed for measurements of: i) temperature dependence of a response to fixed AC and DC applied magnetic field (temperature dependence of the susceptibility); ii) response to field sweep at fixed temperature and AC field (magnetization loops and AC susceptibility); iii) relaxation of a DC magnetic moment (after applied field pulse or step) as a function of time or temperature; iv) frequency dependence at a fixed DC field and temperature. Additional measurement modes require only a software change.

2.5 Data acquisition

The dynamic range of the SQUID is extraordinary, the range of $\pm 500\ \Phi_0$ and spectral flux noise density of $10^{-4}\ \Phi_0\ \mathrm{Hz}^{-1/2}$ represent output voltage range ± 10 V and voltage noise density 10 μV $\mathrm{Hz}^{-1/2}$, a range of 7 orders (140 dB).[6] The frequency response is flat both in a frequency and phase. In slow slew mode the -3 dB point is 100 Hz. The SQUID output signal $m(t)$ falls into an audio range and thus may be easily digitized in "CD" quality as well as the signal of the applied field $H(t)$, recorded on a hard disk, and digitally processed in real time.[7]
Processed data file includes temperature readings.

2.6 AC susceptibility measurement (calculation)

Let the time varying applied AC magnetic field is

$$H(t) = H_{ac} \cos(2\pi f_0 t) = H_{ac} \mathrm{Re} \exp(i 2\pi f_0 t), \tag{5}$$

where H_{ac} is the amplitude and f_0 is the frequency of the applied field. The complex AC susceptibility of the sample is

$$\chi_n = \frac{\mathcal{M}(n f_0)}{H_{ac} V}, \tag{6}$$

[5] CryoCon model 34
[6] This applies to rf-SQUIDs. The flux noise density in DC SQUIDs is lower, $10^{-6}\ \Phi_0\ \mathrm{Hz}^{-1/2}$, corresponding voltage noise density 0.1 μV $\mathrm{Hz}^{-1/2}$, and dynamic range of 9 orders (180 dB).
[7] We use the National Instruments PC cards model PCI-4451 with $\Sigma - \Delta$ digital to analog and analog to digital converters for a digital signal generation and acquisition (two input channels with 16 bit resolution, frequency range from 0 (true DC) to 95 kHz, and sampling rate up to 204.8 kS/s).

where n denotes harmonics and $\mathcal{M}(nf_0)$ are the Fourier components of the magnetic moment $m(t)$. Higher harmonics of the complex susceptibility appear in the case of a nonlinear response to the applied field. Usually the susceptibility is normalized to a volume V (or mass) of the sample. Using the susceptibility, the magnetization loops are

$$M(H(t)) = \sum_n \chi_n H_{ac} \exp(ni2\pi f_0 t), \tag{7}$$

A common way to measure the AC susceptibility is to detect a signal of the magnetic moment using a phase sensitive lock-in amplifier, preferably a two phase instrument indicating both real and imaginary part of the AC susceptibility, and drive the AC field using a signal generator. The conventional analog lock-in amplifier multiplies the input signal $m(t)$ by a square wave $r(t)$ derived from a reference signal $H(t)$ and integrates the product. The DC output are in-phase and out-of-phase components

$$\mathrm{Re}\mathcal{M}(f_0) = \frac{4}{\pi\tau}\int_{t-\tau}^{t}\left[\sum_{n=1}^{\infty}\frac{1}{n}\sin(n\frac{\pi}{2})\cos\left(n2\pi f_0 t'\right)\right] m(t')dt', \tag{8}$$

$$\mathrm{Im}\mathcal{M}(f_0) = \frac{4}{\pi\tau}\int_{t-\tau}^{t}\left[\sum_{n=1}^{\infty}\frac{1}{n}\sin\left(n2\pi f_0 t'\right)\right] m(t')dt', \tag{9}$$

where n is odd and τ is the averaging time constant. Since the reference signal $r(t)$ is a square wave, the DC output is proportional not only to the Fourier component of the first harmonic but also to 1/3 of third, 1/5 of fifth, etc. Evidently, this way of signal processing is not suitable for the measurement of a nonlinear response. One can apply input filters that sufficiently suppress third and higher odd harmonics, but remain unaffected the fundamental frequency. However, suitable tunable filters are complex and expensive.

In the digital signal processor (DSP) lock-in amplifiers the signal is filtered with a simple anti-aliasing filter and digitized by over-sampling ADC with subsequent digital filtering. The DSP chip then synthesizes digital reference sine (and cosine) wave at the reference frequency nf_0 and multiplies the signal by this reference. After multiplication, stages of digital low-pass filtering are applied to average over the signal period. The DSP lock-in amplifier generates the true rms values of the complex Fourier components of $\mathcal{M}(f_0)$ or nth harmonic $\mathcal{M}(nf_0)$:

$$\mathcal{M}(nf_0) = \frac{1}{N\Delta t}\sum_{k=0}^{N-1} m(t_k)\exp(ni2\pi f_0 t_k), \tag{10}$$

where $\Delta t = t_k - t_{k-1}$ is the sampling interval and $N\Delta t$ is averaging time. However, commercial DSP lock-in amplifiers provide only components at single frequency. Hence, unless successive measurements of the harmonics are done, one needs an extra instrument for the each additional harmonic.

With computational power of today's processors in personal computers (PC) and data generation/acquisition hardware the problem as a whole may be solved much more effectively. The single PC card, with essentially the same ADC as are used in the DSP lock-in amplifier, substitutes for the generator and lock-in amplifiers. Since the DACs generating the applied field and ADCs sampling $m(t)$ and $H(t)$ use the same clock, synchronization is guaranteed. In reality, an approach using a direct digital signal generation, acquisition, and processing is more cost effective and less time consuming.

The nth harmonic of the AC susceptibility is given by generalized Eq. 6,

$$\chi_n = \frac{\mathcal{M}(nf_0)}{H_{ac}\exp(ni\varphi)}, \tag{11}$$

where complex $H_{ac}\exp(ni\varphi) \equiv |\mathcal{H}(f_0)|\exp(ni\arg\mathcal{H}(f_0))$ takes into account a phase of the Fourier component of the applied field $\mathcal{H}(f_0)$, i.e. a time shift between a Fourier transformed data segment and cosine field. The $\mathcal{M}(f)$ and $\mathcal{H}(f)$ spectra are computed using a discrete fast Fourier transform (FFT) of real data arrays $m(t_k)$ and $H(t_k)$.

$$\mathcal{M}_l \equiv \sum_{k=0}^{N-1} m(t_k)\exp\left(i2\pi kl/N\right), \tag{12}$$

(the same holds for $H(t) \Leftrightarrow \mathcal{H}(f)$), where N is the transform length (Press et al., 1992). Spectra of the complex amplitudes $\mathcal{M}(f)$ and $\mathcal{H}(f)$ are calculated for frequencies $l\Delta f$, $\mathcal{M}_l \equiv \mathcal{M}(l\Delta f)$. With an applied FFT algorithm N must be a power of 2, FFT is computed in $N\log N$ operations, and $\Delta f = f_s/N$, where $f_s = 1/\Delta t$ is the sampling frequency.[8] Unlike the DSP lock-in amplifiers, where another instrument performing N operations to process $N\Delta t$ long record is need for each measured harmonic, here the whole frequency spectrum from DC to $f/2$ is computed with only $N\log N$ operations using the single instrument. Computation time takes few ms.

Strictly speaking, the measurement of temperature dependence of the susceptibility represents a continuous measurement of magnetization loops at slowly varying temperature. Since the input signals are recorded as well as temperature readings, various time domain and frequency domain filters may be applied thereupon. The magnetization loops may be processed using different time windows (for example to remove a linear trend in $m(t)$) or different averaging times.

3. Critical state in type II superconductors

3.1 Vortex matter

Type II superconductors, ie. those with $\lambda/\xi > 2^{-1/2}$, where λ is the flux penetration length and ξ is the coherence length of a superconducting order parameter, remain superconducting even in a high magnetic field due to lowering of their energy by creating walls between normal and superconducting regions. Consequently, flux lines (vortices) with a normal core of a radius of $\approx \xi$, where the order parameter vanishes, and persistent current circulating around the core and decaying away from the vortex core at distances comparable with λ are created at sample edges and penetrate into an interior of the superconductor. The vortex is a linear (in three dimensions) object which is characterized by a quantized circulation of the phase of the order parameter around its axis and carries a single quantum of the magnetic flux $\Phi_0 = h/2e$. The superconductor penetrated with the flux lines is called to be in a mixed state. A repulsive interaction between the flux lines eventually forms flux line bundles and consecutively a flux

[8] Let us take $N = 2^{14}$ (16 K samples), easy for real time processing on a common PC. With $f_s = 6.4$ kS/s the $\Delta f = 0.390625$ Hz. A right choice for the AC field frequency f_0 is an integer multiple of Δf. For example, with $f_0 = 4\Delta f = 1.5625$ Hz, one period of the AC field is represented by 4 K samples. In this case the 16 K FFT means averaging over 4 periods (2.56 s) of the AC field. If the 16 K data are shifted by 4 K and a void part is replaced with samples of the latest read period, the spectra are averaged over 2.56 s and updated in 0.64 s interval. The index of the nth harmonics amplitude is $l = n4$.

line lattice. In increasing applied field the flux lines enter into the superconductor when the magnetic field exceeds the lower critical field $H_{c1} \approx \Phi_0/\mu_0\lambda^2$. Type II superconductors experience a second-order phase transition into a normal state at the upper critical field $H_{c2} \approx \Phi_0/\mu_0\xi^2$. In type I superconductors this transition is a first-order in a nonzero field.

3.2 Pinning and surface barrier

In a real type II superconductor there are always crystal lattice distortions, voids, interstitials, and impurities with reduced superconducting properties. The superconducting order parameter is either reduced or suppressed completely, just as within a vortex core. That implies that such defects are energetically favorable places for vortices to reside and the vortices will be pinned in the potential of these so-called pinning centers. The efficiency of such a pinning center is at its maximum if its size is of the order of the coherence length ξ. If there is almost no pinning, flux flow occurs (Bardeen, 1965). On the other hand, when there is finite pinning, flux creep of a vortex bundles takes place (Anderson, 1962; 1964). The bundle size is determined by the competition between pinning and the elastic properties of the vortex lattice.

An edge or surface barrier may oppose a flux entry into the sample (Beek et al., 1996). A surface barrier arises as a result of the repulsive force between vortices and the surface shielding current. The first example is Bean-Livingston barrier, which is a feature of flat type II superconductor surfaces in general and is related to a deformation of the vortex at the surface (mirror vortex). The second example is the edge-shape barrier, which is a geometric effect related to the distribution of the Meissner shielding current density in non-ellipsoidal samples.

When an increasing magnetic field is initially applied, flux cannot overcome the barrier, and $M = -H$. At the field of the first flux penetration H_p, the magnetic pressure is sufficiently high to overcome the barrier. If there is no pinning, vortices will now distribute themselves through the sample in such a way that the bulk current is zero and vortex density is homogeneous.

3.3 Flux line dynamics

When the superconductor is carrying a bulk transport or shielding current density $\mathbf{j}$ the flux lines experience a volume density of the driving Lorentz force $\mathbf{f}_L = \mathbf{j} \times \mathbf{B}$, where $\mathbf{B}$ is the flux density inside the flux line. When the Lorentz force acting on the flux lines is exactly balanced by the pinning force density, i.e. $F_L = F_p$, the current density is called the depinning current density, j_c. Under this force the flux lines may move through the crystal lattice and dissipate energy. In this case the electrical losses are no longer zero. In an ideal (homogeneous) type II superconductor there is nothing to hinder the motion of flux lines and the flux lines distribution is homogeneous. The flux lines can move freely, which is equivalent to a vanishing critical depinning current density j_c. On the other hand, the non-dissipative macroscopic currents are the result of the spatial gradients in the density of flux lines or due to their curvature. This is possible only due to the existence of pinning centers, which can compensate the Lorentz force.

The moving flux lines dissipate energy by two effects which give approximately equal contributions: (a) eddy currents that surround each moving flux line and have to pass through the vortex core, which in the model of Bardeen and Stephen is approximated by a normal conducting cylinder (normal currents flowing through the vortex core) (Bardeen, 1962); (b)

Tinkham's mechanism of a retarded recovery of the order parameter at places where the vortex core has passed (Tinkham, 1996).
In general, the current density in type II superconductors can have three different origins: (a) Surface currents within the penetration depth λ. In the Meissner state the current passing through a thick superconductor is restricted to a thin surface layer where the magnetic field can penetrate. Otherwise the magnetic field due to the current would exist inside the superconductor; (b) A gradient of the flux-line density; (c) A curvature of the flux lines.
A flux line motion is discouraged (inhibited) by pinning of individual flux lines, their bundles or lattice. In cases of flux flow and flux creep, the vortices are considered to move in an elastic bundle. With discovery of HTS, however, more complex forms of vortex motion are considered. When the driving force is small, the vortices move in a plastic manner - plastic flow where there are channels in which vortices move with a finite velocity, whereas in other channels the vortices remain pinned (Jensen, 1988). Thus, between moving channels and static channels there are dislocations in the flux lattice. With further increasing driving current, vortices tend to re-order. Through dynamic melting, a stationary flux lattice changes into a moving flux lattice via the plastic flow (Koshelev & Vinokur, 1994).
If pinning is efficient the critical depinning current density j_c becomes high and the material is interesting for applications. The properties of the flux line lattice and the pinning properties are important for applications; on the other hand they are complex and interesting topics of condensed-matter physics and materials science.

3.4 Equation of motion of vector potential

In general, computation of magnetization loops represents a full treatment of a nonlinear 3D problem described by a partial differential equation for a vector potential

$$\frac{\partial \mathbf{A}}{\partial t} = D\nabla^2 \mathbf{A}, \tag{13}$$

where D is the diffusivity. Due to an axial symmetry or for a long sample in a parallel field, the problem may reduce to 2D and the current density $\mathbf{j}$, vector potential $\mathbf{A}$, and electric field $\mathbf{E}$ are parallel to each other and have only a y or ϕ component (applied field is parallel to z axis) (Brandt, 1998). The magnetization loops are obtained solving Eq. 13 using specialized software packages or directly by the time integration of the nonlocal and nonlinear diffusion equation of motion for the azimuthal current density. A long cylinder or slab in parallel field or thin circular disk and strip in an axial field are 1D problems. The flux density and electric field are $\mathbf{B} = \nabla \times \mathbf{A}$ and $\mathbf{E} = -\partial \mathbf{A}/\partial t$, respectively.
In the normal (nonsuperconducting) state with an ohmic conductivity σ is $D = 1/\mu_0\sigma = m/\mu_0 ne^2\tau$. In Meissner state the diffusivity is the pure imaginary $D = i\omega m/\mu_0 n_s e^2$ with a linear frequency dependence, where n_s is the superconducting condensate density.
In an inhomogeneous type II superconductor with flux pinning the electric field is given by nonlinear local and isotropic resistivity $\rho(\mathbf{j})$. A material law $\mathbf{E}(\mathbf{j})$ reflects a flux line pinning. In case of a strong pinning $\mathbf{E}(\mathbf{j})$ is zero up to the critical depinning density j_c at which electric field raises sharply. A power law voltage current relation

$$\mathbf{E}(j) = E_c|j/j_c|^n \mathbf{j}/j = \rho_c|j/j_c|^{n-1}\mathbf{j}, \tag{14}$$

where $j = |\mathbf{j}|$, is observed in numerous experiments (Brandt, 1996). From the theories on (collective) creep, flux penetration, vortex glass picture, and AC susceptibility one obtains the useful general interpolation formula

$$U(J) = U_0 \frac{(j_c/j)^\alpha - 1}{\alpha}. \tag{15}$$

Here $U(j)$ is a current-dependent activation energy for depinning which vanishes at the critical current density j_c, and α is a small positive exponent. In the limit $\alpha \to 0$ one has a logarithmic dependence of the activation energy $U(j) = U_0 \ln(j_c/j)$, which inserted into an Arrhenius law yields

$$\mathbf{E}(j) = E_c \exp\left(-\frac{U(j)}{k_B T}\right) = E_c \left(\frac{j}{j_c}\right)^{U_0/k_B T}. \tag{16}$$

When we compare Eq. 16 with Eq. 14 the exponent is $n = U_0/k_B T$. For $\alpha = -1$ the Eq. 15 coincides with the result of the Kim-Anderson model, $E(j) = E_c \exp[(U_0/k_B T)(1 - j/j_c)]$, (Blatter et al., 1994). For $\alpha = 1$ one gets $E(j) = E_c \exp[(U_0/k_B T)(j_c/j - 1)]$.
In general, the E_c and activation energy U in Eq. 16 depend on the local induction $\mathbf{B}(\mathbf{r})$ and thus also $\alpha(\mathbf{B}, T)$ and $j_c(\mathbf{B}, T)$ depend on $\mathbf{B}$.
With $\mathbf{E} = -\partial\mathbf{A}/\partial t$ and Eq. 14 one obtains for the diffusivity in Eq. 13

$$D(j, j_c, U_0, T) = \frac{1}{\mu_0}\frac{\partial \mathbf{E}}{\partial \mathbf{j}} = \frac{1}{\mu_0}\frac{E_c}{j_c}\left(\frac{j}{j_c}\right)^{U_0/k_B T - 1} = \frac{\rho_c}{\mu_0}\left(\frac{j}{j_c}\right)^{U_0/k_B T - 1}. \tag{17}$$

Power-law electric field versus current density (Eq. 14) induces:
i) An Ohmic conductor behavior with a constant resistivity $\rho = E/j$ for $U_0/k_B T = 1$. This applies also to superconductors in the regime of a linear flux flow or thermally activated flux flow (TAFF) at low frequencies with flux-flow resistivity $\rho_f = \rho_n B/\mu_0 H_{c2}$, known as the Bardeen-Stephen model. The diffusivity D is large and vector potential profiles are time dependent. The magnetization loops have a strong frequency dependence, as well as the susceptibility, and the AC susceptibility has only fundamental component independent on the AC field amplitude (Gömöry, 1997).
ii) Flux creep behavior for $1 \ll U_0/k_B T < \infty$. The magnetization loops have a weak frequency dependence, as well as the AC susceptibility which has higher harmonics and is dependent on the AC field amplitude.
iii) Hard superconductors with strong pinning for $U_0/k_B T \to \infty$. In this case the flux dynamics is quasistatic, described by a Bean model of the critical state with $D = 0$ for $|j| < j_c$ and $D \to \infty$ for $|j| = j_c$. The magnetization loops are frequency independent, as well as the AC susceptibility which has higher harmonics and strongly depends on the AC field amplitude.
A general solution of Eq. 13 represents time dependent vector potential profiles which dynamics covers a viscous flow, diffusion (creep), and quasistatic (sand pile like) behavior. The resistivity generated by the flux creep is Ohmic in the low-driving force limit.

3.5 Analytically solvable models

3.5.1 Normal state with ohmic conductivity and flux flow state

In normal state with an ohmic conductivity $\sigma = ne^2\tau/m$ the diffusion constant is $D = 1/\mu_0\sigma = \omega\delta^2$, where ω is the angular frequency of the applied AC field and $\delta = (2\mu_0\omega\sigma)^{-1/2}$

is the normal skin depth. In this case the analytical solutions to Eq. 13 are known for an infinitely long cylinder and slab in a parallel field, cylinder in a perpendicular field, and sphere (Brandt, 1998; Khoder & Couach, 1991; Lifshitz et al., 1984).
With an increasing ratio δ/R or δ/d, where and R is the radius of the cylinder or sphere and $2d$ id the slab thickness, a sample changes from a diamagnetic (but lossy) at $\delta \ll R$, to absorptive at $\delta \approx R$, and to transparent for applied field at $\delta \gg R$. The magnetization loops $M(H)$ are ellipses which major axis lies on H axis of $H - M$ diagram for transparent medium and gradually turns to $-\pi/4$ direction for diamagnetic medium. The susceptibility as a function of $(\delta/R)^2$ is shown in Fig. 2.
In a limit of low frequencies when the skin depth $\delta \gg R, d$ and the sample is transparent for AC field the first terms in series expansion of the susceptibility are (up to a shape dependent multiplication factor)

$$\mathrm{Re}\chi \approx -\left(R^2\mu\omega\sigma\right)^2 \tag{18}$$

$$\mathrm{Im}\chi \approx \left(R^2\mu\omega\sigma\right), \tag{19}$$

and $\mathrm{Re}\chi \ll \mathrm{Im}\chi$. A measurement of χ yields contactless estimation of the electrical conductivity σ.
In a linear or thermally activated flux flow state as the applied field approaches the upper critical field H_{c2}, the flux density in the superconductor $B \to \mu_0 H_{c2}$ and the flux flow resistivity ρ_f smoothly transforms to $\rho_n = 1/\sigma$

$$\frac{\rho_f}{\rho_n} \approx \frac{B}{\mu_0 H_{c2}} \tag{20}$$

as the phase transition between a mixed state and normal state is of second order (Bardeen Stephen model) (Bardeen, 1965). Flux flow resistivity may be estimated using Eq. 19.

3.5.2 Meissner state

At initial magnetization the superconductor is in Meissner state in field lower that H_{c1}. In this case the diffusivity is pure imaginary $D = i\omega\lambda^2$, where the flux penetration length is $\lambda = (\mu_0 n_s e^2/m)^{-1/2}$. The susceptibility of an infinitely long cylinder and slab in a parallel field, cylinder in a perpendicular field, and sphere is obtained like for normal state but replacing $(1+i)/\delta$ with i/λ (Brandt, 1998; Khoder & Couach, 1991; Lifshitz et al., 1984). The susceptibility as a function of $(\lambda/R)^2$ is shown in Fig. 2.
In a weak field, low temperature part of the susceptibility ($T/T_c < 0.5$) is proportional to the flux penetration length

$$\mathrm{Re}\chi(T) = -1 + a\lambda(T)/R. \tag{21}$$

A measurement of temperature dependence $\lambda(T)$ allows us to distinguish different pairing symmetries. While in conventional superconductors with an isotropic gap the quasiparticle excitations rise with increasing temperature as $\exp(-\Delta/k_B T)$, in nonconventional superconductors, for example HTS, a temperature dependence is power-law. As far as we know, it fails to fit experimental $\chi(T)$ at $T \to T_c$ even for well known $\lambda(T)$, at low temperatures.

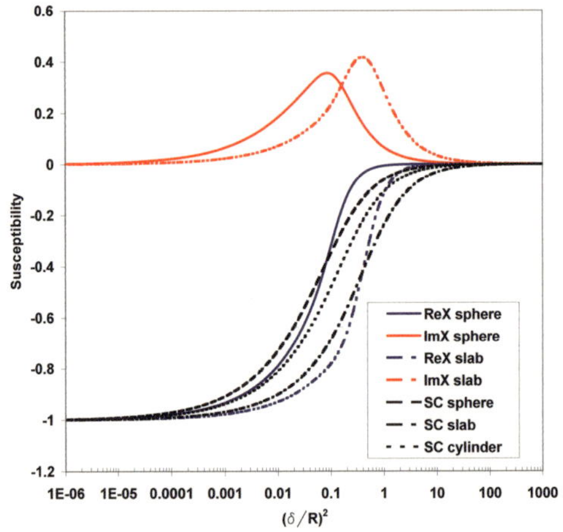

Fig. 2. The dependence of the complex AC susceptibility of a sphere and slab in a normal (ohmic) state in a parallel field on $(\delta/R)^2 \propto \rho_n$ and of the sphere, slab and cylinder in Meissner state on $(\lambda/R)^2 \propto 1/n_s$. In an ohmic state an absorption peak appears on Imχ, the height of which is characteristic of sample shape.

3.5.3 Bean critical state

The Bean model of the critical state is the case of a strong pinning when the flux density variation is quasi-static (frequency independent) in a slowly varying applied magnetic field and the flux density profile changes only when induced shielding current density reaches the critical depinning current density $j = \pm j_c$. An electric field is induced when the flux density changes. In a slab the flux density profile is linear $|\partial B_z(x)/\partial x| = \mu_0 j_c$ in flux penetrated regions and $|\mathbf{B}| = 0$ in untouched regions. The model assumes lower critical field $H_{c1} \to 0$, surface barrier $H_{barrier} \to 0$, and field independent critical depinning current density j_c, i.e. $j_c(B)$ is constant (Bean, 1964).

Analytical solutions for magnetization loops are known for an infinitely long slab or cylinder in a parallel field (Goldfarb, 1991) and thin disk (Clem & Sanchez, 1994; Mikheenko & Kuzovlev, 1993) or strip (Brandt, 1993) in a perpendicular field. In these cases the 3D partial differential equation (PDE) Eq. 13 reduces to a time independent 2D PDE due to sample shape symmetry.

The model to the disks was work out by Clem and Sanches who improved and corrected former model worked out by Mikheenko and Kuzovlev (Clem & Sanchez, 1994). The model is restricted to slow, quasistatic flux changes for which the magnitude of the electric field **E** induced by the moving magnetic flux is small in comparison with $\rho_f j_c$, where ρ_f is the flux flow resistivity. Under these conditions, the magnitude of the induced current density is close to the critical depinning current density. The validity of the model is restricted for $d \ll R$, $d \geq \lambda$ or if $d < \lambda$, that $\Lambda = 2\lambda^2/d \ll R$, where λ is the flux penetration length and Λ is the 2D screening length.

In the case of the infinitely long (or sufficiently long) sample (slab or cylinder) in parallel applied field the shielding current density is at a surface parallel with applied field,

$$\mu_0 j_\phi = -\partial B_z/\partial r \tag{22}$$

while in case of the sufficiently thin sample (disk or strip) in perpendicular applied field

$$\mu_0 j_\phi = \partial B_r / \partial z, \tag{23}$$

the shielding current appears simultaneously everywhere over the sample cross-section upon application of the field, and decreases everywhere simultaneously after a decrease of the field (Beek et al., 1996). The complete magnetic hysteresis loop can be obtained from the first magnetization curve, which is almost the same for the above cases. The hysteresis loop develops from the thin lens-shaped to parallelogram as the H_{ac} is increased or j_c decreases. The lens shape corresponds to partial penetration of the magnetic flux while the parallelogram occurs when the magnetization is saturated.

The component of the magnetization parallel to the applied periodically time varying field $H(\varphi) = H_{ac}\sin\varphi$ is

$$M_{\mp} = \mp\chi_0 H_{ac} \mathrm{S}\left(\frac{H_{ac}}{H_d}\right) \pm \chi_0 \left(H_{ac} \mp H\right) \mathrm{S}\left(\frac{H_{ac} \mp H}{2H_d}\right), \tag{24}$$

where M_- and M_+ are for decreasing and increasing applied field, respectively (Clem & Sanchez, 1994). A characteristic field $H_d = dj_c/2$, where d is the disk thickness and j_c is the critical depinning current density (temperature dependent). The function $\mathrm{S}(x)$ is defined as

$$\mathrm{S}(x) = \frac{1}{2x}\left[\arccos\left(\frac{1}{\cosh x}\right) + \frac{\sinh|x|}{\cosh^2 x}\right]. \tag{25}$$

3.5.4 Mapping of model susceptibility to experimental susceptibility

The model AC susceptibility is calculated for magnetization loops Eq. 24 using Eq. 11, i.e. in the same way as the experimental susceptibility (Youssef et al., 2009). To map the model susceptibility $\chi(H_{ac}/H_d)$ to the experimental temperature dependent susceptibility $\chi(T)$ we use a proportionality of the characteristic field to the critical depinning current density, $H_d = dj_c/2$, and a fact that experimentally observed temperature dependence, $j_c(T) = j_c(0)(1 - T/T_c)^n$, is power-law. Further, we need an inverse function for $j_c(T)$ and insert the amplitude of the applied field. Let us take

$$\frac{j_c(T)}{j_c(0)} = \frac{H_d(T)}{H_d(0)} = \left(1 - \left(\frac{T}{T_c}\right)^m\right)^n. \tag{26}$$

Relation between temperature T and ratio H_d/H_{ac}, i.e. experimental and model susceptibility, is obtained using inverse function for Eq. 26 and multiplying both the numerator and denominator, $H_d/H_d(0)$, by H_{ac}

$$\left(\frac{T}{T_c}\right)_{model} = \left(1 - \left(\frac{H_{ac}}{H_d(0)}\frac{H_d}{H_{ac}}\right)^{1/n}\right)^{1/m}. \tag{27}$$

We have four free parameters $c \equiv H_{ac}/H_d(0)$, n, m, and T_c to match the model and experimental susceptibility

$$\left[\left(1 - \left(c\frac{H_d}{H_{ac}}\right)^{\frac{1}{n}}\right)^{\frac{1}{m}}, \chi\left(\frac{H_d}{H_{ac}}\right)\right] \longleftrightarrow \left[\frac{T}{T_c}, \chi(T)\right]. \tag{28}$$

When we find c, n, m, and T_c, the zero temperature critical depinning current density is

$$j_c(0) = 2H_{ac}/cd \tag{29}$$

and its temperature dependence is given by Eq. 26.

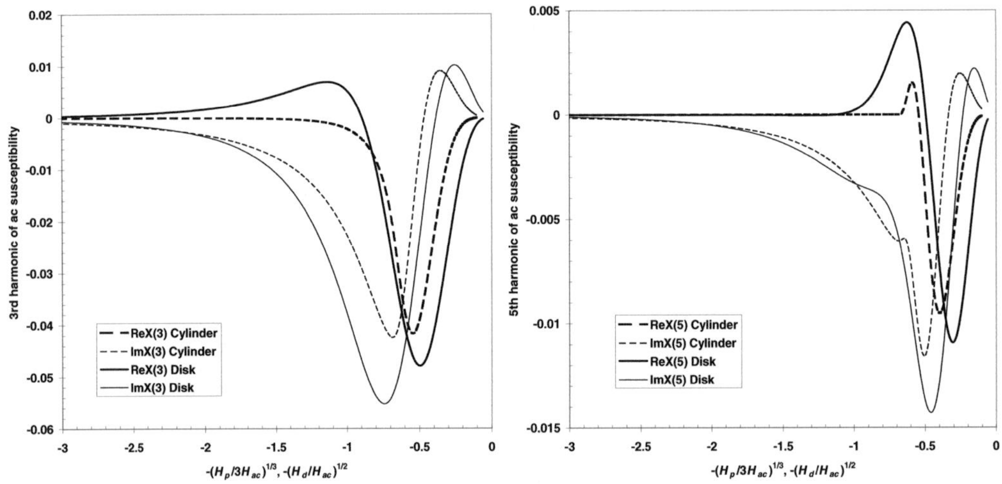

(a) The third harmonic of the AC susceptibility. (b) The fifth harmonic of the AC susceptibility.

Fig. 3. Differences in the harmonics of AC susceptibility for models of cylinders and disks. The susceptibility is plotted versus "model temperature" given by Eq. 27 (Youssef et al., 2009). Here H_p is the characteristic field for a cylinder, $H_p = Rj_c$.

3.5.5 Interpretation of complex AC susceptibility

The real part of the fundamental AC susceptibility represents a magnetic energy of the sample stored in the diamagnetic shielding current. The imaginary part of the fundamental susceptibility is related to losses caused by resistive response (dissipation).

In normal state or in flux flow state the AC susceptibility is a function of applied field frequency, conductivity (resistivity), and temperature but is independent of the field amplitude. On the other hand, in a case of strong pinning the AC susceptibility is a function of the applied field amplitude, critical depinning current density, and temperature but is independent of frequency. Nonlinear dependence of the sample magnetization on applied field amplitude generates harmonics of AC susceptibility. Their behavior is characteristic for a given sample shape. Due to a symmetry of the magnetization loops, $M(H) = -M(-H)$, the coefficients of even harmonics of the AC susceptibility are zero.

4. Experimental results on critical state in type II superconductors

Recently developed second generation of the high temperature superconductor wires on the basis of YBaCuO films and Nb films for superconductor electronics production represent proper materials to study models to the critical state in hard superconductors.

4.1 Materials

The Nb film of thickness of 250 nm was deposited by a dc magnetron sputtering in Ar gas on 400 nm thick silicon-dioxide buffer layer which was grown by a thermal oxidation of a silicon single crystal wafer (May, 1984). The film is polycrystalline with texture of a preferred orientation in the (110) direction and is highly tensile. Grain size is about 100 nm. The square samples of 5×5 mm^2 in dimensions were cut out from the 3-inch wafer.

Second-generation high temperature superconductor wire (2G HTS wire) consists of a 50 μm nonmagnetic nickel alloy substrate (Hastelloy), 0.2 μm of a textured MgO-based buffer stack deposited by an assisting ion beam, 1 μm RE-$Ba_2Cu_3O_x$ superconducting layer SmYBaCuO deposited by metallo-organic chemical vapor deposition, and 2 μm of Ag, with 40 μm total thickness of surround copper stabilizer (20 μm each side) .[9] The sample is cut into 4 mm long segment of 4 mm wide wire.

4.2 Estimation of the critical depinning current density and its temperature dependence

Since the model susceptibility is not given analytically the standard fitting procedures cannot be applied here. A convenient way to map the model susceptibility to the experimental one is to plot the experimental susceptibility as a function of reduced temperature T/T_c and superimpose the model susceptibility by fitting parameters c, n, and m in Eq. 27 and T_c interactively (manually), see Fig. 4. The critical depinning current density estimated using Eq. 29 is $j_c(0) = 3 \times 10^{11}$ A/m^2 in the Nb film with temperature dependence $j_c(T) = j_c(0)[1-(T/T_c)]^{3/2}$. The critical depinning current density found in the YBCO wire is $j_c(0) = 10^{12}$ A/m^2 with steeper temperature dependence, $j_c(T) = j_c(0)[1-(T/T_c)]^2$. This result well agrees with j_c estimated using a four point probe contact measurements (Youssef et al., 2009; 2010).

5. Conclusion

The thin film type II superconductors with a strong pinning allowed us to verify the complete analytical model of a response of a thin disk in the Bean critical state to an applied time varying magnetic field. On the other hand, the application of this model gives a contactless estimation of the critical depinning current density and its temperature dependence.

To observe the characteristic critical state response from an YBCO sample as is shown in Fig. 4 at lower temperatures the applied time varying field has to be of the order of 0.1 T at 77 K and of the order of 1 T at 4.2 K. Such fields may rather be generated using a normal (nonsuperconducting) solenoid that avoids a residual field of flux lines trapped in the superconducting solenoid winding and guaranties a linear $H(I)$ relation. However, dissipated power will be large. Also, since the induced magnetic moment will be large, there is no need for a sensitive superconducting detection system, but a detector with high linearity and flat frequency and phase response is necessary as the maximum amplitude of 3rd harmonic is only 6% and 5th harmonic of only 1% of the real part of the fundamental susceptibility.

The fit to the model reveals an excess of few % of the real part of the susceptibility as temperature decreases to zero. This diamagnetic contribution is due to the temperature

[9] Wire type SCS4050 SuperPower, Inc., Schenectady, NY 12304 USA. The critical current of the wire as estimated using four probe method and 1 μV/cm criterion is from 80 to 110 A at 77 K (97 A for our piece of wire).

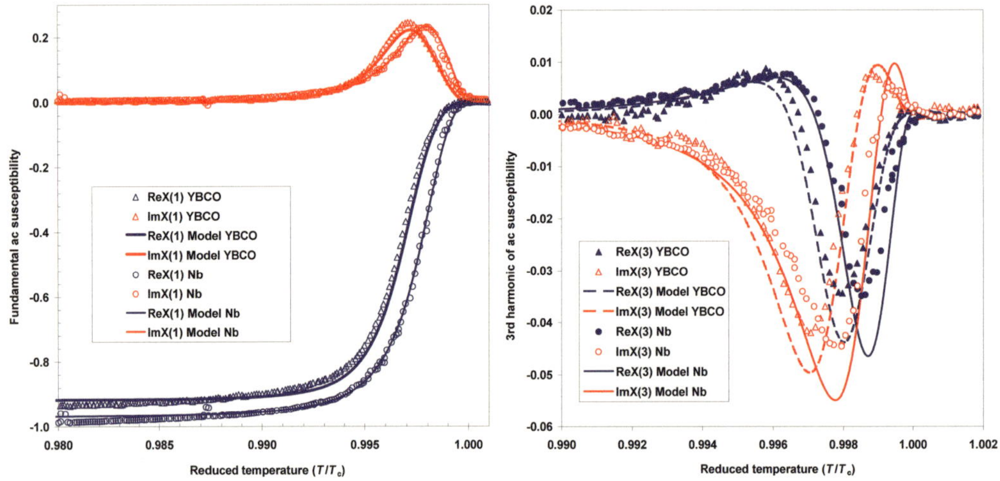

(a) The fundamental AC susceptibility. (b) The third harmonic of the AC susceptibility.

Fig. 4. Temperature dependence of the AC susceptibility of Nb and YBCO films in perpendicular field $\mu_0 H_{ac} = 10\ \mu$T and $f = 1.5625$ Hz (Youssef et al., 2010).

dependent flux penetration length $\lambda(T)$ which depends exponentially on temperature in conventional superconductors (Nb) and obeys a power-law in unconventional ones (YBCO). As was shown by Brandt, the normalized magnetization curves for hard (Bean) superconductors obtained by a numerical treatment differ very little for similar geometries (Brandt, 1996): between strips and circular disks the relative difference is < 0.011, between thin circular and quadratic disks the difference is < 0.002. This makes an application of fully analytical models for contactless estimation of the critical depinning current density and its temperature dependence favorable.

6. Acknowledgements

The authors are grateful to SuperPower, Inc. for providing us with 2G HTS YBCO wire, and to F. Soukup and R. Tichy for technical assistance. This work was supported by Institutional Research Plan AVOZ10100520, Research Project MSM 0021620834 (Ministry of Education, Youth and Sports of the Czech Republic), the Czech Science Foundation under contract No. 202/08/0722, (Javorsky SVV grant 2011-263303) and ESF program NES.

7. References

Anderson, P.W. (1962). Theory of flux creep in hard superconductors, *Phys. Rev. Lett.* Vol. 9:309-311.

Anderson, P.W. & Kim, Y.B. (1962). Hard Superconductivity: Theory of the Motion of Abrikosov Flux Lines, *Rev. Mod. Phys.* Vol. 36:39-43.

Bardeen, J. (1962). Critical fields and currents in superconductors, *Rev. Mod. Phys.* Vol. 34:667-681.

Bean, C.P. (1964). Magnetization of High-Field Superconductors, *Rev. Mod. Phys.* Vol. 36:31-39.

van der Beek, C.J., Indenbom, M.V., D'Anna, G., Benoit, W. (1996). Nonlinear AC susceptibility, surface and bulk shielding, *Physica C* Vol. 258:105-120.

Blatter, G., et al. (1994) Vortices in high-temperature superconductors, *Phys. Mod. Phys.* Vol. 66:1125-1388.

Brandt, E.H., et al. (1993). Type-II Superconducting Strip in Perpendicular Magnetic Field, *Europhys. Lett.* Vol. 22, No. 9: 735 - 740

Brandt, E.H. (1996). Superconductors of finite thickness in a perpendicular magnetic field: Strips and slabs, *Phys. Rev. B* Vol. 54: 4246-4264.

Brandt, E.H. (1998). Superconductor disks and cylinders in an axial magnetic field. I. Flux penetration and magnetization curves, *Phys. Rev. B* Vol. 58: 6506-6522; Superconductor disks and cylinders in an axial magnetic field: II. Nonlinear and linear ac susceptibilities, *Phys. Rev. B* Vol. 58: 6523-6533

Chen, D.-X., et al. (2007). Field dependent alternating current susceptibility of metalorganically deposited YBa2Cu3O7-d films, *J. Appl. Phys.* 101: 073905-.

Clem, J.R. & Sanchez, A. (1994). Hysteretic ac losses and susceptibility of thin superconducting disks, *Phys. Rev. B* Vol. 50: 9355-9362.

deGennes, P. G. (1966), In: *Superconductivity of Metals and alloys* (Benjamin, New York, 1966).

Goldfarb, R.B., Lelenthal, M., Thompson, C.A., Alternating-field susceptometry and magnetic susceptibility of superconductors, In: *Magnetic Susceptibility of Superconductors and Other Spin Systems*, edited by R. A. Hein (Plenum Press 1991), p. 49.

Gömöry, F. (1997). Characterization of high-temperature superconductors by AC susceptibility measurements, *Supercond. Sci. Technol.* Vol. 10: 523-542.

Koshelev A.E., Vinokur V.M. (1994), Dynamic meltig of the vortex lattice, *Phys. Rev. Lett.* Vol. 73: 3580-3583.

Khoder, A.F. and Couach, M. (1991). Early theories of χ' and χ'' of superconductors; the controversial aspects, In: *Magnetic Susceptibility of Superconductors and other Spin Systems*, New York and London: Plenum Press. p 213-228.

Lifshitz, E.M. et al. (1984) In: *Electrodynamics of Continuous Media*, Vol. 8 (Course of Theoretical Physics), Ed. Butterworth-Heinemann.

May T. (1984), Ph.D. Thesis, Institute for Physical High Technology, Jena, Germany 1999.

Mikheenko P. N. & Kuzovlev Yu. E. (1993), Inductance measurements of HTSC films with high critical currents, *Physica C* Vol. 204:229-236.

Press, W.H., et al. (1992). In: *Numerical Recepies in C*, Cambridge University Press, ISBN 0 521 43720 2, Cambridge. p 496 - 536.

Pearl, J. (1964). Current distribution in superconducting films carrying quantized fluxoids, *Appl. Phys. Lett.* 5:65-66.

Sanchez, A. & Navau, C. (1999). X, *IEEE Trans. Appl. Supercond.* Vol. 9:2195-

Tinkham, M. (1996) In: *Introduction to Superconductivity*, (McGraw-Hill, New York, 1996).

Tsoy, G.M. et al. (2000). High-resolution SQUID magnetometer, *Physica B*, Vol. 284, Part 2:2122-2123.

Vrba, J. & Robinson, S.E. (2001). Signal processing in magnetoencephalography. *Methods*, Vol. 25:249-271.

Wellstood, F.C., et al. (1987) Low-frequency noise in dc superconducting quantum interference devices below 1 K *Appl. Phys. Lett.* 50:772-774.

Youssef, A., Svindrych, Z., Janu, Z. (2009) Analysis of magnetic response of critical state in second-generation high temperature superconductor YBa2Cu3Ox wire, *J. Appl. Phys.* Vol. 106: 063901-1-1063901-6.

Youssef, A., et al. (2010). Contactless Estimation of Critical Current Density and Its Temperature Dependence Using Magnetic Measurements, *Acta Physica Polonica A* Vol. 118, No. 5:1036-1037.

13

Current Status and Technological Limitations of Hybrid Superconducting-normal Single Electron Transistors

Giampiero Amato and Emanuele Enrico
The Quantum Research Laboratory, INRIM, Turin
Italy

1. Introduction

Since the original paper from Josephson on tunnel phenomena occurring in superconducting junctions (Josephson, 1962), superconductors have been widely studied by metrologists, because of the quantistic origin of most effects observed in such class of materials. There is, in fact, an intimate relationship between the definition of more accurate and stable standards and Quantum Mechanics. Indeed, the Josephson Voltage Standard (JVS) is believed to be a fundamental quantum physical effect, which is the same everywhere, and at all times.

Tunnel effect has, however, several other implications, one of them being the possibility of localizing a single electron in space. An electric current can flow through the conductor because some electrons are free to move through the lattice of atomic nuclei. The charge transferred through the conductor determines the current. This transferred charge can have practically any value, in particular, a fractional charge value as a consequence of the displacement of the electron cloud against the lattice of atoms. This shift can be changed continuously and thus the transferred charge is a continuous quantity, not quantized at all!

If a discontinuity in space is introduced, e.g. by means of a tunnel junction, electric charge will move through the system by both continuous and discrete processes. Since, from a semi-classical point of view, only discrete electrons can tunnel through junctions, charge will accumulate at the surface of the electrode against the isolating layer, until a high enough bias has built up across the tunnel junction, and one electron will be transferred. This argument, which will be substantiated in a purely quantistic view in the following, led K. Likharev (Likharev, 1988) to coin the term `dripping tap' as an analogy of this process. In other words, if a constant current I is forced to pass through a single tunnel junction, the so called Coulomb oscillations will appear with frequency $f = I/e$ where e is the charge of an electron. The current biased tunnel junction is a very simple circuit able to show the controlled transfer of electrons.

Differently from the JVS, devices capable to control the electron transfer one-by-one are still far to reach the accuracy level necessary for metrological applications. Controlling and counting electrons one-by-one in an electrical circuit will give the possibility of realizing a quantum standard for electrical current. It is important to remember that in the SI system, the base electrical unit is the ampere, but, nowadays, the primary electrical standards are the quantum Hall effect (QHE) resistance standard and the JVS. Both are believed to be fundamental physical effects and widely used in metrological laboratories. The quantum Hall resistance R and Josephson voltage V are given by:

$$R = R_k/i \qquad (R_k = h/e^2) \tag{1}$$

$$V = nf/K_j, \qquad (K_j = 2e/h) \tag{2}$$

where i and n are integers, f is a frequency, h and e are fundamental constants, namely, the Planck's constant and the electron charge.

The QHE ohm and Josephson volt are linked to the ampere via difficult experiments, with a relatively high uncertainty (Flowers, 2004). In consequence, the QHE and JVS are referred to as 'representations' of the SI ohm and volt. To address this inconsistency, the International Committee of Weights and Measures (CIPM) recommended the study of proposals to re-define some of the SI units in 2011.

A quantum electrical standard, based on single electron transport, yields a current given by:

$$I = n'f'e \tag{3}$$

where the current I through the transistor is defined by the number n' of elementary charges (e) injected in one period and f' is the frequency.

There are two basic requirements for a transistor to act as an electron turnstile. The first is that the charging energy for an electron confined into an island of material in between two tunnel junction must be larger than the thermal energy of electrons. This condition can be written as $e^2/2C_\Sigma >> kT$, where C_Σ is the total capacitance of the device. This first condition has two direct technological and physical consequences: to observe Coulomb blockade, junctions with lateral dimension in the 10-100 nm range are required so to have $C_\Sigma \leq 10^{-16}$ F. Of course, the measurement must be carried out at cryogenic temperatures, with typical values in the mK range. What is important to underline here is the need of nano-technologies to realize the device. Standard photolithograpy, widely employed by microelectronic industries for high density package of devices in a single chip, can hardly approach the geometrical limit required, so, Electron Beam Lithography (EBL) is commonly used for the purpose.

The second condition to be fulfilled by an electron turnstile is more related to the basics of Quantum Mechanics. In a classical picture it is clear if an electron is either on an island or not. In other words, the localization is implicitly assumed in a classical formalism. However, in a more precise quantum mechanical description, the number of electrons N localized on an island are in terms of an average value $\langle N \rangle$ which is not necessarily an integer. The so-called Coulomb blockade effect prevents island charging with an extra electron, that is $|N-\langle N \rangle|^2<<1$. Clearly, if the tunnel barriers are not present, or are fairly opaque, no island charging or electrons localization on a quantum dot will be accomplished, because of the absence of confinement for an electron within a certain volume. From a quanto-mechanical point of view, the condition $|N-\langle N \rangle|^2<<1$ requires for the time t which an electron resides on the island, $t >> \Delta t \geq h/\Delta E$. Let us assume that for moderate bias and temperature at most one extra electron resides on the island at any time, so the current cannot exceed e/t. This means that the energy uncertainty on the electron must be $\Delta E<V_b$, where V_b is the applied bias. Trivial calculations lead to the conclusion that the resistance of the tunnel junctions $R_T= V_b/I >> h/e^2$. The last quantity is the von Klitzing constant R_K, known to be $R_K \equiv 25813$ Ω. More rigorous theoretical studies on this issue have supported this conclusion (Zwerger

& Scharpf, 1991). Experimental tests have also shown this to be a necessary condition for observing single-electron charging effects (Geerligs et al., 1989).

An important experiment, in which all the three electrical standards are joined together, is the Metrological Triangle. We can describe this experiment like a sort of quantum validation of the Ohm's Law. Joining eqs. (1), (2) and (3), we will yield the product $R_k K_j e$. This is expected to be exactly 2. Any discrepancy from this value will indicate a flaw in our understanding of one or more of these quantum effects. This experiment will be an important input into the CIPM deliberations on the future of the SI. It is one of the higher priorities in fundamental metrology today.

Current pumps based on mesoscopic metallic tunnel junctions have been proposed in the past (Geerligs et al., 1990; Pothier et al., 1992) and demonstrated to drive a current with a very low uncertainty (Keller et al., 1996). Unfortunately, these systems are difficult to control and relatively slow (Zimmerman & Keller, 2003). Amongst the various attempts to overcome these limitations by using e.g. surface-acoustic-wave driven one-dimensional channels (Talyanskii et al. 1997), superconducting devices (Vartiainen et al., 2007; Niskanen et al., 2003; Lotkhov, 2004; Governale et al., 2005; Kopnin et al. ,2006; Mooij and Nazarov, 2006, Cholascinski & Chhajlany, 2007) and semiconducting quantum dots (Blumenthal et al., 2007), a system based on hybrid superconducting-metal assemblies and capable of higher accuracy has been recently proposed (Pekola et al., 2008).

2. Theorethical background

2.1 The Orthodox theory

In the present chapter, we will review the Orthodox (Averin & Likharev, 1991) theory for the Normal-metal Single Electron Transistor (n-SET) with the aim of extending it to the case of hybrid Superconductor/Normal structures. This model, which will be discussed in a following section, enables to predict the h-SET performances when different superconductors are employed.

For clarity purposes, we will give a heuristic treatment for the n-SET but without any lack of generality, while a more detailed discussion will be devoted to the hybrid case.

The energy that determines the transport of electrons through a single-electron device is Helmholtz's free energy which is defined as difference between total energy E_Σ stored in the device and work done by power sources. The total energy stored includes all the before mentioned energy components that have to be considered when charging an island with an electron. The change in Helmholtz's free energy a tunnel event causes is a measure of the probability of this tunnel event. The general fact that physical systems tend to occupy lower energy states, is apparent in electrons favoring those tunnel events which reduce the free energy.

In the framework of the Orthodox theory (Averin & Likharev, 1991) the tunneling rate Γ across a single junction between two normal metal electrodes can be extracted using the Golden Rule as:

$$\Gamma^{+} = e^2 R_T^{-1} \int_{-\infty}^{+\infty} f(E,T)[1 - f(E - \Delta F,T)]dE$$

$$\Gamma^{-} = e^2 R_T^{-1} \int_{-\infty}^{+\infty} f(E + \Delta F,T)[1 - f(E,T)]dE \tag{4.1}$$

where ΔF is the variation in the Helmholtz free energy of the system. Integration of (4.1) yields:

$$\Gamma = -\Delta F e^2 R_T^{-1} \left[1 - \exp(\Delta F / k_B T) \right] \tag{4.2}$$

It can be easily concluded that, in the low T limit, Γ= 0 when ΔF > 0, whereas:

$$\Gamma = -\Delta F e^2 R_T^{-1} \quad \Delta F < 0 \tag{4.3}$$

The quantity ΔF for a n-SET with i junctions can be written in the following way:

$$\Delta F_i^{\pm} = e(e / 2C_{\Sigma} \pm V_i) \tag{5}$$

where i=1,2 in a single-island n-SET, V_i is the voltage bias across the junctions. Here, we are dealing with 4 different equations, which consider the possibility for one electron to enter in (+) or to exit from (-) the island both from junctions 1 or 2.
Eq. (5) gives a perspicuous representation of the Helmoltz free energy for an island limited by two tunnel junctions. The energy $E_c=e^2/2C_\Sigma$ is clearly the energy stored in the device, whereas $\pm eVi$ represents the work done by the power sources.

2.2 The Normal-Insulator-Normal SET

In Fig. 1 a SET equivalent circuit is displayed. First, it is helping to write the equations for a double junction system, and then to correct them when a gate contact is added.
The charge q_i at the i-th junction can be written as $q_i=C_iV_i$, so, the total charge into the island is $q= q_2-q_1+q_0=-ne+q_0$ where q_0 is the background charge inside the island and $n=n_1-n_2$ is an integer number indicating the electrons in excess.

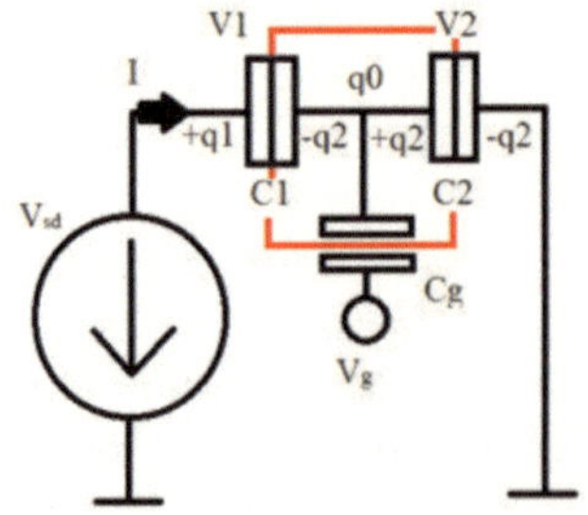

Fig. 1. Equivalent circuit of a single-island, two-junctions SET

The voltage bias across the i-th junction is then:

$$V_i = \left[C_{(3-i)} V_{SD} + (-1)^i (q_0 - ne) \right] C_{\Sigma}^{-1} \tag{6}$$

where V_{SD} is the bias across the device ($V_{SD}= \Sigma V_i$) and $C_\Sigma=\Sigma C_i$.
To add the contribution of the gate contact in the device, we can simply take into account for effect of the gate electrode on the background charge q_0. This quantity can be changed at will, because the gate additionally polarizes the island, so that the island charge becomes:

$$q = -ne + q_0 + C_g\left(V_g - V_2\right) \tag{7}$$

with V_g the gate voltage.
Now, after, some trivial calculations, one can write the final relationship giving the voltage bias across the i-th junction in Single Electron Transistor (SET) composed by 1 island surrounded by 2 tunnel junctions:

$$V_i = C_\Sigma^{-1}\left\{\left[C_{(3-i)} + C_g\delta_{i,1}\right]V_{SD} + (-1)^i C_g V_g + (-1)^{i+1} ne + q_0\right\} \tag{8}$$

where $\delta_{i,1}$ is the Dirac's function and $C_\Sigma=\Sigma C_i+C_g$.
By combining (8) and (5) it is possible to explicitly write the equations governing the free energy change in a system with two tunnel junctions and a gate electrode. For example, under the particular conditions: $q_0 = 0$, $R_1 = R_2$ and $C_1 = C_2 = C >> C_g$, one gets:

$$\Delta F_1^{\pm} = 2E_c\left(\pm n \mp n_g + 1/2\right) \pm V_{SD}/2$$

$$\Delta F_2^{\pm} = 2E_c\left(\mp n \pm n_g + 1/2\right) \pm V_{SD}/2 \tag{9}$$

where $n_g=C_g(V_g-V_2)$ and $E_c=e^2/2C$.
In order to model the behavior of such a complex system, some simplifying assumptions are needed. First of all, we consider the tunneling events as instantaneous and uncorrelated, say, one is occurring at a time. Since any single-electron tunneling event changes the charge state of the island, at least two states are required for current transport.
Having the rates of tunneling through the two junctions at hand we can now define the rates of elementary charge variation for the island as:

$$\Gamma_{n+1,n} = \overleftarrow{\Gamma}_1(n) + \overrightarrow{\Gamma}_2(n) \qquad \Gamma_{n-1,n} = \overleftarrow{\Gamma}_2(n) + \overrightarrow{\Gamma}_1(n) \tag{10}$$

With the aid of the above considerations it is possible to define a master equation that governs the behavior of the system, whose solution is (Ingold & Nazarov, 1992):

$$\Gamma_{n,n+1}P_{n+1} = \Gamma_{n+1,n}P_n \tag{11}$$

where P_n is the probability distribution for the island charge state.
Taking as a starting state that one with no excess charge in the island and considering that only the nearest neighbors states are connected by non-null rates, the probability distribution can be derived from eq. (11) as:

$$P_n = P_0\prod_{m=0}^{n-1}\Gamma_{m+1,m}/\Gamma_{m,m+1} \qquad n>0$$

$$P_n = P_0\prod_{m=n+1}^{0}\Gamma_{m-1,m}/\Gamma_{m,m-1} \qquad n<0 \tag{12}$$

where the free parameter P_0 can be extracted from the normalization condition $\sum_{-\infty}^{+\infty} P_n = 1$.

Being the steady-state currents through the two junctions equal to I we can write:

$$I = e\sum_{-\infty}^{+\infty} P_n\left[\overrightarrow{\Gamma_1}(n) - \overleftarrow{\Gamma_1}(n)\right] = e\sum_{-\infty}^{+\infty} P_n\left[\overrightarrow{\Gamma_2}(n) - \overleftarrow{\Gamma_2}(n)\right] \tag{13}$$

It's trivial to note that in the T=0 limit the terms ($\overrightarrow{\Gamma_1} - \overleftarrow{\Gamma_1}$ and $\overrightarrow{\Gamma_2} - \overleftarrow{\Gamma_2}$) of eq. (13) are identically null for some values of V_{SD} and n_g. In these states it is also noted that the probability distribution P_n=1 for a well defined value of n. This means that these regions are stable in terms of the number of charges on the island and both tunnel junctions are in the so-called Coulomb Blockade state.
In the zero temperature limit, by imposing $\Delta F_i^{\pm} = 0$, one is able to write down the equations providing the dependence of V_{SD} on n_g at the boundaries between the regions in which tunneling is allowed ($\Delta F_i^{\pm} < 0$) and forbidden ($\Delta F_i^{\pm} > 0$). Without going into details on this rather simple calculation, we can easily observe that such dependence is linear, with slopes given by $C_g / \left[C_{(3-i)} + C_g \delta_{i,1}\right]$ and intercepts related to the number n of excess electrons into the island. These lines give rise to the well-known Stability Diagram for a n-SET depicted in Fig. 2.
Diamonds in the Stability Diagram are representative for the region where tunneling is inhibited ($\Delta F_i^{\pm} > 0$). They are defined by two families of parallel lines having positive (1st junction) and negative (2nd junction) slopes, respectively. Outside such regions, current can flow freely across the device, whereas the control of the charging state at single-electron level can be obtained only when the working point with coordinates n_g ,V_{SD} lies inside a stable diamond.

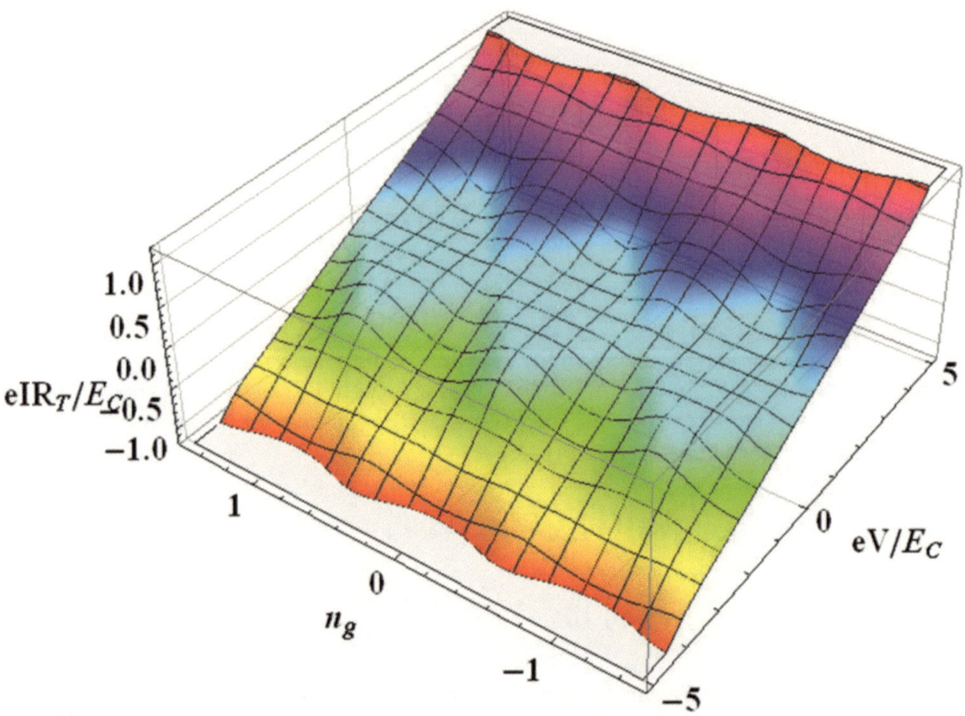

Fig. 2. Stability Diagram for a n-SET

It is important to stress that the stable states in the case of n-SETs have a single degeneracy point in which the the states with n or n +1 are equiprobable (Fig. 2).
The only location on the stability diagram, and therefore the only set of coordinates n_g ,V_{SD} which allows the system to switch from one stable state to another passes through the degeneracy point where the bias voltage V_{SD} is nil in any circuit configuration. Then, the reader can understand how a simple n-SET can control the number of elementary charges in excess on the island, solely, but not the flow of single electrons from source to drain electrodes. This because the system switch from n to n +1 can occur either through the forward tunneling in the first junction or the backward tunneling in the second junction, with the same probability. In other words, V_{SD}=0 implies that no directionality for the events is defined, that is, the n-SET cannot work as a turnstile.
For $V_{SD} \neq 0$, the current can freely flow across the device in well-defined V_g intervals. The so-called SET oscillations can then be observed (Fig. 3).
Investigators have tried to circumvent this problem by using multi-island electron pumps (Zimmerman & Keller, 2003). In such devices some islands are in series and driven by their own gate contact. Sinusoidal waveforms for each of these gates are shifted in phase, so to ensure that successive tunnel process occur from the first to the last junction. The relatively complicated experiment with such a slow device yields an output value for the singular current much lower than the limit (10^{-10}-10^{-9} A) necessary for carrying out the Metrological Triangle experiment with the required accuracy.

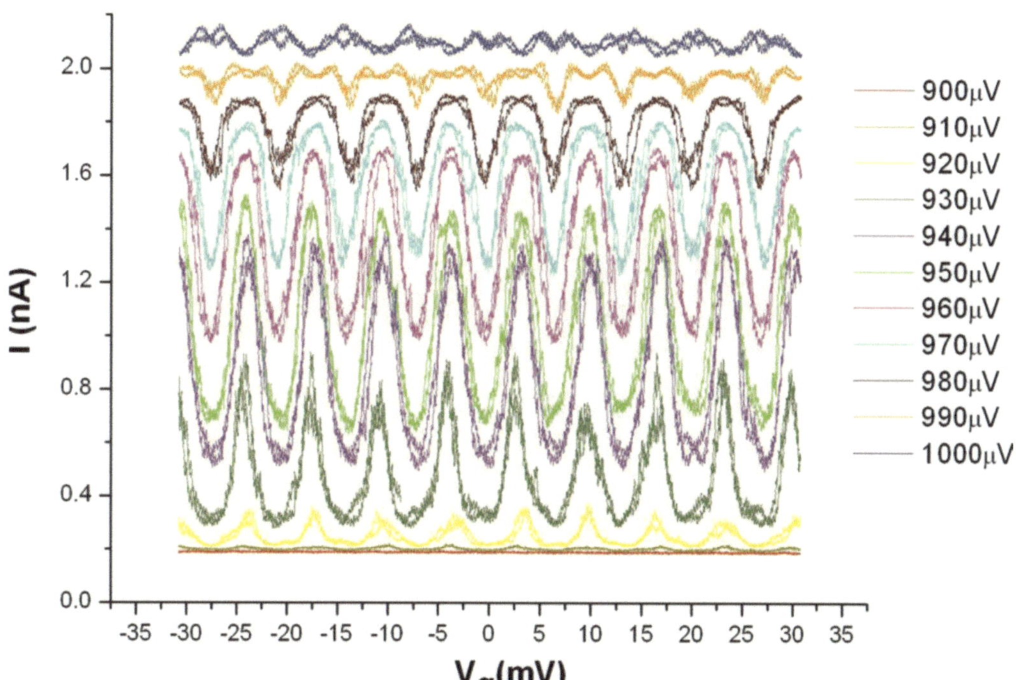

Fig. 3 The SET oscillations occurring when $V_{SD} \neq 0$. Values for V_{SD} are given on the right side of the Fig. With scanning V_g, we find peaks representing the current flow through the device, when $\Delta F_i^{\pm} < 0$ (outside of the diamonds in Fig. 2), and minima related to $\Delta F_i^{\pm} > 0$.

2.3 The hybrid SET

Hybrid superconducting-metal assemblies have been recently proposed and shown to be capable of higher accuracy (Pekola et al., 2008). From a technological point of view, this assembly is composed by a normal-metal island sandwiched by two superconducting electrodes (SNS), or the reverse (NSN) scheme. For the purpose of this chapter, the theoretical description is the same for both the arrangements.
In the following chapter, eqs. (4) will be rewritten in the case of NIS junction and applied to h-SET.

2.3.1 Tunneling in a S-I-N junction

Typical applications of SIN junctions are microcoolers (Nahum et al., 1994; Clark et al., 2005; Giazotto et al., 2006) and thermometers (Nahum & Martinis, 2003; Schmidt et al. 2003; Meschke et al. 2006; Giazotto et al., 2006). In these applications, SIN junctions are usually employed in the double-junction (SINIS) geometry. The opposite NISIN geometry has gathered less attention. Recently, there has been interest in SINIS structures with considerable charging energy. They have been proposed for single-electron cooler applications (Pekola et al., 2007; Saira et al., 2007) that are closely related to the quantized current application (Pekola et al., 2008). Thermometry in the Coulomb-blocked case has been considered, too (Koppinen et al., 2009).
In the case where the superconductor in study is well below its transition temperature ($T_S<T_c$) it can be assumed for the superconducting gap Δ that $\Delta(T_S)=\Delta(0)=\Delta$.
Because the number of particles for a given amount of energy must be the same in the superconducting state (quasi-particles) and in the normal one (free electrons), the relationship:

$$g_s(E)dE = g_n(\varepsilon)d\varepsilon \tag{14}$$

must hold, and then:

$$g_s(E) = g_n(\varepsilon)E\left(E^2-\Delta^2\right)^{-1/2}\theta(E-\Delta) \tag{15}$$

where θ is the Heaviside's step function.
Then, the density of states in a superconductor can be written as:

$$g_s(E) \cong g_n(0)E\left(E^2-\Delta^2\right)^{-1/2}\theta(E-\Delta) \tag{16}$$

by considering that:

1. all the energy terms at low temperatures have significant values of the order of k_BT (which is several orders of magnitude less than the Fermi energy, $k_BT<<E_F$);
2. the energies are measured with respect to the Fermi level ($\varepsilon=0$ at E_F);
3. $g_n(\varepsilon)\approx g_n(0)$.

A further assumption is that the electrons in the metal and the quasiparticles in superconductor are weakly interacting and at thermal equilibrium due to the high potential barrier of the dielectric layer. It is then possible, to consider tunneling as a perturbation and to apply the Golden Rule approach. The dominating current transport mechanism in a NIS junction is single-electron tunneling between the normal metal and the quasi-particle states of the superconductor.

The equations governing the rate of tunneling back and forth in a NIS can be written in a similar way to that for the NIN system by simply adding a term proportional to the superconductor density of states:

$$\Gamma^{+} = e^{2}R_{T}^{-1}\int_{-\infty}^{+\infty} n_{S}(E)\,f(E,T_{S})\left[1-f(E-\Delta F,T_{N})\right]dE$$

$$\Gamma^{-} = e^{2}R_{T}^{-1}\int_{-\infty}^{+\infty} n_{S}(E)\,f(E+\Delta F,T_{N})\left[1-f(E,T_{S})\right]dE \quad (17)$$

where T_s and T_N are temperatures for the superconductor and normal electrodes, respectively and $n_s=g_s(E)/g_N(0)$.
For T = 0, the corresponding of eq. (4.3) is found for the SIN case:

$$\Gamma = e^{-2}R_{T}^{-1}\sqrt{\Delta F^{2}-\Delta^{2}} \qquad \Delta F \leq -\Delta \quad (18)$$

whereas $\Gamma = 0$ for $\Delta F > -\Delta$ (the other two solutions cannot be considered because transitions are allowed only for negative free energy variations).
Finding a solution for eq. (17) is not a trivial task and will not be reported here, but some words are deserved to the tunneling effects occurring in the SIN junction at voltages values below gap. Here, if the condition $K_bT_N \ll \Delta F$ (i.e., if $T_N \ll 1.76\ T_c$) is fulfilled, the rate of tunneling through the SIN junction is given by:

$$\Gamma_{T}(V,T_{N}) = \Gamma_{0}\exp\left[(\Delta F-\Delta)/k_{b}T_{N}\right] \quad (19)$$

the quantity $\Gamma_{0} = \Delta e^{-2}R_{T}^{-1}\sqrt{\pi k_{b}T_{N}/2\Delta}$ being called the characteristic rate and approximately representing the tunneling rate when the free energy variation approaches the gap.
From eq. (19) it can be seen that for free energy variations below the gap the tunneling rate strongly depends on temperature. This opens the possibility of using this type of junction as a thermometer at low temperature. As a drawback, limits in the accuracy of electron counting for metrological applications of h-SETs can arise, as discussed in the next chapters.

2.3.2 Stability diagram for h-SET

Following the same n-SET master equation approach for the SINIS system, it is now possible to combine eqs. (9) and (17) in order to consider the case in which the mesoscopic tunnel junctions charging energy is not negligible and the central island is coupled to a gate electrode.
Results from calculations of the electrical characteristics for the previous ideal system are shown in Fig. 4. Using a similar procedure for the n-SET device we can study the h-SET behavior at temperature T->0 K in order to extract the modified stability diagram.
It is observed from eq. (18) that when $\Delta F > -\Delta$, the tunneling rate is nil (in principle) and the junction does not allow for the electron flow. The areas in the stability diagram in which such conditions hold identify the stable regions with a defined number of elementary charges on the island (n).

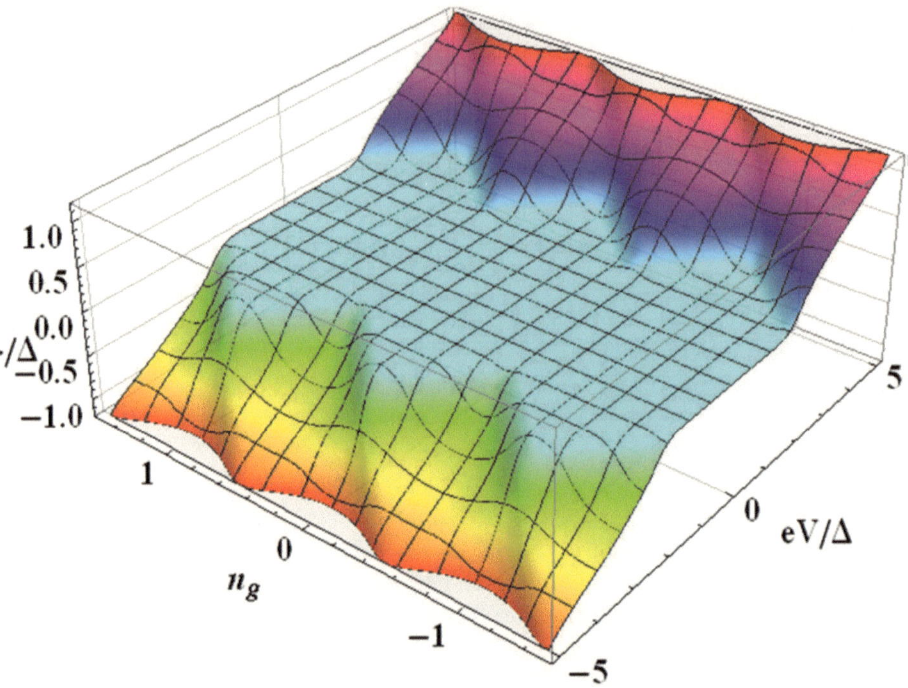

Fig. 4. 3-dimensional view of the Stability Diagram for a h-SET

The present formalism allows us to treat the hybrid assembly in the same way as the normal system. Then, very trivial calculations let us to extract the equations for the two families of straight lines defining the boundaries between regions of allowed and forbidden tunneling, with an offset value with respect to the lines in the normal case of 2Δ (Fig. 5).

The dashed lines in Fig. 5 define the regions of inhibited tunneling, as in the case of the n-SET (purple areas). In the hybrid system, each of them is shifted by an offset 2Δ, defining the blue regions where the mechanism of tunnel inhibition is the band offset at the SIN junction.

When the Current across a Single Electron Tunneling device is recorded as a function of the V_{SD}, for different n_g values, a family of Current-Voltage characteristics is obtained. This means we are moving along parallel vertical pathways on the Stability Diagram. The extension of the Coulomb gap, obviously depends on the n_g value: in a Normal SET it periodically oscillates from 0 to e/2C, with a periodicity of one unit (see Fig. 6).

It is then interesting to compare the Current-Voltage characteristics of the Normal and Hybrid SETs. This comparison, reported in Fig. 6, clearly indicates a broadening of the conduction gap in the hybrid structure. The gap oscillates with tuning V_g from Δ to Δ+ e/2C. In a few words, the presence of the superconducting gap broadens the region of inhibited tunneling, whose width never equals to zero.

In this configuration for the h-SET, the degeneracy point linking the stable states is suppressed by a region in the $V_{SD}-n_g$ space where the pathway from point A to point B occurs with negligible backward tunneling at both junctions (V_{SD}> 0) and without departing from the stable regions. E.g., with V_g oscillating between the states A and B, one can move a single electron per cycle from source electrode to drain with a well defined directionality given by the sign of V_{SD}.

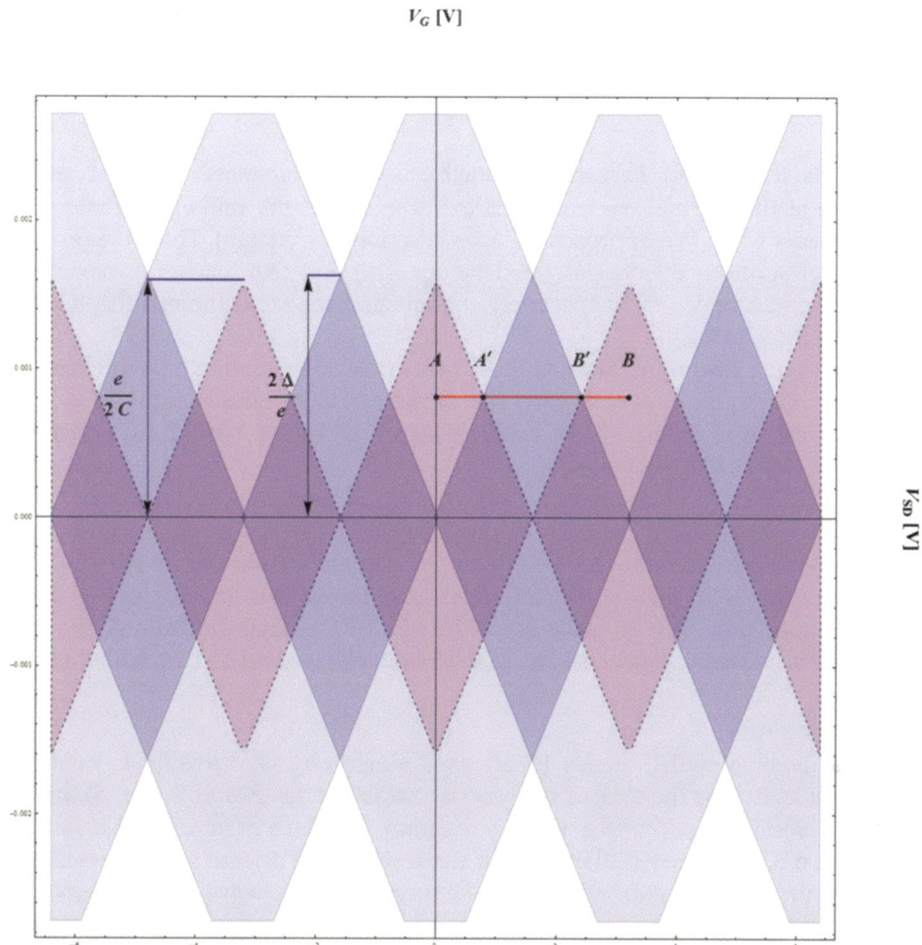

Fig. 5. Plan view of the Stability Diagram for a h-SET. For clarity purposes, Δ is given in eV, so to compare directly with V_{SD}.

In the next section we will analyze some of the possible effects that can alter the process of controlled transport of electric charges in a h-SET turnstile configuration.
The extension of the Coulomb blocked region to V_{SD} values $\neq 0$ is the peculiar feature of the hybrid assembly. This opens the possibility for such a device to operate as a turnstile. In fact, we can operate the device along the pathway between points A and B with $V_{SD} \neq 0$ (Fig. 5). From points A and A' (B' and B) tunneling inhibition is accomplished thanks to the Coulomb Energy $e^2/2C$ ($\Delta F<0$), whereas in the intermediate region A'B', the presence of the superconducting gap is the limiting mechanism ($0<\Delta F<\Delta$).

2.3.3 Error sources in hybrid SET

The following treatment on the error sources in Hybrid SET will not be exhaustive, since second-order (e.g. co-tunneling), and technology-related (e.g. Adreev's reflections at the oxide pinholes) effects, will not be discussed. We will focus on a sort of "ideal" h-SET, in order to determine the optimal conditions for turnstile operation.

It is rather intuitive that small V_{SD} values lead to an increased probability for tunneling events in the backward direction, according to the relationship (Pekola et al., 2008):

$$\Gamma_b \propto \exp(-eV_{SD} / k_B T_N) \tag{20}$$

where Γ_b is the rate of backward tunneling k_B the Boltzmann constant and T_N the temperature at the Normal electrode. On the other hand, the rate of unwanted intra-gap events increases when V_{SD} approaches 2Δ, as described by eq. (19). Thus, the probability of both kinds of spurious events described by eqs. (19) and (20) reaches a maximum value either for $V_{SD}= 2\Delta$ or $V_{SD}=0$, respectively. Minimizing the contributions displayed in eqs. (19) and (20) leads to $V_{SD}=\Delta$.
It could seem, at a first sight, that the incorporation of superconductors with larger Δ is, at first sight advisable, if a drastic suppression of thermal error rates is required as in the case of metrological applications. This because larger Δ values would in principle allow operating the device at higher V_{SD} bias.
Examples of h-SETs in literature generally employ Al as the superconductive component ($\Delta \approx 170$ μeV). Apart from the ease of producing efficient dielectric junction barriers by means of simple Al oxidation, the Δ value for Al is relatively low, if compared e.g with Nb ($\Delta \approx 1.4$ meV). As a matter of fact, there are limitations in employing larger gap superconductors (Pb, Nb) in state-of-art hybrid SETs. Such limitations are either of fundamental or of technological nature. In the followings we will discuss both these aspects.

2.3.4 A scaling rule

The capability of a h-SET device to act as a single elecron turnstile is related to the possibility of switching the system between two stable states A and B (Fig. 5), keeping the system in a blocked region of the stability diagram. All paths at nonzero V_{SD} values which connect A and B, necessarly contain a set of states where the current is suppressed by means of the superconducting gap, solely. In the present chapter we consider the simplest theoretical and experimental setup for a turnstile with dc bias and ac gate voltage: in this framework the system switches between two blocked states, the first related to the Coulomb blockade in analogy with the n-SET and depicted by means of the AA' and BB' segments, the second represented by the A'B' segment in which the tunnel current is suppressed by the superconducting gap. As previously discussed the superconductive gap cannot be considered as a perfect barrier and the transition in the A'B' segment is a potential source of current leakage inside the tunnel junctions, then some considerations are needeed in order to minimize this effect mantaining the advantages of h-SET turnstile configuration.
Minimizing the resident time t_Δ in this region is then an important issue in order to reduce errors related to leakage effects. Authors (Pekola et al., 2008), suggested a squared waveform for the V_g signal, even if the sinusoidal signal can be more easily handled during a turnstile experiment.
Evaluation of such resident time is easily obtained in the case of sinusoidal waveform, by considering the extension of the A'B' region in Fig. 5. We consider a value for $V_{SD}=\Delta$ (with Δ in eV), say, we assume the SET as working in the optimal conditions according to eqs. (19) and (20). From geometrical considerations, as can be evident when observing Fig. 5, the condition:

$$\Delta < E_c. \tag{21}$$

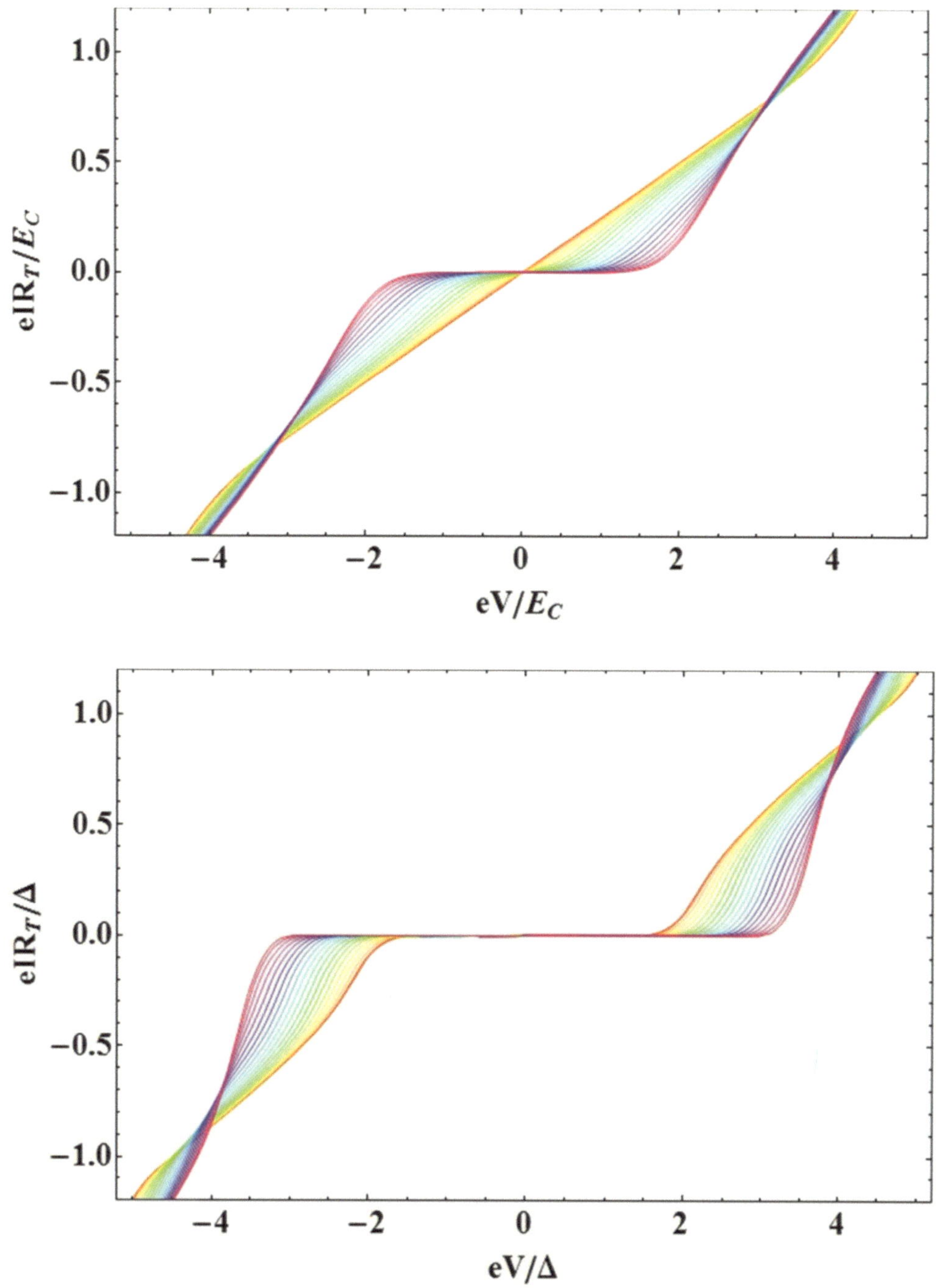

Fig. 6. Comparison between the Current-Voltage characteristics of a Normal (top) and hybrid (bottom) SET, taken at different gate voltage values. According to the Stability Diagram of Fig. 2 we observe the broadening of the Coulomb gap with varying V_g. In the hybrid assembly the contribution from Δ broadens the region of inhibited tunneling.

must hold, otherwise the system will never reach a stable Coulomb-blocked state. Such simple relationship provides an important scaling rule for designing h-SETs. It says that employing high gap superconductors (Nb is a key example) into a hybrid assembly does not guarantee better device performances. That is, the Charging Energy E_C must be increased, too. As an example, if we envisage to replace Al with Nb ($2\Delta \approx 340$ µeV vs. $2\Delta \approx 3$ meV), we have to find a way to increase the E_C value by a factor of ~10; this can be accomplished by decreasing the tunnel capacitance values, solely.
The ratio between t_Δ, the time interval in which the system is blocked only by the superconductive gap during a cycle, and the cycle half-period T/2, can be written as:

$$2t_\Delta / T = \frac{1}{\pi}\left[ar\cos(-\Delta / E_c) - ar\cos(\Delta / E_c)\right] \quad (22)$$

and displayed as a function of the junction capacitance and the superconducting gap Δ(Fig. 7). The $2t_\Delta/T$ ratio is <1 (indicative for the presence of a Coulomb Blockade region in the Stability Diagram, see Fig. 5) only in the portion of the Δ-C plane in which the values of Δ, and/or C are low. As a comparison, Δ-values for typical low-T_c superconductors are indicated together with the reasonnable lower limits for junction capacitance with the most common SET technologies, the SAIL (Self Aligning In-Line) (Götz et al., 1996) and the Shadow evaporation (Dolan, 1977).

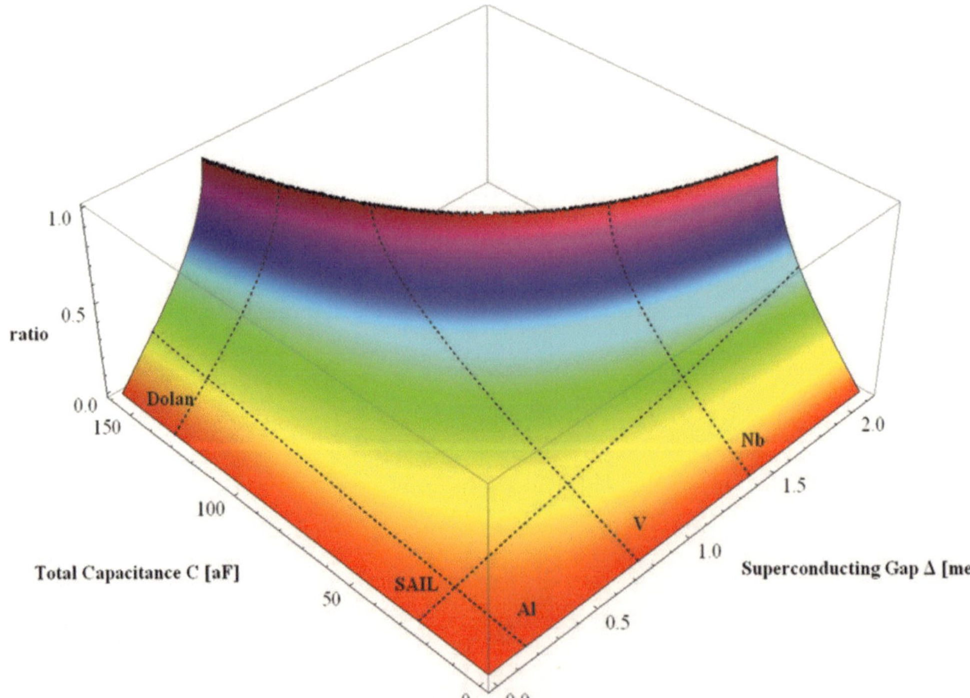

Fig. 7. The graph displays the calculated dependence of The $2t_\Delta/T$ on the superconductor gap Δ and the junction capacitance C. Lines perpendicular to the Δ-axis show the typical gap values for most common low-T_c superconductors, whereas the lines across the C-axis represent the limit of two typical techniques for producing SETs (see text for details).

The following chapter will review the technological approaches to realize SET devices, with the purpose of identifying the most promising ones as far as the capacitance reduction issue is concerned.

3. SET Technologies

3.1 The Shadow evaporation technique

The shadow evaporation technique (Dolan, 1977) was the first to be used for the fabrication of single-electron devices based on metallic systems and is currently the most widespread. This technology takes advantage from a shadow effect, implying that the deposition techniques must be highly non-conformal. The typical deposition process is then thermal or, better, e-beam evaporation: this dramatically limits the choice of materials to be deposited (Nb, for example, being a refractory material, is hardly evaporated).

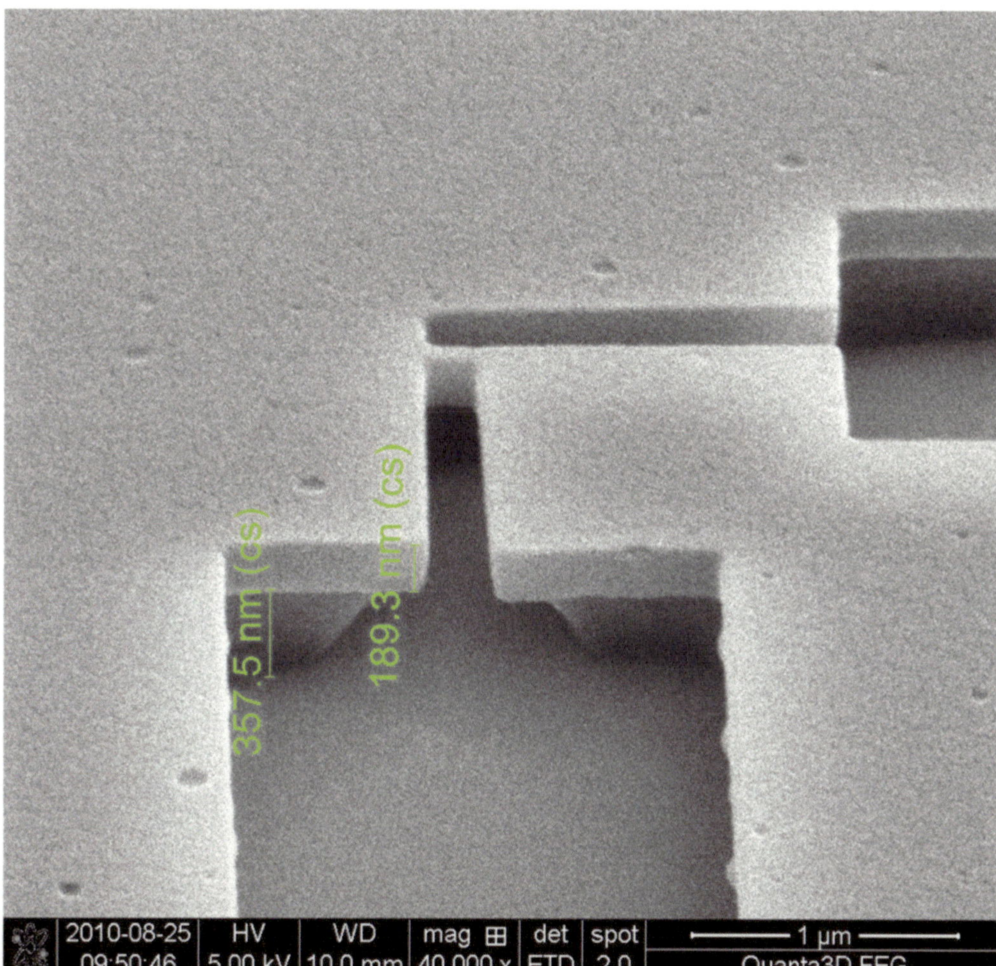

Fig. 8. SEM image of a suspended mask for Shadow evaporation.

The critical step for the success of the process is to fabricate suspended segments of electron beam resist at a certain distance from the substrate. In common lift-off process, the films are defined by evaporating the metal through the openings in the mask at normal incidence substrate, so as to ensure the break between the parts of the layer on the substrate and those on the mask.

The creation of masks with suspended bridges is possible thanks to the use of two different types of resists for electron beam lithography, the lower with greater sensitivity to electron beam than the upper one. During the development step, the exposed resist region is chemically removed in a selective way, with a wider pattern in the polymer underneath. In this way, using the so-called proximity effect, typical of electron beam lithography, it is possible to obtain suspended bridges structures.

Fig. 8 shows the SEM tilted view of the mask we are dealing with: it consists of a support resist layer of thickness δ_1~350 nm, on which the layer that define the structures, with thickness δ_2~200 nm is lying.

If the mask is suspended one no longer needs to deposit the metal at normal incidence to guarantee the successful lift-off and can vary the angle of deposition thus obtaining different patterns on the substrate. From simple geometrical considerations we can see that creating an opening of width W_0 in the top layer of resist and carrying out the evaporation at an angle Θ respect to the normal will produce a deposided feature of width:

$$W_0 = \delta_2 \tan(\Theta) \quad (22)$$

If the angle of incidence is greater than the critical one:

$$\Theta > \Theta_0 = \arctan(W_0 / \delta_2) \quad (23)$$

the opening in the mask appears as "closed" and the deposition does not reach the substrate.

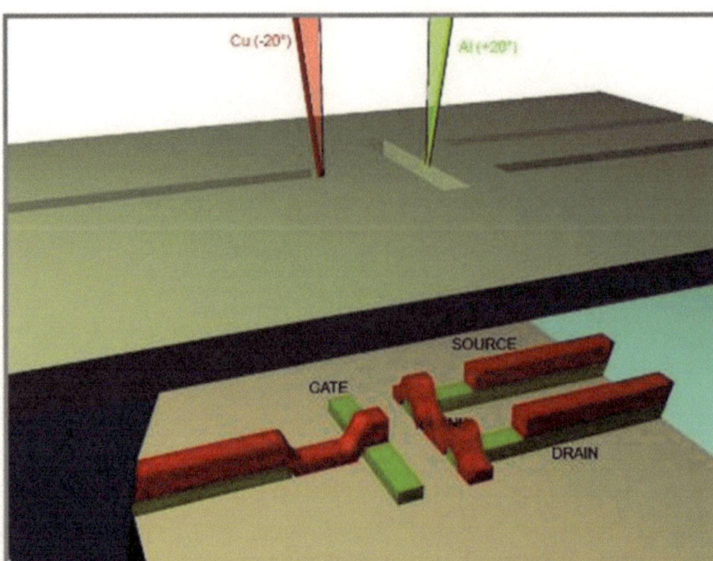

Fig. 9. Schematics of the angled Shadow evaporation process

The practical realization of this effect depends on the ability to produce shadow masks similar to the ideal ones presented so far. To apply this calculation it is important that the experimental values of δ_1 and δ_2 are reliable, and that the cross section of the top resist layer is rectangular.
For the construction of tunnel junctions, first a pattern mask with two, very tight openings must be created. A bridge in the top layer resist between them is then defined.
One can then proceed to the fabrication of tunnel junctions with a deposition-oxidation-deposition sequence, which occurs in the same vacuum cycle. After the first evaporation performed to an angle δ_1, (Al in Fig. 9) the deposited film is oxidized in O_2 atmosphere then growing an insulating layer, commonly Al oxide, ~1nm thick. After pumping down, the second layer is then deposited at angle δ_2 (Cu in Fig. 9).

3.2 The Self Aligning In Line Process (SAIL)

The principle of the SAIL technique (Koch, 1987) is to fabricate the tunnel junctions at the two sides of the island, so that the size of the junctions is determined by the thickness and width of metal thin films: in this way one gets a planar configuration with vertical barriers.
In this section we will discuss the basic steps of the process originally created and provide some hints on how it could be used for manufacturing h-SETs.
The SAIL process, as presented by Gotz (Gotz et al., 1995) consists of the following steps:

i. Preparation of a narrow and thin metal film on the substrate (Fig. 10 (a)).
ii. Fabrication of a resist mask which leaves the area open for the following counter electrode deposition step (Fig. 10 (b)).
iii. Anisotropic etching of the film in order to define the island (Fig. 10 (c)).
iv. Formation of a dielectric barrier on the exposed surface of the island (Fig. 10 (d))
v. Deposition of the second metal film (Fig. 10 (e)).
vi. Lift off (Fig. 10 (f)).

There are no particular requirements for the island deposition technique, e.g. sputtering or evaporation, while the subsequent transfer of the pattern can be accomplished with lift off or anisotropic etching.
The mask generated in the second step defines the location and size of the island and that of source and drain electrodes. The process is self-aligned along the length of the island, while mismatches in the cross direction can be easily compensated by choosing one of the two metal strips wider: then one can realize an island sandwiched with two wide electrodes (WNW), as shown in Figure 9, or a large island between two narrow electrodes (NWN), obtaining in both cases the same junction area.
Difficulties could arise from the use of the same mask for etching and lift off: in fact, the resist must remain soluble and thick enough to allow reliable lift off, even after the ion beam bombardment. One will then need to tune the thickness of the resist or the metal depending on the etching selectivity. The solution may be to replace the ion beam etching, barely physical, with Reactive Ion Etching (RIE), taking advantage from the chemical selectivity of the gas employed.
An alternate solution is the use of a multi-layered mask, e.g. two layers of resist with an intermediate layer with lower etching rate. In this way, the lower resist layer is protected against the ion bombardment, and can be used as lift off mask.

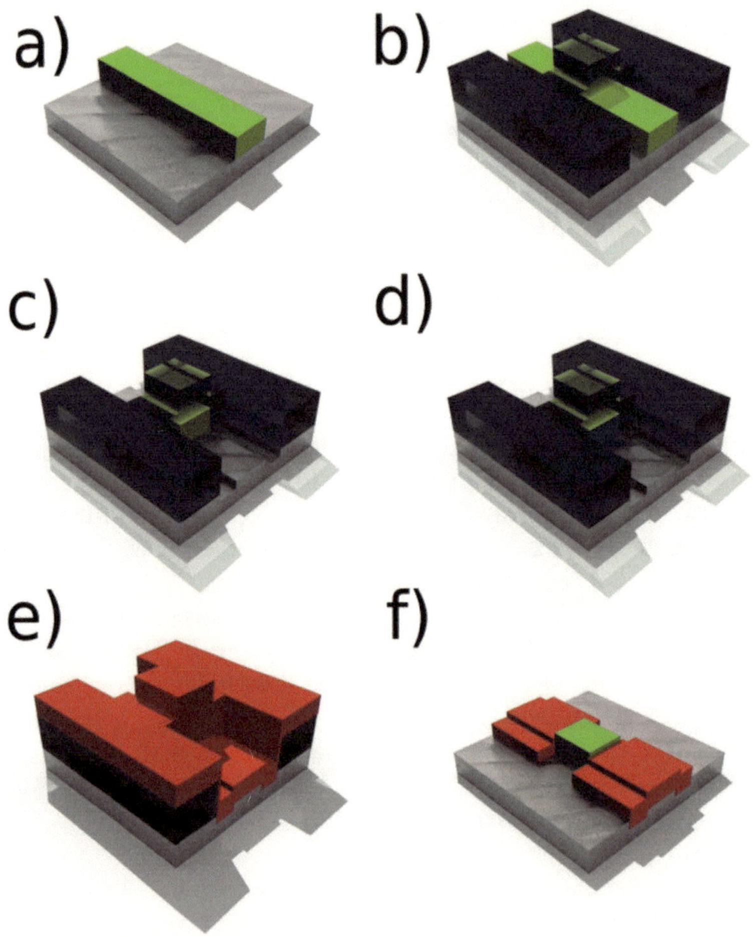

Fig. 10. The main technological steps for the SAIL technique. See text for details.

In order to be used for lift off, the resist mask should show in section walls with negative slope. The generation of a suitable mask is the crucial step and more complicated in the SAIL technique than in the shadow evaporation one.
The creation of the barrier after the anisotropic etching of the first mask avoids its damage due to high-energy ions.
Over-etching in the substrate during step iii. can lead to re-deposition of substrate material on the exposed sides of the island, and then serious barrier uniformity problems can arise. To improve the quality of the barrier as well as to minimize the over-etching, it is possible to choose as substrate the same material of the barrier to be fabricated: in fact, the barrier dielectrics usually have lower etching rates than the corresponding pure metals, and therefore can excellently act as etch-stop layers.
A further technological complication is that the formation of reliable contacts requires a more anisotropic etching (step iii.) than the second metal deposition step (step vi.).
Apart from these difficulties, the SAIL process has several advantages if compared to the shadow evaporation technique.

As mentioned above, there is complete freedom in the choice of the deposition process of metal layer, e.g. evaporation can be replaced by sputtering. It is worth noting, for instance, that the latter technique is more suitable for depositing a robust and reliable superconductor like Nb. Moreover, one can get rid of fragile structures like suspended bridges necessary for the shadow evaporation. Finally, since the tunnel junction is obtained at the sides of the island, the electrodes overlapping is absent, and the junction capacitance is lower than in devices realized by the shadow evaporation.
The first SET made with the SAIL technique was reported by M. Gotz (Gotz et al., 1995). The device is based on the system $Al/AlO_x/Al$. The island, with thickness and width of 50 nm and 80-150 nm, respectively, is defined by EBL and subsequent lift off on a single layer of AR-P610 resist. The metal was deposited by sputtering. The second mask was made with a double resist layer composed by AR-P671 and AR-P 641. The thickness of the second metal layer was 100nm. The anisotropic etching was carried out with Ar^+ ions. Immediately after the etching, the dielectric barrier has been created by means of oxidation step in dry air. The reported yield is 40%.
From the width of the Coulomb Blockade areas, the junction capacitance was estimated to be 0.5 fF, a value in agreement with the calculations for a tunnel junction area of 50 x 150 nm^2, and a barrier thickness of the order of 1 nm.

4. Conclusions

The employment of the Shadow evaporation technique dramatically limits the choice for superconductors to use, either from a merely technical (materials to be evaporated) or from a more fundamental (difficulties in reducing junction areas) points of view. As a matter of fact, h-SETs made by Al/Cu assemblies have been recently produced and characterized (Pekola et al., 2008). The SAIL technique seems promising, since it allows for a wider choice of superconducting materials. It is possible, for example, to envisage the employment of In:Pb alloys (with improved electrical and thermal properties with respect to the unalloyed elements) in SAIL SETs by taking advantage from composition-related gap tunability. In this case, however, technological problems related to deposition of continuous, ~10 nm thick, films from metals with low fusion temperature require solution. It is noteworthy that such alloys were used years ago in the first generation Josephson junctions (Lacquaniti et al., 1982). V or Ta could be interesting alternatives, but the best candidate for the realization of stable and robust turnstiles should obviously be Nb. Indeed, the graph in Fig. 7 shows that the inclusion of such material in a h-SET arrangement still requires to overcome the technological limitations of the SAIL technique.
The possibility of device biasing, offered by the hybrid arrangement can improve the accuracy of electron pumping process, but care must be taken in reducing leakage through the superconducting gap. Optimizing between these opposite effects requires the increase of both the superconducting gap and the charging energy.

5. Acknowlegdments

The work has been carried out at Nanofacility Piemonte supported by Compagnia di San Paolo.

6. References

Altshuler, B. L.; Lee, P. A. & Webb, R. A. (1991). *Mesoscopic phenomena in solids*. North Holland, 1991, 978-044-4884-54-1

Averin, D. V.; Korotkov, A. N. & Likharev, K. K. (1991). "Theory of single-electron charging of quantum wells and dots," *Physical Review B*, vol. 44, n°. 12, pag. 6199, 1991.

Averin, D. V. & Pekola, J. P. (2008). "Nonadiabatic Charge Pumping in a Hybrid Single-Electron Transistor," *Physical Review Letters*, vol. 101, n°. 6, pag. 066801, 2008.

Blumenthal, M. D.;Kaestner, B.;Li, L.;Gibling, S.;Janssen, T.J.B.M.;Pepper, M.;Anderson, D.;Jones, G. & Ritchie, D.A. "Gigahertz quantized charge pumping," *Nat Phys*, vol. 3, n°. 5,(2007)

Cholascinski, M. & Chhajlany, R. W. (2007). "Stabilized Parametric Cooper-Pair Pumping in a Linear Array of Coupled Josephson Junctions," *Physical Review Letters*, vol. 98, n°. 12, pag. 127001, Mar. 2007.

Clark, A. M. (2005).;Miller, N.A.;Williams, A.;Ruggiero, S.T.;Hilton, G.C.;Vale, L.R.;Beall, J.A.; Irwin, K.D. & Ullom, J.N. , "Cooling of bulk material by electron tunneling refrigerator ". *Appl. Phys. Lett.* 86, 173508 (2005).

Dolan, G. J. "Offset masks for lift-off photoprocessing". *Appl. Phys. Lett.* 31, 337 (1977).

Flowers, J. (2004). "The Route to Atomic and Quantum Standards," *Science*, vol. 306, n°. 5700, pagg. 1324-1330, Nov. 2004.

Geerligs, L. J. ; Anderegg, V. F.; van der Jeugd, C. A.; Romijn, J. & Mooij, J. E. (1989). "Influence of Dissipation on the Coulomb Blockade in Small Tunnel Junctions" *Europhys. Lett.* 10, 79, 1989.

Geerligs, L. J.; Anderegg, V.F.;Holweg, P.A.M.; Mooij, J.E.; Poithier, H.; Esteve, D.; Urbina, C. & Devoret, M.H. (1990) "Frequency-locked turnstile device for single electrons," *Physical Review Letters*, vol. 64, n°. 22, pag. 2691, Mag. 1990.

Giazotto, F.; Heikkilä, T. T.; Luukanen, A.; Savin, A. M. & Pekola, J. P. 2006). "Opportunities for mesoscopics in thermometry and refrigeration: Physics and applications" *Rev. Mod. Phys.* 78, 217 (2006).

Governale, M.; Taddei, F.; Fazio, R. & Hekking, F. W. J. (2005). "Adiabatic Pumping in a Superconductor-Normal-Superconductor Weak Link," *Physical Review Letters*, vol. 95, n°. 25, pag. 256801, Dic. 2005.

Götz, M.; Bluthner, K.; Krech, W.,Nowack, A.; Fuchs, H.J.; Kley, E.B.; Thieme, P.; Wagner, Th. Eska, G.; Hecker, K. & Hegger, H. (1996) "Self-aligned in-line tunnel junctions for single-charge electronics," *Physica B: Condensed Matter*, vol. 218, n°. 1, pagg. 272-275, Feb. 1996.

Ingold, G. L. & Nazarov, Y. V. (1992) in Single charge tunneling, Vol. 294 of NATO ASI Series B, edited by H. Grabert & M. H. Devoret, Plenum Press, New York, 1992, ISBN 0-306-44229-9.

Josephson, B. D. (1962). "Possible new effects in superconductive tunneling," *Physics Letters*, vol. 1, n°. 7, pagg. 251-253, Lug. 1962.

Keller, M. W.; Martinis, J. M.; Zimmerman, N. M. Steinbach, & A. H. (1996). "Accuracy of electron counting using a 7-junction electron pump," *Applied Physics Letters*, vol. 69, n°. 12, pag. 1804, 1996.

Koch H. "Self-Aligned In-Line Junction - Fabrication and Application to DC-SQUIDS", *Int. Supercond. Electr. Conf.*, Tokyo August 1987 (Extended Abstracts) pp. 281-284.

Kopnin, N. B.; Mel'nikov, A. S. & Vinokur, V. M. (2006). "Resonance Energy and Charge Pumping through Quantum SINIS Contacts," *Physical Review Letters*, vol. 96, n°. 14, pag. 146802, Apr. 2006.

Koppinen, P.; Kühn, T. & Maasilta, I. (2009), "Effects of Charging Energy on SINIS Tunnel Junction Thermometry " *J. Low Temp. Phys.* 154, 179 (2009).

Lacquaniti, V; Battistoni, C.; Paparazzo, E.; Cocito, M.; Palumbo, S. (1982). "Tunnelling barrier structure of Nb/Pb and Nb/(Pb-In) thin film Josephson junctions studied by Auger electron spectroscopy and x-ray photoelectron spectroscopy analysis, " *Thin Sol. Films,* vol. 94, n°. 4, pagg. 331-339. 1982

Likharev, K. K. (1988) "Correlated discrete transfer of single electrons in ultrasmall tunnel junctions," *IBM Journal of Research and Development*, 1988.

Lotkhov, S. V. (2004)"Radio-frequency-induced transport of Cooper pairs in superconducting single electron transistors in a dissipative environment," *Journal of Applied Physics*, vol. 95, n°. 11, pag. 6325, 2004.

Meschke, M.; Guichard, W. & Pekola, J. P. (2006). "Single-mode heat conduction by photons" *Nature* 444, 187 (2006).

Mooij, J. E. & Nazarov, Y. V. (2006). "Superconducting nanowires as quantum phase-slip junctions," *Nature Physics*, vol. 2, n°. 3, pagg. 169-172, Mar. 2006.

Nahum, M.; Eiles, T. M. & Martinis, J. M. (1994). "Electronic microrefrigerator based on a normal-insulator-superconductor tunnel junction" *Appl. Phys. Lett.* 65, 3123 (1994).

Nahum, M.; & Martinis, J. M. (1993). "Ultrasensitive-hot-electron microbolometer" *Appl. Phys. Lett.* 63, 3075 (1993).

Niskanen, A. O.; Pekola, J. P. & Seppä, H. (2003). "Fast and Accurate Single-Island Charge Pump: Implementation of a Cooper Pair Pump," *Physical Review Letters*, vol. 91, n°. 17, pag. 177003, Ott. 2003.

Pekola, J. P.; Giazotto, F. & Saira, O.-P. (2007). "Radio-Frequency Single-Electron Refrigerator" *Phys. Rev. Lett.* 98, 037201 (2007).

Pekola, J. P.; Vartiainen, J. J.; Mottonen, M.; Saira, O.-P.; Meschke, M. & Averin, D. V. (2008). "Hybrid single-electron transistor as a source of quantized electric current," *Nat Phys*, vol. 4, n°. 2, pagg. 120-124, 2008.

Pothier, H.; Lafarge, P.; Urbina, C.; Esteve, D. & Devoret, M. H. (1992). "Single-Electron Pump Based on Charging Effects," *Europhysics Letters (EPL)*, vol. 17, n°. 3, pagg. 249-254, 1992.

Saira, O.-P.; Meschke, M.; Giazotto, F.; Savin, A. M.; Möttönen, M. & Pekola, J. P. (2007). *Phys. Rev. Lett.* 99, 027203 (2007).

Schmidt, D. R.; Yung, C. S. & Cleland, A. N. (2003). "Nanoscale radio-frequency thermometry" *Appl. Phys. Lett.* 83, 1002 (2003).

Talyanskii, V. I.; Shilton, J.M.; Pepper, M.; Smith, C.G.; Ford, C.J.B.; Linfield, E.H.; Ritchie, D.A. & Jones, G.A.C. (1997). "Single-electron transport in a one-dimensional channel by high-frequency surface acoustic waves," *Physical Review B*, vol. 56, n°. 23, pag. 15180, Dic. 1997.

Vartiainen, J. J.; Möttönen, M.; Pekola, J. P. & Kemppinen, A. (2007). "Nanoampere pumping of Cooper pairs," *Applied Physics Letters*, vol. 90, n°. 8, pag. 082102, 2007.

Zimmerman, N. M. & Keller, M. W. (2003). "Electrical metrology with single electrons," *Measurement Science and Technology*, vol. 14, n°. 8, pagg. 1237-1242, 2003.

Zwerger, W. & Scharpf, M. (1991). "Crossover from Coulomb-blockade to ohmic conduction in small tunnel junctions," *Zeitschrift für Physik B Condensed Matter*, vol. 85, n° 3.

14

Photonic Band Structure and Transmittance of the Superconductor Photonic Crystal

Ting-Hang Pei and Yang-Tung Huang
National Chiao Tung University
Taiwan,
R.O.C.

1. Introduction

The photonic crystal (PhC) is formed with a dielectric periodic structure and exhibits new electromagnetic phenomena (John, 1987). It shows some properties analog to the semiconductor, such as the photonic band structure (PBS) including photonic passing bands and photonic band gaps (PBGs), and complicated dispersion relations. In analogous to the electron transport in the semiconductor, the Bloch theorem is also applied to describe electromagnetic waves propagating in the PhC very well.

The PBS strongly depends on refracted indices of constituent materials and the geometry of the PhC. Once the materials and geometry structure of a PhC are constructed, the possible way to change its PBS is tuning the refracted indices of its constituent materials utilizing the temperature effect, the external electric field effect, or the external magnetic field effect, etc (Busch & John, 1999; Kee & Lim, 2001; Kee et al., 2000, 2001; Figotin et al., 1998; Takeda & Yoshino, 2003a, 2003b, 2003c, 2003d, 2004). For PhCs composed of ferroelectric or ferromagnetic materials, PBSs can be tuned by the external electric field effect and the external magnetic field effect (Busch & John, 1999; Figotin et al.). On the other hand, the variation on the PBS of the liquid-crystal PhC controlled by the external electric field or the temperature has also been investigated (Kee & Lim, 2001b; Takeda & Yoshino, 2003a, 2003b, 2003c, 2003d, 2004). Another potential material that can be used to tune the PBS is the superconductor by varying the temperature and the external magnetic field (Lee et al., 1995; Raymond Ooi et al., 2000; Takeda & Yoshino, 2003e).

In our previous works, we have designed a tunable PhC Mach-Zehnder interferometer composed of copper oxide high-temperature superconductors (HTSCs) utilizing the temperature modulation to reach the on and off states (Pei & Huang, 2007a). The Mach-Zehnder interferometer, whose path-length difference of two arms is fixed after designed, can be realized as an optical switching device or sensor due to the temperature effect. In the output, the signals from two arms interfere with each other, and the phases of these two signals can be modulated by HTSCs. Besides, we also discussed the superprism effect in the superconductor PhC (Pei & Huang, 2007b). The superprism effect was demonstrated experimentally by Kosaka *et al.* in 1998 (Kosaka *et al.*, 1998). They found that the refracted angle of a light beam in a PhC is very sensitive to the incident angle and wavelength. The basic explanation of the superprism effect is based on the anomalous dispersion characteristics of the PBS. The propagation direction of light in the PhC is the same as the direction of the group velocity, which is determined by the equifrequency surfaces (EFS). The group velocity is normal to the EFS at a certain wave vector and is defined as $\vec{v}_g = grad_{\vec{k}}\omega$, where $\vec{k}$ and ω are the wave vector and the frequency, respectively. Notomi has published a detailed study on the superprism effect (Notomi, 2000). In our work, we not only study the transmission of light propagating through the superconductor PhC, but also pay lots of attentions on the refraction. The result shows that the refraction can be changed sensitively by the temperature of the superconductor.

In this chapter, we deduce the way to calculate the PBS of the superconductor PhC based on the plane wave expansion method first. It is not like the way to calculate the PBS of the PhC only composed of dielectric materials. Second, the finite-difference time-domain (FDTD) method for the PhC composed of dispersive materials such as superconductors are derived carefully. The time-domain auxiliary differential equations (ADEs) are introduced to represent effects of currents in dispersive materials. The ADE-FDTD algorithm can be used to calculate the transmission of the finite superconductor PhC. It has also been used in our previous works to discuss the tunability of the PhC Mach-Zehnder interferometer composed of HTSCs and the superprism effect in the superconductor PhC.

Finally, the internal-field expansion method developed by Sakoda is also introduced (Sakoda, 1995a, 1995b, 2004). This method is used to calculate the transmission of the two-dimensional PhC composed of air cylinders embedded in certain background medium. It is much like the grating theory that describes the scattering waves as Bragg waves. He successively calculated the transmission and the Bragg reflection spectra using this method, and also mentioned that the existences of the uncoupled modes (Sakoda, 1995a, 1995b). However, this method has not been yet verified on the superconductor PhC. We use this method to calculate the transmission of the finite superconductor PhC and compare the result of it with that of the ADE-FDTD method.

2. The plane wave expansion method for calculating the photonic band structure of the superconductor photonic crystal

The superconductivity of the superconductor is strongly sensitive to the temperature and the external magnetic field. We only discuss the temperature effect in this chapter. The PhC structure is composed of superconductor cylinders with triangular lattice in air as shown in Fig. 1. The two-fluid model is used to describe the electromagnetic response of a typical superconductor without an additional magnetic field (Tinkham, 2004), and it describes that the electrons occupy two states. One is the superconducting state, in which the superconducting electrons of density $N_s(x, y)$ are paired and transport with no resistance. The definition of the superconducting state under the temperature and magnetic field effects is shown in Fig. 2. The other is the normal state, in which the normal conducting electrons of density $N_n(x, y)$ act like electrons in general materials with a nonzero resistance. Both superconducting and normal conducting electrons coexist in the superconductor when the temperature is lower than the critical temperature. This model also characterizes the performance of high-frequency superconductive devices very well (Van Duzer & Truner, 1998).

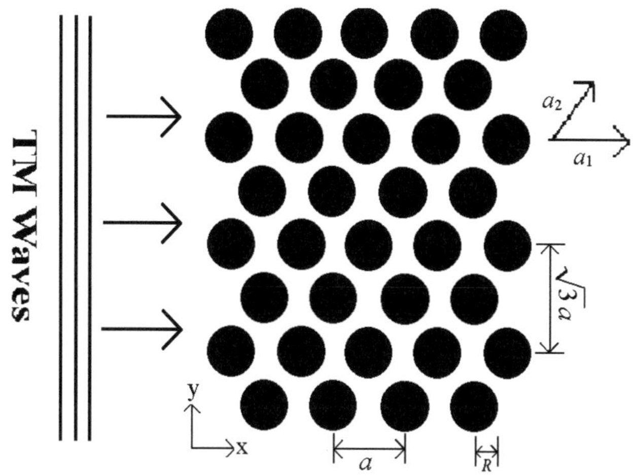

Fig. 1. The cross-section of the two-dimensional PC formed in a triangular array.

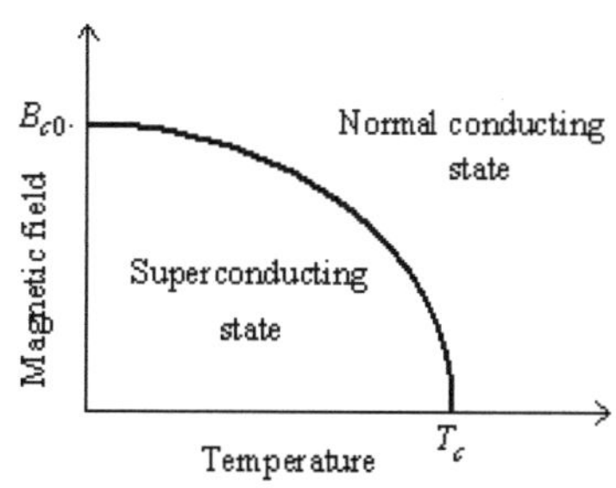

Fig. 2. The definition of the superconducting state under the effects of the temperature and the magnetic field.

Utilizing this model, the E-polarized light with its electric field parallel to the z-axis (TM mode) is incident on a two-dimensional PhC lying in the x-y plane. In the presence of the external electric field, superconducting and normal conducting current densities J_{sz} and J_{nz} flowing along the z-axis can be expressed as the following equations (Tinkham, 2004):

$$\frac{\partial J_{sz}(x,y)}{\partial t} = \varepsilon_0 \varepsilon_1(x,y)\omega_p^s(x,y)^2 E_z(x,y) \tag{1}$$

$$\tau \frac{\partial J_{nz}(x,y)}{\partial t} + J_{nz}(x,y) = \tau\varepsilon_0 \varepsilon_1(x,y)\omega_p^n(x,y)^2 E_z(x,y), \tag{2}$$

where $\omega_p^s(x,y)$ and $\omega_p^n(x,y)$ are the plasma frequencies of the superconducting and normal conducting electrons given by

$$\omega_p^s(x,y) = \sqrt{N_s(x,y)e^2 / \varepsilon_0\varepsilon_1(x,y)m} = c / \lambda(x,y)\sqrt{\varepsilon_1(x,y)}, \tag{3}$$

$$\omega_p^n(x,y) = \sqrt{N_n(x,y)e^2 / \varepsilon_0\varepsilon_1(x,y)m} \tag{4}$$

$\lambda(x, y)$ is the London penetration depth, $\varepsilon_1(x, y)$ is the distribution of the dielectric constant, τ is the relaxation time, c is the wave velocity in free space, and m is the mass of the electron. Because the incident electric field is harmonic with frequency ω, the induced J_{sz} and J_{nz} also have the same oscillating period. Eqs. (1) and (2) can be further expressed as follows:

$$J_{sz}(x,y) = -i\varepsilon_0\varepsilon_1(x,y)\frac{\omega_p^s(x,y)^2}{\omega}E_z(x,y) \tag{5}$$

$$J_{nz}(x,y) = \varepsilon_0\varepsilon_1(x,y)\frac{\omega_p^n(x,y)^2\tau}{(1+i\omega\tau)}E_z(x,y) \quad . \tag{6}$$

Substituting Eqs. (5) and (6) into the E-polarized wave equation results in the following equation (Pei & Huang, 2007a):

$$\frac{\partial^2 E_z(x,y)}{\partial x^2} + \frac{\partial^2 E_z(x,y)}{\partial y^2} = -\frac{\omega^2\varepsilon_1(x,y)}{c^2}E_z(x,y) + i\mu_0\omega\{J_{sz}(x,y) + J_{nz}(x,y)\}$$

$$= -\frac{\omega^2\varepsilon_1(x,y)}{c^2}\left\{1 - \frac{\omega_p^s(x,y)^2}{\omega^2} - \frac{\omega_p^n(x,y)^2\tau}{\omega(1+i\omega\tau)}\right\}E_z(x,y)$$

$$= -\frac{\omega^2}{c^2}\varepsilon_s(x,y,\omega)E_z(x,y), \tag{7}$$

where $\varepsilon_s(x, y, \omega)$ is the effective dielectric function given by

$$\varepsilon_s(x,y,\omega) = \varepsilon_1(x,y)\left\{1 - \frac{\omega_p^s(x,y)^2}{\omega^2} - \frac{\omega_p^n(x,y)^2\tau}{\omega(1+i\omega\tau)}\right\}. \tag{8}$$

For HTSCs, optical characteristics show the anisotropic properties (Takeda & Yoshino, 2003e). The electric fields parallel and perpendicular to the c-axis feel different dielectric indices. However, Eq. (7) is still valid even for anisotropic materials (Lee et al., 1995). When the electric fields are parallel to the c-axis, plasma frequencies are in the microwave and far-infrared regions (Takeda & Yoshino, 2003e). In our study, the z-axis is chosen as the c-axis.

In the superconducting state, the electromagnetic wave can propagate in the range of the London penetration depth. The London penetration depth is dependent on the temperature T, which can be expressed as $\lambda(T) = \lambda_0 \big/ \sqrt{1-(T/T_c)^4}$ (Zhou, 1999), where T_c and λ_0 are the critical temperature and the London penetration depths at the absolute zero temperature, respectively. When the temperature is above about 0.8 times the critical temperature, the London penetration depth increases rapidly and then approaches infinity as the temperature is close to T_c. Besides, $\omega_p^s(x,y)$ strongly depends on the London penetration depth as well as the temperature. Based on the experimental results (Shibata & T. Yamada, 1996; Matsuda, 1995) in the far-infrared region, the small contribution of the normal conducting electrons can be neglected and the plasma frequency $\omega_p^s(x,y)$ can be assumed to be uniform within the rods. Then the third term on the right side of Eq. (6) can be dropped and then simplified as

$$\varepsilon_s(x,y,\omega) \approx \varepsilon_1(x,y)\left(1-\frac{\omega_p^s(x,y)^2}{\omega^2}\right). \tag{9}$$

Eq. (9) is known as the Drude model (Grosso & Parravicini, 2000) which can also be applied to the kind of PhCs constituting metallic components.
Kuzmiak *et al.* (Kuzmiak *et al.*, 1994) has dealt with the two-dimensional PhC containing metallic components. We use the same method based on the plane-wave expansion to calculate the PBSs of the superconductor PhCs. In this method, the dielectric function of the PhC is directly expanded in a Fourier series. The dielectric constant of the PhC can be written explicitly in the form

$$\varepsilon_s(\vec{r},\omega) = 1+\left[\varepsilon_s(\omega)-1\right]\sum_{\vec{a}} S(\vec{r}-\vec{a}) \quad , \tag{10}$$

where the function $S(\vec{r})=1$ and $\varepsilon_s(\vec{r},\omega)=\varepsilon_s(\omega)$ if $\vec{r}$ is inside the cylinder, and $S(\vec{r})=1$ and $\varepsilon_s(\vec{r},\omega)=1$ if $\vec{r}$ is outside the cross section. The expansion of $\varepsilon_s(\vec{r},\omega)$ in a Fourier series on reciprocal lattice vectors $\vec{G}$ is

$$\varepsilon_s(\vec{r},\omega) = \sum_{\vec{G}} \varepsilon_s(\vec{G},\omega) e^{i\vec{G}\cdot\vec{r}} \quad , \tag{11}$$

where the Fourier coefficient $\varepsilon_s(\vec{G},\omega)$ is given by

$$\varepsilon_s(\vec{G},\omega) = \frac{1}{a_c}\int_{V_0^{(2)}} \varepsilon_s(\vec{r},\omega) e^{-i\vec{G}\cdot\vec{r}} d\vec{r}$$

$$= \delta_{\vec{G}0} + \left[\varepsilon_s(\omega)-1\right]\frac{1}{a_c}\int_{V_0^{(2)}} S(\vec{r}) e^{-i\vec{G}\cdot\vec{r}} d\vec{r} \quad , \tag{12}$$

where the integral is now over the 2D unit cell $V_0^{(2)}$ and a_c is the area of the unit cell in the PhC. Eq. (12) can be expressed as

$$\varepsilon_s(\vec{G}=\vec{0},\omega) = 1+\left[\varepsilon_1\left(1-\frac{\omega_p^{s2}}{\omega^2}\right)-1\right]f, \tag{13}$$

$$\varepsilon_s(\vec{G}\neq\vec{0},\omega) = \varepsilon_1\left(\frac{\varepsilon_1-1}{\varepsilon_1}-\frac{\omega_p^{s2}}{\omega^2}\right)\frac{2J_1(GR)}{GR}f, \tag{14}$$

where *f* is the filling fraction. For the triangular superconductor PhC, the filling fraction of the superconductor rod in a unit cell is $f=(2\pi/\sqrt{3})R^2/a^2$. According to the Bloch theory, the electric field can be expanded in the form

$$E_z(x,y) = \sum_{\vec{G}} E_z(\vec{k}\,|\,\vec{G}) e^{i(\vec{k}+\vec{G})\cdot\vec{r}} \quad , \tag{15}$$

where $\vec{k} = k_x\hat{i} + k_y\hat{j}$ is the wave vector of the electromagnetic waves propagating inside the PhC. Substituting Eqs. (11) and (15) into Eq. (7), we obtain a set of equations for the coefficients $E_z(\vec{k}\,|\,\vec{G})$. It is a standard eigenvalue problem of a real and symmetric matrix with respect to the frequency ω. The set of equations for coefficients $E_z(\vec{k}\,|\,\vec{G})$ shows as follows:

$$\sum_{\vec{G}'}\left[\left(\vec{k}+\vec{G}\right)^2\delta_{\vec{G},\vec{G}'} + \varepsilon_1\left(\frac{\omega_p^s}{c}\right)^2\frac{1}{a_c}\int S(\vec{r})e^{-i(\vec{G}-\vec{G}')\cdot\vec{r}}d\vec{r}\right]E_z(\vec{k}\,|\,\vec{G}')$$

$$=\left(\frac{\omega}{c}\right)^2\sum_{\vec{G}'}\left[\left(1-f+\varepsilon_1 f\right)\delta_{\vec{G},\vec{G}'} + \frac{\varepsilon_1-1}{a_c}\int S(\vec{r})e^{-i(\vec{G}-\vec{G}')\cdot\vec{r}}d\vec{r}\right]E_z(\vec{k}\,|\,\vec{G}'). \tag{16}$$

Rearranging Eq. (16) that we have

$$\sum_{\vec{G}'}\left\{\left(\vec{k}+\vec{G}\right)^2\delta_{\vec{G},\vec{G}'} + f\left(\frac{\omega_s^p}{c}\right)^2\delta_{\vec{G},\vec{G}'} + f\left(\frac{\omega_s^p}{c}\right)^2\frac{2J_1\left(\left|\vec{G}-\vec{G}'\right|R\right)}{\left(\left|\vec{G}-\vec{G}'\right|R\right)}\right\}E_z(\vec{k}\,|\,\vec{G}')$$

$$=\frac{\omega^2}{c^2}\sum_{\vec{G}'}\left\{\left[\varepsilon_0(1-f)+\varepsilon_1 f\right]\delta_{\vec{G},\vec{G}'} - (\varepsilon_0-\varepsilon_1)\frac{2J_1\left(\left|\vec{G}-\vec{G}'\right|R\right)}{\left(\left|\vec{G}-\vec{G}'\right|R\right)}\right\}E_z(\vec{k}\,|\,\vec{G}) \tag{17}$$

To solve this matrix eigenvalue problem, the frequencies can be determined at a certain wave vector and the whole PBS can be obtained.

3. The E-polarized photonic band structure

In our designed device, we used high-T_c superconductor $Bi_{1.85}Pb_{0.35}Sr_2Ca_2Cu_{3.1}O_y$ (Takeda & Yoshino, 2003). Previous study (Takeda & Yoshino, 2003) utilized parallel copper oxide HTSCs rods to form PhCs with square lattices repeating in two-dimensional directions (*x-y* plane). The authors theoretically investigated the tunability of the photonic band gap (PBG) of the two-dimensional PhC by changing temperatures of superconductors and external magnetic fields. The PhC structure we discuss here is composed of superconductor cylinders with triangular lattice in air as shown in Fig. 1. The *E*-polarized electromagnetic wave with the electric field parallel to the extended direction of the rod propagates in the *x-y* plane. Adjusting the temperature of the superconductor can control the refracted index of the superconductor as well as the PBS of the superconductor PhC. When $T \le T_c$ is satisfied, the dependence of the plasma frequency on the temperature is given by (Zhou, 1999)

$$\omega_p^s(T) = \omega_p^s(0)\Big/\sqrt{1-(T/T_c)^4}. \tag{18}$$

For this superconductor, the London penetration depth of the copper oxide HTSCs is $\lambda = 23$ μm at $T = 5$ K, the critical temperature $T_c = 107$ K, and the dielectric constant is $\varepsilon_1 = 12$ (Shibata & Yamada, 1996). When $T = 5$ K, we obtain $\omega_p^s / c \approx 1.3\times10^4$ cm^{-1}.

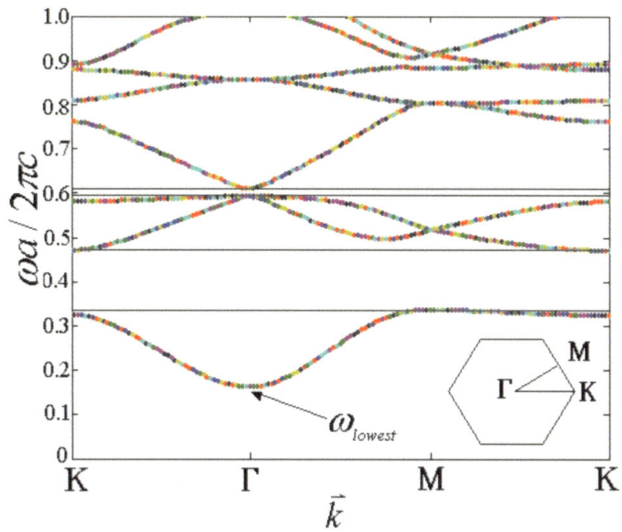

Fig. 3. The PBS of the PhC composed of superconductor cylinders at $T = 5$ K with radius of cylinders $r = 0.2a$ and lattice constant $a = 100$ μm. The first Brillouin zone shows at the right lower corner. In this figure, the region between two paired horizontal lines is the PBG.

The theory discussed in Section 2 is used to calculate the PBS along the three directions MΓ, ΓK, and KM in the reduced Brillouin zone when the periodic lattice constant of the PhC is $a = 100$ μm, the radius of cylinders is $r = 0.2a$, and the overall temperature is fixed at 5 K. The PBS is shown in Fig. 3 and the first Brillouin zone at the right lower corner. The reduced Brillouin zone is denoted as the triangle ΓKM. From Eq. (9), we can see that the optical response of the superconductor under the *E*-polarized wave is the same as that of the metal described by the Drude model. The lowest point of the first band for a metal or metal-like material is above zero frequency, which is not like a non-dispersive material whose lowest point of the first band is at zero frequency. A PBG exists from zero to a certain frequency ω_{lowest}, which means that the light can propagate in the PhC only in the frequency range above ω_{lowest}. In Fig. 3, the region between two paired horizontal lines is the PBG region. The PhC has a large second PBG, which is located in the frequency range from 0.33 to 0.47 ($2\pi c/a$). The third PBG is located in the frequency range from 0.595 to 0.605 ($2\pi c/a$).

4. The finite-difference time-domain method for the photonic crystal composed of dispersive materials

In 1966, K. S. Yee first provided the FDTD method to solve electromagnetic scattering problems (Yee, 1966). The Yee's equations are obtained to discretize Maxwell's equations in time and space. The fields on the nodal points of the space-time mesh can be calculated in an iteration process when the source is excited. Because the finite resource of the hardware limits the size of simulation domain, an absorbing boundary condition (ABC) needs to be set on the outer surface of the computational domain. In 1994, Berenger proposed a perfectly matched layer (PML), which is an artificial electromagnetic wave absorber with electric conductivity σ and magnetic conductivity σ^* (Berenger, 1994). The PML absorbs outgoing waves very well, so it can simulate the electromagnetic wave propagating in free space. Therefore, we apply the PML as the absorbing layer used in the FDTD method.

In the FDTD method, Maxwell's equations are solved directly in time domain via finite differences and time steps without any approximations or theoretical restrictions. The basic approach is relatively easy to understand and is an alternative to more usual frequency-domain approaches, so this method is widely used as a propagation solution technique in integrated optics. Imagine a region of space where no current flows and no isolated charge exists. Maxwell's curl equations can be written in Cartesian coordinates as six simple scalar equations. Two examples are:

$$\frac{\partial E_z}{\partial t} = -\frac{1}{\varepsilon}\left(\frac{\partial H_y}{\partial x} - \frac{\partial H_x}{\partial y}\right), \tag{19}$$

$$\frac{\partial H_z}{\partial t} = -\frac{1}{\mu_0}\left(\frac{\partial E_x}{\partial y} - \frac{\partial E_y}{\partial x}\right). \tag{20}$$

The most common method to solve these equations is based on Yee's mesh and computes the E and H field components at points on a grid with grid points spaced Δx, Δy, and Δz apart, which are named grid sizes. The E and H field components are then interlaced in all three spatial dimensions. Furthermore, time is broken up into discrete steps of Δt. The E field components are then computed at times $t = n\Delta t$ and the H at times $t = (n + 1/2)\Delta t$, where n is an integer representing the computing step. For example, the E field at a time $t = n\Delta t$ is equal to the E field at $t = (n - 1)\Delta t$ plus an additional term computed from the spatial variation, or curl of the H field at time t. This method results in six equations that can be used to compute the field at a given mesh point, denoted by integers i, j, k

$$E_z\Big|_{i,j}^{n+1} = E_z\Big|_{i,j}^{n} + \frac{\Delta t}{(\varepsilon_0\varepsilon_1)_{i,j}}\left(\frac{H_y\Big|_{i+1/2,j}^{n+1/2} - H_y\Big|_{i-1/2,j}^{n+1/2}}{\Delta x} - \frac{H_x\Big|_{i,j+1/2}^{n+1/2} - H_x\Big|_{i,j-1/2}^{n+1/2}}{\Delta y}\right), \tag{21}$$

$$H_z\Big|_{i,j}^{n+1} = H_z\Big|_{i,j}^{n} + \frac{\Delta t}{\mu_0}\left(\frac{E_x\Big|_{i,j}^{n} - E_x\Big|_{i,j-1}^{n}}{\Delta y} - \frac{E_y\Big|_{i,j}^{n} - E_y\Big|_{i-1,j}^{n}}{\Delta x}\right). \tag{22}$$

These equations are iteratively solved in a leapfrog manner, alternating between computing the E and H fields at subsequent $\Delta t/2$ intervals. The grid sizes and time step in 2D simulations are set $\Delta x = \Delta y$ and $\Delta t = \Delta x/2$.

The method for implementing FDTD models of dispersive materials utilizes ADE equations which describe the time variation of the electric current densities (Taflove & Hagness, 2005). These equations are time-stepped synchronously with Maxwell's equations. ADE-FDTD method is a second-order accurate method.

Consider a dispersive medium whose Ampere's Law can be expressed as

$$\nabla \times \vec{H}(t) = \varepsilon_0\varepsilon_\infty \frac{\partial \vec{E}(t)}{\partial t} + \sigma\vec{E}(t) + \sum_p \vec{J}_p(t), \tag{23}$$

where $\vec{J}_p(t)$ is the polarization current. The goal of the ADE technique is to develop a simple time-stepping scheme for $\vec{J}_p(t)$. In our superconductor system, $\vec{J}_{sz}$ and $\vec{J}_{nz}$ contribute to $\sigma\vec{E}$ and $\vec{J}_p$, respectively, so Eq. (23) can be rewritten as

$$\nabla\times\vec{H}(x,y,t)=\varepsilon_0\varepsilon_1(x,y)\frac{\partial\vec{E}_z(x,y,t)}{\partial t}+\vec{J}_{sz}(x,y,t)+\vec{J}_{nz}(x,y,t). \tag{24}$$

Another time-dependent Maxwell's curl equation is

$$\nabla\times\vec{E}(x,y,t)=-\mu(x,y)\frac{\partial\vec{H}(x,y,t)}{\partial t}, \tag{25}$$

where $\mu(x,y)$ is the position dependent permeability of the material. Eqs. (24) and (25) can be discretized in two-dimensional space and time by the Yee-cell technique(Yee, 1966). Eqs. (1) and (2) are the required ADEs for $\vec{J}_{sz}$ and $\vec{J}_{nz}$. They both can be easily and accurately implemented in an FDTD code using the semi-implicit scheme where fields at time-step n+1 are created and updated by fields known at time-step n. Then, we implement Eqs. (1) and (2) in an FDTD code by finite differences, centered at time-step n+1/2:

$$\frac{J_{sz}|_{i,j}^{n+1}-J_{sz}|_{i,j}^{n}}{\Delta t}=\left(\varepsilon_0\varepsilon_1\omega_p^s\right)_{i,j}^2\frac{E_z|_{i,j}^{n+1}+E_z|_{i,j}^{n}}{2}, \tag{26}$$

$$\frac{J_{nz}|_{i,j}^{n+1}-J_{nz}|_{i,j}^{n}}{\Delta t}+\frac{J_{nz}|_{i,j}^{n+1}+J_{nz}|_{i,j}^{n}}{2\tau}=\left(\varepsilon_0\varepsilon_1\omega_p^{n2}\right)_{i,j}\frac{E_z|_{i,j}^{n+1}+E_z|_{i,j}^{n}}{2}. \tag{27}$$

Solving Eqs. (26) and (27) for $J_{sz}|_{i,j}^{n+1}$ and $J_{nz}|_{i,j}^{n+1}$, we obtain

$$J_{sz}|_{i,j}^{n+1}=J_{sz}|_{i,j}^{n}+\frac{\Delta t}{2}\left(\varepsilon_0\varepsilon_1\omega_p^s\right)_{i,j}^2\left(E_z|_{i,j}^{n+1}+E_z|_{i,j}^{n}\right) \tag{28}$$

$$J_{nz}|_{i,j}^{n+1}=\frac{(1-\Delta t/2\tau)}{(1+\Delta t/2\tau)}J_{nz}|_{i,j}^{n}+\frac{\Delta t}{2(1+\Delta t/2\tau)}\left(\varepsilon_0\varepsilon_1\omega_p^{n2}\right)_{i,j}\left(E_z|_{i,j}^{n+1}+E_z|_{i,j}^{n}\right). \tag{29}$$

Then we can evaluate Eq. (24) at time-step $n+1/2$:

$$\nabla\times\vec{H}\Big|_{i,j}^{n+1/2}=\left(\varepsilon_0\varepsilon_1\right)_{i,j}\left(\frac{\vec{E}_z\Big|_{i,j}^{n+1}-\vec{E}_z\Big|_{i,j}^{n}}{\Delta t}\right)+\frac{1}{2}\left(\vec{J}_{sz}\Big|_{i,j}^{n+1}+\vec{J}_{sz}\Big|_{i,j}^{n}\right)+\frac{1}{2}\left(\vec{J}_{nz}\Big|_{i,j}^{n+1}+\vec{J}_{nz}\Big|_{i,j}^{n}\right). \tag{30}$$

Applying Eq. (30) into the implementation of Eq. (25) in an FDTD code by finite differences, we obtain the E fields at time-step n+1:

$$E_z|_{i,j}^{n+1}=E_z|_{i,j}^{n}+\frac{\Delta t}{\left(\varepsilon_0\varepsilon_1\right)_{i,j}}\left(\frac{H_y|_{i+1/2,j}^{n+1/2}-H_y|_{i-1/2,j}^{n+1/2}}{\Delta x}-\frac{H_x|_{i,j+1/2}^{n+1/2}-H_x|_{i,j-1/2}^{n+1/2}}{\Delta y}\right). \tag{31}$$

Thus, the ADE-FDTD algorithm for calculating dispersive media has three processes. Starting with the assumed known values of $\vec{E}_z$, $\vec{J}_{sz}^n$, $\vec{J}_{nz}^n$, and $\vec{H}^{n+1/2}$, we first calculate the new E_z^{n+1} components using Eq. (31). Next, we calculate the new J_{sz}^{n+1} and J_{nz}^{n+1} components using Eqs. (28) and (29). Finally, $H_x^{n+3/2}$ is obtained from $H_x^{n+1/2}$ and E_z^{n+1} by using Eq. (30). $H_y^{n+3/2}$ is updated as $H_x^{n+3/2}$ be done.

In the end of this section, let us return to discuss the numerical stability. We choose the two-dimensional cell space steps, i.e. Δx and Δy, and the time step Δt based on the required accuracy. The space step is usually chosen less one twentieth of the smallest wavelength in order to avoid the non-physical oscillation. The time step must satisfy the well-known "Courant Condition":

$$\Delta t \le \frac{1}{V_{\max}}\left[\frac{1}{(\Delta x)^2}+\frac{1}{(\Delta y)^2}\right]^{-1}, \tag{32}$$

where V_{max} is the maximum wave velocity in the computational domain.

5. The transmission of the finite photonic crystal composed of the superconductor

In this section, the ADE-FDTD method is used to calculate the transmission of the finite thickness PhC from the frequency 0.01 to 1.00 ($2\pi c/a$). As we know, the PBS represents the existing mode with photon energy inside the infinite PhC; but in practice, the thickness of a PhC is always finite. So it is necessarily to calculate the transmission and further compare to the PBS in the previous section. This also verifies calculations of the PBS through the ADE-FDTD method. The triangular PhC is shown in Fig. 1 in which the interface is along the ΓM direction (x-direction). Light is normally incident and propagates along the ΓK direction (y-direction). The numbers of layers along the x- and y-directions are 40 and 30, respectively. The lattice constant along the x-direction is a_1 and that along y-direction a_2. We choose $a_2 = a$ = 100 μm and $a_1 = \sqrt{3}\, a_2$. For simplicity, the square unit cell $\Delta x \Delta y$ are used in the ADE-FDTD calculations where $\Delta x = \Delta y = a/30$. The time increment is $\Delta t = \Delta x / 2c$. The Gaussian wave is supposed to be incident from air on the PhC. The transmissions from 0.01 to 1.00 ($2\pi c/a$) are shown in Fig. 4. The increment of the frequency is 0.005 ($2\pi c/a$). In this simulation, we consider both currents $\vec{J}_{sz}$ and $\vec{J}_{nz}$. The contribution of the current $\vec{J}_{sz}$ in the calculations is dominant and that of the current $\vec{J}_{nz}$ is very small. It can be seen that almost zero transmission below frequency 0.16 ($2\pi c/a$) matches the prediction of the PBS. All transmissions are more than 0.80 at frequencies from 0.16 to 0.33 ($2\pi c/a$). This frequency region just corresponds to the first photonic band, in which the highest transmission is close to 1.0 at 0.28 ($2\pi c/a$). The second PBG occurs at frequencies between 0.33 and 0.47 ($2\pi c/a$). It can be seen that the transmission dramatically drops to nearly zero at 0.33 ($2\pi c/a$) and then continues almost zero until 0.47 ($2\pi c/a$). The second and third bands both occupy the frequency region from 0.47 to 0.59 ($2\pi c/a$), so the transmission becomes larger in this frequency region. From the PBS, we can predict that another sharp drop should take place in a narrow region between 0.59 and 0.61 ($2\pi c/a$), which is just the third PBG. A sharp drop after 0.59 ($2\pi c/a$) is indeed investigated and then rapid rise after 0.61 ($2\pi c/a$) from the ADE-

FDTD calculations. The frequency region above 0.61 ($2\pi c/a$) and below 1.00 ($2\pi c/a$) is occupied by several bands. Hence, the most parts of this frequency region should have non-zero transmission.

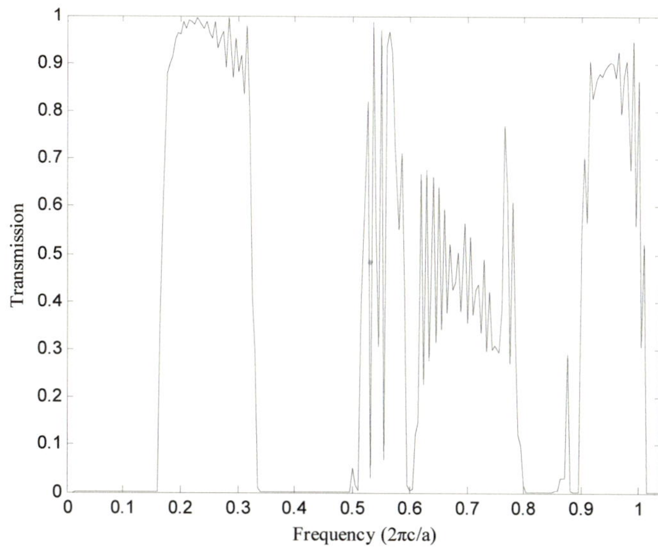

Fig. 4. Transmissions calculated by the FDTD method when the Gaussian wave is incident from air into the PhC. The radius of cylinders is $0.2a$ and the lattice constant a is 100 μm. We set $\Delta x = a/30$ and 30 layers in the propagation direction.

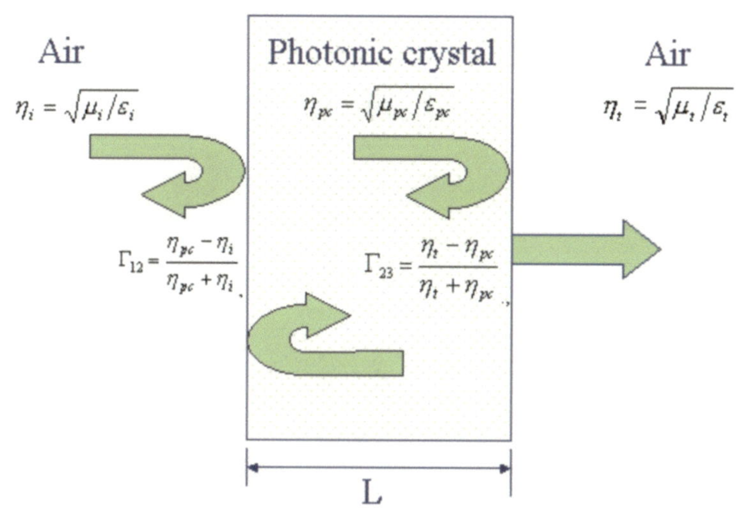

Fig. 5. The transmission and reflection when the electromagnetic wave propagates through three media including two interfaces.

Another possible low transmission predicted by the PBS occurs in the vicinity of the intersection between the fifth and sixth bands. In Fig. 4, these two bands intersect at the Γ point of the first Brillouin zone when the frequency is 0.86 ($2\pi c/a$). Because the modes in the

PhC only occupy k=0 states, the density of states (DOS) is very small at the intersection. On the other hand, from the viewpoint of the effectively refracted index of the PhC, they have very small effectively refracted indices in the vicinity of the intersection. Hence, the multilayer model instead of DOS can be used to explain the extremely low transmissions. In our calculations, the PhC is sandwiched between two homogeneous media. If the PhC is replaced with an effective homogeneous medium (Pei et al., 2011a, 2011b), it can define the effectively dielectric constant. ε_{pc}, effectively magnetic permeability μ_{pc}, effectively refracted index n_{pc}, and effective impedance η_{pc} in the normally incident case. The effective refracted index and effective impedance are defined as $n_{pc}=\sqrt{\varepsilon_{pc}\mu_{pc}}$ and $\eta_{pc}=\sqrt{\mu_{pc}/\varepsilon_{pc}}$ for the positive refraction, respectively. For the negative refraction, the refracted index and effective impedance are defined as $n_{pc}=\sqrt{(-\varepsilon_{pc})(-\mu_{pc})}$ and $\eta_{pc}=\sqrt{(-\mu_{pc})/(-\varepsilon_{pc})}$ (Engheta & Ziolkowski, 2006). The relation between the effectively refracted index and effective impedance is

$$n_{pc}=\varepsilon_{pc}\eta_{pc}. \tag{33}$$

The transmission and reflection become the problem of multiple scattering by interfaces as shown in Fig. 5 (Moreno, 2002). The total transmitted coefficient of the system consisting of one finite and two semi-infinite media with two interfaces is (Yariv & Yeh, 2002)

$$t=\frac{t_{12}t_{23}e^{-ik_{pc}L}}{1+r_{12}r_{23}e^{-i2k_{pc}L}}, \tag{34}$$

where $k_{pc}=n_{pc}\omega/c$ and L is the length of the PhC along the y direction. The r_{12}, r_{23}, t_{12}, and t_{23} are defined as

$$r_{12}=\frac{\eta_{pc}-\eta_i}{\eta_{pc}+\eta_i}, \tag{35}$$

$$r_{23}=\frac{\eta_t-\eta_{pc}}{\eta_t+\eta_{pc}}, \tag{36}$$

$$t_{12}=\frac{2\eta_{pc}}{\eta_{pc}+\eta_i}, \tag{37}$$

$$t_{23}=\frac{2\eta_t}{\eta_t+\eta_{pc}}, \tag{38}$$

where $\eta_i=\sqrt{\mu_i/\varepsilon_i}$ and $\eta_t=\sqrt{\mu_t/\varepsilon_t}$ are the impedances of the incident region and transmitted region, respectively. The incident and transmitted regions are both air here, so the $\eta_i=\eta_t$. If the effectively refracted index of the PhC is zero in Eq. (33), either ε_{pc} or μ_{pc} has to be zero. It deduces that η_{pc} is zero if μ_{pc}=0, and η_{pc} is infinite if ε_{pc}=0. By calculations, the effectively refracted index at the intersection of two bands is zero as well as the transmission. We can further check this point of view by using the ADE-FDTD method at 0.86 ($2\pi c/a$). The result shows that most of the electromagnetic wave cannot pass through the first interface between the incident region and the PhC. According to the above discussion of transmission and Eq. (37), the zero t_{12} deduces zero effective impedance η_{pc}. Due to $\eta_{pc}=0$, we have a non-zero ε_{pc} and a zero μ_{pc} here.

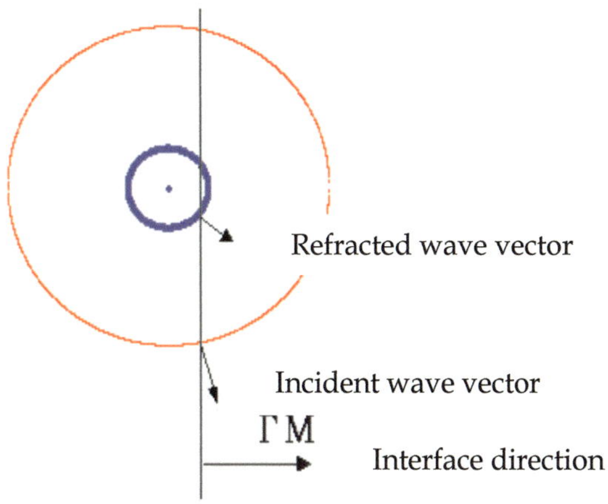

Fig. 6. The determination of the refracted wave vector and refracted angle by the conservation of wave vector parallel to the interface. The outer circle represents the EFS with frequency ω in air, and the inner one the EFS with frequency ω in the PhC. The line is perpendicular to the interface along the ΓM direction.

From the ADE-FDTD calculation in Fig. 4, we find out that the extremely low transmission not only takes place at the intersection, but also extends to the vicinity. They locate at frequencies between 0.80 and 0.88 ($2\pi c/a$). To explain it we should calculate the effectively refracted indices in this frequency region. The effectively refracted indices can be determined from EFSs. According to the conservation rule, the incident and the refracted wave vectors are continuous for the tangential components parallel to the interface. Given the incident wave vector and an incident angle, the refracted wave vector and the refracted angle will be determined. How to determine the refracted wave vector and refracted angle by the conservation rule is shown in Fig. 6. The incident and refracted waves are on different sides of the normal line, so it is the positive refraction. By applying Snell's law, the effectively refracted index can be further determined. But traditional Snell's law cannot be applied if the EFSs move inward with an increasing frequency (Notomi, 2000). It needs to add minus sign on the effectively refracted index. Figs. 7–10 show the 3D EFSs of the fourth to seventh photonic bands in the first Brillouin zone. It can be seen that the frequency range of 3D EFSs matches the calculation in Fig. 3. They all form a bell shape. Some erect upward and some erect downward. The upward bell usually corresponds to the positive refraction and the downward bell usually corresponds to the negative refraction.

Crosscutting the 3D EFS at a certain frequency reduces to a two-dimensional contour. Thus, we obtain a lot of k_x and k_y at the same frequency drawn in the two-dimensional plane. Each point on the contour is the allowed propagating mode in the PhC for the chosen frequency. In the following, we further discuss the extremely low transmission at frequencies from 0.80 to 0.88 ($2\pi c/a$). EFSs of frequencies 0.81, 0.83, and 0.85 ($2\pi c/a$) for discussions are shown in

Figs. 11–13. A large circle drawn at the center in each figure represents the FES of air at the same frequency. All EFSs of air in these three figures have larger radius than those in the PhC. If we use the conservation rule mentioned before, they all result in the same conclusion that the refracted angle is larger than the incident angle. It also indicates that the absolute value of the effectively refracted index is smaller than 1.0. Because each EFS shrinks with an increasing frequency, the effectively refracted index is negative. Therefore, the negative refraction takes place here. When the frequency is higher, the shape of the EFS of the fifth photonic band is closer to a circle. The circular EFS means that the PhC can be considered as a homogeneous medium at this frequency. The relations between the incident and refracted angles for these three frequencies are shown in Figs. 15(a)-(c). On the one hand, the lower curve of each figure shows the negative refraction, where the refracted angle is defined as negative for convenience. The negative angles are calculated from lines intersecting with the EFSs in the first Brillouin zone as shown in Figs. 11-13. It can be seen that the relation between the incident angle and refracted angle is much like that in a homogeneous medium. On the other hand, the upper curves for a larger incident angle in Fig. 15 (a) and (b) show the normal refraction with positive refracted angle. They are calculated from lines intersecting with the EFSs in the right repeated Brillouin zone as shown in Figs. 11-13.

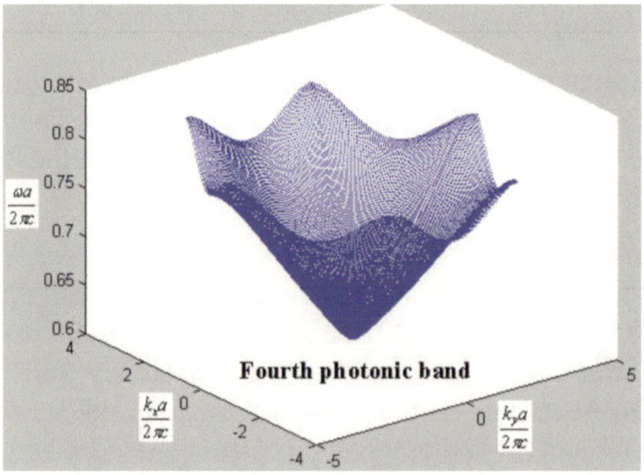

Fig. 7. The EFS of the fourth photonic band in the first Brillouin zone.

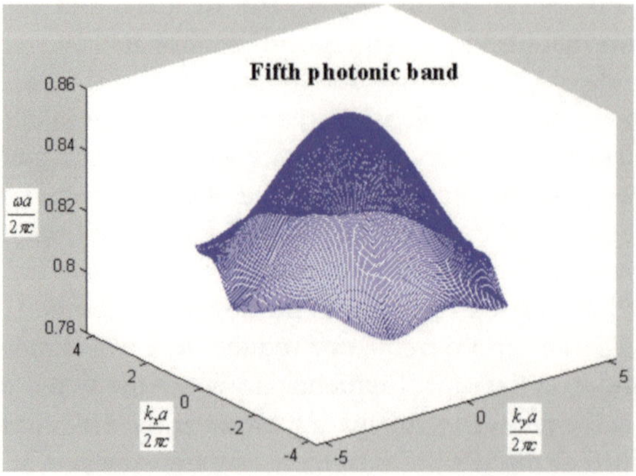

Fig. 8. The EFS of the fifth photonic band in the first Brillouin zone.

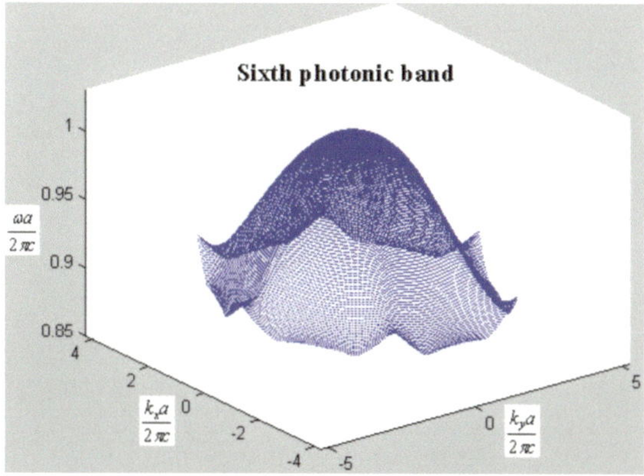

Fig. 9. The EFS of the sixth photonic band in the first Brillouin zone.

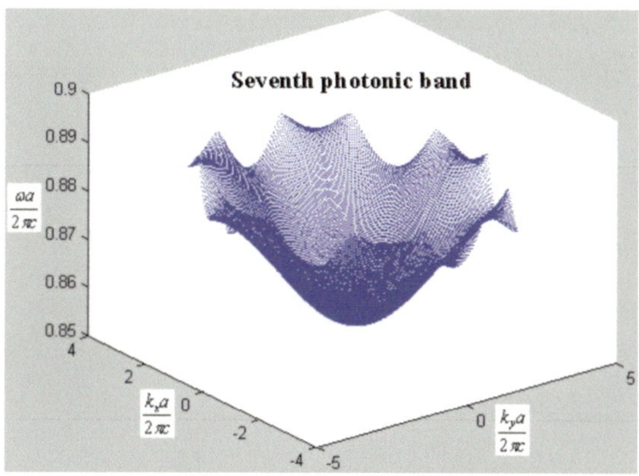

Fig. 10. The EFS of the seventh photonic band in the first Brillouin zone.

Next, we calculate the effectively refracted index varying with the incident angle only for negative refraction as shown in Fig. 16. From Figs. 15(a)-(c), the normally incident case belongs to negative refraction. It can be found out that the effectively refracted indices of three frequencies 0.81, 0.83, and 0.85 ($2\pi c/a$) at incident angle 0° are -0.31, -0.30, and -0.16, respectively. Using Eq. (33), we can calculate these three corresponding effective impedances. But we do not explicitly know the effectively dielectric constants for these three frequencies at normal incidence. However, according to the previous discussion about the normal incidence at 0.86 ($2\pi c/a$), we have the conclusion that the effective impedance η_{pc} is zero with a zero μ_{pc} and a non-zero ε_{pc}. Utilizing the similar explanation and a little correction, the effective impedance at 0.85 ($2\pi c/a$) should be very close to zero with non-zero μ_{pc} and ε_{pc}. The conclusion can also be applied to frequencies 0.81 and 0.83 ($2\pi c/a$). As a result, the effective impedances in the frequency range from 0.81 to 0.85 ($2\pi c/a$) are very small. Using Eqs. (34)-(38), we obtain extremely low transmissions at frequencies from 0.81 to 0.85 ($2\pi c/a$).

6. The internal-field expansion method

In this section, we introduce the internal-field expansion method (IFEM) to calculate the transmission of the finite thickness PhC (Sakoda, 1995a, 1995b, 2004). This method is based on

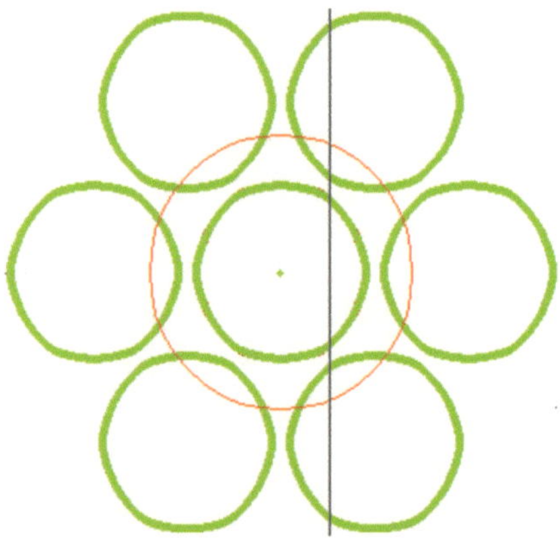

Fig. 11. The EFS of the fifth photonic band at 0.81 ($2\pi c/a$) in the repeated Brillouin zone. The largest circle at the center represents the FES of air with the same frequency.

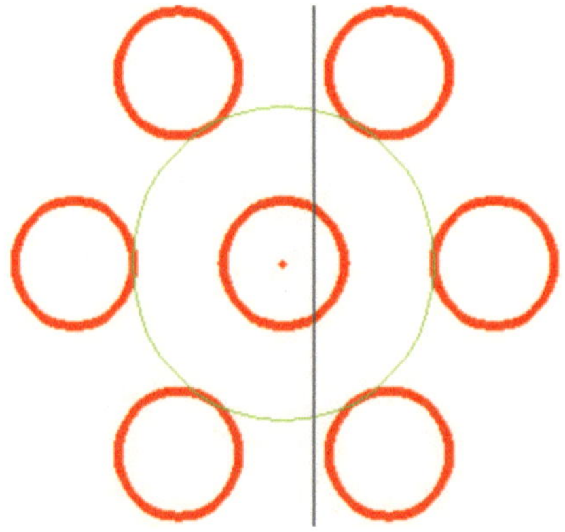

Fig. 12. The EFS of the fifth photonic band at 0.83 ($2\pi c/a$) in the repeated Brillouin zone. The largest circle at the center represents the FES of air with the same frequency.

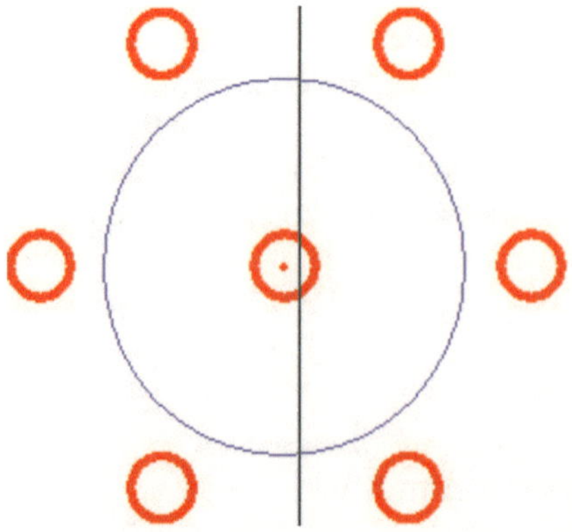

Fig. 13. The EFS of the fifth photonic band at 0.85 ($2\pi c/a$) in the repeated Brillouin zone. The largest circle at the center represents the FES of air with the same frequency.

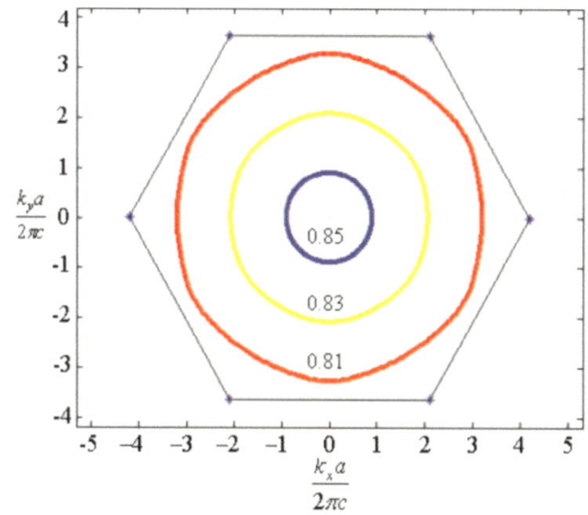

Fig. 14. EFSs of the fifth photonic band when frequencies are 0.81, 0.83, and 0.85 ($2\pi c/a$).

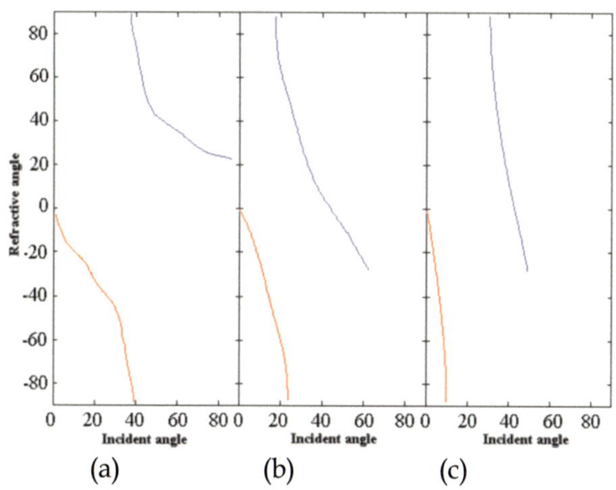

Fig. 15. Refracted angles vs. incident angles calculated from Figs. 11–13 for T = 5 K at (a) 0.81, (b) 0.83, and (c) 0.85 ($2\pi c/a$).

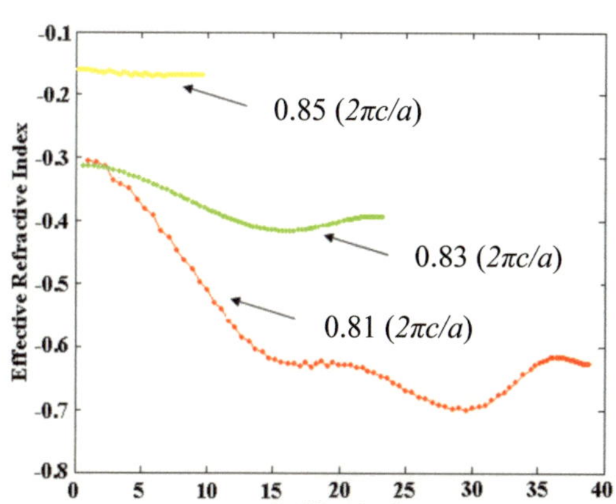

Fig. 16. Effectively refracted indices versus incident angles at 0.81, 0.83, and 0.85 ($2\pi c/a$).

the internal fields expanded in Fourier series. We consider a two-dimensional PhC composed of a triangular array of air cylinders with radius of r in a dielectric background. The dielectric constants of the cylinders and the background are ε_a and ε_b, respectively. The infinitely extended direction of air holes is parallel to the z-axis. The PhC is infinitely extended in the x-direction and the width of the PhC in the y-direction is finite. Therefore, two dielectric-PhC interfaces exist at the left and right sides of the PhC. a_1 and a_2 are the lattice periods along the x- and y- directions, respectively. The region from the interface to the edge of the nearest cylinder is called the edge region, in which the width is d. The PhC has two edge regions at the left and right sides. The other region including all the cylinders is called the middle region. The total width L of the PhC in the y-direction is $L = 2(r + d) + (N_L - 0.5)a_2$, where N_L is the periodic number. So the total layers of cylinders are $2N_L$. The configuration of the PhC is shown in Fig. 17. The first Brillouin zone is shown at the up-right corner. The region at the left-handed side of the PhC is called the incident region, and that at the right-handed side of the PhC is called the transmitted region. The plane wave in the incident region is incident on the left interface. After propagating through the PhC, the transmitted wave is through the right interface and into the transmitted region.

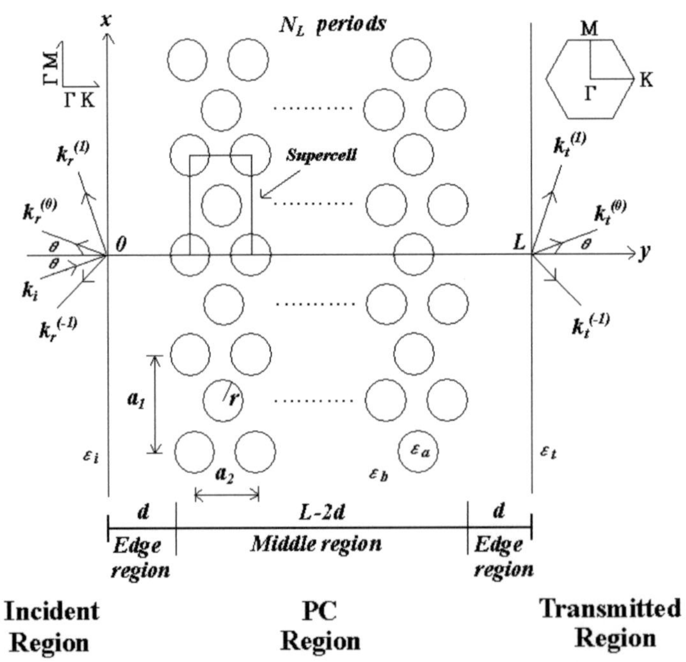

Fig. 17. The PhC structure with finite length in the y-direction and infinite length in the x-direction.

Because the two-fluid model is only suitable for the currents flowing along the cylinder direction, we only discuss an E-polarized plane wave incident upon the superconductor PhC here. Two interfaces are along the ΓM direction of the PhC. The 2D wave vector of the incident wave is denoted by $\vec{k}_i = (k_i \sin\theta, k_i \cos\theta) = (k_{i,x}, k_{i,y})$, where θ is the incident angle and $k_i = \sqrt{\varepsilon_i}\omega/c$. ε_i is the dielectric constant of the incident region, ω is the angular frequency of the incident wave, and c is the light velocity in vacuum. The wave in the incident region consists of the incident plane wave and the reflected Bragg waves. In the transmitted region, the wave is composed of the transmitted Bragg waves. The reflected and transmitted Bragg waves are represented as space harmonics with the wave vector

$$k_{r,x}^{n} = k_{t,x}^{n} = k_{x}^{n} = k_{i,x} + G_{n}, \tag{39}$$

where $k_{r,x}^{n}$ and $k_{t,x}^{n}$ are the wave-vector components parallel to the interface for the Bragg reflected and transmitted waves of order n, respectively, $G_n = 2n\pi / a_1$ is the reciprocal lattice vector corresponding to the periodicity a_1, and n is an integer. Each component of the Eq. (39) is called the nth order phase matching condition. It means that the periodicity along the ΓM direction is like a diffraction grating. The wave-vector components of the nth order Bragg reflected and transmitted waves normal to the interface are

$$k_{r,y}^{n} = \begin{cases} -\sqrt{k_i^2 - \left(k_x^n\right)^2} & if \;\; k_i \geq \left|k_x^n\right| \\ -i\sqrt{\left(k_x^n\right)^2 - k_i^2} & Otherwise \end{cases} \tag{40}$$

$$k_{t,y}^{n} = \begin{cases} \sqrt{k_t^2 - \left(k_x^n\right)^2} & if \;\; k_i \geq \left|k_x^n\right| \\ i\sqrt{\left(k_x^n\right)^2 - k_t^2} & Otherwise \end{cases} \tag{41}$$

Here, $k_t = \sqrt{\varepsilon_t}\omega / c$ and ε_t are the transmitted wave vector and dielectric constant of the transmitted region, respectively. The electric fields in the incident region and the transmitted region are given by

$$E_{iz}(x,y) = E_0 \exp(i\vec{k}_i \cdot \vec{r}) + \sum_{n=-\infty}^{\infty} R_n \exp(i\vec{k}_r^n \cdot \vec{r}), \tag{42}$$

$$E_{tz}(x,y) = \sum_{n=-\infty}^{\infty} T_n \exp(i\vec{k}_t^n \cdot \vec{r}), \tag{43}$$

where E_0, R_n, and T_n are the amplitudes of the electric field of the incident wave, the reflected Bragg wave, and the transmitted Bragg wave, respectively. The electric field inside the PhC satisfies the following equation derived from Maxwell's equations:

$$L_E\left[E_{PC}(x,y)\right] = \left[\frac{1}{\varepsilon(x,y,\omega)}\left(\frac{\partial^2}{\partial x^2} + \frac{\partial^2}{\partial y^2}\right) + \frac{\omega^2}{c^2}\right] E_{PC}(x,y) = 0. \tag{44}$$

Now, we introduce a boundary value function $f_E(x,y)$:

$$f_E(x,y) = \sum_{n=-\infty}^{\infty} \left\{\frac{y}{L} T_n + \left(1 - \frac{y}{L}\right)\left(\delta_{n0} E_0 + R_n\right)\right\} \exp(ik_x^n x), \tag{45}$$

where δ_{nm} is the Kronecker's δ. The boundary value function $f_E(x,y)$ satisfies the boundary conditions at each interface:

$$f_E(x,0) = E_{iz}(x,0) \text{ and } f_E(x,L) = E_{tz}(x,L) \quad (46)$$

Moreover, we define

$$\psi_E(x,y) = E_{PhC}(x,y) - f_E(x,y) \quad (47)$$

If we substitute Eq. (47) into Eq. (44), we have

$$L_E[\psi_E(x,y)] = -L_E[f_E(x,y)], \quad (48)$$

$$\psi_E(x,0) = \psi(x,L) = 0. \quad (49)$$

The problem of unknown E_{pc} becomes to deal with the internal field. We have to solve Eq. (48) to obtain E_{pc} field in the PhC. If we expand $\psi_E(x,y)$ and $\varepsilon^{-1}(x,y,\omega)$ in Fourier series, we have

$$\psi_E(x,y) = \sum_{n=-\infty}^{\infty}\sum_{m=1}^{\infty} A_{nm}\exp(ik_x^n x)\sin\frac{m\pi}{L}y, \quad (50)$$

$$\frac{1}{\varepsilon(x,y,\omega)} = \sum_{n=-\infty}^{\infty}\sum_{m=1}^{\infty} \kappa_{nm}\exp(iG_n x + k_m y). \quad (51)$$

Then the electric field in the PhC is expressed as

$$E_{PC}(x,y) = \sum_{n=-\infty}^{\infty}\left\{\frac{y}{L}T_n + \left(1-\frac{y}{L}\right)(\delta_{n0}E_0 + R_n)\right\}\exp(ik_x^n x)$$

$$+\sum_{n=-\infty}^{\infty}\sum_{m=1}^{\infty} A_{nm}\exp(ik_x^n x)\sin\frac{m\pi}{L}y. \quad (52)$$

If we substitute Eqs. (45), (50), and (51) into Eq. (48) and compare the independent Fourier components, the equation about coefficients R_n, T_n, and A_{nm} are obtained as follows:

$$\frac{\omega^2}{c^2}A_{nm} + \sum_{n'=-\infty}^{\infty}\sum_{m'=1}^{\infty}\left\{\left(k_x^{n'}\right)^2 + \left(\frac{m'\pi}{L}\right)^2\right\}\left(\kappa_{n-n',m+m'} - \kappa_{n-n',|m-m'|}\right)A_{n'm'}$$

$$= -\frac{2}{\pi}\sum_{n'=-\infty}^{\infty}\left(k_x^{n'}\right)^2\sum_{m'=1}^{\infty}\left(\kappa_{n-n',m+m'} - \kappa_{n-n',|m-m'|}\right)\times\frac{(-1)^{m'-1}T_{n'} + R_{n'} + \delta_{n'0}E_0}{m'}$$

$$-\frac{2}{\pi}\frac{\omega^2}{c^2}\frac{(-1)^{m-1}T_n + R_n + \delta_{n0}E_0}{m}, \quad (53)$$

where $\kappa_{n,m}$ is the Fourier coefficient of the inverse of $\varepsilon(x,y,\omega)$. Next, we calculate the Fourier coefficients of the configuration shown in Fig. 17. In our case, we have

$$\frac{1}{\varepsilon(x,y,\omega)}=\frac{1}{\varepsilon_b}+\left(\frac{1}{\varepsilon_a}-\frac{1}{\varepsilon_b}\right)\sum_{j=1}^{4}\sum_{l=-\infty}^{\infty}S\left(\vec{r}-\vec{u}\left(j,l,l'\right)\right), \tag{54}$$

where $\varepsilon_a=\varepsilon_s(\omega)$ and $\vec{u}(j,l,l')$ is the center of each cylinder, which are

$$\vec{u}(1,l,l')=(la_1,l'a_2+R+d), \tag{55}$$

$$\vec{u}(2,l,l')=((l+1/2)a_1,(l'+1/2)a_2+R+d), \tag{56}$$

$$\vec{u}(3,l,l')=(la_1,-l'a_2-R-d), \tag{57}$$

$$\vec{u}(4,l,l')=((l+1/2)a_1,-(l'+1/2)a_2-R-d). \tag{58}$$

The S here is the spatial function of the cylinder. The inverse of $\varepsilon(x,y,\omega)$ now is extended symmetrically to the negative y region ($-L\leqq y\leqq 0$) to calculate the Fourier coefficients. Then, we obtain

$$\kappa_{n,m}=\frac{1}{2a_1L}\int_0^{a_1}dx\int_{-L}^{L}\frac{1}{\varepsilon(x,y,\omega)}\exp\left[-i\left(G_nx+\frac{m\pi}{L}y\right)\right]dy$$

$$=\frac{1}{\varepsilon_b}\delta_{n0}\delta_{m0}+\frac{1}{2a_1L}\left(\frac{1}{\varepsilon_a}-\frac{1}{\varepsilon_b}\right)\sum_{j=1}^{4}\sum_{l=-\infty}^{\infty}\sum_{l'=0}^{\infty}\int_0^{a_1}dx\int_{-L}^{L}S\left(\vec{r}-\vec{u}\left(j,l,l'\right)\right)e^{-i\vec{G}_{nm}\cdot\vec{r}}dy, \tag{59}$$

where $\vec{G}_{nm}=(G_n,m\pi/L)$. After calculating the integration, we obtain

$$\kappa_{0,0}=\frac{f}{\varepsilon_a}+\frac{1-f}{\varepsilon_b}, \tag{60}$$

$$\kappa_{2n,2m}=\begin{cases}\dfrac{2f}{N_L}\left(\dfrac{1}{\varepsilon_a}-\dfrac{1}{\varepsilon_b}\right)(-1)^m\cos\dfrac{a_2m\pi}{2L}\dfrac{\sin(a_2mN_L\pi/L)}{\sin(a_2m\pi/L)}\dfrac{J_1(G_{2n.2m}R)}{G_{2n.2m}R} & (m\neq 0)\\ 2f\left(\dfrac{1}{\varepsilon_a}-\dfrac{1}{\varepsilon_b}\right)\dfrac{J_1(G_{2n.2m}R)}{G_{2n.2m}R} & (n\neq 0,m=0)\end{cases}, \tag{61}$$

$$\kappa_{2n-1,2m-1}=\frac{2f}{N_L}\left(\frac{1}{\varepsilon_a}-\frac{1}{\varepsilon_b}\right)(-1)^{m-1}\sin\frac{a_2(2m-1)\pi}{4L}\frac{\sin\left[a_2(2m-1)N_L\pi/2L\right]}{\sin\left[a_2(2m-1)\pi/2L\right]}\frac{J_1(G_{2n-1.2m-1}R)}{G_{2n-1.2m-1}R}, \tag{62}$$

where f is the filling fraction of the superconductor rods in the calculation domain:

$$f=2N_L\pi R^2/a_1L. \tag{63}$$

Finally, we want to solve the unknown coefficients, A_{nm}, R_n, and T_n. Eq. (53) is not enough to solve all unknown coefficients because the number of equations is less than the number of all unknown coefficients. We need other boundary conditions to solve all A_{nm}, R_n, and T_n. The reminder boundary conditions is the continuity of the x components of the magnetic field, which leads to

$$\pi\sum_{m=1}^{\infty} mA_{nm} = \left(iLk_{r,y}^{n}+1\right)R_n - T_n + \delta_{n0}E_0\left(iLk_{i,y}\right), \tag{64}$$

$$\pi\sum_{m=1}^{\infty} m(-1)^{m-1}A_{nm} = R_n + \left(iLk_{t,y}^{n}-1\right)T_n + \delta_{n0}E_0. \tag{65}$$

Follow the calculation processes and consider the boundary conditions for the E-polarized mode, we can determine the unknown coefficients, A_{nm}, R_n, and T_n. In practical calculation, we restrict the Fourier expansion up to $n = \pm N$ and $m = M$. So there are $(2N + 1)M$ terms in the Fourier expansion. The total number of the unknown coefficients is $(2N + 1)(M + 2)$. From Eqs. (42) and (43) and the boundary conditions, we also obtain $(2N + 1)(M + 2)$ independent equations. Solving these independent equations can obtain these coefficients.
To discuss the reflection and transmission along the y-direction, we can sum up the y-components of the Poynting vectors of all waves and consider the energy flow conservation across these two interfaces. The y component of the wave vector with an imaginary value represents the evanescent wave in the incident region or the transmitted region, so it's not necessary to consider this kind of wave in the summation. Then, we obtain the following relations for the E-polarized mode:

$$\sum_{n}\frac{c}{4\pi}\frac{\left|k_{r,y}^{n}\right|}{k_i}\sqrt{\varepsilon_i}\left|R_n\right|^2 + \sum_{n'}\frac{c}{4\pi}\frac{k_{t,y}^{n'}}{k_t}\sqrt{\varepsilon_t}\left|T_{n'}\right|^2 = \frac{c\cos\theta}{4\pi}\sqrt{\varepsilon_i}\left|E_0\right|^2, \tag{66}$$

where n and n' represent the summation over the Bragg waves with real wave vectors. Then, we can use the Eq. (52) to define the transmission and reflection:

$$T = \sum_{n'}\frac{k_{t,y}^{n'}}{k_t}\sqrt{\frac{\varepsilon_t}{\varepsilon_i}}\left|T_n'\right|^2 \Big/ \cos\theta\left|E_0\right|^2, \tag{67}$$

$$R = \sum_{n}\frac{k_{r,y}^{n}}{k_i}\left|R_n\right|^2 \Big/ \cos\theta\left|E_0\right|^2. \tag{68}$$

Eqs. (67) and (68) will be used in Section 7.

7. The transmission calculated by internal-field expansion method

In previous Section, we have introduced the internal-field expansion method to calculate the finite thickness PhC. This method used to calculate the transmission of the electromagnetic wave propagating through the PhC is faster than the FDTD method if the size of the $(2N + 1)(M + 2) \times (2N + 1)(M + 2)$ matrix is not very large. In the original references (Sakoda, 1995a, 1995b, 2004), the author concludes that this method can be used for the general two-dimensional PhC. In the following, we use this method to calculate transmissions of the superconductor PhC. Obviously, the boundary conditions of the magnetic field in Eqs. (64) and (65) are no more suitable for the superconductor PhC if superconductor rods are embedded in air. It is the factor that the boundary conditions of the magnetic field in this method are dealt with at the interface between two homogeneous media but not between cylinders and a homogeneous medium. In the latter half part of this section, we try to overcome this problem by adding a virtual edge region. At the beginning, transmissions are directly calculated without adding a virtual edge region. Then we investigate the effect on transmissions after adding it.

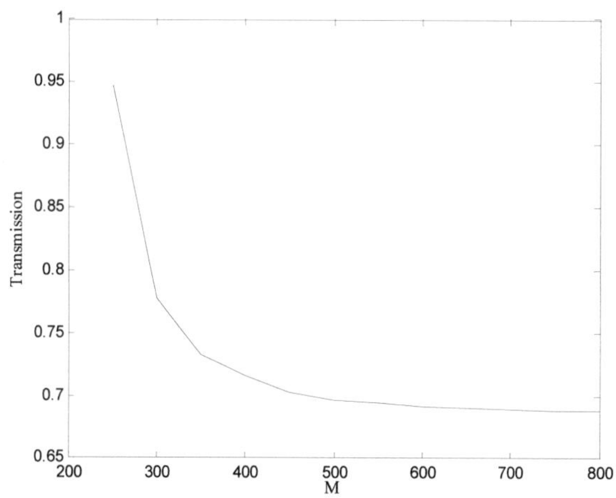

Fig. 18. The transmission versus the M value when the frequency is 0.54 ($2\pi c/a$) without a virtual edge region.

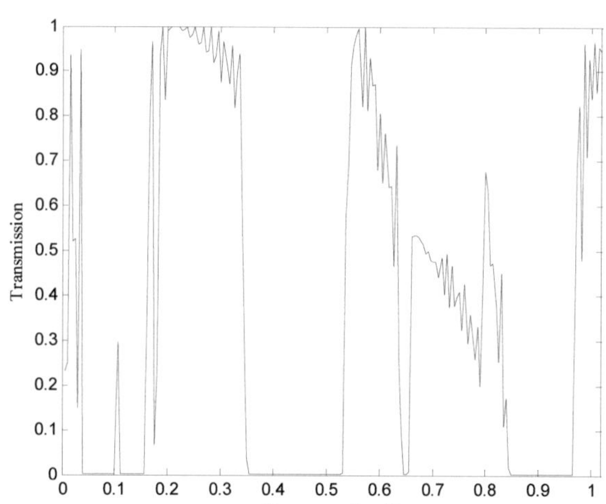

Fig. 19. Transmissions versus frequency with N=5 and M=600 without a virtual edge region.

The same parameters as those in the previous section are used. The final results of this method are compared with those of the ADE-FDTD method. The wave is supposed to be normally incident on the superconductor PhC, and the propagation direction is along the ΓK direction (y-direction) perpendicular to the interface which is along the ΓM direction (x-direction). The number of layers along the x-direction is assumed to be infinite. The number of layers along the y-direction is still 30. The lattice constant along the x-direction is a_1 and that along the y-direction is a_2. We choose $a_2 = a = 100$ μm and $a_1 = \sqrt{3}\, a_2$. The radius of all superconductor cylinders is $0.2a$.

First, the width of the edge region d is considered to be zero. N is fixed at 5 and M is determined at the situation when the transmission is convergent. The frequency is chosen at 0.54 ($2\pi c/a$). In Fig. 18, it is found out that M=600 is enough for calculation. Then N=5 and M=600 are used to calculate transmissions from 0.01 to 1.00 ($2\pi c/a$). The transmissions of the internal-field expansion method have some differences with those of the ADE-FDTD method shown in Fig. 4. The transmissions at frequencies below 0.17 ($2\pi c/a$) are not all close to zero. They are more than 0.1 when the frequencies are below 0.035 ($2\pi c/a$) and at 0.105 ($2\pi c/a$). These results are not coincident with the results of the PBS and the ADE-FDTD method. From the calculations of the PBS before, no propagation modes exist below 0.16 ($2\pi c/a$). The calculations of the ADE-FDTD method also confirm this conclusion even if the thickness of the PhC is finite. It means that the internal-field expansion method has some errors at the low frequency region. In the frequency region from 0.17 to 0.33 ($2\pi c/a$), transmissions of the internal-field expansion method and the ADE-FDTD method almost match each other except for the value at 0.175 ($2\pi c/a$).

The region from 0.33 to 0.47 ($2\pi c/a$) is the PBG region. It is found out that this region shifts to the right in the internal-field expansion method. The region extends to 0.53 ($2\pi c/a$) in Fig. 19. After the PBG region, the PBS displays two photonic bands existing between 0.47 and 0.595 ($2\pi c/a$), and a narrow PBG region between 0.595 and 0.605 ($2\pi c/a$). However, the transmissions in Fig. 19 show that high values exist between 0.53 and 0.64 ($2\pi c/a$) and nearly zero between 0.64 and 0.65 ($2\pi c/a$). In this region, they show a shift about 0.035 ($2\pi c/a$) forward higher frequency. From 0.65 to 0.845 ($2\pi c/a$), the trend of the transmissions in Fig. 19 is much similar to that of the ADE-FDTD method but the frequency region shifts to the right about 0.045 ($2\pi c/a$). In frequencies from 0.80 to 0.895 ($2\pi c/a$), the transmissions of the ADE-FDTD method show the third zero-transmission region. This region exists between 0.845 and 0.955 ($2\pi c/a$) in Fig. 19, which is 0.02 ($2\pi c/a$) larger than that of the ADE-FDTD method.

Next, we try to extend the boundary away from the edge of the cylinder by increasing the width d of the edge region. It is an imaginary boundary between air and the superconductor PhC because the background medium of the superconductor PhC is also air. In fact, such edge region doesn't exist. The nonzero edge region implies that the results should have something to do with the width of it. Several values of d=0.5a, 1.0a, 1.5a, and 2.0a are calculated and all of them are shown in Figs. 5-20(a)-(d). After comparing all results, we find out that the nonzero edge region only affects transmissions below 0.17 ($2\pi c/a$), where the dielectric function in Eq. (9) is negative. The transmissions above 0.17 ($2\pi c/a$) are almost unchanged. So it explicitly reveals that this method is not suitable for negative dielectric function.

To summarize, some transmissions of the internal-field expansion method are close to those of the ADE-FDTD method, and some frequency regions have relative shifts between two methods. Roughly speaking, the shift is about 0.06 multiplying the frequency, so it is obvious that all zero-transmission regions below 1.00 ($2\pi c/a$) broaden in the internal-field expansion method. In Fig. 21, both results of the internal-field expansion method and the ADE-FDTD method are shown, in which the frequency scale of the ADE-FDTD method is

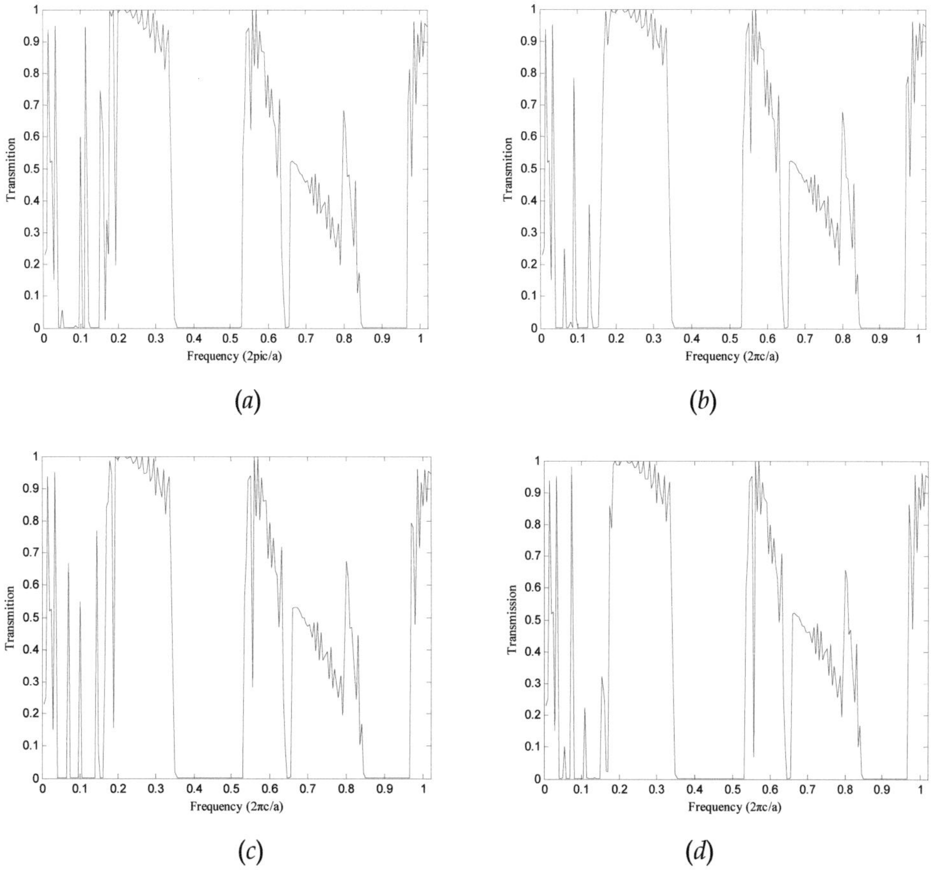

Fig. 20. Transmissions at (a) $d = 0.5a$, (b) $d = 1.0a$, (c) $d = 1.5a$, and (d) $d = 2.0a$, respectively.

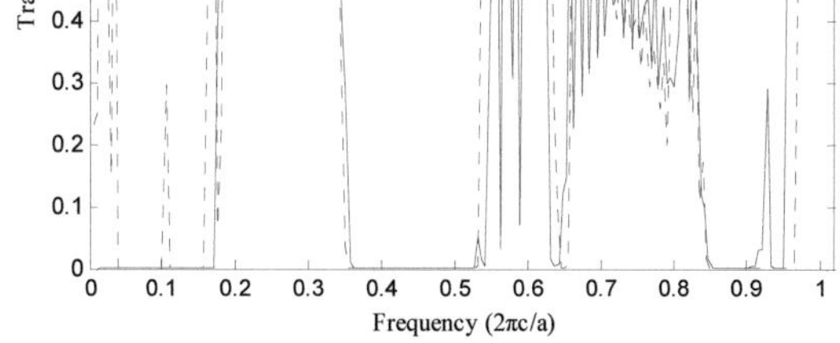

Fig. 21. The transmission versus the M value when the width of the edge region and frequencies are $10.6a$ and 0.1 ($2\pi c/a$), respectively.

multiplied by 1.06. It can be seen that most results of two methods can match each other much better after 0.17 (*2πc/a*). We find that the calculations of the internal-field expansion method exists some errors, which need to be overcome. It still cannot solve the problem, even the edge region is added in calculations. In order to match the results of the PBS and the ADE-FDTD method, the internal-field expansion method needs to be modified in some way.

8. Conclusion

This study focuses on the transmissions of the two-dimensional superconductor PhC. The PhC, composed of copper oxide high-temperature superconductor rods in a triangular array in air, can be tunable utilizing the temperature modulation. We use the plane wave expansion method introduced in Section 2 to calculate the PBS of it, which is much like a metallic PhC system described by the Drude's model if the normal conducting current is ignored. The frequency of the fundamental mode in the superconductor PhC is far above zero. It is the reason that the dielectric function is positive when frequency is more than $\omega_p^s(x,y)$, the plasma frequency of the superconducting electron. Because both the electric susceptibility and magnetic permeability have to be either positive or negative, light has the ability to propagate through the superconductor.

Then we use the ADE-FDTD method introduced in Section 4 to calculate the transmission when light is normally incident from air into the superconductor PhC. The results of the ADE-FDTD method are consistent with the PBS and also verify the frequency of the fundamental mode is more than $\omega_p^s(x,y)$, which is 0.17 (*2πc/a*) in our demonstrated case. It can be seen that the extremely low transmissions correspond to the PBG regions. Some extremely low transmissions exist at the fifth and sixth bands. They can be explained by treating the superconductor PhC as an effective medium sandwiched between two air regions.

Finally, we use the internal-field expansion method developed by Sakoda to calculate the transmission when light is also normally incident from air into the superconductor PhC. It can be found out that transmissions below 0.17 (*2πc/a*) are not all close to zero. These non-zero transmissions can't reach convergent values even we use large M in calculations. The results point out that this method can't be directly applied on the negative dielectric constant media. We try to increase the width of the edge region to overcome this problem, but it is useless. Transmissions above 0.17 (*2πc/a*) can reach stable values as long as M is large enough. However, the frequency scale has to reduce 1.06 times in order to match the results of the ADE-FDTD method. To sum up, this method is successful to calculate the transmission of the PhC with air cylinders embedded in the homogeneous medium but not suitable very well for the superconductor PhC. One reason is that the boundary conditions between the superconductor PhC and air are not correct. So this method needs modification to obtain correct transmissions.

9. References

Berenger, J. P. (1994). A perfectly matched layer for the absorption of electromagnetic waves. *Journal of Computational Physics*, Vol. 114, No.2, (October 1994), pp. 185-200, ISSN 0021-9991

Busch, K. & John, S. (1999). Liquid-Crystal Photonic-Band-Gap Materials: The Tunable Electromagnetic Vacuum. *Physical Review Letters*, Vol. 83, No. 5, (August 1999), pp. 967–970, ISSN 0031-9007

Engheta, N. & Ziolkowski, R. W. (2006). *Electromagnetic Metamaterials: Physics and Engineering Explorations*, John Wiley & Sons, ISBN 978-0471761020, New York, USA

Figotin, A.; Godin, Y. A. & Vitebsky, I. (1998). Two-dimensional tunable photonic crystals. *Physical Review B*, Vol. 57, No. 5, (February 1998), pp. 2841–2848, ISSN 1098-0121

Grosso, G. & Parravicini, G. P. (2000). *Solid State Physics*, Academic Press, ISBN 978-0123044600, Burlington, USA

John, S. (1987). Strong localization of photons in certain disordered dielectric superlattices. *Physical Review Letters*, Vol. 58, No. 23, (June 1987), pp. 2486–2489, ISSN 0031-9007

Kee, C.-S.; Kim, J.-E.; Park, H.-Y.; Park, I. & Lim, H. (2000). Two-dimensional tunable magnetic photonic crystals. *Physical Review B*, Vol. 61, No. 23, (June 2000), pp. 15523-1 - 15523-3, ISSN 1098-0121

Kee, C.-S. & Lim, H. (2001). Tunable complete photonic band gaps of two-dimensional photonic crystals with intrinsic semiconductor rods. *Physical Review B*, Vol. 64, No. 12, (September 2001), pp. 121103-1 - 121103-4, ISSN 1098-0121

Kee, C.-S.; Lim, H.; Ha, Y.-K.; Kim, J.-E. & Park, H.-Y. (2001). Two-dimensional tunable metallic photonic crystals infiltrated with liquid crystals. *Physical Review B*, Vol. 64, No. 8, (August 2001), pp. 085114-1 - 085114-7, ISSN 1098-0121

Lee, W. M.; Hui, P. M. & Stroud, D. (1995). Propagating photonic modes below the gap in a superconducting composite. *Physical Review B*, Vol. 51, No. 13, (April 1995), pp. 8634 -8637, ISSN 1098-0121

Matsuda, Y.; Gaifullin, M. B.; Kumagai, K.; Kadowaki, K. & Mochiku, T. (1995). Collective Josephson Plasma Resonance in the Vortex State of $Bi_2Sr_2CaCu_2O_{8+\delta}$. *Physical Review Letters*, Vol. 75, No. 24, (December, 1995), pp. 4512-4515, ISSN 0031-9007

Moreno, E.; Erni, D. & Hafner, C. (2002). Modeling of discontinuities in photonic crystal waveguides with the multiple multipole method. *Physical Review E*, Vol. 66, No. 3, (September, 2002), pp. 036618-1 - 036618-12, ISSN 1539-3755

Notomi, M. (2000). Theory of light propagation in strongly modulated photonic crystals: Refractionlike behavior in the vicinity of the photonic band gap. *Physical Review B*, Vol. 62, No. 16, (October 2000), pp. 10696-1 - 10696-12, ISSN 1098-0121

Pei, T.-H. & Huang, Y.-T. (2007a). A temperature modulation photonic crystal Mach-Zehnder interferometer composed of copper oxide high-temperature superconductor. *Journal of Applied Physics*, 101, (April 2007), pp. 084502-1-084502-5, ISSN 0021-8979.

Pei, T.-H. & Huang, Y.-T. (2007b). Temperature Modulation of the Superprism Effect in Photonic Crystals Composed of the Copper Oxide High-Temperature Superconductor. *Japanese Journal of Applied Physics*, Vol. 46, No. 24, (June 2007), pp. L593–L595, ISSN 0021-4922.

Pei, T.-H. & Huang, Y.-T. (2011a). The high-transmission photonic crystal heterostructure Y-branch waveguide operating at photonic band region. Journal of Applied Physics, 109, (February 20011), pp. 034504-1-034504-8, ISSN 0021-8979.

Pei, T.-H. & Huang, Y.-T. (2011b). The equivalent structure and some optical properties of the periodic-defect photonic crystal. Journal of Applied Physics, 101, (April 2011), pp. 073104-1-073104-9, ISSN 0021-8979.

Raymond Ooi, C. H.; Au Yeung, T. C.; Kam, C. H. & Lim, T. K. (2000). Photonic band gap in a superconductor-dielectric superlattice. *Physical Review B*, Vol. 61, No. 9, (March 2000), pp. 5920-5923, ISSN 1098-0121

Sakoda, K. (1995a). Optical transmittance of a two-dimensional triangular photonic lattice. *Physical Review B*, Vol. 51, No. 7, (October 1995), pp. 4672 - 4675, ISSN 1098-0121

Sakoda, K. (1995b). Transmittance and Bragg reflectivity of two-dimensional photonic lattices. *Physical Review B*, Vol. 52, No. 12, (April 1995), pp. 8992-9002, ISSN 1098-0121

Sakoda, K. (2004). *Optical Properties of Photonic Crystals*, 2nd ed., Springer, ISBN 978-3540206828, Berlin, Germany

Shibata, H. & Yamada, T. (1996). Superconducting-plasma resonance along the *c* axis in various copper oxide superconductors. *Physical Review B*, Vol. 54, No. 10, (September 1996), pp. 7500-7504, ISSN 1098-0121

Taflove, A. & Hagness, S. C. (2005). *Computational Electrodynamics: The Finite-Difference Time-Domain Method, 3rd Ed.*, Artech House, ISBN 978-1580538329, Norwood, USA

Takeda, H. & Yoshino, K. (2003a). Tunable refraction effects in two-dimensional photonic crystals utilizing liquid crystals. *Physical Review B*, Vol. 67, No. 5, (May 2003), pp. 056607-1 - 056607-5, ISSN 1098-0121

Takeda, H. & Yoshino, K. (2003b). Disappearances of uncoupled modes in two-dimensional photonic crystals due to anisotropies of liquid crystals. *Physical Review B*, Vol. 67, No. 5, (May 2003), pp. 056612-1 - 056612-5, ISSN 1098-0121

Takeda, H. & Yoshino, K. (2003c). Tunable light propagation in Y-shaped waveguides in two-dimensional photonic crystals utilizing liquid crystals as linear defects. *Physical Review B*, Vol. 67, No. 7, (February 2003), pp. 073106-1 - 073106-4, ISSN 1098-0121

Takeda, H. & Yoshino, K. (2003d). Coupling of the TE and TM modes of electromagnetic waves in two-dimensional photonic crystals with surface defects of liquid crystals. *Physical Review B*, Vol. 68, No. 4, (October 2003), pp. 046602-1 - 046602-5, ISSN 1098-0121

Takeda, H. & Yoshino, K. (2003e). Tunable photonic band schemes in two-dimensional photonic crystals composed of copper oxide high-temperature superconductors. *Physical Review B*, Vol. 67, No. 24, (June 2003), pp. 245109-1 - 245109-6, ISSN 1098-0121

Takeda, H. & Yoshino, K. (2004). TE-TM mode coupling in two-dimensional photonic crystals composed of liquid-crystal rods. *Physical Review B*, Vol. 70, No. 2, (August 2004), pp. 026601-1 - 026601-7, ISSN 1098-0121

Tinkham, M. (2004). *Introduction to Superconductivity*, 2nd ed., Dover Publications, ISBN 978-0486435039, New York, USA.

V. Kuzmiak, A. A. Maradudin, & F. Pincemin, (1994). Photonic band structures of two-dimensional systems containing metallic components. *Physical Review B*, Vol. 50, No. 23, (December 1994), pp. 16835-16844, ISSN 1098-0121

Van Duzer, T. & Truner, C. W. (1998). *Principle of superconductive device and Circuits*, 2nd ed., Arnold, ISBN 978-0132627429, London, the United Kingdom

Yariv, A. & Yeh, P. (2002). *Optical Waves in Crystals: Propagation and Control of Laser Radiation*, John Wiley & Sons, ISBN 978-0471430810, New York, USA

Yee, K. S. (1966). *IEEE Transactions on Antennas and Propagation*, Vol. AP-14, (May 1966), pp. 302-307, ISSN 0018-926X

Zhou, S.-A. (1999). *Electrodynamics of solid and Microwave Superconductivity*, John Wiley & Sons, ISBN 978-0471354406, New York, USA

15

Electrodynamics of High Pinning Superconductors

Klimenko E.Yu.
National Research Nuclear University "MEPhI"
Russian Federation

1. Introduction

High pinning superconductors (HPSC) hold much promise for power engineering. There are some annoying consequences of up-to-date unequal state of these materials electrodynamics. The main one is for lack of superconducting winding engineering environment. Applied superconductivity is still a field of empirical activity. There are two ways for a reliable design developing, either using of steady state stabilized conductor with a lot of normal metal, or protracted trial-and-error manufacturing of full-scale model windings. Both ways are too expensive. In the narrow sense, new approaches are called for effective computation of real conductor stability, loss, etc. Probably in the more wide sense, putting the fundamentals in order will provide more stability to the whole applied superconductivity for opposition from new wasteful initiatives.The state of HPSC physics gives another cause for concerning. Now it is a vast collection of incoherent effects and several options of mezoscopic models for each one. The analysis of the situation has uncovered the following probable causes of the modern state of the electrodynamics: 1. the most popular constitutive law so called thermal activation model [Anderson 1962], is not only inconvenient but most likely unequal; 2 a good deal of discovered effects is conditioned by incorrect data processing.
A detailed examination [Klimenko et al., 2005] had shown that in fact the thermal activation hypothesis had neither grounds nor verifications. However, taking into account the fabulous popularity of the hypothesis we considered necessary to revise all the points from the beginning [Klimenko et al., 2005].
This paper contains an alternative constitutive law and reasoning in its favour, general setting up the problem of quasisteady HPSC electrodynamics, and discussion of some features of HPSC which must be kept in mind during experiment setting up and data processing.

2. HPSC constitutive law (Nb-Ti)

There were several reasons for selecting a niobium-titanium as the material for constitutive law revising. It is analogous to niobium-zirconium wires widely used in basic experiments favouring establishment of the thermal activation model. However, this is not the only ground for this choice: we believe niobium-titanium wire and foil to be the most suitable materials for studying the transition characteristics of superconductors with high pinning as well as the problems of their electrodynamics. The high level of commercial technology of niobium-titanium alloys provides relatively homogeneous materials exhibiting uniform properties along the wire length and the foil area. The niobium–titanium conductors are convenient for pinning anisotropy studying. Sufficiently high anisotropic pinning is not complicated with critical field anisotropy. No other commercially available material offers these advantages. These advantages make it possible to study the general laws of electrodynamics in technical superconductors [Klimenko et al., 1997], which are almost not masked by specific features of particular samples. Experiments with niobium–titanium wires are not complicated by brittleness and high sensitivity to straining; those are typical features of intermetallic compounds and HTSC. On the other hand, niobium–titanium wires are by no means a simple material. These wires were displaced from the focus of research, not even having been exhaustively studied. Later, it was found that niobium–titanium alloys are two-component (as manifested by a difference in the critical fields of the grain body and boundaries [Klimenko et al., 2001a]) and are characterized by anisotropic pinning in the cross section of a wire [Klimenko et al., 2001b].

We used a commercial monofilament copper coated Nb-50wt%Ti wire 0.15 mm in diameter. Superconducting core diameter was 0.12 mm. Several dozens of voltage-current curves were recorded in magnetic field range from zero up to B_{c2} [Klimenko et al., 2005]. The curves were converted into ohm-ampere ones. The latter were linear in semi-logarithmic scale at several

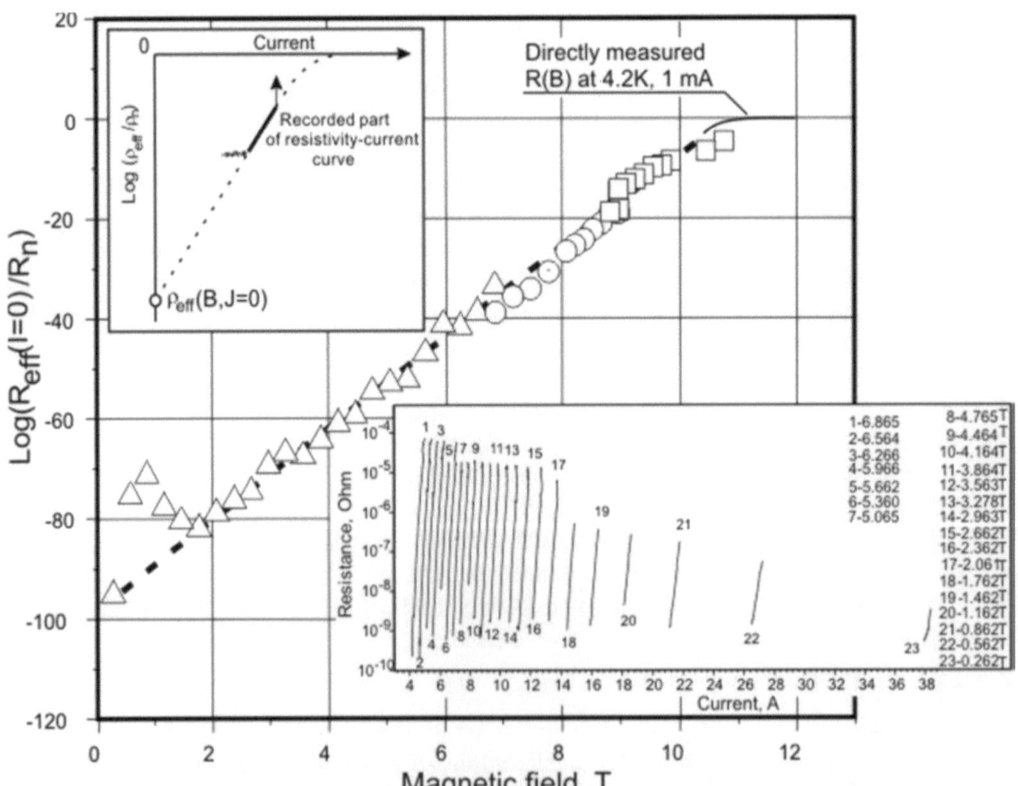

Fig. 1. Dependence of reduced resistivity of Nb-Ti wire on magnetic field at zero current [11]. The resistivity was obtained by the extrapolation of ohm-ampere curves to zero current. Some of the used ohm-ampere curves are shown on a panel. Another panel illustrates the extrapolation procedure.

orders of resistance values. There were two reasons to use just ohm-ampere curves. Firstly, a finite value of resistance being obtained by the curve extrapolation to zero current (Fig.1) was quite natural contrary to a finite voltage value at zero current. The later self-contradiction of thermal activation model is avoidable by no workarounds [Ketterson&Song]. Secondly, this approach corresponded to usual way of critical values determination, say, T_c and B_{c2}.
The experimental results were compared with a simplest constitutive law proposed thirty years ago [Dorofeev et al., 1980]:

$$\sigma(T,B,j) = \sigma_n\{1+\exp[\left(-1+\frac{T}{T_c}+\frac{|B|}{B_{c2}}+\frac{|j|}{\frac{j_c}{2}}\right)\frac{1}{\delta}]\} \tag{1}$$

The law was derived from assumption that a critical surface ($\frac{T}{T_c}+\frac{B}{B_{c2}}+\frac{j}{j_c}=1$) in its traditional form should be replaced with a smooth transition layer. The comparison is not complicated because expression (1) contains only one fitting parameter: critical current corresponding to effective resistivity equal to a half of normal one at zero temperature and zero magnetic field. In the strict sense, T_c and B_{c2} are also fitting parameters, but they must be closely alligned to the critical values measured at small current. Fig.2 shows that this critical current value is a good constant, as well as parameter δ describing transition width. Corresponding to another measurement series [Dorofeev et al., 1979] Fig.3 shows rather a vast tablelands of the parameters in dependence on temperature and magnetic field. Their rise at low magnetic field governs with pinning anisotropy. It will be discussed in part 5 of the paper.
It follows from (1) that nothing as "true critical current" exists. More likely a certain insignificant seed resistivity exists at zero temperature and magnetic fields. The resistivity increases exponentially with raising any parameter *T, B* or *j*. Fig.1 is a kind of this conception confirming. However, the extremely low resistivity is more likely statistics of local resistive barriers than the homogeneous property. Anybody will be wrong calculating, for example, skin depth by using this resistivity. This value may not be less than λ (length of field penetration depth into superconductor).
An isotropic HPSC is a media which conductivity depends on magnetic field direction relative to current. In this case the conductivity must be a tensor. Current density and electric field are related through the material equation:

$$j_\alpha = \sigma_{\alpha\beta}E_\beta \tag{2}$$

Where $\sigma_{\alpha\beta}$ just is the electrical conductivity tensor.
According to the general principles, the electrical conductivity tensor for high pinning superconductors in magnetic field obeys the condition

$$\sigma_{\alpha\beta}(\mathbf{B}) = \sigma_{\beta\alpha}(-\mathbf{B}) \tag{3}$$

The part of the conductivity tensor odd with respect to magnetic field is antisymmetric with respect to transposition of indices and determines the physical phenomena such as Hall Effect etc. This part does not contribute to the heat generation and is small in superconductors. Therefore, we restrict ourselves only with the symmetric part of this tensor. For isotropic superconductors in magnetic field the symmetric part may be presented in the form

$$\sigma_{\alpha\beta} = \sigma_t(\delta_{\alpha\beta} - b_\alpha b_\beta) + \sigma_l b_\alpha b_\beta \quad (4)$$

Here $\delta_{\alpha\beta}$ is Kronecker delta and b is the unit vector along the magnetic field B. In the normal state the electrical conductivity is isotropic, $\sigma_t = \sigma_l = \sigma_n$, and $j = \sigma_n E$. The first term in the right-hand side of (4) refers to the dissipative motion of vortices under the action of Lorentz force. The longitudinal part σ_l in (4) is related to the dissipation processes induced when the current flows along magnetic field. Although the longitudinal conductivity σ_l is well-established, its influence, as well as the tensor character of HPSC electrical conductivity, is commonly neglected in the applied researches. Such approximation holds only when current and magnetic field are mutually perpendicular. That is not the case, for example, in twisted multifilament wires.

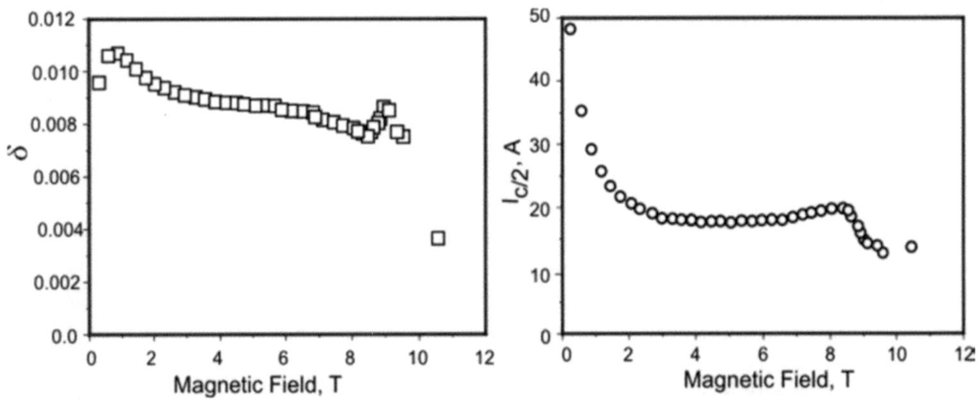

Fig. 2. The parameters describing Nb-Ti wire transition are surprisingly stable

In the critical state model, it is assumed that the transition from the superconducting state to normal one occurs when physical parameters attain the critical surface

$$K(T, |B|, |j|) = 0 \quad (5)$$

where |B| and |j| are modules of a vector. Due to intrinsic textural inhomogeneity of HPSC the actually observed transition to resistive state is smoothed at the vicinity of critical state. The tensor approach requires generalising of (1) for the transverse and longitudinal parts of electrical conductivity in superconductors:

$$\sigma_t = \sigma_n[1 + \exp\left(\frac{K_t}{\delta}\right)] \quad (6)$$

$$K_t = 1 - \frac{T}{T_c} - \frac{|\mathbf{B}|}{B_{c2}} - \frac{|\mathbf{j}_t|}{j^t_{c/2}} \quad (7)$$

$$j_{t\alpha} = (\delta_{\alpha\beta} - b_\alpha b_\beta)j_\beta \quad (8)$$

here $\delta_{\alpha\beta} = \{\begin{matrix}1, & \alpha = \beta \\ 0, & \alpha \neq \beta\end{matrix}$ is Kronecker delta. It has nothing common with δ in (1, 6 and 9). By analogy with (6-8) longitudinal conductivity may be written

$$\sigma_l = \sigma_n[1 + \exp\left(\frac{K_l}{\delta}\right)] \quad (9)$$

$$K_l = 1 - \frac{T}{T_c} - \frac{|\mathbf{B}|}{B_{c2}} - \frac{|\mathbf{j}_l|}{j_{c/2}^l} \quad (10)$$

$$j_{l,\alpha} = b_\alpha b_\beta j_\beta \quad (11)$$

Here σ_n is the electrical conductivity in the normal state. In this notation the critical surface defined by (5) corresponds to $\sigma_{eff} = 2\sigma_n$ and parameters T_c, B_{c2} and $j_{c/2}$ are the intercepts of the critical surface with reference axes. Parameter δ characterizes the width of a gradual kinetic transition from superconducting to normal state. In the limit $\delta \to 0$ our scheme tends to the critical state model [Bean, 1962]. As δ varies in the range 0.05–0.005, the values of T_c and $B_{c2}(0)$ appear to be close to the related thermodynamic quantities (generally, these parameters may be redefined if needed).
There is no trace of thermal activation process in the experimentally approved HPSC constitutive law. An alternative model is statistic one. [Baixeras&Fournet, 1967].

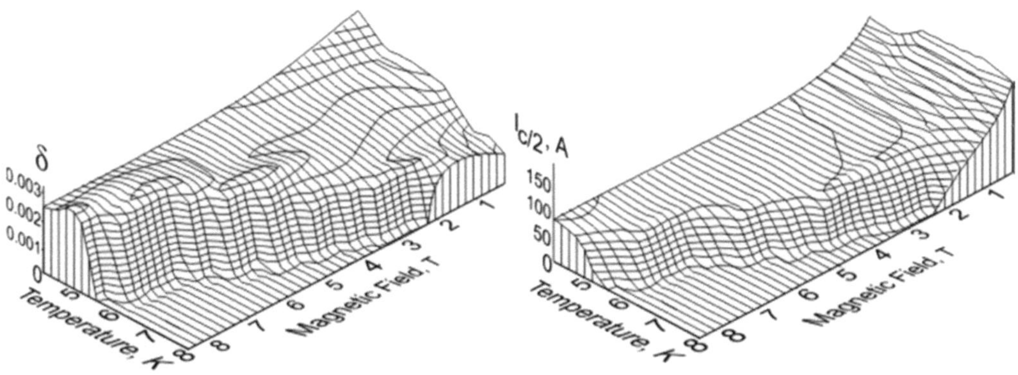

Fig. 3. These parameters weakly depends on temperature as well as on magnetic field.

It appeared that (1) described well transition of a model multilinked network consisting of superconducting elements if their critical temperatures had been normally distributed around certain mean value (Fig.4a) [Klimenko, 1983, 1985]. This model bears relation to "bulk inhomogeneous" superconductor such as monofilament wire. Another type of wires was current important in eighties. It was multifilament wire with broken filaments. This type needed another approach. We proposed regarding it as "longitudinally inhomogeneous" [Dorofeev et al., 1980].

$$V = R_m \int_0^I \frac{I - i_c}{\sigma\sqrt{2\pi}} \exp\left[\frac{(I_c - i_c)^2}{2\sigma^2}\right] di_c \quad (12)$$

This carrying back makes sense due to far-reaching analogy between voltage-current curves of longitudinally inhomogeneous wires and modern HTSC. It would be absurd to look for any physical reasons of the analogy. Statistical reason seems more probable. This hypothesis helps discussing some features of HTSC, though, certainly, no analogy has evidential force.
A group of voltage-current curves according to (12) is presented at Fig.4b,c. In semi-logarithmic scale the curves are parabolas (Fig4b), as it was shown and experimentally approved in [Dorofeev et al., 1980]. However, the SC community had preferred straightening curves by using logarithmic scale at both coordinates (Fig.4c). This transformation had given power behaviour to rather long parts of the curves. This explanation looks less naive than a model known as "logarithmic potential well" [Zeldov et al., 1990].

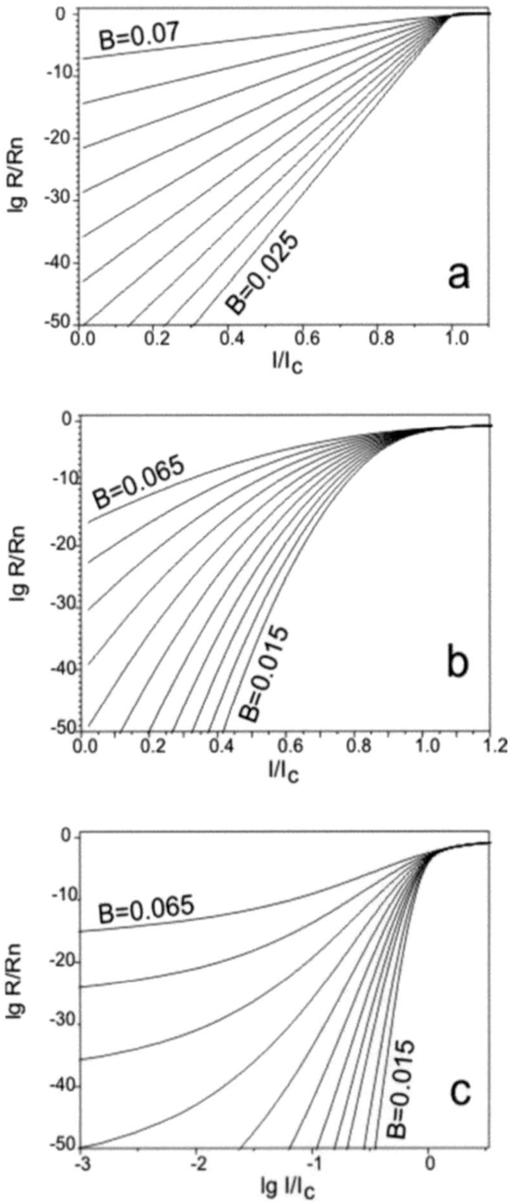

Fig. 4. Om-ampere curves of bulk inhomogeneous (a) and longitudinal inhomogeneous (b,c) superconductors.

An irreversible line is used to consider as one of the important characteristic of HTSC. It is assumed that this line separates voltage-current curves with positive and negative curvatures, the negative curvature considering as an evidence of the true superconducting condition. Fig.4c hints that irreversible line, perhaps, has no real sense. In fact, every curve changes its curvature from negative value to positive one at certain resistance level. The positive curvature itself arises due to stretching the current coordinate by logarithmic scale at low currents.

3. Electrodynamics equations

The electrodynamics of isotropic HPSC may be described in terms of the general quasistationary electrodynamics of a continuous media. In fact all the known HPSC are anisotropic. However, isotropic electrodynamics is ever considered as a necessary step [Klimenko et al., 2010].

$$rot\mathbf{E} = -\frac{\partial \mathbf{B}}{\partial t} \tag{13}$$

$$div\mathbf{B} = \tag{14}$$

$$rot\mathbf{B} = \mu_0 \mathbf{j} \tag{15}$$

Here **E** and *B* are the electric and magnetic fields, respectively, **j** is the current density, and μ_0 is the magnetic constant. At a boundary of superconductor 1 and normal metal 2 the following components must be continuous

$$\mathbf{B}^{(1)} = \mathbf{B}^{(2)} - \mathbf{M}(B^{(2)}) \tag{16}$$

$$\mathbf{nE}^{(1)} = \mathbf{nE}^{(2)} \tag{17}$$

$$\mathbf{nj}^{(1)} = \mathbf{nj}^{(2)} \tag{18}$$

here **n** is the unit vector normal to a boundary. **M**-is magnetic moment of ideal type II superconductor. Ch. Bean [Bean, 1962] was the first, who had neglected this value. In fact, it is rather small, if $\kappa = \frac{\lambda}{\xi} \gg 1$: ($M(B) = \frac{B_{c2} - B}{(2\kappa^2 - 1)\beta_A}$)
Instead of electric and magnetic fields, it is more conveniently to use the scalar and vector potentials,

$$\mathbf{B} = rot\mathbf{A} \tag{19}$$

$$\mathbf{E} = -\nabla\varphi - \frac{\partial \mathbf{A}}{\partial t} \tag{20}$$

In terms of potentials (19) and (20) the equations of electrodynamics are given by

$$\nabla^2 A_\alpha = \mu_0 \sigma_{\alpha\beta}(\nabla_\beta \varphi + \frac{\partial A_\beta}{\partial t}) \tag{21}$$

$$\nabla_\alpha \left(\sigma_{\alpha\beta}\left(\nabla_\beta \varphi + \frac{\partial A_\beta}{\partial t}\right)\right) = 0 \tag{22}$$

The potentials ϕ and **A** should be continuous and satisfy boundary conditions (16-18).
Due to high sensitivity SC conductivity to temperature the electrodynamics equations must be supplemented by the heat transfer equation

$$C(T)\frac{\partial T}{\partial t} = \nabla_\alpha\left(\kappa_{\alpha\beta}(T)\nabla_\beta T\right) + G(T) \tag{23}$$

with Neumann boundary conditions. Here C(T) is the specific heat per unit volume, $\kappa_{\alpha\beta}(T)$ is thermal conductivity, and the heat generation G(T) is determined as

$$G(T) = \mathbf{jE} \tag{24}$$

The constitutive law (6,9) permits to enclose the set of equations.
A package of computer codes was developed on the basis of Eq(21-24) for real geometry and heat exchange condition. It provides a possibility of stability and AC loss computation for arbitrary cycles of external magnetic field and current. The results will be soon published on behalf of the whole team.

4. 2D voltage-current curves

The introduced above tensor conductivity is in contradiction with a widespread belief that current and electric field are collinear in isotropic superconductor: $\mathbf{j} = j_C \frac{\mathbf{E}}{E}$. [Carr, 1983].
W.J. Carr had supposed it as an intuitive generalization of Bean model. However the generalisation has appeared wrong. It is right only for the case of magnetic field perpendicular to current, as well as Bean model. Indeed moving vortices generate electric field in a plane normal to magnetic field. This field must be tilted to the current, if later does not lie in the plane.

$$\mathbf{E} = [\mathbf{Bv}] = \rho_t[\mathbf{j} - \mathbf{b}(\mathbf{jb})] \qquad (25)$$

The normal to current electric field component was called "satellite electric field" [Klimenko 2001a]. Fig.5 brings it clearly. It is significant that the tilt disappears, when the external magnetic field exceeds critical value, and the field components become independent. It may be the most convincing demonstration of electric field generating by vortices movement in superconductors.

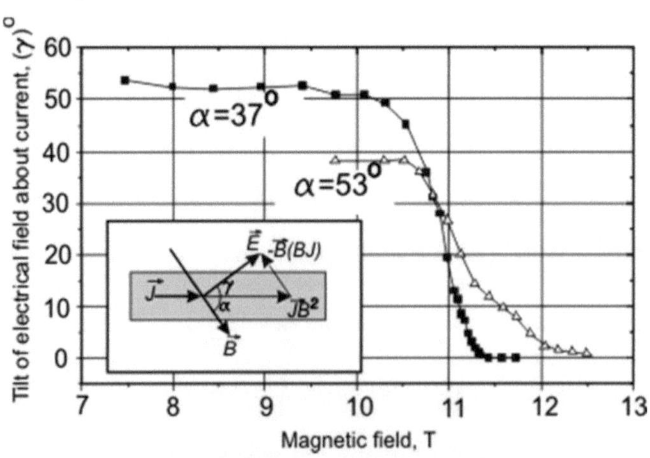

Fig. 5. The satellite electric field observation at 10 μm Nb-Ti foil covered with 1 μm layer of copper (ϕ=37°) and -the same foil without copper layer (ϕ=53°).

Another significant feature is quite large field interval (~2T) in which the tilt falls to zero. It is evidence of inhomogeneity of the superconductor. In all likelihood, the critical fields of elongated grains bodies and their borders are different. This assumption is supported with results of the similar experiment with the same foil samples cut at various angles to rolling direction. (Fig.6). One can see pikes in the transition interval. The current flows at angle to the superconducting borders and must to cross normal bodies generating transversal electric field.

5. Pinning anisotropy

There are no isotropic HPSC actually. Their electrodynamics is mainly of theoretical interest. Real anisotropic pinning brings to a wide variety of phenomena and will provide a lot of new "discoveries" if the proper electrodynamics is not developed in the nearest future.

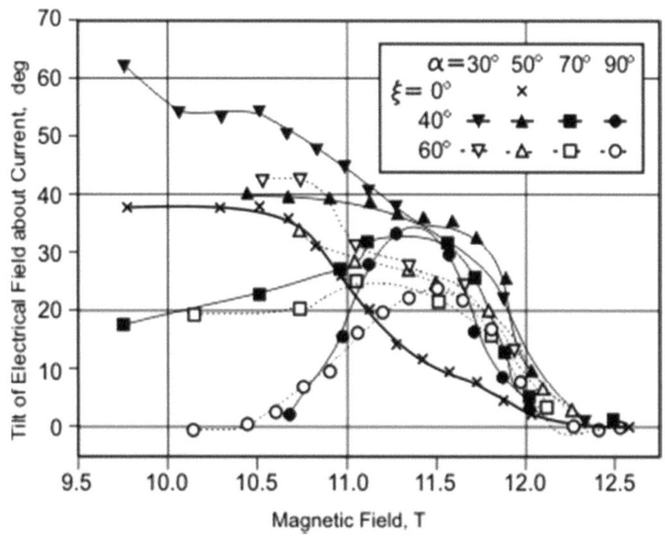

Fig. 6. The satellite electric field observation at the same foil samples cut at angle ξ to the rolling direction.

We cannot yet offer something being equivalent to the theory described in part 3. An approach [Klimenko et al., 1997] was developed in frames of critical state model. That time we used Critical Lorentz Force for critical state description. Critical Lorentz force (scalar) related to unit superconductor volume is a radius of certain closed surface (called "pinning surface") in a space of Lorentz forces. Notice, subsequent reasoning always relates to unit superconducting volume. In the case of isotropic superconductor critical Lorentz force doesn't depend on Lorentz force direction and the pinning surface is 3D sphere. In the case of anisotropic pinning the critical Lorentz force depends not only on Lorentz force direction but also on magnetic field direction. So the pinning surface must be constructed in 5D space: three components of magnetic field plus only two components of Lorentz force because the Lorentz force is always normal to magnetic field.

The following procedure was proposed as a simplest option of the pinning 5D-surface construction. It is well known that energy of pinned array of vortices is less than energy of free one. It means that the pinning array forms a potential well, which may be described with three parameters: height, width and steepness of a potential barrier. The height is an energy gain of pinned magnetic flux at rest. The width is a distance between nearby positions of the flux with the same energy gain. It is obvious that the width equals to the least of two mean values: distance between pinning centers or between vortices. The barrier steepness is a maximum derivative of the flux energy with respect to coordinate when the flux is displaced from the rest position. If shapes or distribution of pinning centers are anisotropic, the barrier parameters are described with tensors corresponding to certain ellipsoids which main diameters are aligned with the main directions of the pinning centers array.

Tensor U corresponds to the barrier height:

$$U = \begin{vmatrix} \frac{1}{U_x^2} & 0 & 0 \\ 0 & \frac{1}{U_y^2} & 0 \\ 0 & 0 & \frac{1}{U_z^2} \end{vmatrix} \quad (26)$$

The flux energy gain U_b at rest depends on magnetic field direction. It equals U-ellipsoid radius collinear to the magnetic field.

$$U_b^2 = (\mathbf{b}\mathrm{U}\mathbf{b})^{-1} \quad (27)$$

here $\mathbf{b} = \frac{\mathbf{B}}{|B|}$ is a unit vector in magnetic field direction.

Tenzor L will help to calculate critical Lorentz force in direction $\boldsymbol{l}$, which is $F_{Ll}^c = \max(\frac{\partial u}{\partial l})$,

$$L = \begin{vmatrix} \frac{1}{l_{ex}^2} & 0 & 0 \\ 0 & \frac{1}{l_{ey}^2} & 0 \\ 0 & 0 & \frac{1}{l_{ez}^2} \end{vmatrix} \quad (28)$$

here $l_{ei} = \frac{U}{\max(\frac{\partial u}{\partial l_i})}$. This effective barrier half-width allows critical Lorentz force calculating:

$$F_L^c = \frac{U(B)}{l_e(f)} \quad (29)$$

$\mathbf{f} = \frac{\mathbf{F_L}}{|F_L|}$ is a unit vector in Lorentz force direction. and

$$l_e^2 = (\mathbf{f}\mathrm{L}\mathbf{f})^{-1} \quad (30)$$

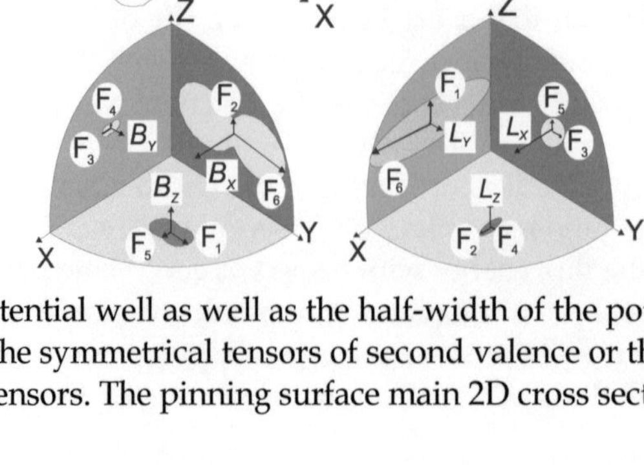

Fig. 7. The depth of potential well as well as the half-width of the potential barrier are described either with the symmetrical tensors of second valence or the ellipsoids corresponding to the tensors. The pinning surface main 2D cross sections are shown.

The effective half-width was assumed a geometrical parameter independent on U_b. Experimental data treatment must show, if the assumption was correct. The complete pinning surface may be constructed by division all radii of U-ellipsoid by L-ellipsoid radii in its cross section normal to the field directions. Fig.7 shows some 2D-cross-sections of the 5D pinning surface. Fig. 8. shows an example of 3D-cross section built for varying magnetic field directions.
The model doesn't allow getting all the six main diameters of the ellipsoids from critical Lorentz forces measurements. It is possible to write six values:

$$F_{Lij}^{c} = \frac{U_i}{l_{eff}^{j}}, \qquad i \neq j \tag{31}$$

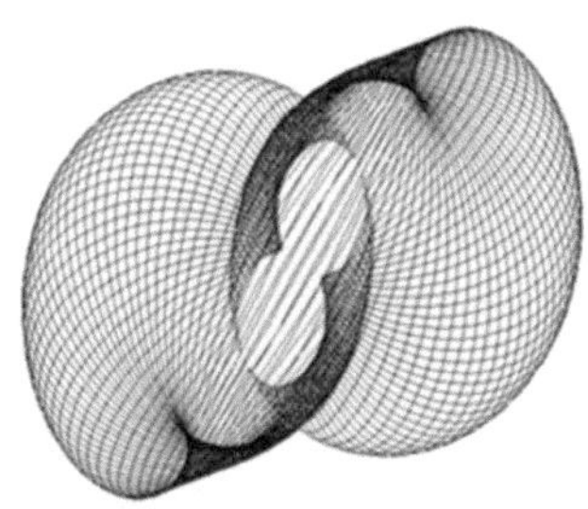

Fig. 8. 3D- cross section of a pinning surface. (U- ellipsoid main radii are related as 1:2:3, L-ellipsoid ones - as 1:2;4. Magnetic fields vectors are laying in the U-central plane with radii related as 1:3).

It is easy to see that $F_{L13}^{c}F_{L21}^{c}F_{L32}^{c} = F_{L12}^{c}F_{L23}^{c}F_{L31}^{c}$. Thus, only five of them are independent on one another.
We have studied a large series of samples made from cold deformed Nb-Ti foil. They were cut at various angles to rolling direction and tested in magnetic fields tilted at various angles both to the sample plane and current direction. Fig.9 shows the main radii of the ellipsoids, the barrier half-width L_y normal to the foil plane being accepted as unity. The pinning centers density in this direction was maximum, and the half-width didn't change while magnetic field increased in contrary with L_x aligned to the rolling direction.

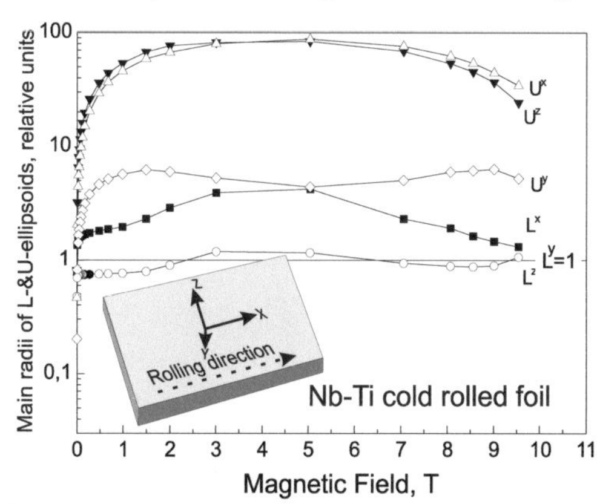

Fig. 9. The main radii of L- and U- ellipsoids of the cold rolled Nb-Ti 10 μ foil. The data are extracted from a set of experiments with various orientations of magnetic fields and currents.

The degree of the foil anisotropy is seen from Figs.10 and 12. It allows estimating of agreement between experimental data and model predictions.
There are two causes of transverse electric field origin. The above-mentioned satellite field arises due to movement of vortices tilted to current direction. Another one is known as guided vortices motion [Niessen&Weijsenfeld, 1969]. It arises due to vortices movement at an angle to Lorentz force direction. Fig. 11a explains this phenomenon. Due to the special shape of a cross section of the pinning surface normal to the magnetic field, a certain projection of the Lorentz force vector pierces the pinning surface in point 'd', whereas the vector itself does not reach point 'c' at the surface. So the magnetic flux moves in the projection direction. Fig.11b compares the prediction with our experimental data.

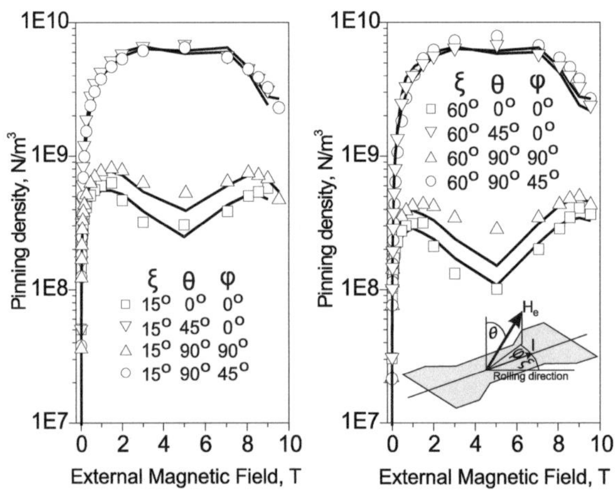

Fig. 10. A comparison of the experimental data on pinning density with predictions (solid curves) calculated with the main radii of L- and U- ellipsoids. The dependence of the pinning anisotropy on both the magnetic field and Lorentz force directions can be seen.

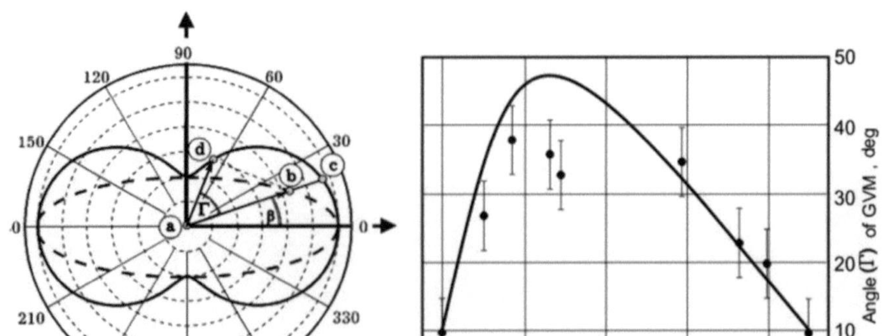

(a) (b)

Fig. 11. A scheme of guiding vortices motion arising (left) and comparison of experimental points and predicted curve (right) obtained by magneto-optical method in low magnetic field.

A problem of critical current in longitudinal magnetic field was very exciting for a long time due to nontrivial process of vortices reconnecting. There were tested four foil samples in magnetic field aligned to current direction with accuracy better than 0.2°. The samples were cut at different angles ξ to the rolling direction. Fig.13 shows results of foil samples testing compared with model calculations made on the following assumptions: a. The vortices reconnection is free at pinning centers, b. The vortices array breaks virtually up into longitudinal and transverse ones moving in opposite directions, c. pinning centers number is sufficient for independent pinning of both virtual arrays. The semiquantitative agreement is obvious. The model predicts correctly nontrivial dependence of longitudinal critical currents on pinning.

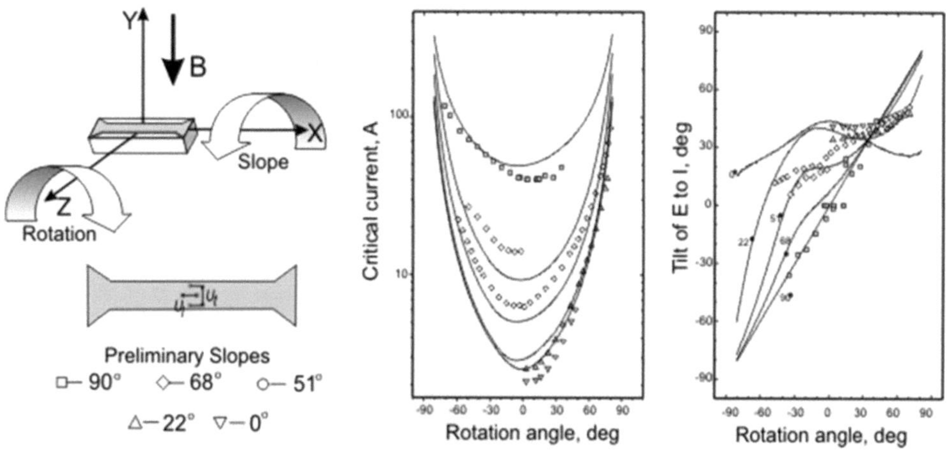

Fig. 12. Results of studying critical currents and tilts of electrical field to current directions in dependence on preliminary slopes and rotation angles.

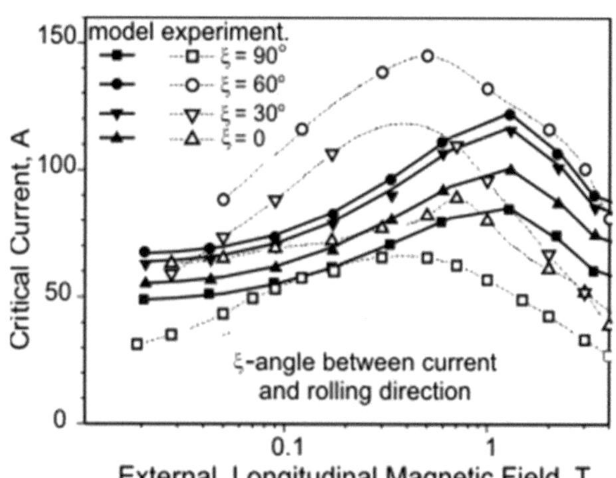

Fig. 13. The critical currents in the longitudinal magnetic fields. The experimental values obtained with the samples (1.3 mm width) cut from a piece of Nb-Ti 10 μm foil at various tilts to the rolling direction are compared with predictions (curves) calculated with the main radii of L- and U- ellipsoids (Fig.9)

The foil anisotropy arises due to the rolling process. The wire drawing process has certain features in common with rolling. It also forms the anisotropic structure. Significant difference in critical current values for axial and azimuth currents is well known [Jungst, 1977]. It appeared that significant pinning anisotropy existed in a wire cross section [Klimenko et al., 2001b]. It was found out on trials of a Nb-Ti wire 0.26 mm in diameter with cross section reduced by grinding into segment shape (segment height was 0.21 of the wire diameter).

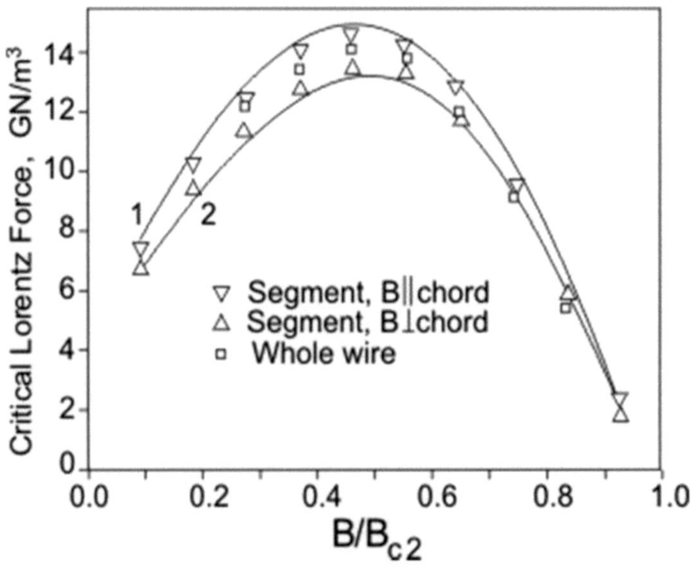

Fig. 13. Critical Lorentz Force anisotropy in Nb-Ti wire cross section. 1. The critical value for azimuthally aligned vortices, 2. The critical value for radial aligned vortices.

Maximum and minimum critical Lorentz Forces (curves 1 and 2 at Fig.13) were derived from results of segment tests in magnetic fields of orthogonal directions. The anisotropy affects the wire critical current and the magnetic moment. Figs.14 and 15 show these effects, the foil anisotropy parameters being used for the calculations to make the effects more pronounced. The results differ in dependence on prevalence of radial or azimuth pinning.

The anisotropy affects critical currents in low magnetic field, where azimuth component of the current self field becomes dominant (Fig.14), as it is seen from current distributions shown at the left pictures. When the azimuth aligned vortices pinning is higher than one of radial vortices the critical current rises steeply up as the field decreases (curve 2 at Fig.14). The Nb-Ti wire demonstrates just this type of $I_c(B)$ curve. A material with opposite ratio of pinning forces would show a plateau in this field region (curve 1).

There is a large range of magnetic fields where critical currents don't depend on the type of anisotropy. Current distributions in this range are similar (right pictures of Fig.14 This independence allowed the constitutive law (part 2 of this paper) deducing under the assumption that the averaged current density had a definite physical meaning (part 6 of the paper).

The type of anisotropy influences on the wire magnetic moment in the whole range of magnetic fields due to difference in distances of current density maxima from the cross section symmetry lines (Fig 15).

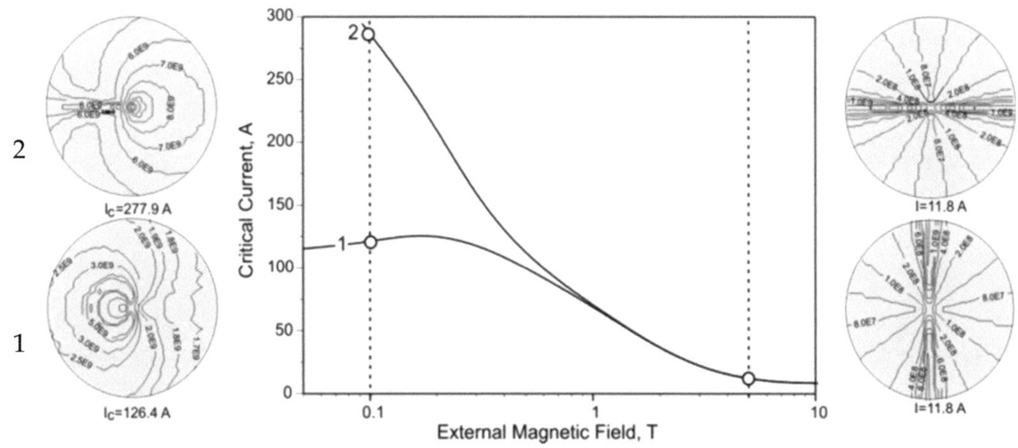

Fig. 14. Comparison of field dependences of the critical current of wires on the type of anisotropy. 1. Pinning of radial aligned vortices prevails. 2. pinning of azimuth aligned vortices prevails. Current density distributions in low and high magnetic fields are shown on left and right sides of the picture.

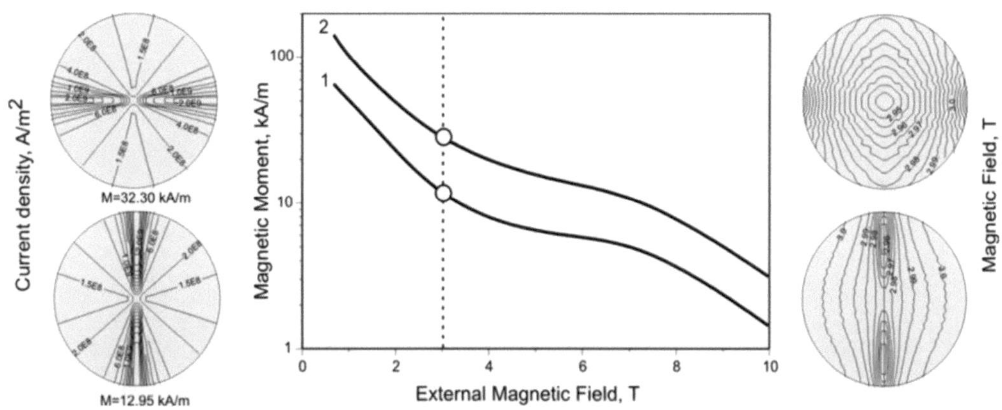

Fig. 15. Comparison of field dependences of the magnetic moments of wires on the type of anisotropy. 1. Pinning of radial aligned vortices prevails. 2. Pinning of azimuth aligned vortices prevails. Current density and magnetic field distributions are shown on left and right sides of the picture.

6. Self-consistent distributions of magnetic field and current density

The most of important problems of applied superconductivity, such as conductor stability, AC loss, winding quench, require nonsteady equations solving. There is, may be, only one situation which needs steady state analyzing. That is testing of a conductor, namely voltage-current curve registration. There is a crafty trap in this seemingly simplest procedure. The point is that this procedure gives an integral result that is dependence of the curves on external magnetic field or, less appropriately, dependence of critical current on the external magnetic field ($I_c(H_e)$). This result is sufficient for a winding designer. A material researcher needs differential result that is dependence of critical current density on internal magnetic field ($j_c(B)$). It is considered usually that

$$j_c(\mu_0 H_e) = \frac{I_c(H_e)}{S} \tag{32}$$

Firstly, it is not trivial because current distribution is not homogeneous in conductor cross section due to current self field. There was shown [Klimenko&Kon, 1977] that in high fields

$$I_c(H_e) = j_c(\mu_0 H_e)\pi r_0^2 \left(1 - 0.031 \frac{j_c(\mu_0 H_e) r_0}{H_e}\right) \tag{33}$$

here r_0 – wire radius, $j_c(B)\sim B^{-0.5}$ was assumed. Taken from the same paper Fig.16 shows that (32) may not be used in low external fields due to the current self field becomes more than the external field. An example of habitual mistake [Kim et al., 1963]: the dependence

$$j_c(B) = \frac{\alpha_c}{B + B_0} \tag{34}$$

by no means follows from more or less acceptable approximation : $I_c(B_e) = \frac{C}{B_e + B_0}$

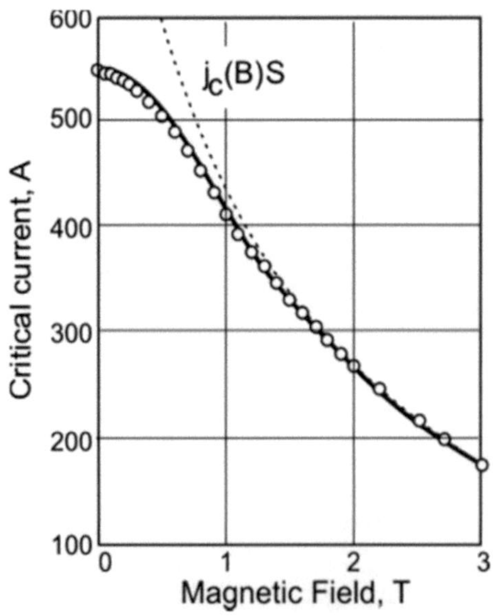

Fig. 16. Critical current dependence on external magnetic field calculated and measured for the case wire with Nb-Ti core 0.22 mm in diameter(Critical current density was assumed 1.06 1010B-0.5 A/m2)

If the constitutive law is known, the self consisted distributions of current density and inner magnetic field can be found by iterations for any external magnetic and electric fields. In the case of anisotropic pinning results of the solution seem to be unexpected. Fig.17 shows calculated critical currents of a tape 4 mm wide (*a*) and 2 μm thick (*b*) for two anisotropy directions. The constitutive law was used in the form (1). It is seen that non-monotone run of the current curves is a macroscopic effect that follows from quite monotone critical current density falling with magnetic field rising.

The critical current corresponding to zero external magnetic field is the presently accepted standard of HTSC conductor evaluating. The insufficient information is not a main drawback of the standard. Sometimes it provokes false conclusions. Fig.18 suggests that HTSC layer thickness increasing uses to spoil the material properties; in fact the current density goes down due to current self field increasing.

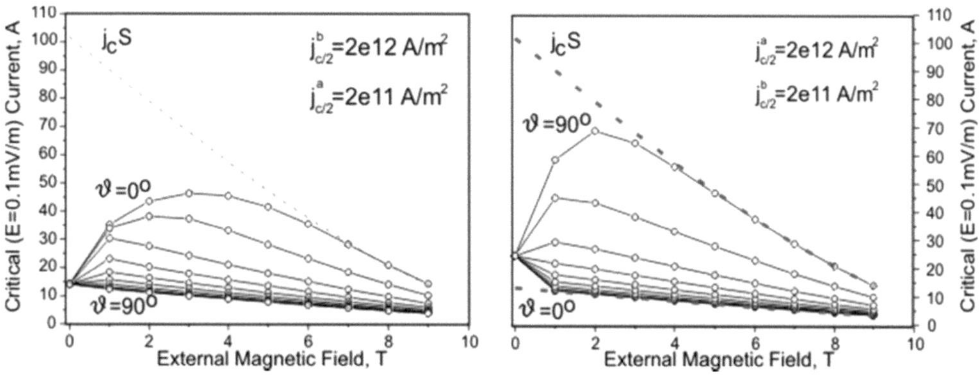

Fig. 17. Calculated $I_c(B)$ curves depending on magnetic field tilt (ϑ) in respect to the normal to the tape surface for the cases when maximum critical Lorentz force direction aligns to the tape width (left) and to the thickness (right).

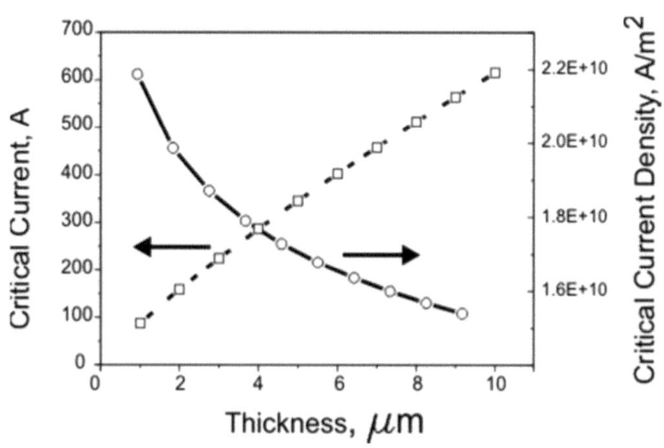

Fig. 18. Calculated dependence of critical current and averaged critical current density on the HTSC layer thickness.

7. Conclusion

There are countless numbers of complete phenomena and characteristics of HPSC discovered during last half century and last quarter in particular. We hope that the completeness is not inherent property of the HPSC but it is consequence of superposition of several quite simple features: nonlinear constitutive law, inhomogeneity, various types of anisotropy, self consistent distributions of magnetic field and current density and may be something else.

8. References

Anderson P. W., (1962). Theory of Flux Creep in Hard Superconductors, *Phys. Rev. Lett.* 9, pp.309-311

Baixeras J.and Fournet G., J. (1967).Pertes par deplacement de vortex dans un supraconducteur de type II non ideal Phys. Chem. Solids 28, pp.1541-1545

Bean C.P. (1964), Magnetization of High-Field Superconductors, *Rev.Mod.Phys.* 36,31-39

Carr W.J. (1983) *AC Loss and Macroscopic Theory of Superconductors*, Gordon &Beach, ISBN 0-677-05700-8, London, New York, Paris.

Dorofeev G. L., Imenitov A. B., and Klimenko E. Yu., (1980)Voltage-current characteristics of type III superconductors, *Cryogenics* 20, 307-310

Dorofeev G. L., Imenitov A. B., and Klimenko E. Yu., (1978),Voltage-current curves of deformed SC wires of III type *Preprint* No. 2987, IAÉ (Inst. of Atomic Energy, Moscow)

Jungst K.-P., (1975), Anisotropy of pinning forces in NbTi, *IEEE Transaction on Magnetics*, v.MAG-11, N2, 340-343

Kim Y.B., Hempstead C.F., Strnad A.R., (1965), Flux-Flow Resistance in Type-II Superconductors, *Phys.Rev.*, v.139, N4A, A1163-A1172

Klimenko E. Yu. and Kon V. G., (1977), On critical state of real shape superconducting samples in low magnetic field., in « Superconductivity »:*Proceedings of Conference on Technical Applications of Superconductivity*, Alushta-75 Atomizdat, Vol. 4, pp. 114-121

Klimenko E. Yu. and Trenin A. E., (1983), Numerical calculation of temperature dependent Superconducting Transition in inhomogenions Superconductor, *Cryogenics* 23, 527-530

Klimenko E. Yu. and Trenin A. E., (1985), Applicability of the Normal distribution for calculation voltage- current characteristics of superconductors *Cryogenics* 25, pp. 27-28

Klimenko E. Yu., Shavkin S. V., and Volkov P. V., (1997), Anisotropic Pinning in macroscopic electrodynamics of superconductors *JETP* 85, pp. 573-587

Klimenko E. Yu., Shavkin S. V., and Volkov P. V., (2001), Manifestation of Macroscopic Nonuniformities in Superconductors with Strong Pinning in the dependences of the transverse Current-Voltage Curves on the magnetic Field near Hc2.*Phys. Met. Metallogr.*92, pp. 552-556

Klimenko E. Yu., Novikov M. S., and Dolgushin A. N., (2001), Anizotropy of Pinning in the Cross Section of a Superconducting Wire. *Phys. Met. Metallogr.* 92, pp. 219-224.

Klimenko E. Yu., Imenitov A.B., Shavkin S. V., and Volkov P. V., (2005), Resistance-Current Curves of High Pinning Superconductors, *JETP* 100, n.1, pp. 50-65

Klimenko E.Yu., Chechetkin V.R. , Khayrutdinov R.R. , (2010), Solodovnikov S.G., Electrodynamics of multifilament superconductors, Cryogenics 50, pp. 359-365.

Ketterson J.B.&Song S.N. (1999). Superconductivity,CUP, ISBN 0-521-56295-3, UK Niessen A.K., Weijsenfeld C.H., (1969), Anisotropic Pinning and guided Motion of Vortices in Type II Superconductors, *J.Appl.Phys.*, 40, pp.384-393

E. Zeldov, N. M. Amer, G. Koren, et al., (1990), Flux Creep Characteristics in High-Temperature Superconductors Appl. Phys. Lett. 56, pp. 680-682, ISSN: 0003-6951